APPLIED
ABSTRACT ALGEBRA
WITH
MAPLE™ AND MATLAB®

THIRD EDITION

TEXTBOOKS in MATHEMATICS

Series Editors: Al Boggess and Ken Rosen

TEXTBOOKS in MATHEMATICS

APPLIED ABSTRACT ALGEBRA
WITH
MAPLE™ AND MATLAB®
THIRD EDITION

Richard Klima
Appalachian State University
Boone, North Carolina, USA

Neil Sigmon
Radford University
Radford, Virginia, USA

Ernest Stitzinger
North Carolina State University,
Raleigh, North Carolina, USA

CRC Press
Taylor & Francis Group
Boca Raton   London   New York

CRC Press is an imprint of the
Taylor & Francis Group an **informa** business
A CHAPMAN & HALL BOOK

CRC Press
Taylor & Francis Group
6000 Broken Sound Parkway NW, Suite 300
Boca Raton, FL 33487-2742

© 2016 by Taylor & Francis Group, LLC
CRC Press is an imprint of Taylor & Francis Group, an Informa business

No claim to original U.S. Government works

Printed on acid-free paper
Version Date: 20151020

International Standard Book Number-13: 978-1-4822-4823-4 (Hardback)

Library of Congress Cataloging-in-Publication Data

Klima, Richard E.
  [Applications of abstract algebra with Maple and MATLAB]
  Applied abstract algebra with Maple and MATLAB / Richard Klima, Neil P. Sigmon, and Ernest Stitzinger. -- Third edition.
       pages cm. -- (Textbooks in mathematics ; 34)
    "A CRC title."
    Previous edition: Applications of abstract algebra with Maple and MATLAB, 2007.
    Includes bibliographical references and index.
    ISBN 978-1-4822-4823-4 (alk. paper)
    1. Algebra, Abstract--Data processing. 2. Maple (Computer file) 3. MATLAB. I. Sigmon, Neil. II. Stitzinger, Ernest. III. Title.

QA162.K65 2016
512'.02--dc23   2015026385

Visit the Taylor & Francis Web site at
http://www.taylorandfrancis.com

and the CRC Press Web site at
http://www.crcpress.com

Printed and bound by CPI Group (UK) Ltd, Croydon, CR0 4YY

Contents

Preface

About the Course

Several years ago we created an applied algebra course at North Carolina State University for students who had successfully completed semesters of linear and abstract algebra. We intended for the course to give students more exposure to basic algebraic concepts, and show some practical uses of these concepts. The course was received enthusiastically by both students and faculty at NC State, where it remains a popular elective.

When deciding on material for the course, we knew we wanted to include several topics from coding theory, cryptology, and graph theory. With this in mind, at Michael Singer's suggestion we used George Mackiw's book *Applications of Abstract Algebra*, and supplemented as we saw fit. After a few years, Mackiw's book went out of print. Rather than searching for a new book, we decided to write our own notes and teach the course from a coursepack. Around this same time, NC State incorporated the software package Maple™ into its calculus sequence, and so we decided to use it in our course as well. The use of Maple[1] played a central role in the development of the course, because it provided a way for students to see realistic examples without having to struggle with extensive computations. With additional notes detailing the use of Maple, our coursepack evolved into the first edition of this book. MATLAB® and three new chapters were added for the second edition, and an additional new chapter was added for the third edition. Besides the topics presented in this book, we have included a number of other topics in the course. However, the topics presented in this book have become the core for the course, which we have now had the pleasure of teaching at several institutions, as Klima and Sigmon, who were graduate students at NC State under Stitzinger when the course was created, have established their own careers in academia.

Our philosophy concerning the use of technology in the course is that it be a useful tool and not present new problems or frustrations. As such,

---

[1]Maple (2015). Maplesoft, a division of Waterloo Maple Inc., Waterloo, Ontario.

we have included very detailed instructions regarding the use of Maple and MATLAB[2], and posted all of our Maple and MATLAB worksheets and functions at http://www.radford.edu/npsigmon/algebrabook.html. With the exception of only the user-written Maple function **midichords** in Section 13.4, all of our Maple worksheets and functions will work going back to at least version 13, and all of our MATLAB worksheets and functions will work going back to at least version 7.7. Many of our MATLAB worksheets and functions require the MATLAB Symbolic Math Toolbox, which uses the MuPAD computer algebra system. We hope that our Maple and MATLAB presentations are thorough enough to allow these packages to be used without outside aid. If outside aid is needed, we would suggest one of Maple's manuals, which are available at http://www.maplesoft.com/documentation_center/, or MATLAB's online tutorial at http://www.mathworks.com/academia/student_center/tutorials/launchpad.html.

We generally do not require students to produce computer code used in the course. However, we do expect students to obtain a level of proficiency such that they can make basic changes to provided worksheets in order to complete exercises. We have included a few exercises in which we ask readers to produce computer code, although we tend to save these for midterm or final projects or for exceptional students. Also, in order that this book can be used in courses in which advanced technology is not incorporated, we have separated all computer material and exercises into sections that are clearly labeled. All computer material and exercises can be omitted without any loss of continuity, and all other non-research exercises in the book can be completed with at most a hand-held calculator. Having said that, we should also be clear that we do not think it would be best to teach a course covering the material in this book without the use of some kind of advanced technology to help make the material more accessible to students.

About the Book

When teaching the course, we discuss the material in Chapter 1 as needed rather than reviewing it all at once. More specifically, we discuss the topics

---

[2]MATLAB and Simulink are registered trademarks of The MathWorks, Inc. For product information, please contact:
The MathWorks, Inc.
3 Apple Hill Drive
Natick, MA 01760-2098 USA
Tel: 508-647-7000
Fax: 508-647-7001
E-mail: info@mathworks.com
Web: www.mathworks.com

in Chapter 1 the first time they are needed in the applications that follow. Some of the material in Chapter 1 does not apply directly to the applications that follow. However, Chapter 1 does provide a comprehensive and concise review of all prerequisite advanced mathematics.

In Chapter 2 we present block designs. Then in Chapters 3–5 we present some topics from coding theory. In Chapter 3 we introduce error-correcting codes and present Hadamard, Reed-Muller, and Hamming codes. Then in Chapters 4 and 5 we give detailed presentations of BCH and Reed-Solomon codes, respectively. Each of these chapters is dependent in part on the preceding ones. The dependency of Chapter 3 on Chapter 2 can be avoided by omitting Sections 3.2–3.5 on Hadamard and Reed-Muller codes. In Chapters 6–10 we present some topics from cryptology. In Chapter 6 we introduce algebraic cryptology and present shift, affine, and Hill ciphers. Then in Chapter 7 we give a detailed presentation of Vigenère ciphers. In Chapter 8 we introduce public-key cryptology and present the RSA cryptosystem and some related topics, including the Diffie-Hellman key exchange. Then in Chapters 9 and 10 we give detailed presentations of elliptic curve cryptography and the Advanced Encryption Standard, respectively. There is a slight dependency of Chapters 7–10 on Chapter 6, and of Chapter 9 on Chapter 8. Chapters 11–13 are stand-alone from the rest of the applications in the book. In Chapter 11 we introduce the Pólya counting techniques, including Burnside's Theorem and the Pólya Enumeration Theorem. Then in Chapters 12 and 13 we use these ideas and others in applications involving graph and music theory, respectively. Chapters 12 and 13 are each dependent on Chapter 11, although Chapter 13 is not also dependent on Chapter 12.

This book contains more than enough material for a semester course, and perhaps contains enough material for a two-course sequence if the courses were supplemented with student projects. From our experience, Chapters 2–10 make a nice course on mathematical coding theory and cryptology, from which Chapter 2 and Sections 3.2–3.5 could be omitted if time constraints dictate.

New for the Third Edition

Notable additions for the third edition of this book include Section 6.7, Chapter 13, and more than 100 new exercises. Several other parts of the book were also rewritten, including many of the MATLAB sections due to their adoption of the MuPAD computer algebra system since the second edition. In addition, a complete solutions manual was produced for the first time and is available through the publisher. Countless cosmetic changes and wording updates were also made, as well as a few corrections.

Acknowledgments

We wish to thank everyone who had a part in the development of the course and this book. Pete Hardy taught from the book and improved it with his suggestions, as did Michael Singer, who has been especially invaluable at recommending and documenting topics. Many students have completed theses or capstone projects based on material in the book, including Diana Alexander, Gus Miraglia, Shiloh Moore, Phillip Bare, and Karen Klein, whose master's project on elliptic curves was particularly interesting. We also wish to thank Vicky Klima, who wrote the first draft of Chapter 13, and taught us a great deal about applications of algebra in music theory. Finally, we wish to thank Jack Levine, our mentor and friend, for his interest in our projects, his guidance as we learned about applications of algebra, and his many contributions to the subject, especially in cryptology. Jack passed away during the writing of the second edition of this book, and is sorely missed.

We welcome comments, questions, corrections, and suggestions for improving future editions of this book, and hope you enjoy using it.

Richard Klima
klimare@appstate.edu

Neil Sigmon
npsigmon@radford.edu

Ernest Stitzinger
stitz@math.ncsu.edu

Chapter 1

Preliminary Mathematics

There are two purposes to this chapter. We very quickly and concisely review some of the basic algebraic concepts that are probably familiar to many readers, and also introduce some topics for specific use in later chapters. We will generally not pursue topics any further than necessary to obtain the material needed for the applications that follow. Topics reviewed in this chapter include permutation groups, the ring of integers, polynomial rings, finite fields, and examples that incorporate these topics using the philosophies of concepts covered in later chapters.

1.1 Permutation Groups

Suppose a set G is *closed* under an operation $*$. That is, suppose $a * b \in G$ for all $a, b \in G$. Then $*$ is called a *binary* operation on G. We will use the notation $(G, *)$ to represent the set G with this operation. Suppose $(G, *)$ also satisfies the following three properties.

1. $(a * b) * c = a * (b * c)$ for all $a, b, c \in G$.

2. There exists an *identity* element $e \in G$ for which $e * a = a * e = a$ for all $a \in G$.

3. For each $a \in G$, there exists an *inverse* element $b \in G$ for which $a * b = b * a = e$. The inverse of a is usually denoted by a^{-1} if $*$ is a general operation or multiplication, and $-a$ if $*$ is addition.

Then $(G, *)$ is called a *group*. For example, it can easily be verified that for the set $\mathbb{Z}$ of integers, $(\mathbb{Z}, +)$ is a group with identity element 0, but $(\mathbb{Z}, \cdot)$ with normal integer multiplication is not a group.

Let S be a set, and let $B(S)$ be the collection of all bijections (i.e., one-to-one and onto mappings) on S. Then any $\alpha \in B(S)$ can be uniquely expressed by its action $\alpha(s)$ on the elements $s \in S$.

Example 1.1   If $A = \{1, 2, 3\}$, then $B(A)$ contains six elements. One $\alpha \in B(A)$ can be expressed as $\alpha(1) = 2$, $\alpha(2) = 3$, and $\alpha(3) = 1$.   □

Let $\circ$ represent the composition operation on $B(S)$. Specifically, if $\alpha, \beta \in B(S)$, then define $\alpha \circ \beta$ by the action $(\alpha \circ \beta)(s) = \alpha(\beta(s))$ for $s \in S$. Since the composition of two bijections on S is also a bijection on S, then $\alpha \circ \beta \in B(S)$. Thus, $\circ$ is a binary operation on $B(S)$. It can easily be verified that $(B(S), \circ)$ is a group.

A group $(G, *)$ is said to be *abelian* or *commutative* if $a * b = b * a$ for all $a, b \in G$. For example, since $m + n = n + m$ for all $m, n \in \mathbb{Z}$, then the group $(\mathbb{Z}, +)$ is abelian. However, for a set S with more than two elements, there do exist $\alpha, \beta \in B(S)$ such that $\alpha \circ \beta \neq \beta \circ \alpha$. Thus, for a set S with more than two elements, the group $(B(S), \circ)$ is not abelian.

For a set S, we will represent the number of elements in S by $|S|$, and call this number the *order* of S. Suppose $A = \{1, 2, 3, \ldots, n\}$. Then $(B(A), \circ)$ is denoted by S_n, and called the *symmetric* group on n elements. It can easily be shown that $|S_n| = n!$. Suppose $\alpha \in S_n$. Then α can be represented as follows by listing the elements in A in a row with their images under α listed immediately below.

$$\alpha : \begin{pmatrix} 1 & 2 & \cdots & n \\ \alpha(1) & \alpha(2) & \cdots & \alpha(n) \end{pmatrix}$$

Example 1.2 Consider $\alpha \in S_3$ given by $\alpha(1) = 2$, $\alpha(2) = 3$, and $\alpha(3) = 1$. Then α can be represented as follows.

$$\alpha : \begin{pmatrix} 1 & 2 & 3 \\ 2 & 3 & 1 \end{pmatrix}$$

□

An element $\alpha \in S_n$ is called a *permutation*. Note that for permutations $\alpha, \beta \in S_n$, we can represent $\alpha \circ \beta$ as follows.

$$\begin{pmatrix} 1 & \cdots & n \\ \alpha(1) & \cdots & \alpha(n) \end{pmatrix} \begin{pmatrix} 1 & \cdots & n \\ \beta(1) & \cdots & \beta(n) \end{pmatrix} = \begin{pmatrix} 1 & \cdots & n \\ \alpha(\beta(1)) & \cdots & \alpha(\beta(n)) \end{pmatrix}$$

For example, consider $\alpha \in S_4$ given by $\alpha(1) = 2$, $\alpha(2) = 4$, $\alpha(3) = 3$, and $\alpha(4) = 1$, and $\beta \in S_4$ given by $\beta(1) = 4$, $\beta(2) = 3$, $\beta(3) = 2$, and $\beta(4) = 1$. Then $\alpha \circ \beta$ can be represented as follows.

$$\begin{pmatrix} 1 & 2 & 3 & 4 \\ 2 & 4 & 3 & 1 \end{pmatrix} \begin{pmatrix} 1 & 2 & 3 & 4 \\ 4 & 3 & 2 & 1 \end{pmatrix} = \begin{pmatrix} 1 & 2 & 3 & 4 \\ 1 & 3 & 4 & 2 \end{pmatrix}$$

We now present another way to express elements in S_n. Let $i_1, i_2, \ldots, i_k$ be distinct elements in the set $A = \{1, 2, \ldots, n\}$. Then $(i_1\, i_2\, i_3\, \cdots\, i_{k-1}\, i_k)$ is called a *cycle* of length k, or a *k-cycle*, and represents the element in S_n that maps $i_1 \mapsto i_2$, $i_2 \mapsto i_3$, $\ldots$, $i_{k-1} \mapsto i_k$, $i_k \mapsto i_1$, and maps every other element in A to itself. For example, consider the following permutation $\alpha \in S_6$.

$$\alpha : \begin{pmatrix} 1 & 2 & 3 & 4 & 5 & 6 \\ 3 & 4 & 5 & 1 & 6 & 2 \end{pmatrix}$$

This permutation α can be expressed as the 6-cycle (135624). Note that this expression of α as a cycle is not unique, as α can also be expressed as (356241) and (562413), among others.

Next, consider the following permutation $\beta \in S_6$.

$$\beta : \begin{pmatrix} 1 & 2 & 3 & 4 & 5 & 6 \\ 3 & 4 & 5 & 6 & 1 & 2 \end{pmatrix}$$

To express β using cycle notation, we must use more than one cycle. For example, we can express β as the "product" (135)(246) of two 3-cycles. Since these two 3-cycles contain no elements in common, they are said to be *disjoint*, and the order in which they are written does not matter. The permutation β can also be expressed as (246)(135).

Every permutation in S_n can be expressed as either a single cycle or a product of disjoint cycles. When a permutation is expressed as a product of disjoint cycles, cycles of length one are customarily not included. For example, consider the following permutation $\gamma \in S_6$.

$$\gamma : \begin{pmatrix} 1 & 2 & 3 & 4 & 5 & 6 \\ 3 & 4 & 5 & 2 & 1 & 6 \end{pmatrix}$$

The fact that γ maps 6 to itself would be expressed as the 1-cycle (6). However, this cycle would customarily not be included in the expression of γ as a product of disjoint cycles. That is, γ would normally be expressed as (135)(24) or (24)(135), and the absence of 6 from these expressions would indicate that γ maps 6 to itself.

In an expression of a permutation as a product of cycles, the cycles need not be disjoint. For example, the permutation $\alpha = (135624)$ that we defined previously can also be expressed as the product (14)(12)(16)(15)(13) of 2-cycles. Because these 2-cycles are not disjoint, the order in which they are listed does matter.

A 2-cycle is also called a *transposition*. Any permutation can be expressed as a product of transpositions in the way illustrated previously for α. Specifically, the cycle $(i_1\, i_2\, i_3\, \cdots\, i_{k-1}\, i_k)$ can be expressed as the product $(i_1\, i_k)(i_1\, i_{k-1})\, \cdots\, (i_1\, i_3)(i_1\, i_2)$ of transpositions. If a permutation can be expressed as a product of more than one disjoint cycle, then each cycle can be considered separately when expressing the permutation as a product of transpositions. For example, the permutation $\beta = (135)(246)$ that we defined previously can be expressed as $(15)(13)(26)(24)$, and the permutation $\gamma = (135)(24)$ that we defined previously can be expressed as $(15)(13)(24)$.

There are many ways to express a permutation as a product of transpositions, and the number of transpositions in these expressions can vary. However, the number of transpositions in the expressions of a permutation as a product of transpositions must be always even or always odd. A permutation is said to be *even* if it can be expressed as the product of an even number of transpositions, and *odd* otherwise. Thus, the product of two even permutations is even, as is the product of two odd permutations.

The inverse of the cycle $(i_1\ i_2\ i_3\ \cdots\ i_{k-1}\ i_k)$ is $(i_k\ i_{k-1}\ \cdots\ i_3\ i_2\ i_1)$. Suppose $\alpha = t_1 t_2 \cdots t_m \in S_n$, where each t_j is a transposition. Then $\alpha^{-1} = t_m^{-1} t_{m-1}^{-1} \cdots t_1^{-1} = t_m t_{m-1} \cdots t_1$, since $t_j^{-1} = t_j$ for each transposition t_j. Thus, the inverse of an even permutation is even. Also, because the identity permutation is even, the subset of even permutations in S_n forms a group. This group is denoted by A_n, and called the *alternating* group on n elements. Since A_n is a group and also a subset of the group S_n, then A_n is called a *subgroup* of S_n.

Definition 1.1 *Let $(G, *)$ be a group, and suppose H is a nonempty subset of G. If $(H, *)$ is a group, then H is called a subgroup of G.*

Consider now a regular polygon P, such as for example an equilateral triangle or a square. Any movement of P that preserves the general shape of P is called a *rigid motion*. There are two types of rigid motions: rotations and reflections. For a regular polygon P with n sides, there are $2n$ distinct rigid motions. These include the n rotations of P through $\frac{360j}{n}$ degrees for $j = 0, 1, 2, \ldots, n - 1$. The remaining n rigid motions are reflections. If n is even, these are reflections of P across the lines that connect opposite vertices or bisect opposite sides of P. If n is odd, these are reflections of P across the perpendicular bisectors of the sides of P. Since all rigid motions of P preserve the general shape of P, they can be viewed as permutations of the vertices (i.e., corners) of P. The set of all rigid motions of P with the composition operation forms a group called the *symmetries* of P.

Example 1.3 Consider the group of symmetries of a square. To express these symmetries as permutations of the vertices of a square, consider the following general figure.

The 8 symmetries of a square can be expressed as permutations of the vertices of this general figure as follows (rotations are counterclockwise).

Rigid Motion	Permutation
0° rotation (identity)	(1)
90° rotation	(1234)
180° rotation	(13)(24)
270° rotation	(1432)
reflection across horizontal	(12)(34)
reflection across vertical	(14)(23)
reflection across 1–3 diagonal	(24)
reflection across 2–4 diagonal	(13)

Note that expressing these rigid motions as permutations of the vertices of the preceding general figure yields a subgroup of S_4. □

When the symmetries of a regular polygon with n sides are expressed as permutations of the set $\{1, 2, \ldots, n\}$, the resulting subgroup of S_n is denoted by D_n, and called the *dihedral* group on n elements. The subgroup of S_4 in Example 1.3 is D_4.

A group G (with any operation) is said to be *cyclic* if there is an element $x \in G$ for which $G = \{x^i \mid i \in \mathbb{Z}\}$. For a cyclic group $G = \{x^i \mid i \in \mathbb{Z}\}$, x is called a *generator* for G. More generally, suppose a is an element in a group G, and let $H = \{a^i \mid i \in \mathbb{Z}\}$. Then H is a subgroup of G called the cyclic subgroup generated by a. Suppose $a^j = a^k$ for some integers $0 \le j < k$. Then $a^{k-j} = a^k(a^j)^{-1} = e$, where e is the identity element in G. Thus, since $k - j$ is a positive integer, there must be a smallest positive integer m for which $a^m = e$. Now, suppose $a^t = e$ for some positive integer t. Since $t = mq + r$ for some integer r with $0 \le r < m$, and $a^t = a^{mq+r} = (a^m)^q a^r = a^r$, it follows that $r = 0$, and m divides evenly into t. Since $a^j = a^k$ for integers $0 \le j < k$ forces $a^{k-j} = e$, a contradiction if $0 < k - j < m$, the set $\{a^i \mid 0 \le i < m\}$ must consist of m distinct elements. Furthermore, for any integer s, we can write $s = mq + r$ for some integer r with $0 \le r < m$ and $a^s = a^r$. Thus, $H = \{a^i \mid 0 \le i < m\}$, and H contains exactly m distinct elements. We summarize this discussion as the following theorem.

Theorem 1.2 *Suppose a is an element in a group G, and there is some positive integer k for which $a^k = e$ (where e is the identity element in G). If m is the smallest positive integer for which $a^m = e$, then the cyclic subgroup of G generated by a contains exactly m distinct elements.*

The value of m in Theorem 1.2 is called the *order* of a. Thus, the order of an element a in a group G is equal to the order of the cyclic subgroup of G generated by a. We will show in Theorem 1.4 that for an element of order m in a group of order n, it must be the case that m divides evenly into n. Thus, in a group G of order n, it must be true that $a^n = e$ for all $a \in G$, where e is the identity element in G. We summarize this as the following corollary.

Corollary 1.3 *Suppose a is an element in a group G, and G has order n. Then $a^n = e$ (where e is the identity element in G).*

Example 1.4 Consider the dihedral group D_n, which has order $2n$. Recall that the elements in D_n can be viewed as the rigid motions of a regular polygon P with n sides. Each of the n reflections of P has order 2. Also, the rotations of P through $\frac{360}{n}$ and $\frac{360(n-1)}{n}$ degrees have order n (as do, possibly, some other rotations). Note that each of these orders divides evenly into the order of D_n.   □

1.2   Cosets and Quotient Groups

Let H be a subgroup of a group G. For an element $g \in G$, we define $gH = \{gh \mid h \in H\}$, called a *left coset* of H in G. Since $gh_1 = gh_2$ implies $h_1 = h_2$ for all $h_1, h_2 \in H$, there is a one-to-one correspondence between the elements in gH and H. Thus, if H is finite, $|gH| = |H|$. Now, suppose $g_1, g_2 \in G$. If $x \in g_1 H \cap g_2 H$ for some $x \in G$, then $x = g_1 h_1 = g_2 h_2$ for some $h_1, h_2 \in H$, and $g_1 = g_2 h_2 h_1^{-1} \in g_2 H$. Then for any $y \in g_1 H$, it follows that $y = g_1 h_3$ for some $h_3 \in H$, and so $y = g_1 h_3 = g_2 h_2 h_1^{-1} h_3 \in g_2 H$. Thus, $g_1 H \subseteq g_2 H$. Similarly, $g_2 H \subseteq g_1 H$, and so $g_1 H = g_2 H$. The preceding discussion implies that if $g_1, g_2 \in G$, then either $g_1 H = g_2 H$, or $g_1 H$ and $g_2 H$ are disjoint. As a result, G is the union of pairwise disjoint left cosets of H in G.[1]

Example 1.5 Consider the subgroup A_n of S_n. If α is an odd permutation in S_n, then αA_n and A_n will be disjoint. If β is also an odd permutation in S_n, then $\beta^{-1}\alpha$ will be even. Thus, $\beta^{-1}\alpha \in A_n$, and $\alpha A_n = \beta A_n$. From this we can conclude that there are exactly two distinct left cosets of A_n in S_n, one consisting of the even permutations in S_n, and the other consisting of the odd permutations in S_n.   □

For a finite group G with subgroup H, the following theorem is a fundamental algebraic result regarding the number of left cosets of H in G.

---

[1] We began this section by defining the left cosets gH of a subgroup H in a group G. Similarly, the sets $Hg = \{hg \mid h \in H\}$ are called the *right cosets* of H in G, and results analogous to those in this section for left cosets hold for right cosets as well.

Theorem 1.4 (Lagrange's Theorem) *Let G be a group of order n with a subgroup H of order k, and suppose that there are t distinct left cosets of H in G. Then $n = kt$.*

Proof. Exercise. □

As a consequence of Lagrange's Theorem, the order of any subgroup of a finite group G must divide evenly into the order of G. As an example of this, consider the dihedral group D_4 of order 8, which is a subgroup of the symmetric group S_4 of order 24.

Next, we will describe how cosets can be used to construct new groups from existing ones. Suppose H is a subgroup of a group G. Then for any $x \in G$, we define $x^{-1}Hx = \{x^{-1}hx \mid h \in H\}$. If $x^{-1}Hx \subseteq H$ for all $x \in G$, then we call H a *normal* subgroup of G, and the set of distinct left cosets of H in G forms a group with the operation $(xH)(yH) = (xy)H$. To see this, note first that if H is a normal subgroup of G, then it will be the case that $x^{-1}Hx \subseteq H$ for all $x \in G$, including if we replace x with x^{-1}. That is, it will also be the case that $xHx^{-1} \subseteq H$ for all $x \in G$. Thus, for any $h \in H$, it follows that $h = x^{-1}(xhx^{-1})x = x^{-1}h_1x \in x^{-1}Hx$ for some $h_1 \in H$. As a result, $H \subseteq x^{-1}Hx$, and it follows that $x^{-1}Hx = H$. Therefore, a subgroup H in a group G is normal in G if and only if it satisfies $x^{-1}Hx = H$ for all $x \in G$.

To see that the operation that we defined previously for the left cosets of H in G is well defined, suppose $xH = x_1H$ and $yH = y_1H$ for some $x, x_1, y, y_1 \in G$. Since $xH = x_1H$ and $yH = y_1H$, it follows that $x = x_1h_1$ and $y = y_1h_2$ for some $h_1, h_2 \in H$. Also, since H is normal in G, then $y_1^{-1}h_1y_1 = h_3$ for some $h_3 \in H$, or, equivalently, $h_1y_1 = y_1h_3$ for some $h_3 \in H$. Then $xy = x_1h_1y_1h_2 = x_1y_1h_3h_2 \in x_1y_1H$. Thus, $xy \in x_1y_1H$, and so $xyH = x_1y_1H$. Therefore, the operation that we defined previously for the left cosets of H in G is well defined.

We can now easily show that if H is a normal subgroup of a group G, then the set of left cosets of H in G forms a group with the operation $(xH)(yH) = (xy)H$. This group is denoted G/H (read G *mod* H) and called a *quotient* group.

Theorem 1.5 *Suppose H is a normal subgroup of a group G. Then the set G/H of all distinct left cosets of H in G forms a group with the operation $(xH)(yH) = (xy)H$.*

Proof. If e is the identity element in G, then $eH = H$ is the identity in G/H, since $(eH)(xH) = (ex)H = xH$ and $(xH)(eH) = (xe)H = xH$ for all $x \in G$. Also, the inverse of the element xH in G/H is $x^{-1}H$, since $(x^{-1}H)(xH) = (x^{-1}x)H = eH = H$. Finally, associativity in G/H can easily be verified. □

Note also that if G is an abelian group, then any subgroup H of G will be normal, and G/H will be abelian.

Example 1.6 Let $G = (\mathbb{Z}, +)$. Choose an integer $n \in \mathbb{Z}$, and let H be the cyclic subgroup of G generated by n. Since the operation on this group is addition, then $H = \{kn \mid k \in \mathbb{Z}\}$, and additive notation $x + H$ is used for the cosets of H in G. That is, the cosets of H in G are the sets $x + H = \{x + h \mid h \in H\} = \{x + kn \mid k \in \mathbb{Z}\}$ for all $x \in \mathbb{Z}$. The distinct left cosets of H in G are the sets $H, 1 + H, 2 + H, \ldots, (n-1) + H$. Thus, G/H consists of these sets with the operation $(x + H) + (y + H) = (x + y) + H$. Note that if we would perform this operation without including H in the notation, we would simply be doing integer addition modulo n. Note also that G/H is cyclic with generator $1 + H$.   □

Suppose H is a normal subgroup of a group G, and define the mapping $\phi : G \to G/H$ by $\phi(x) = xH$. For this mapping, it can easily be verified that $\phi(xy) = \phi(x)\phi(y)$ for all $x, y \in G$. Since ϕ satisfies this property, ϕ is called a *homomorphism*.

Definition 1.6 *Let G and H be groups. A mapping $\phi : G \to H$ that satisfies $\phi(xy) = \phi(x)\phi(y)$ for all $x, y \in G$ is called a homomorphism.*

Example 1.7 Let H be the group $H = \{odd, even\}$ with identity element *even*. Define $\phi : S_n \to H$ by $\phi(x) = even$ if x is an even permutation, and $\phi(x) = odd$ otherwise. Then ϕ is a homomorphism.   □

Example 1.8 Let G be the group of nonsingular $n \times n$ matrices over the set $\mathbb{R}$ of real numbers (i.e., with entries in $\mathbb{R}$) under ordinary matrix multiplication. Then the determinant function is a homomorphism from G onto the group $\mathbb{R}^*$ of nonzero real numbers under multiplication.   □

Let G_1, G_2 be groups, and suppose $\phi : G_1 \to G_2$ is a homomorphism. We define the *kernel* of ϕ to be the set $\text{Ker}(\phi) = \{g \in G_1 \mid \phi(g) = e\}$, where e is the identity element in G_2. It can easily be verified that $\text{Ker}(\phi)$ is a normal subgroup of G_1. Also, for any normal subgroup H of any group G, if we define the mapping $\phi : G \to G/H$ by $\phi(x) = xH$, then $\text{Ker}(\phi) = H$. Thus, every normal subgroup of a group G is the kernel of a homomorphism with domain G, and the kernel of every homomorphism with domain G is a normal subgroup of G.

1.3   Rings and Euclidean Domains

Let R be a set with two binary operations, an "addition" $+$ and a "multiplication" $*$. Suppose R also satisfies the following three properties.

1. $(R, +)$ is an abelian group, with identity we will denote by 0.

2. $(a * b) * c = a * (b * c)$ for all $a, b, c \in R$.

3. $a * (b + c) = (a * b) + (a * c)$ and $(a + b) * c = (a * c) + (b * c)$ for all $a, b, c \in R$.

Then R is called a *ring*. If it is also true that $a * b = b * a$ for all $a, b \in R$, then R is said to be a *commutative* ring. Also, if R contains a multiplicative identity element (i.e., an element, usually denoted by 1, that satisfies $1 * a = a * 1 = a$ for all $a \in R$), then R is said to be a ring *with identity*. As is customary (and as we have already done frequently when dealing with multiplication in groups), we will usually suppress the $*$ from the notation when performing the multiplication operation in rings.

All of the rings that we will use in this book are commutative rings with identity. A commutative ring R with identity is called an *integral domain* if $a, b \in R$ with $ab = 0$ implies either $a = 0$ or $b = 0$. For example, $\mathbb{Z}$ with ordinary addition and multiplication is an integral domain. A commutative ring R with identity is called a *field* if every nonzero element in R has a multiplicative inverse in R. For example, $\mathbb{R}$ with ordinary addition and multiplication is a field. Also, since all fields are integral domains, then $\mathbb{R}$ is an integral domain.

In addition to $\mathbb{Z}$, we will make extensive use of the ring $F[x]$ of polynomials in x with coefficients in a field F, with the operations of ordinary addition and multiplication. Like $\mathbb{Z}$, the ring $F[x]$ is an integral domain but not a field.

Suppose now that B is a nonempty subset of a commutative ring R. If $(B, +)$ is a subgroup of $(R, +)$, and if $rb \in B$ for all $r \in R$ and $b \in B$, then B is called an *ideal* of R. For an ideal B of R, if there exists an element $b \in B$ for which $B = \{rb \mid r \in R\}$, then B is called a *principal* ideal of R. In this case, B is denoted by (b), and called the ideal *generated by* b.

If $f(x) \in F[x]$ for some field F, then $(f(x))$ consists of all multiples of $f(x)$ over F. That is, $(f(x))$ consists of all polynomials in $F[x]$ that have $f(x)$ as a factor. A similar result holds for integers $n \in \mathbb{Z}$. We will show in Theorem 1.9 that all ideals of $F[x]$ and $\mathbb{Z}$ are principal ideals. Since $F[x]$ and $\mathbb{Z}$ are integral domains in which every ideal is principal, they are called *principal ideal domains*.

Ideals play a role in ring theory similar to the role played by normal subgroups in group theory. For example, we can use an ideal to construct a new ring from an existing one. Suppose B is an ideal of a commutative ring R. Then since $(B, +)$ is a subgroup of the abelian group $(R, +)$, it follows that $R/B = \{r + B \mid r \in R\}$ is an abelian group with addition operation $(r + B) + (s + B) = (r + s) + B$. As it turns out, R/B is also a commutative ring with multiplication operation $(r + B)(s + B) = (rs) + B$. To see that this multiplication operation is well defined, suppose $r + B = r_1 + B$ and $s + B = s_1 + B$ for some $r, r_1, s, s_1 \in R$. Since $r + B = r_1 + B$ and $s+B = s_1+B$, it follows that $r = r_1+b_1$ and $s = s_1+b_2$ for some $b_1, b_2 \in B$. Then $rs = (r_1 + b_1)(s_1 + b_2) = r_1s_1 + r_1b_2 + b_1s_1 + b_1b_2 \in r_1s_1 + B$. Thus, $rs \in r_1s_1 + B$, and so $rs + B = r_1s_1 + B$. Therefore, the multiplication operation that we defined previously for R/B is well defined. The ring R/B is called a *quotient* ring.

Suppose B is an ideal of a commutative ring R, and define the mapping $\phi : R \to R/B$ by $\phi(x) = x + B$. For this mapping, it can easily be verified that $\phi(rs) = \phi(r)\phi(s)$ and $\phi(r + s) = \phi(r) + \phi(s)$ for all $r, s \in R$. Since ϕ satisfies these properties, ϕ is called a *ring homomorphism*.

Definition 1.7 *Let R and S be rings. A mapping $\phi : R \to S$ that satisfies $\phi(rs) = \phi(r)\phi(s)$ and $\phi(r + s) = \phi(r) + \phi(s)$ for all $r, s \in R$ is called a ring homomorphism. Also, we define the kernel of ϕ to be the set $Ker(\phi) = \{r \in R \mid \phi(r) = 0\}$.*

Proposition 1.8 *Suppose R and S are commutative rings, and ϕ is a ring homomorphism from R onto S. Then the following statements are true.*

1. *If B_1 is an ideal of R, then the set $\phi(B_1) = \{\phi(r) \in S \mid r \in B_1\}$ is an ideal of S.*

2. *If B_2 is an ideal of S, then the set $\phi^{-1}(B_2) = \{r \in R \mid \phi(r) \in B_2\}$ is an ideal of R.*

Proof. Exercise.   □

Suppose D is an integral domain with nonzero elements D^*, and let $\mathbb{N}$ be the set of nonnegative integers. If there is a mapping $\delta : D^* \to \mathbb{N}$ such that for all $a \in D$ and $b \in D^*$, there exist $q, r \in D$ with $a = bq + r$ and either $r = 0$ or $\delta(r) < \delta(b)$, then D is called a *Euclidean domain*. Two examples of Euclidean domains are $\mathbb{Z}$ with $\delta(n) = |n|$, and the ring $F[x]$ of polynomials over a field F with $\delta(f(x)) = \deg(f(x))$.

Theorem 1.9 *Every Euclidean domain is a principal ideal domain.*

Proof. Suppose B is an ideal of a Euclidean domain D, and B contains at least one nonzero element. Choose nonzero $b \in B$ such that $\delta(b) \leq \delta(x)$ for all nonzero $x \in B$. Then for any $a \in B$, there must exist $q, r \in D$ with $a = bq + r$ and either $r = 0$ or $\delta(r) < \delta(b)$. However, since $r = a - bq$ and B is an ideal, then $r \in B$. By the choice of b, it follows that $r = 0$. Therefore, $a = bq$, and $a \in (b)$. Thus, $B \subseteq (b)$. Since certainly $(b) \subseteq B$, it follows that $B = (b)$.   □

If an element a in an integral domain D has a multiplicative inverse in D, then a is called a *unit*. We will denote the set of units in an integral domain D by $U(D)$. Thus, for example, $U(\mathbb{Z}) = \{1, -1\}$, and for a field F, the set $U(F[x])$ of units in $F[x]$ is the set of nonzero constants in F. For an element a in an integral domain D, an element $b \in D$ is called an *associate* of a if and only if $b = au$ for some unit $u \in D$. The only associates of an element $n \in \mathbb{Z}$ are n and $-n$. The associates of an element $f(x) \in F[x]$ are $cf(x)$ for all nonzero $c \in F$.

For elements a and b in an integral domain D, if there exists $x \in D$ for which $ax = b$, then a is said to *divide* b, written $a|b$.

Proposition 1.10 *Suppose a, b, and c are elements in an integral domain D. Then the following statements are true.*

1. *If $a|b$ and $b|c$, then $a|c$.*

2. *$a|b$ and $b|a$ if and only if a and b are associates of each other in D.*

3. *$a|b$ if and only if $(b) \subseteq (a)$.*

4. *$(a) = (b)$ if and only if a and b are associates of each other in D.*

Proof. Exercise. □

Let a be an element in a Euclidean domain D, and suppose that the only elements in D that divide a are the units in D and the associates of a in D. Then a is said to be *irreducible*. Let M be an ideal of a Euclidean domain D with $M \neq D$, and suppose that for all ideals B of D with $M \subseteq B \subseteq D$, it follows that either $M = B$ or $B = D$. Then M is said to be a *maximal* ideal in D.

Theorem 1.11 *An element a in a Euclidean domain D is irreducible if and only if the ideal (a) in D is maximal.*

Proof. Suppose first that a is irreducible. If $(a) \subseteq (b) \subseteq D$ for some $b \in D$, then b must divide a. It follows that either b is a unit in D, in which case $(b) = D$, or b is an associate of a, in which case $(a) = (b)$. Thus, (a) is maximal. Now, suppose (a) is maximal. If b divides a, then $(a) \subseteq (b)$. It follows that either $(b) = D$, in which case b is a unit (since there must exist $x \in D$ for which $bx = 1$), or $(b) = (a)$, in which case b is an associate of a. Thus, a is irreducible. □

Theorem 1.12 *An ideal M in a Euclidean domain D is maximal if and only if the quotient ring D/M is a field.*

Proof. Suppose first that M is maximal. Choose $r + M \in D/M$ such that $r + M \neq M$, and let $B = (r+M) \subseteq D/M$. Then for the ring homomorphism ϕ from D onto D/M defined by $\phi(x) = x + M$, let $C = \phi^{-1}(B)$. Since B is an ideal of D/M, we know by Proposition 1.8 that C is an ideal of D. Therefore, $M \subseteq C \subseteq D$. However, since M is maximal and $r + M \neq M$, it follows that $C = D$, and so $B = D/M$. Thus, there exists an element $s + M \in D/M$ for which $(r + M)(s + M) = 1 + M$, and so $r + M$ has an inverse in D/M. Therefore, D/M is a field. Now, suppose D/M is a field. Let B be an ideal of D for which $M \subseteq B \subseteq D$. We know by Proposition 1.8 that $\phi(B)$ is an ideal of D/M, and since the only ideals in a field are the field itself and $\{0\}$, it follows that either $\phi(B) = M$ or $\phi(B) = D/M$. Thus, either $B = M$ or $B = D$, and M is maximal. □

By combining the results of Theorems 1.11 and 1.12, we obtain the following theorem.

Theorem 1.13 *Suppose a is an element in a Euclidean domain D. Then the following statements are equivalent.*

1. *a is irreducible in D.*

2. *(a) is maximal in D.*

3. *$D/(a)$ is a field.*

1.4   Finite Fields

Finite fields play an important role in several of the applications that we will present in this book. In this section, we will describe the theoretical basis of constructing finite fields.

It can easily be shown that the ring $\mathbb{Z}_p = \{0, 1, 2, \ldots, p-1\}$ for prime p is a field with the usual operations of addition and multiplication modulo p (i.e., divide the result by p and take the remainder). This shows that there are finite fields of order p for every prime p. In the following discussion, we show how the fields $\mathbb{Z}_p$ can be used to construct finite fields of order p^n for every prime p and positive integer n.

Suppose m is an irreducible element in a Euclidean domain D, and let $B = (m)$. Then by Theorem 1.13, we know that D/B must be a field. If D is the ring $\mathbb{Z}$ of integers and $m > 0$, then m must be a prime p. Note that if we perform the addition and multiplication operations in D/B without including B in the notation, these operations will be exactly the addition and multiplication operations in $\mathbb{Z}_p$. Thus, we can view D/B as $\mathbb{Z}_p$.

Now, suppose D is the integral domain $\mathbb{Z}_p[x]$ of polynomials over $\mathbb{Z}_p$ for prime p, and let $B = (f(x))$ for some irreducible polynomial $f(x)$ of degree

n in D. Then again by Theorem 1.13, we know that D/B must be a field. Each element in D/B is a coset of the form $g(x) + B$ for some $g(x) \in \mathbb{Z}_p[x]$. Since $\mathbb{Z}_p[x]$ is a Euclidean domain, there exists $r(x) \in \mathbb{Z}_p[x]$ for which $g(x) + B = r(x) + B$ with either $r(x) = 0$ or $\deg(r(x)) < n$. Therefore, each element in D/B can be expressed as $r(x) + B$ for some $r(x) \in \mathbb{Z}_p[x]$ with either $r(x) = 0$ or $\deg(r(x)) < n$. Since a polynomial $r(x) \in \mathbb{Z}_p[x]$ with either $r(x) = 0$ or $\deg(r(x)) < n$ can contain up to n terms, and each of these terms can have any of p coefficients (the p elements in $\mathbb{Z}_p$), there are p^n polynomials $r(x) \in \mathbb{Z}_p[x]$ with either $r(x) = 0$ or $\deg(r(x)) < n$. Thus, the field D/B will contain p^n distinct elements. The operations on this field are the usual operations of addition and multiplication modulo $f(x)$ (i.e., divide the result by $f(x)$ and take the remainder). For convenience, when we write elements and perform the addition and multiplication operations in D/B, we will not include B in the notation. That is, we will express the elements $r(x) + B$ in D/B as just $r(x)$.

Because it is possible to find an irreducible polynomial of degree n over $\mathbb{Z}_p$ for every prime p and positive integer n, the comments in the preceding paragraph indicate that there are finite fields of order p^n for every prime p and positive integer n. It is also true that all finite fields have order p^n for some prime p and positive integer n (see Theorem 1.14).

Example 1.9 Suppose $D = \mathbb{Z}_3[x]$, and let $B = (f(x))$ for the irreducible polynomial $f(x) = x^2 + x + 2 \in \mathbb{Z}_3[x]$. (Note: We could very easily verify that $f(x)$ is irreducible in $\mathbb{Z}_3[x]$ by showing that $f(a) \neq 0$ for all $a \in \mathbb{Z}_3$.) Then the field D/B will contain the $3^2 = 9$ polynomials of degree less than 2 in $\mathbb{Z}_3[x]$. That is, $D/B = \{\, 0,\, 1,\, 2,\, x,\, x+1,\, x+2,\, 2x,\, 2x+1,\, 2x+2 \,\}$. To add elements in D/B, we simply reduce the coefficients in $\mathbb{Z}_3$. For example, $(2x + 1) + (2x + 2) = 4x + 3 = x$. To multiply elements in D/B, we can use several methods. One method is to divide the product by $f(x)$ and take the remainder. For example, to multiply $2x + 1$ and $2x + 2$ in D/B, we could form $(2x+1)(2x+2) = 4x^2 + 6x + 2 = x^2 + 2$. Dividing $x^2 + 2$ by $f(x)$ yields a quotient of 1 and a remainder of $-x = 2x$. Thus, $(2x + 1)(2x + 2) = 2x$ in D/B. Another method for multiplying elements in D/B uses the fact that $x^2 + x + 2 = 0$ in D/B. Thus, $x^2 = -x - 2 = 2x + 1$ in D/B, an identity that can be used to reduce powers of x in D/B. For example, we can also compute the product of $2x + 1$ and $2x + 2$ in D/B by forming $(2x + 1)(2x + 2) = 4x^2 + 6x + 2 = x^2 + 2 = (2x + 1) + 2 = 2x$. We will describe a third method for multiplying elements in D/B next, and then illustrate this method in Example 1.10. □

A fundamental fact regarding finite fields is that the nonzero elements in every finite field form a cyclic group under multiplication (see Theorem 1.15). Suppose $D = \mathbb{Z}_p[x]$ for some prime p, and let $B = (f(x))$ for some irreducible polynomial $f(x) \in D$. For the field $F = D/B$, if x is a cyclic generator for the nonzero elements F^* in F, then $f(x)$ is said to be *primitive*. Thus, if $f(x)$ is primitive, then F^* can be completely generated by constructing powers of x modulo $f(x)$. This is useful because it allows products of elements in F^* to be formed by converting the elements into their representations as powers of x, multiplying the powers of x, and then converting the result back into an element in F^*. We illustrate this in the following example.

Example 1.10 Consider the field D/B in Example 1.9. We can use the identity $x^2 = 2x + 1$ to construct the elements in this field that correspond to powers of x. For example, we can construct the field element that corresponds to x^3 as follows.

$$x^3 = x^2 x = (2x + 1)x = 2x^2 + x = 2(2x + 1) + x = 5x + 2 = 2x + 2$$

Thus, $x^3 = 2x + 2$ in D/B. Also, we can construct the field element that corresponds to x^4 as follows.

$$x^4 = x^3 x = (2x + 2)x = 2x^2 + 2x = 2(2x + 1) + 2x = 6x + 2 = 2$$

Thus, $x^4 = 2$ in D/B. The field elements that correspond to subsequent powers of x can be constructed similarly. We list the field elements that correspond to the first 8 powers of x in the following table.

Power	Field Element
x^1	x
x^2	$2x + 1$
x^3	$2x + 2$
x^4	2
x^5	$2x$
x^6	$x + 2$
x^7	$x + 1$
x^8	1

Note that the only element in D/B not listed in this table is 0. Since all nonzero elements in D/B are generated by computing powers of x, then $f(x) = x^2 + x + 2$ is primitive in $\mathbb{Z}_3[x]$. This table is useful for computing products in D/B. For example, we can compute the product of $2x + 1$ and $2x + 2$ in D/B as follows.

$$(2x + 1)(2x + 2) = x^2 x^3 = x^5 = 2x$$

Note that this result is identical to the product we obtained for the same elements by two other methods in Example 1.9. We can also compute the product of $2x$ and $x + 2$ in D/B as follows.

$$(2x)(x+2) = \dot{x}^5 x^6 = x^{11} = x^8 x^3 = 1x^3 = 2x + 2$$

Other products of nonzero elements in D/B can be computed similarly. □

Example 1.11 Suppose $D = \mathbb{Z}_3[x]$, and let $B = (f(x))$ for the irreducible polynomial $f(x) = x^2 + 1 \in \mathbb{Z}_3[x]$. Since $f(x)$ is irreducible of degree 2 in $\mathbb{Z}_3[x]$, then D/B is a field of order $3^2 = 9$, with the exact same elements as the field in Example 1.9. However, note that $x^2 = -1 = 2$ in D/B, and so $x^4 = 4 = 1$ in D/B. Thus, it follows that computing powers of x will not generate all 8 nonzero elements in D/B. Therefore, $f(x) = x^2 + 1$ is not primitive in $\mathbb{Z}_3[x]$, and we cannot compute all possible products of nonzero elements in D/B using the method illustrated in Example 1.10. However, we can still compute all possible products in D/B using either of the two methods illustrated in Example 1.9. □

We close this section by proving a pair of fundamental results we have mentioned regarding finite fields.

Theorem 1.14 *Suppose F is a finite field. Then $|F| = p^n$ for some prime p and positive integer n.*

Proof. Let H be the additive subgroup of F generated by 1. Suppose that $|H| = mn$ for some positive integers m, n with $m \neq 1$ and $n \neq 1$. Then $0 = (mn)1 = (m1)(n1)$. However, since $m1 \neq 0$ and $n1 \neq 0$, this contradicts the fact that F is a field. Thus, $|H| = p$ for some prime p. That is, $H = \mathbb{Z}_p$ for some prime p. The field F can be viewed as a vector space over H with scalar multiplication given by the field multiplication, and so F has a basis with a finite number of elements, say n. The order of F is then the number p^n of linear combinations of these basis elements over $\mathbb{Z}_p$. □

Theorem 1.15 *Suppose F is a finite field. Then the nonzero elements F^* in F form a cyclic multiplicative group.*

Proof. Clearly, F^* is an abelian multiplicative group. To show that F^* is cyclic, we use the first of the well known Sylow Theorems, which states that for any finite group G of order n, if p^k divides n for some prime p and positive integer k, then G contains a subgroup of order p^k. Suppose $|F^*|$ has prime factorization $p_1^{n_1} p_2^{n_2} \cdots p_t^{n_t}$, and let S_i be subgroups of order $p_i^{n_i}$ in F^* for each $i = 1, 2, \ldots, t$. Also, let $k_i = p_i^{n_i - 1}$ for each $i = 1, 2, \ldots, t$. Then if S_i is not cyclic for some i, it follows that $a^{k_i} = 1$ for all $a \in S_i$. Thus, $f(x) = x^{k_i} - 1$ has $p_i^{n_i}$ roots in F, a contradiction. Therefore, each S_i must have a cyclic generator a_i. Let $b = a_1 a_2 \cdots a_t$. Since b has order $|F^*|$, then b is a cyclic generator for F^*. □

1.5   Finite Fields with Maple

In this section, we will show how Maple can be used to construct the nonzero elements as powers of x in a finite field $\mathbb{Z}_p[x]/(f(x))$ for prime p and primitive polynomial $f(x) \in \mathbb{Z}_p[x]$. We will consider the field used in Examples 1.9 and 1.10.

We will begin by defining the polynomial $f(x) = x^2 + x + 2 \in \mathbb{Z}_3[x]$ used to construct the field elements.

```
>   f := x -> x^2+x+2;
```
$$f := x \to x^2 + x + 2$$

We can use the Maple **Irreduc** function as follows to verify that $f(x)$ is irreducible in $\mathbb{Z}_3[x]$. The following command will return *true* if $f(x)$ is irreducible in $\mathbb{Z}_3[x]$, and *false* otherwise.

```
>   Irreduc(f(x)) mod 3;
```
$$true$$

Thus, $f(x)$ is irreducible in $\mathbb{Z}_3[x]$, and $\mathbb{Z}_3[x]/(f(x))$ is a field. However, in order for us to be able to construct all of the nonzero elements in this field as powers of x, it must be the case that $f(x)$ is also primitive. We can use the Maple **Primitive** function as follows to verify that $f(x)$ is primitive in $\mathbb{Z}_3[x]$. The following command will return *true* if $f(x)$ is primitive in $\mathbb{Z}_3[x]$, and *false* otherwise.

```
>   Primitive(f(x)) mod 3;
```
$$true$$

Thus, $f(x)$ is primitive in $\mathbb{Z}_3[x]$.

To construct the nonzero elements in $\mathbb{Z}_3[x]/(f(x))$ as powers of x, we can use the Maple **Powmod** function. For example, the following command returns the field element that corresponds to x^6 in $\mathbb{Z}_3[x]/(f(x))$.

```
>   Powmod(x, 6, f(x), x) mod 3;
```
$$x + 2$$

The operation performed as a consequence of entering the preceding command is the polynomial x given in the first parameter raised to the power 6 given in the second parameter, with the output displayed after the result is reduced modulo the third parameter $f(x)$ defined over the specified coefficient modulus 3. The fourth parameter is the variable used in the first and third parameters.

We will now use a Maple **for** loop to construct and display the field elements that correspond to each of the first 8 powers of x in $\mathbb{Z}_3[x]/(f(x))$. Note that since $f(x)$ is primitive and $\mathbb{Z}_3[x]/(f(x))$ only has a total of eight nonzero elements, this will cause each of the nonzero elements in $\mathbb{Z}_3[x]/(f(x))$ to be displayed exactly once. In the following commands, we store the results returned by **Powmod** for each of the first 8 powers of x in the variable *temp*, and display these results using the Maple **print** function. Note where we use colons and semicolons in this loop, and note also that we use back quotes (') in the **print** statement.

```
>   for i from 1 to 8 do
>       temp := Powmod(x, i, f(x), x) mod 3:
>       print(x^i, ' Field Element: ', temp);
>   od:
```

$$x, \quad \textit{Field Element: }, x$$
$$x^2, \quad \textit{Field Element: }, 2x+1$$
$$x^3, \quad \textit{Field Element: }, 2x+2$$
$$x^4, \quad \textit{Field Element: }, 2$$
$$x^5, \quad \textit{Field Element: }, 2x$$
$$x^6, \quad \textit{Field Element: }, x+2$$
$$x^7, \quad \textit{Field Element: }, x+1$$
$$x^8, \quad \textit{Field Element: }, 1$$

Note that these results are identical to those listed in Example 1.10 for the nonzero elements in $\mathbb{Z}_3[x]/(f(x))$.

1.6   Finite Fields with MATLAB

In this section, we will show how MATLAB can be used to construct the nonzero elements as powers of x in a finite field $\mathbb{Z}_p[x]/(f(x))$ for prime p and primitive polynomial $f(x) \in \mathbb{Z}_p[x]$. We will consider the field used in Examples 1.9 and 1.10.

We will begin by declaring the variable x as symbolic, and defining the polynomial $f(x) = x^2 + x + 2 \in \mathbb{Z}_3[x]$ used to construct the field elements.

```
>> syms x
>> f = @(x) x^2+x+2

f =

    @(x) x^2 + x + 2
```

To verify that $f(x)$ is irreducible in $\mathbb{Z}_3[x]$, we will use the user-written function **Irreduc**, which we have written separately from this MATLAB session and saved as the M-file Irreduc.m. The following command illustrates how the function **Irreduc** can be used. The function will return TRUE if $f(x)$ is irreducible in $\mathbb{Z}_3[x]$, and FALSE otherwise.

```
>> Irreduc(f(x), 3)

ans =

TRUE
```

Thus, $f(x)$ is irreducible in $\mathbb{Z}_3[x]$, and $\mathbb{Z}_3[x]/(f(x))$ is a field. However, in order for us to be able to construct all of the nonzero elements in this field as powers of x, it must be the case that $f(x)$ is also primitive. To verify that $f(x)$ is primitive in $\mathbb{Z}_3[x]$, we will use the user-written function **Primitive**, which we have written separately from this MATLAB session and saved as the M-file Primitive.m. The following command will return TRUE if $f(x)$ is primitive in $\mathbb{Z}_3[x]$, and FALSE otherwise.

```
>> Primitive(f(x), 3)

ans =

TRUE
```

Thus, $f(x)$ is primitive in $\mathbb{Z}_3[x]$.

To construct the nonzero elements in $\mathbb{Z}_3[x]/(f(x))$ as powers of x, we will use the user-written function **Powmod**, which we have written separately from this MATLAB session and saved as the M-file Powmod.m. For example, the following command returns the field element that corresponds to x^6 in $\mathbb{Z}_3[x]/(f(x))$.

```
>> Powmod(x, 6, f(x), x, 3)

ans =

x + 2
```

The operation performed as a consequence of entering the preceding command is the polynomial x given in the first parameter raised to the power 6 given in the second parameter, with the output displayed after the result is reduced modulo the third parameter $f(x)$ defined over the modulus 3 given in the fifth parameter. The fourth parameter is the variable used in the first and third parameters.

We will now use a MATLAB **for** loop to construct and display the field elements that correspond to each of the first 8 powers of x in $\mathbb{Z}_3[x]/(f(x))$. Note that since $f(x)$ is primitive and $\mathbb{Z}_3[x]/(f(x))$ only has a total of eight nonzero elements, this will cause each of the nonzero elements in $\mathbb{Z}_3[x]/(f(x))$ to be displayed exactly once. In the following commands, we store the results returned by **Powmod** for each of the first 8 powers of x in the variable *temp*, and display these results as a vector. The MATLAB **pretty** function causes the results to be displayed in a form somewhat resembling typeset mathematics. Also, note where we use a semicolon to suppress the printing of *temp* after each passage through the loop.

```
>> for i = 1:8
        temp = Powmod(x, i, f(x), x, 3);
        pretty([x^i 'Field_Element' temp])
   end


+-                  -+
| x, Field_Element, x |
+-                  -+


+- 2                    -+
| x , Field_Element, 2 x + 1 |
+-                    -+


+- 3                    -+
| x , Field_Element, 2 x + 2 |
+-                    -+


+- 4               -+
| x , Field_Element, 2 |
+-               -+


+- 5                -+
| x , Field_Element, 2 x |
+-                -+


+- 6                -+
| x , Field_Element, x + 2 |
+-                -+
```

```
+- 7                              -+
| x , Field_Element, x + 1 |
+-                               -+


+- 8                        -+
| x , Field_Element, 1 |
+-                         -+
```

Note that these results are identical to those listed in Example 1.10 for the nonzero elements in $\mathbb{Z}_3[x]/(f(x))$.

1.7   The Euclidean Algorithm

Suppose a and b are nonzero elements in a Euclidean domain D, and consider an element $d \in D$ for which $d|a$ and $d|b$. Suppose that d also has the property that for all $x \in D$, if $x|a$ and $x|b$, then $x|d$. Then d is called a *greatest common divisor* of a and b, denoted $\gcd(a, b)$.

Greatest common divisors do not always exist for pairs of nonzero elements in rings, although as we will show in Theorem 1.16, greatest common divisors do always exist for pairs of nonzero elements in Euclidean domains. Also, greatest common divisors, when they exist, need not be unique. For example, in the Euclidean domain $\mathbb{Z}$, both 1 and -1 are greatest common divisors of any pair of distinct primes. However, it can easily be verified that if both d_1 and d_2 are greatest common divisors of a pair of nonzero elements in a Euclidean domain D, then d_1 and d_2 will be associates of each other in D.

Theorem 1.16 *Suppose a and b are nonzero elements in a Euclidean domain D. Then there exists a greatest common divisor d of a and b that can be expressed as $d = au + bv$ for some $u, v \in D$.*

Proof. Let B be an ideal of D of smallest order that contains both a and b. It can easily be shown that $B = \{ar + bs \mid r, s \in D\}$. Since D is a Euclidean domain, we know by Theorem 1.9 that D must also be a principal ideal domain, and $B = (d)$ for some $d \in D$. Since d generates B, and $a, b \in B$, then $d|a$ and $d|b$. Also, since $d \in B = \{ar+bs \mid r, s \in D\}$, then $d = au+bv$ for some $u, v \in D$. Now, if $x \in D$ with $x|a$ and $x|b$, then $a = xr$ and $b = xs$ for some $r, s \in D$. Thus, $d = au + bv = xru + xsv = x(ru + sv)$, and so $x|d$.   □

When considering specific rings, it is often convenient to place restrictions on greatest common divisors to make them unique. For example, for nonzero elements a and b in $\mathbb{Z}$, there is a unique positive greatest common divisor of a and b. Also, for nonzero polynomials a and b in the ring $F[x]$ of polynomials over a field F, there is a unique monic (i.e., with a leading coefficient of 1) greatest common divisor of a and b. Since these are the only two rings that we will use extensively in this book, for the remainder of this book we will assume that greatest common divisors are defined uniquely with these restrictions. We should also note that even though the greatest common divisor of two integers or polynomials a and b is uniquely defined with these restrictions, the u and v that yield $\gcd(a,b) = au + bv$ need not be unique.

In several of the applications in this book, we will need to determine not only the greatest common divisor of two integers or polynomials a and b, but also u and v that yield $\gcd(a,b) = au + bv$. We will use the Euclidean algorithm to do this. In order to describe how the Euclidean algorithm works, suppose D is a Euclidean domain with nonzero elements D^*, and let $\mathbb{N}$ be the set of nonnegative integers. Then for any $a, b \in D^*$, since D is a Euclidean domain, there is a mapping $\delta : D^* \to \mathbb{N}$ such that we can find $q_1, r_1 \in D$ with $a = bq_1 + r_1$ and either $r_1 = 0$ or $\delta(r_1) < \delta(b)$. Say $r_1 \neq 0$. Then we can find $q_2, r_2 \in D$ with $b = r_1 q_2 + r_2$ and either $r_2 = 0$ or $\delta(r_2) < \delta(r_1)$. Say $r_2 \neq 0$. Then we can find $q_3, r_3 \in D$ with $r_1 = r_2 q_3 + r_3$ and either $r_3 = 0$ or $\delta(r_3) < \delta(r_2)$. Suppose we continue this process until the first time $r_i = 0$, which is guaranteed to happen eventually since the $\delta(r_i)$ form a strictly decreasing sequence of nonnegative integers. This yields the following list of equations.

$$
\begin{array}{rcll}
a &=& bq_1 + r_1 & (r_1 \neq 0) \\
b &=& r_1 q_2 + r_2 & (r_2 \neq 0) \\
r_1 &=& r_2 q_3 + r_3 & (r_3 \neq 0) \\
&\vdots& & \vdots \\
r_{n-2} &=& r_{n-1} q_n + r_n & (r_n \neq 0) \\
r_{n-1} &=& r_n q_{n+1} + 0 &
\end{array}
$$

By working up this list of equations from the bottom, we can see that r_n will divide both a and b. By working down the list from the top, we can see that any $x \in D$ that divides both a and b will also divide r_n. Thus, $\gcd(a,b) = r_n$. This technique for determining $\gcd(a,b)$ is the *Euclidean algorithm*.

We have now shown a technique for determining the greatest common divisor of two integers or polynomials a and b. We must still show a technique for determining u and v that yields $\gcd(a,b) = au + bv$. To do this, we will consider the following table constructed using the quotients q_i and remainders r_i from the divisions in the Euclidean algorithm, and u_i, v_i that we will describe after the table. We will call this table a *Euclidean algorithm table*.

Row	Q	R	U	V
-1	$-$	$r_{-1} = a$	$u_{-1} = 1$	$v_{-1} = 0$
0	$-$	$r_0 = b$	$u_0 = 0$	$v_0 = 1$
1	q_1	r_1	u_1	v_1
2	q_2	r_2	u_2	v_2
$\vdots$	$\vdots$	$\vdots$	$\vdots$	$\vdots$
n	q_n	r_n	u_n	v_n

The entries in each row $i = 1, 2, \ldots, n$ of this table are constructed as follows. The q_i, r_i are from the following equation that results from the ith division in the Euclidean algorithm.

$$r_{i-2} \quad = \quad r_{i-1} q_i + r_i \tag{1.1}$$

Note that if we solve for r_i in (1.1), we obtain the following equation.

$$r_i \quad = \quad r_{i-2} - r_{i-1} q_i \tag{1.2}$$

The u_i, v_i in the Euclidean algorithm table are then constructed by following this exact same pattern for constructing r_i from q_i. Specifically, we construct u_i, v_i from q_i as follows.

$$u_i \quad = \quad u_{i-2} - u_{i-1} q_i \tag{1.3}$$
$$v_i \quad = \quad v_{i-2} - v_{i-1} q_i \tag{1.4}$$

Many useful relationships exist between the entries in a Euclidean algorithm table. For example, the following equation is true for all rows $i = -1, 0, 1, 2, \ldots, n$ in the table.

$$r_i \quad = \quad a u_i + b v_i \tag{1.5}$$

Clearly, this equation is true for $i = -1$ and 0. To see that it is true for all subsequent rows, suppose it is true for all rows i from -1 through $k - 1$. Then using (1.2), (1.3), and (1.4), we obtain the following.

$$
\begin{aligned}
r_k &= r_{k-2} - r_{k-1} q_k \\
&= (a u_{k-2} + b v_{k-2}) - (a u_{k-1} + b v_{k-1}) q_k \\
&= a(u_{k-2} - u_{k-1} q_k) + b(v_{k-2} - v_{k-1} q_k) \\
&= a u_k + b v_k
\end{aligned}
$$

Specifically, $r_n = a u_n + b v_n$. We have already noted that $r_n = \gcd(a, b)$. Thus, for $u = u_n$ and $v = v_n$, we will have $\gcd(a, b) = au + bv$.

Another useful relationship between the entries in a Euclidean algorithm table is the following equation for all rows $i = -1, 0, 1, 2, \ldots, n - 1$.

$$r_i u_{i+1} - u_i r_{i+1} \quad = \quad (-1)^i b \tag{1.6}$$

Clearly, this equation is true for $i = -1$. To see that it is true for all subsequent rows, suppose it is true for $i = k - 1$. Then using (1.2), (1.3), and the fact that adding a multiple of one row of a matrix to another row does not change the determinant of the matrix, we obtain the following.

$$
\begin{aligned}
r_k u_{k+1} - u_k r_{k+1} &= \begin{vmatrix} r_k & u_k \\ r_{k+1} & u_{k+1} \end{vmatrix} \\
&= \begin{vmatrix} r_k & u_k \\ r_{k-1} - r_k q_{k+1} & u_{k-1} - u_k q_{k+1} \end{vmatrix} \\
&= \begin{vmatrix} r_k & u_k \\ r_{k-1} & u_{k-1} \end{vmatrix} \\
&= r_k u_{k-1} - u_k r_{k-1} \\
&= -(r_{k-1} u_k - u_{k-1} r_k) \\
&= -(-1)^{k-1} b \\
&= (-1)^k b
\end{aligned}
$$

Two additional useful relationships that exist between the entries in a Euclidean algorithm table are the following pair of equations for all rows $i = -1, 0, 1, 2, \ldots, n - 1$.

$$
\begin{aligned}
r_i v_{i+1} - v_i r_{i+1} &= (-1)^{i+1} a && (1.7) \\
u_i v_{i+1} - u_{i+1} v_i &= (-1)^{i+1} && (1.8)
\end{aligned}
$$

These equations can be verified similarly to how we verified (1.6).

We close this section with two examples in which we use the Euclidean algorithm to find $\gcd(a, b)$, and then use a Euclidean algorithm table to find u and v such that $\gcd(a, b) = au + bv$.

Example 1.12 Let $a = 81$ and $b = 64$ in $\mathbb{Z}$. To use the Euclidean algorithm to find $\gcd(a, b)$, we form the following equations.

$$
\begin{aligned}
81 &= 64 \cdot 1 + 17 \\
64 &= 17 \cdot 3 + 13 \\
17 &= 13 \cdot 1 + 4 \\
13 &= 4 \cdot 3 + 1 \\
4 &= 1 \cdot 4 + 0
\end{aligned}
$$

Thus, $\gcd(81, 64) = 1$. It can then easily be verified that these equations yield the following Euclidean algorithm table.

Row	Q	R	U	V
−1	−	81	1	0
0	−	64	0	1
1	1	17	1	−1
2	3	13	−3	4
3	1	4	4	−5
4	3	1	−15	19

Therefore, $u = -15$ and $v = 19$ satisfy $\gcd(81, 64) = 81u + 64v$. □

Example 1.13 Let $a = x^6 + x^2 + x$ and $b = x^4 + x^2 + x$ in $\mathbb{Z}_2[x]$. To use the Euclidean algorithm to find $\gcd(a, b)$, we form the following equations.

$$
\begin{aligned}
a &= b(x^2 + 1) + x^3 \\
b &= x^3(x) + (x^2 + x) \\
x^3 &= (x^2 + x)(x + 1) + x \\
x^2 + x &= x(x + 1) + 0
\end{aligned}
$$

Thus, $\gcd(a, b) = x$. The u_i and v_i for the resulting Euclidean algorithm table are constructed as follows (with all coefficients expressed in $\mathbb{Z}_2$).

$$
\begin{aligned}
u_1 &= u_{-1} - u_0 q_1 &= 1 - 0(x^2 + 1) & &= 1 \\
v_1 &= v_{-1} - v_0 q_1 &= 0 - 1(x^2 + 1) & &= x^2 + 1 \\
u_2 &= u_0 - u_1 q_2 &= 0 - 1x & &= x \\
v_2 &= v_0 - v_1 q_2 &= 1 - (x^2 + 1)x & &= x^3 + x + 1 \\
u_3 &= u_1 - u_2 q_3 &= 1 - x(x + 1) & &= x^2 + x + 1 \\
v_3 &= v_1 - v_2 q_3 &= (x^2 + 1) - (x^3 + x + 1)(x + 1) & &= x^4 + x^3
\end{aligned}
$$

Therefore, the Euclidean algorithm table is the following.

Row	Q	R	U	V
−1	−	$x^6 + x^2 + x$	1	0
0	−	$x^4 + x^2 + x$	0	1
1	$x^2 + 1$	x^3	1	$x^2 + 1$
2	x	$x^2 + x$	x	$x^3 + x + 1$
3	$x + 1$	x	$x^2 + x + 1$	$x^4 + x^3$

Thus, $u = x^2 + x + 1$ and $v = x^4 + x^3$ satisfy $\gcd(a, b) = au + bv$. □

Exercises

1. For the set $\mathbb{Z}$ of integers, show that $(\mathbb{Z}, +)$ is a group with identity element 0, but $(\mathbb{Z}, \cdot)$ with normal integer multiplication is not a group.

2. Prove that the composition of two bijections on S is a bijection on S.

3. For the set $B(S)$ of all bijections on a set S and composition operation $\circ$ on $B(S)$, prove that $(B(S), \circ)$ is a group.

4. Show that $|S_n| = n!$.

5. Consider the following elements α, β, and γ in S_6.

$$\alpha : \begin{pmatrix} 1 & 2 & 3 & 4 & 5 & 6 \\ 4 & 3 & 6 & 2 & 1 & 5 \end{pmatrix}$$

$$\beta : \begin{pmatrix} 1 & 2 & 3 & 4 & 5 & 6 \\ 4 & 1 & 6 & 2 & 3 & 5 \end{pmatrix}$$

$$\gamma : \begin{pmatrix} 1 & 2 & 3 & 4 & 5 & 6 \\ 4 & 1 & 6 & 2 & 5 & 3 \end{pmatrix}$$

(a) Express $\alpha \circ \beta$ and $\beta \circ \gamma$ in array form (i.e., like how α, β, and γ are displayed), where $\circ$ represents the composition operation.

(b) Express α, β, and γ as a cycle or product of disjoint cycles.

(c) Is $\alpha \circ \gamma$ even or odd? Explain or show how you know.

(d) Find the inverses of α, β, and γ.

(e) Express α, β, and γ as a product of transpositions.

6. List the elements in A_4.

7. List the elements in D_5.

8. List the elements in $D_5 \cap A_5$.

9. Show that A_3 is cyclic.

10. Prove that if a group G is cyclic, then G must also be abelian.

11. Prove that if H is a subgroup of a cyclic group, then H must be cyclic.

12. Find the cyclic subgroups generated by and the orders of each of the following five group elements.

(a) The $144°$ rotation in D_5.

(b) The $144°$ rotation in D_{10}.

(c) Reflection across horizontal in D_{10}.

(d) The element α in Exercise 5.

(e) The element $(123)(45)(67)$ in A_7.

13. List the elements in each of the distinct left cosets of A_4 in S_4.

14. Prove Lagrange's Theorem (Theorem 1.4).

15. Prove that if H is a subgroup of a cyclic group G, then G/H must also be cyclic.

16. Show that the subgroup D_4 of S_4 is not normal.

17. Prove that the subgroup A_n of S_n is normal for all integers $n \geq 1$.

18. Prove that if G is an abelian group, then any subgroup H of G must be normal, and G/H must be abelian.

19. Let $G = (\mathbb{Z}, +)$, and let H be the cyclic subgroup of G generated by 4. List the elements in each of the distinct left cosets of H in G.

20. Verify that the function ϕ in Example 1.7 is a homomorphism, and find the kernel of ϕ.

21. Find the kernel of the homomorphism in Example 1.8.

22. Let G_1 and G_2 be groups, and suppose $\phi : G_1 \to G_2$ is a homomorphism. Prove that $\text{Ker}(\phi)$ is a normal subgroup of G_1.

23. Prove that all fields are integral domains.

24. Show that $\mathbb{Z}$ with ordinary addition and multiplication is an integral domain but not a field.

25. Prove that the only ideals of a field F are F and $\{0\}$.

26. Let F be a field, and suppose a is a fixed element in F. Prove that the mapping $\phi : F[x] \to F$ defined by $\phi(f(x)) = f(a)$ is a ring homomorphism, and find the kernel of ϕ.

27. Prove Proposition 1.8.

28. Show that $\mathbb{Z}$ is a Euclidean domain with $\delta(n) = |n|$.

29. Prove that the ring $F[x]$ of polynomials over a field F is a Euclidean domain with $\delta(f(x)) = \deg(f(x))$.

30. Prove Proposition 1.10.

31. Find all irreducible elements in the ring $\mathbb{Z}$ of integers.

32. Perform the following calculations:

    (a) $(x + 2) + (2x + 2)$ in the field D/B in Examples 1.9 and 1.10.

(b) $(x+2)(2x+2)$ in the field D/B in Examples 1.9 and 1.10.

(c) $(x+2)+(2x+2)$ in the field D/B in Example 1.11.

(d) $(x+2)(2x+2)$ in the field D/B in Example 1.11.

33. For each of the following polynomials $f(x)$, decide if $f(x)$ is primitive in $\mathbb{Z}_3[x]$. For each $f(x)$ that is primitive, construct the field elements that correspond to powers of x in $\mathbb{Z}_3[x]/(f(x))$. (Note: We will use the fields resulting from the $f(x)$ that are primitive later in this book.)

(a) $f(x) = x^2 + 2x + 2$

(b) $f(x) = x^2 + 2x + 1$

(c) $f(x) = 2x^2 + 2$

(d) $f(x) = 2x + 2$

34. For each of the following polynomials $f(x)$, decide if $f(x)$ is primitive in $\mathbb{Z}_2[x]$. For each $f(x)$ that is primitive, construct the field elements that correspond to powers of x in $\mathbb{Z}_2[x]/(f(x))$. (Note: We will use the fields resulting from the $f(x)$ that are primitive later in this book.)

(a) $f(x) = x^3 + x + 1$

(b) $f(x) = x^3 + x^2 + 1$

(c) $f(x) = x^4 + x + 1$

(d) $f(x) = x^4 + x^3 + 1$

(e) $f(x) = x^4 + x^3 + x^2 + 1$

(f) $f(x) = x^4 + x^3 + x^2 + x + 1$

35. Let $f(x) = x^2 + x + 2$.

(a) Show that $f(x)$ is primitive in $\mathbb{Z}_5[x]$ by constructing the field elements that correspond to powers of x in $\mathbb{Z}_5[x]/(f(x))$. (Note: We will use this field later in this book.)

(b) Show that $f(x)$ is not primitive in $\mathbb{Z}_{11}[x]$ by showing that $f(x)$ is not irreducible in $\mathbb{Z}_{11}[x]$.

36. Prove that if both d_1 and d_2 are greatest common divisors of a pair of nonzero elements in a Euclidean domain D, then d_1 and d_2 will be associates of each other in D.

37. Suppose a and b are nonzero elements in a Euclidean domain D, and let B be an ideal of D of smallest order that contains both a and b. Prove that $B = \{ar + bs \mid r, s \in D\}$.

38. Verify (1.7).

39. Verify (1.8).

40. Let $a = 448$ and $b = 153$ in $\mathbb{Z}$. Use the Euclidean algorithm to find $\gcd(a, b)$, and then use a Euclidean algorithm table to find u and v such that $\gcd(a, b) = au + bv$. (Note: You will need this Euclidean algorithm table in the Chapter 8 Exercises.)

41. Let $a = 2272$ and $b = 716$ in $\mathbb{Z}$. Use the Euclidean algorithm to find $\gcd(a, b)$, and then use a Euclidean algorithm table to find u and v such that $\gcd(a, b) = au + bv$.

42. Let $a = x^5 + x^4 + x^3 + x^2$ and $b = x^4 + x^3 + x + 1$ in $\mathbb{Z}_2[x]$. Use the Euclidean algorithm to find $\gcd(a, b)$, and then use a Euclidean algorithm table to find u and v such that $\gcd(a, b) = au + bv$.

Computer Exercises

43. For each of the following polynomials $f(x)$, all of which are primitive in $\mathbb{Z}_2[x]$, construct the field elements that correspond to powers of x in $\mathbb{Z}_2[x]/(f(x))$. (Note: We will use the fields resulting from these $f(x)$ later in this book.)

    (a) $f(x) = x^5 + x^3 + 1$
    (b) $f(x) = x^6 + x^5 + 1$
    (c) $f(x) = x^7 + x + 1$
    (d) $f(x) = x^8 + x^4 + x^3 + x^2 + 1$

44. For each of the following polynomials $f(x)$, both of which are primitive in $\mathbb{Z}_5[x]$, construct the field elements that correspond to powers of x in $\mathbb{Z}_5[x]/(f(x))$. (Note: We will use the fields resulting from these $f(x)$ later in this book.)

    (a) $f(x) = x^5 + 4x + 2$
    (b) $f(x) = 3x^7 + 4x + 1$

45. Find a primitive polynomial of degree 4 in $\mathbb{Z}_3[x]$, and use this polynomial to construct the nonzero elements in a finite field. (Note: You will need a field of this size in the Chapter 2 Exercises.)

46. Find a primitive polynomial of degree 2 in $\mathbb{Z}_{11}[x]$, and use this polynomial to construct the nonzero elements in a finite field. (Note: You will need a field of this size in the Chapter 2 Exercises.)

47. Use a primitive polynomial to construct the nonzero elements in a finite field of order 127.

Chapter 2

Block Designs

Suppose a magazine editor wishes to compare seven different automobiles by assessing the opinions of seven test-drivers with regard to a variety of topics such as handling and comfort. One way for the editor to obtain a fair comparison of the automobiles would be to have each of the test-drivers evaluate each of the vehicles. However, because of time or monetary constraints, it may not be reasonable for the editor to have each of the test-drivers evaluate each of the vehicles. A more reasonable way for the editor to obtain a comparison of the automobiles would be to have each of the test-drivers evaluate just one of the vehicles, but due to potential differences between the test-drivers, this might not yield a fair comparison of the vehicles. In this chapter, we will present some techniques that the editor could use to ensure a scheme for comparing the automobiles that is reasonable but still yields a fair comparison of the vehicles.

2.1 General Properties

Let $B_1, B_2, \ldots, B_b$ be subsets of a set $S = \{a_1, a_2, \ldots, a_v\}$. We will refer to the elements a_i as *objects* and to the subsets B_j as *blocks*. Suppose this collection of objects and blocks also satisfies the following three conditions.

1. Each object appears in the same number of blocks.

2. Each block contains the same number of objects.

3. Every possible pair of objects appears together in the same number of blocks.

Then this collection of objects and blocks is called a *balanced incomplete block design*. For convenience, we will refer to balanced incomplete block designs as simply *block designs*. We will describe a block design using the parameters (v, b, r, k, λ) if the design has v objects and b blocks, each object appears in r blocks, each block contains k objects, and every possible pair of objects appears together in λ blocks.

In all of the (v, b, r, k, λ) block designs that we will consider, we will assume that $k < v$ and $\lambda > 0$. It is reasonable to make these assumptions. Clearly $k \leq v$, and $k = v$ corresponds to the case in which each block contains all of the objects. In the example in the introduction to this chapter, this would represent the potentially unreasonable case in which each of the test-drivers (represented by the blocks) evaluates each of the vehicles (represented by the objects). Also, clearly $\lambda \geq 0$, and $\lambda = 0$ corresponds to the case in which each block contains just one of the objects. In the example in the introduction to this chapter, this would represent the potentially unfair case in which each of the test-drivers evaluates just one of the vehicles.

Example 2.1 Suppose a magazine editor wishes to obtain a fair and reasonable comparison of seven automobiles by assessing the opinions of seven test-drivers. If we represent the automobiles by the objects in the set $S = \{1, 2, 3, 4, 5, 6, 7\}$, then each test-driver can be represented by a block containing the vehicles to be evaluated by that test-driver. For example, the subsets $\{1, 2, 4\}$, $\{2, 3, 5\}$, $\{3, 4, 6\}$, $\{4, 5, 7\}$, $\{1, 5, 6\}$, $\{2, 6, 7\}$, and $\{1, 3, 7\}$ of S are the blocks in a $(7, 7, 3, 3, 1)$ block design. The blocks in this design indicate that the first test-driver should evaluate vehicles 1, 2, and 4, the second test-driver should evaluate vehicles 2, 3, and 5, and so forth. Notice that in this block design, each vehicle is evaluated three times, each test-driver evaluates three vehicles, and every possible pair of vehicles is evaluated by the same test-driver exactly once. Thus, this design yields a fair comparison of the automobiles while requiring a total of only 21 evaluations, as opposed to 49 evaluations if each of the test-drivers were to evaluate each of the vehicles. $\qquad\square$

In this chapter, we will present several techniques for constructing block designs, including a technique that yields the design in Example 2.1. Before presenting any of these techniques, we will first mention some general properties of block designs.

Theorem 2.1 *The parameters (v, b, r, k, λ) in any block design satisfy $vr = bk$ and $(v - 1)\lambda = r(k - 1)$.*

Proof. To show that the equation $vr = bk$ is true, we consider the set $T = \{(a, B) \mid a \text{ is an object in block } B\}$, and count $|T|$ in two different ways. First, since a (v, b, r, k, λ) block design has v objects that each appear in r blocks, then $|T| = vr$. However, since the design also has b blocks that each contain k objects, then $|T| = bk$. Thus, $vr = bk$. To show that the equation $(v - 1)\lambda = r(k - 1)$ is true, we consider a particular object a_i in the design, and for the set $U = \{(x, B) \mid x$ is an object with a_i in block $B\}$, we count $|U|$ in two different ways. First, since there are $v - 1$ objects in the design that each appear in λ blocks with a_i, then $|U| = (v - 1)\lambda$. However, since there are also r blocks in the design that each contain a_i and $k - 1$ other objects, then $|U| = r(k - 1)$. Thus, $(v - 1)\lambda = r(k - 1)$. □

For a block design with objects $a_1, a_2, \ldots, a_v$ and blocks $B_1, B_2, \ldots, B_b$, let A be the $v \times b$ matrix whose entry in the ith row and jth column is 1 if $a_i \in B_j$, and 0 otherwise. Then A is called an *incidence matrix* for the design.

Example 2.2 The following is the incidence matrix for the block design in Example 2.1, with the objects and blocks considered in the orders in which they are written in Example 2.1.

$$
\begin{bmatrix}
1 & 0 & 0 & 0 & 1 & 0 & 1 \\
1 & 1 & 0 & 0 & 0 & 1 & 0 \\
0 & 1 & 1 & 0 & 0 & 0 & 1 \\
1 & 0 & 1 & 1 & 0 & 0 & 0 \\
0 & 1 & 0 & 1 & 1 & 0 & 0 \\
0 & 0 & 1 & 0 & 1 & 1 & 0 \\
0 & 0 & 0 & 1 & 0 & 1 & 1
\end{bmatrix}
$$

□

In this chapter, we will use incidence matrices for two purposes. In the next section, we will use them to construct some specific block designs. In the remainder of this section, we will use them to prove some general results about block designs.

Let A be an incidence matrix for a (v, b, r, k, λ) block design, with objects $a_1, a_2, \ldots, a_v$ considered in this order. Note that the dot product of any row i of A with itself will be equal to the number r of blocks in the design that contain a_i. Also, note that the dot product of any row i of A with any other row j of A will be equal to the number λ of blocks in the design that contain both a_i and a_j. Since the entry in the ith row and jth column of the matrix AA^T can be viewed as the dot product of the ith row of A with the jth row of A, then AA^T is as follows, where I is the $v \times v$ identity matrix, and J is a $v \times v$ matrix containing a one in every position.

$$
AA^T = \begin{bmatrix}
r & \lambda & \cdots & \lambda \\
\lambda & r & \cdots & \lambda \\
\vdots & \vdots & & \vdots \\
\lambda & \lambda & \cdots & r
\end{bmatrix} = (r - \lambda)I + \lambda J
$$

Lemma 2.2 *Suppose B is a $v \times v$ matrix with $B = (r - \lambda)I + \lambda J$, where I is the $v \times v$ identity, and J is a $v \times v$ matrix containing a one in every position. Then the determinant of B is $det(B) = (r - \lambda)^{v-1}(r + (v - 1)\lambda)$.*

Proof. Note first that B will have the following form.

$$B = \begin{bmatrix} r & \lambda & \lambda & \cdots & \lambda & \lambda \\ \lambda & r & \lambda & \cdots & \lambda & \lambda \\ \lambda & \lambda & r & \cdots & \lambda & \lambda \\ \vdots & \vdots & \vdots & & \vdots & \vdots \\ \lambda & \lambda & \lambda & \cdots & r & \lambda \\ \lambda & \lambda & \lambda & \cdots & \lambda & r \end{bmatrix}$$

Now, if we subtract the first column of B from each of the other columns of B, we obtain the following matrix B_1.

$$B_1 = \begin{bmatrix} r & \lambda - r & \lambda - r & \cdots & \lambda - r & \lambda - r \\ \lambda & r - \lambda & 0 & \cdots & 0 & 0 \\ \lambda & 0 & r - \lambda & \cdots & 0 & 0 \\ \vdots & \vdots & \vdots & & \vdots & \vdots \\ \lambda & 0 & 0 & \cdots & r - \lambda & 0 \\ \lambda & 0 & 0 & \cdots & 0 & r - \lambda \end{bmatrix}$$

Then if we add to the first row of B_1 each of the other rows of B_1, we obtain the following matrix B_2.

$$B_2 = \begin{bmatrix} r + (v - 1)\lambda & 0 & 0 & \cdots & 0 & 0 \\ \lambda & r - \lambda & 0 & \cdots & 0 & 0 \\ \lambda & 0 & r - \lambda & \cdots & 0 & 0 \\ \vdots & \vdots & \vdots & & \vdots & \vdots \\ \lambda & 0 & 0 & \cdots & r - \lambda & 0 \\ \lambda & 0 & 0 & \cdots & 0 & r - \lambda \end{bmatrix}$$

Since B_2 is triangular, then $det(B_2)$ will be equal to the product of the diagonal entries of B_2. Thus, $det(B_2) = (r - \lambda)^{v-1}(r + (v - 1)\lambda)$. Since $det(B) = det(B_2)$, this establishes the result.   □

Theorem 2.3 *The parameters (v, b, r, k, λ) in any block design satisfy $v \leq b$ and $k \leq r$.*

Proof. Let A be an incidence matrix for a (v, b, r, k, λ) block design. Since we are assuming that $k < v$ in all block designs, by Theorem 2.1 it follows that $\lambda < r$. Then by Lemma 2.2, it follows that $det(AA^T) \neq 0$. However, since the rank of a product of matrices is at most the smallest rank of the factors, then rank$(A) \geq$ rank$(AA^T) = v$. Thus, since A is of size $v \times b$, it follows that $v \leq b$. By Theorem 2.1, it follows that $k \leq r$.   □

A block design is said to be *symmetric* if the number of objects in the design is the same as the number of blocks in the design. That is, a (v, b, r, k, λ) block design is symmetric if $v = b$, which, by Theorem 2.1, implies $r = k$. The block design in Example 2.1 is symmetric.

Theorem 2.4 *In a (v, v, r, r, λ) block design, every pair of blocks will contain exactly λ objects in common.*

Proof. Let A be an incidence matrix for a (v, v, r, r, λ) block design. By Lemma 2.2, it follows that A must be nonsingular. Also, for the $v \times v$ matrix J containing a one in every position, it will be the case that $AJ = JA$, since each entry in both products will be equal to r. Now, since $AA^T = (r - \lambda)I + \lambda J$ for the $v \times v$ identity matrix I, and $AJ = JA$, then the following shows that $AA^T A$ will be equal to AAA^T.

$$AA^T A = ((r - \lambda)I + \lambda J)A = A((r - \lambda)I + \lambda J) = AAA^T$$

Since A is nonsingular, it can be canceled from the left of both sides of the equation $AA^T A = AAA^T$, yielding $A^T A = AA^T = (r - \lambda)I + \lambda J$. Thus, the dot product of any two distinct columns of A (the off-diagonal entries in $A^T A$) will be equal to λ. From this, it follows that every pair of blocks in the design will contain exactly λ objects in common.   □

Theorem 2.4 states that in any symmetric block design, there will be a specific number of objects common to every pair of blocks, and this number will be equal to the number of blocks common to every pair of objects. Thus, in the block design in Example 2.1, every possible pair of test-drivers will evaluate the same vehicle exactly once, adding an extra dimension of fairness to the comparison scheme given by the design.

2.2   Hadamard Matrices

In this section, we will show how Hadamard matrices can be used to construct block designs. An $n \times n$ matrix H is called a *Hadamard matrix* if the entries in H are all 1 or -1, and $HH^T = nI$, where I is the $n \times n$ identity matrix.

For an $n \times n$ Hadamard matrix H, since $\frac{1}{n}H^T = H^{-1}$, then it is also true that $H^T H = nI$. As a result, since $HH^T = H^T H = nI$, we can see that the dot product of any row or column of H with itself will be equal to n, and the dot product of any two distinct rows or columns of H will be equal to 0. Thus, changing the sign of every entry in a row or column of H will yield another Hadamard matrix. Also, a Hadamard matrix H is said to be *normalized* if both the first row and first column of H contain only positive ones. Every Hadamard matrix can be converted into a normalized Hadamard matrix by changing the sign of each of the entries in necessary rows and columns. Because both the first row and first column of a normalized Hadamard matrix H will contain only positive ones, each of the other rows and columns of H will contain the same number of positive and negative ones. Thus, for a Hadamard matrix H of order n (i.e., of size $n \times n$), if $n > 1$, then n must be even. In fact, if $n > 2$, then n must be a multiple of 4. To see this, note that for $H = (h_{ij})$, the following holds.

$$\sum_j (h_{1j} + h_{2j})(h_{1j} + h_{3j}) = \sum_j h_{1j}^2 = n$$

Since $(h_{1j} + h_{2j})(h_{1j} + h_{3j}) = 0$ or 4 for each j, the result is apparent.

The only normalized Hadamard matrices of orders one and two are $H_1 = \begin{bmatrix} 1 \end{bmatrix}$ and $H_2 = \begin{bmatrix} 1 & 1 \\ 1 & -1 \end{bmatrix}$. Also, $H_4 = \begin{bmatrix} H_2 & H_2 \\ H_2 & -H_2 \end{bmatrix}$ is a normalized Hadamard matrix of order four. This construction of H_4 from H_2 can be generalized. Specifically, if H is a normalized Hadamard matrix, then so is $\begin{bmatrix} H & H \\ H & -H \end{bmatrix}$. This shows that there are Hadamard matrices of order 2^n for every nonnegative integer n.

We are interested in Hadamard matrices because they provide us with a method for constructing block designs. For a normalized Hadamard matrix H of order $4t \geq 8$, if we delete both the first row and first column from H, and change all of the negative ones in H into zeros, the resulting matrix will be an incidence matrix for a $(4t - 1, 4t - 1, 2t - 1, 2t - 1, t - 1)$ block design. We state this as the following theorem.

Theorem 2.5 *Suppose H is a normalized Hadamard matrix of order $4t \geq 8$. If both the first row and first column of H are deleted, and all of the negative ones in H are changed into zeros, the resulting matrix will be an incidence matrix for a $(4t - 1, 4t - 1, 2t - 1, 2t - 1, t - 1)$ block design.*

Proof. Delete both the first row and first column from H, and change all of the negative ones in H into zeros. Call the resulting matrix A. Every row and column of H except the first will contain $2t$ ones. Thus, every row and column of A will contain $2t - 1$ ones. As a result, the dot product of any row or column of A with itself will be equal to $r = 2t - 1$. Furthermore, in any pair of distinct rows of H excluding the first, there will be $2t$ positions in which the rows differ, t positions in which the rows both contain a one, and t positions in which the rows both contain a negative one. Thus, in the corresponding pair of rows of A, there will be $t - 1$ positions in which both rows contain a one. As a result, the dot product of any pair of distinct rows of A will be equal to $\lambda = t - 1$. Therefore, $AA^T = (r - \lambda)I + \lambda J$, where I is the $(4t - 1) \times (4t - 1)$ identity matrix, and J is a $(4t - 1) \times (4t - 1)$ matrix containing a one in every position. Since also $JA = rJ$, then A is an incidence matrix for a $(4t - 1, 4t - 1, 2t - 1, 2t - 1, t - 1)$ block design. $\square$

Example 2.3 Consider the following normalized Hadamard matrix of order 8, where H_4 is the normalized Hadamard matrix of order four that we constructed previously.

$$H_8 = \begin{bmatrix} H_4 & H_4 \\ H_4 & -H_4 \end{bmatrix}$$

Using H_8 as H, Theorem 2.5 states that the following matrix is an incidence matrix for a $(7, 7, 3, 3, 1)$ block design.

$$A = \begin{bmatrix} 0 & 1 & 0 & 1 & 0 & 1 & 0 \\ 1 & 0 & 0 & 1 & 1 & 0 & 0 \\ 0 & 0 & 1 & 1 & 0 & 0 & 1 \\ 1 & 1 & 1 & 0 & 0 & 0 & 0 \\ 0 & 1 & 0 & 0 & 1 & 0 & 1 \\ 1 & 0 & 0 & 0 & 0 & 1 & 1 \\ 0 & 0 & 1 & 0 & 1 & 1 & 0 \end{bmatrix}$$

Note that this incidence matrix is not the same as the incidence matrix in Example 2.2 for the $(7, 7, 3, 3, 1)$ block design in Example 2.1. $\square$

2.3 Hadamard Matrices with Maple

In this section, we will show how Maple can be used to construct the Hadamard matrices H_{2^n} and corresponding block designs. We will consider the design resulting from the incidence matrix in Example 2.3.

Because some of the functions that we will use are in the Maple **LinearAlgebra** package, we will begin by including this package.

```
>  with(LinearAlgebra):
```

Next, we will define the Hadamard matrix $H_1 = \begin{bmatrix} 1 \end{bmatrix}$.

```
>  H1 := Matrix([[1]]);
```

$$H1 := \begin{bmatrix} 1 \end{bmatrix}$$

Recall that the Hadamard matrix H_{2^k} can be constructed as a block matrix from the Hadamard matrix $H_{2^{k-1}}$. Thus, we can construct the Hadamard matrices H_2, H_4, and H_8 by using the Maple **Matrix** function as follows.

```
>  H2 := Matrix([[H1, H1], [H1, -H1]]);
```

$$H2 := \begin{bmatrix} 1 & 1 \\ 1 & -1 \end{bmatrix}$$

```
>  H4 := Matrix([[H2, H2], [H2, -H2]]);
```

$$H4 := \begin{bmatrix} 1 & 1 & 1 & 1 \\ 1 & -1 & 1 & -1 \\ 1 & 1 & -1 & -1 \\ 1 & -1 & -1 & 1 \end{bmatrix}$$

```
>  H8 := Matrix([[H4, H4], [H4, -H4]]);
```

$$H8 := \begin{bmatrix} 1 & 1 & 1 & 1 & 1 & 1 & 1 & 1 \\ 1 & -1 & 1 & -1 & 1 & -1 & 1 & -1 \\ 1 & 1 & -1 & -1 & 1 & 1 & -1 & -1 \\ 1 & -1 & -1 & 1 & 1 & -1 & -1 & 1 \\ 1 & 1 & 1 & 1 & -1 & -1 & -1 & -1 \\ 1 & -1 & 1 & -1 & -1 & 1 & -1 & 1 \\ 1 & 1 & -1 & -1 & -1 & -1 & 1 & 1 \\ 1 & -1 & -1 & 1 & -1 & 1 & 1 & -1 \end{bmatrix}$$

In the preceding three **Matrix** commands, the Hadamard matrices are constructed by listing their rows in order surrounded by brackets and separated by commas, with the individual blocks within each row also separated by commas. The normalized Hadamard matrices H_{2^k} for $k \geq 4$ can be constructed similarly.

We will now construct the incidence matrix shown in Example 2.3 that results from the Hadamard matrix H_8. We begin by deleting the first row and first column from H_8 by applying the Maple **DeleteRow** and **DeleteColumn** functions as follows.

```
>  A := DeleteRow(H8, 1):
```

```
>  A := DeleteColumn(A, 1):
```

We can then obtain the incidence matrix by changing each of the negative ones in A into zeros. To do this, we will use the following function f.

```
>  f := x -> if x = -1 then 0 else 1 fi:
```

To apply this function f to all of the entries in A at the same time, we can use the Maple **map** function as follows.

```
>  A := map(f, A);
```

$$A := \begin{bmatrix} 0 & 1 & 0 & 1 & 0 & 1 & 0 \\ 1 & 0 & 0 & 1 & 1 & 0 & 0 \\ 0 & 0 & 1 & 1 & 0 & 0 & 1 \\ 1 & 1 & 1 & 0 & 0 & 0 & 0 \\ 0 & 1 & 0 & 0 & 1 & 0 & 1 \\ 1 & 0 & 0 & 0 & 0 & 1 & 1 \\ 0 & 0 & 1 & 0 & 1 & 1 & 0 \end{bmatrix}$$

The resulting matrix A is the incidence matrix in Example 2.3.

Finally, we will list the objects that are contained within each of the blocks in the design. To do this, we first assign the values of the block design parameters v, b, and k.

```
>  v := 7:
```

```
>  b := 7:
```

```
>  k := 3:
```

Since each block in the design will contain k objects, we will create the following vector *block* of length k in which to store the objects that are contained within each block. By default, this vector will initially contain a zero in every position.

```
>  block := Vector[row](k);
```

$$block := \begin{bmatrix} 0 & 0 & 0 \end{bmatrix}$$

We can then use the following nested **for** loop to list the objects that are contained within each block. In these commands, the outer loop spans the columns of A, while the inner loop spans the rows.

```
>  for j from 1 to b do
```

```
>        bct := 0:
```

```
>        for i from 1 to v do
```

```
>            if A[i, j] = 1 then
>                bct := bct+1;
>                block[bct] := i;
>            fi;
>        od:
>        print('Block  ', j, '  contains objects  ', block);
>    od:
```

$$Block\ , 1,\ contains\ objects\ , \begin{bmatrix} 2 & 4 & 6 \end{bmatrix}$$

$$Block\ , 2,\ contains\ objects\ , \begin{bmatrix} 1 & 4 & 5 \end{bmatrix}$$

$$Block\ , 3,\ contains\ objects\ , \begin{bmatrix} 3 & 4 & 7 \end{bmatrix}$$

$$Block\ , 4,\ contains\ objects\ , \begin{bmatrix} 1 & 2 & 3 \end{bmatrix}$$

$$Block\ , 5,\ contains\ objects\ , \begin{bmatrix} 2 & 5 & 7 \end{bmatrix}$$

$$Block\ , 6,\ contains\ objects\ , \begin{bmatrix} 1 & 6 & 7 \end{bmatrix}$$

$$Block\ , 7,\ contains\ objects\ , \begin{bmatrix} 3 & 5 & 6 \end{bmatrix}$$

Note that in the preceding commands, we use colons after both **od** statements. This causes the output to be suppressed after each passage through the loops (except the output that results from the **print** function). Note also that, as in Section 1.5, we use back ticks in the **print** command.

2.4   Hadamard Matrices with MATLAB

In this section, we will show how MATLAB can be used to construct the Hadamard matrices H_{2^n} and corresponding block designs. We will consider the design resulting from the incidence matrix in Example 2.3.

We will begin by defining the Hadamard matrix $H_1 = \begin{bmatrix} 1 \end{bmatrix}$.

```
>> H1 = [1]

H1 =

     1
```

Recall that the Hadamard matrix H_{2^k} can be constructed as a block matrix from the Hadamard matrix $H_{2^{k-1}}$. Thus, we can construct the Hadamard matrices H_2, H_4, and H_8 as follows.

```
>> H2 = [H1 H1; H1 -H1]

H2 =

        1     1
        1    -1

>> H4 = [H2 H2; H2 -H2]

H4 =

        1     1     1     1
        1    -1     1    -1
        1     1    -1    -1
        1    -1    -1     1

>> H8 = [H4 H4; H4 -H4]

H8 =

        1     1     1     1     1     1     1     1
        1    -1     1    -1     1    -1     1    -1
        1     1    -1    -1     1     1    -1    -1
        1    -1    -1     1     1    -1    -1     1
        1     1     1     1    -1    -1    -1    -1
        1    -1     1    -1    -1     1    -1     1
        1     1    -1    -1    -1    -1     1     1
        1    -1    -1     1    -1     1     1    -1
```

The parameters in the preceding three commands are an ordered list of the blocks that form the resulting matrices, with each row terminated by a semicolon. The normalized Hadamard matrices H_{2^k} for $k \geq 4$ can be constructed similarly.

We will now construct the incidence matrix shown in Example 2.3 that results from the Hadamard matrix H_8. We begin by assigning the Hadamard matrix H_8 as A.

```
>> A = H8;
```

In the following two commands, we delete the first row and first column from A.

```
>> A(1, :) = [];
>> A(:, 1) = [];
```

In the preceding two commands, the colon refers to all of the elements in the first row and first column of A, respectively. The first row and first column of A are deleted by assigning each to an empty array, represented by the empty brackets.

We can now obtain the incidence matrix by changing each of the negative ones in A into zeros. To do this, we will use the MATLAB **find** function, which is used to determine the indices of array elements that meet a given criterion. In the following command, we create and store an array f that represents the indices where A contains negative values.

```
>> f = find(A < 0);
```

Next, we will use f to assign values of zero to the negative entries in A.

```
>> A(f) = 0

A =

     0     1     0     1     0     1     0
     1     0     0     1     1     0     0
     0     0     1     1     0     0     1
     1     1     1     0     0     0     0
     0     1     0     0     1     0     1
     1     0     0     0     0     1     1
     0     0     1     0     1     1     0
```

The resulting matrix A is the incidence matrix in Example 2.3.

Finally, we will list the objects that are contained within each of the blocks in the design. To do this, we first assign the value of the block design parameter b.

```
>> b = 7;
```

We can then use the following **for** loop to list the objects that are contained within each block.

```
>> for j = 1:b
        i = find(A(:, j) == 1)';
        fprintf('%s %2.0f %s', 'Block', j, ...
        ' contains objects')
        fprintf('%3.0f', i)
        fprintf('\n')
    end
Block  1  contains objects  2  4  6
```

```
Block  2  contains objects  1  4  5
Block  3  contains objects  3  4  7
Block  4  contains objects  1  2  3
Block  5  contains objects  2  5  7
Block  6  contains objects  1  6  7
Block  7  contains objects  3  5  6
```

In the preceding commands, the loop spans the columns of A. Then for each column, the **find** function determines the numbers of the rows in A where A contains a one, which shows the numbers of the objects that occur in the corresponding block. The single quote at the end of the **find** command is the transpose function, which converts i from a column vector into a row vector. Since we terminated the **find** command with a semicolon, the resulting output is suppressed. To display the results, we use the MATLAB **fprintf** function, which is designed to do formatted printing. In the first **fprintf** command, the first parameter is a string enclosed in single quotes that contains the format specifiers that will be used when the parameters that follow are displayed. The first %s instructs MATLAB to display the parameter *Block* as a string of characters. The %2.0f instructs MATLAB to display the parameter j as a fixed point number with a width of two spaces and no digits displayed to the right of the decimal. The second %s instructs MATLAB to also display *contains objects* as a string. In the second **fprintf** command, the %3.0f instructs MATLAB to display all of the elements in the vector i as fixed point numbers with widths of three spaces and no digits displayed to the right of the decimal. In the final **fprintf** command, the \n instructs MATLAB to end the current line and begin future printing on the next line.

2.5 Difference Sets

In this section, we will show how to construct difference sets, and how difference sets can be used to construct block designs. As we will show, difference sets yield block designs with more of a variety of parameters than designs that result from Hadamard matrices. Specifically, block designs that result from difference sets do not need to be symmetric.

Suppose G is an abelian group of order v with identity 0, and let D be a subset of order k in G. If every nonzero element in G can be expressed as the difference of two elements in D in exactly λ ways with $\lambda < k$, then D is called a *difference set* in G, described by the parameters (v, k, λ).

Example 2.4 The set $D = \{0, 1, 2, 4, 5, 8, 10\}$ is a $(15, 7, 3)$ difference set in $\mathbb{Z}_{15}$. $\qquad\square$

Example 2.5 The set $D = \{1, 2, 4\}$ is a $(7, 3, 1)$ difference set in $\mathbb{Z}_7$. Also, note that if we take the elements in $\mathbb{Z}_7$ one at a time, and add these elements to each of the elements in D (i.e., if we form the sets $i + D$ for $i = 0, 1, \ldots, 6$), the seven resulting sets are the blocks in the block design in Example 2.1 (with 0 represented by 7 in the blocks in Example 2.1). Thus, the $(7, 3, 1)$ difference set $D = \{1, 2, 4\}$ in $\mathbb{Z}_7$ can be used to construct the $(7, 7, 3, 3, 1)$ block design in Example 2.1. $\square$

The fact that a block design results from adding each of the elements in $\mathbb{Z}_7$ to each of the elements in the difference set D in Example 2.5 is guaranteed by the following theorem.

Theorem 2.6 Let $D = \{d_1, d_2, \ldots, d_k\}$ be a (v, k, λ) difference set in a group $G = \{g_1, g_2, \ldots, g_v\}$. Consider $g_i + D$ for $i = 1, 2, \ldots, v$ defined as follows.

$$g_i + D = \{g_i + d_1, g_i + d_2, \ldots, g_i + d_k\}$$

These sets are the blocks in a (v, v, k, k, λ) block design.

Proof. Clearly there are v objects in the design. Also, the v blocks $g_i + D$ for $i = 1, 2, \ldots, v$ are distinct, for if $g_i + D = g_j + D$ for some $i \neq j$, then $(g_i - g_j) + D = D$. We can then find k differences of elements in D that are equal to $g_i - g_j$, contradicting the assumption that $\lambda < k$. Now, if we add an element in D to each of the elements in G, the result will be G. Thus, each element in G will appear exactly k times among the elements $g_i + d_j$ for $i = 1, 2, \ldots, v$ and $j = 1, 2, \ldots, k$. Therefore, each element in G will appear in exactly k blocks. Also, by construction, each block will contain exactly k objects. It remains to be shown only that each pair of elements in G appears together in exactly λ blocks. Let $x, y \in G$ be distinct. If x and y appear together in some block $g + D$, then $x = g + d_i$ and $y = g + d_j$ for some i, j. Then $x - y = d_i - d_j$, and so $x - y$ is the difference of two elements in D. Since D is a (v, k, λ) difference set in G, then $x - y$ can be written as the difference of two elements in D in exactly λ ways. Since $x = g + d_i = h + d_i$ implies $g = h$, the difference $d_i - d_j$ cannot come from more than one block. Thus, x and y cannot appear together in more than λ blocks. On the other hand, suppose $x - y = d_i - d_j$ for some i, j. Then $x = g + d_i$ for some $g \in G$, and $y = x - (d_i - d_j) = (x - d_i) + d_j = g + d_j$. So x and y appear together in the block $g + D$. Therefore, x and y must appear together in at least λ blocks. With our previous result, this implies that x and y must appear together in exactly λ blocks. $\square$

As illustrated in Example 2.5, Theorem 2.6 provides us with a very easy method for constructing block designs from difference sets. Of course, in order to use Theorem 2.6, we must first have a difference set. Also,

Theorem 2.6 only yields block designs that are symmetric, at least some of which we could obtain by using Hadamard matrices (including the designs that result from each of the difference sets in Examples 2.4 and 2.5). So before discussing how to construct difference sets, we will first generalize them in a way that will allow us to use them to construct block designs that do not need to be symmetric.

Suppose G is an abelian group of order v with identity 0, and let $D_0, D_1, \ldots, D_{t-1}$ be subsets of order k in G. If every nonzero element in G can be expressed as the difference of two elements from the same subset D_i in exactly λ ways with $\lambda < k$, then $D_0, D_1, \ldots, D_{t-1}$ are called the *initial blocks* in a *generalized difference set* in G, described by the parameters (v, k, λ). Note that for $t = 1$, this definition reduces exactly to our prior definition of a difference set.

The following theorem generalizes the method given in Theorem 2.6 for constructing block designs from difference sets.

Theorem 2.7 *Let $D_0, D_1, \ldots, D_{t-1}$ be the initial blocks in a (v, k, λ) generalized difference set in a group $G = \{g_1, g_2, \ldots, g_v\}$. Then the sets $g_i + D_j$ for $i = 1, 2, \ldots, v$ and $j = 0, 1, \ldots, t - 1$ are the blocks in a (v, vt, kt, k, λ) block design.*

Proof. Exercise.   □

Example 2.6 The sets $D_0 = \{1, 7, 11\}$, $D_1 = \{2, 3, 14\}$, and $D_2 = \{4, 6, 9\}$ are the initial blocks in a $(19, 3, 1)$ generalized difference set in $\mathbb{Z}_{19}$. Theorem 2.7 states that if we take the elements in $\mathbb{Z}_{19}$ one at a time, and add these elements to each of the elements in D_0, D_1, and D_2, the 57 resulting sets will be the blocks in a $(19, 57, 9, 3, 1)$ block design.   □

As illustrated in Example 2.6, Theorem 2.7 provides us with a very easy method for constructing a (v, vt, kt, k, λ) block design, given that we are first able to find t initial blocks in a (v, k, λ) generalized difference set. However, it is not obvious how to find initial blocks in generalized difference sets. The following two propositions give us methods for doing exactly this.

Proposition 2.8 *Suppose that $v = 6t + 1 = p^n$ for some prime p and for some positive integers n and t. Let F be a finite field of order p^n, and choose $a \in F$ such that a is a cyclic generator for F^*. Then the sets $D_i = \{a^i, a^{2t+i}, a^{4t+i}\}$ for $i = 0, 1, \ldots, t - 1$ are the initial blocks in a $(6t + 1, 3, 1)$ generalized difference set in F.*

Proof. Exercise. (Hint: See the proof of Proposition 2.9.)   □

Example 2.7 We can use Proposition 2.8 to construct the initial blocks in Example 2.6 as follows. Let $F = \mathbb{Z}_{19}$, and choose cyclic generator $a = 2$ for $\mathbb{Z}_{19}^*$. Since $19 = 6t + 1$ implies $t = 3$, Proposition 2.8 yields three initial blocks. These initial blocks are $D_0 = \{2^0, 2^6, 2^{12}\} = \{1, 7, 11\}$, $D_1 = \{2^1, 2^7, 2^{13}\} = \{2, 14, 3\}$, and $D_2 = \{2^2, 2^8, 2^{14}\} = \{4, 9, 6\}$. $\square$

Proposition 2.9 *Suppose that* $v = 4t + 1 = p^n$ *for some prime* p *and for some positive integers* n *and* t. *Let* F *be a finite field of order* p^n, *and choose* $a \in F$ *such that* a *is a cyclic generator for* F^*. *Then the sets* $D_i = \{a^i, a^{t+i}, a^{2t+i}, a^{3t+i}\}$ *for* $i = 0, 1, \ldots, t-1$ *are the initial blocks in a* $(4t + 1, 4, 3)$ *generalized difference set in* F.

Proof. Since a is a cyclic generator for F^*, the order of a is $4t$. Thus, $a^{4t} = 1$, and $a^{2t} \neq 1$. Also, $a^{4t} - 1 = (a^{2t} - 1)(a^{2t} + 1) = 0$ implies $a^{2t} = -1$. Furthermore, $a^t - 1 \neq 0$. As a result, $a^t - 1 = a^s$ for some s between 1 and $4t$. Forming all possible differences from the sets $\pm a^i(a^t - 1)$, $\pm a^i(a^{2t} - 1)$, $\pm a^i(a^{2t} - a^t)$, $\pm a^i(a^{3t} - 1)$, $\pm a^i(a^{3t} - a^t)$, and $\pm a^i(a^{3t} - a^{2t})$, we obtain the following.

$$
\begin{array}{lcl}
\pm a^i(a^t - 1) & = \pm a^i(a^s) & = a^{i+s},\ a^{i+s+2t} \\
\pm a^i(a^{2t} - 1) & = \pm a^i(2a^{2t}) & = 2a^{i+2t},\ 2a^i \\
\pm a^i(a^{2t} - a^t) & = \pm a^i a^t(a^t - 1) & = a^{i+t+s},\ a^{i+3t+s} \\
\pm a^i(a^{3t} - 1) & = \pm a^i a^{3t}(1 - a^t) & = a^{i+t+s},\ a^{i+3t+s} \\
\pm a^i(a^{3t} - a^t) & = \pm a^i a^t(2a^{2t}) & = 2a^{i+3t},\ 2a^{i+t} \\
\pm a^i(a^{3t} - a^{2t}) & = \pm a^i a^{2t}(a^s) & = a^{i+2t+s},\ a^{i+s}
\end{array}
$$

Multiplication by a^s and 2 are bijections, and so these elements can be canceled from the preceding expressions. The only remaining elements are a^i, a^{t+i}, a^{2t+i}, and a^{3t+i} for $i = 0, 1, \ldots, t-1$ repeated three times each. Since these are all of the elements in F^*, then $\lambda = 3$. $\square$

Example 2.8 Suppose we wish to obtain a fair and reasonable comparison of 9 automobiles by assessing the opinions of 18 test-drivers. We can use Proposition 2.9 to construct a block design for this comparison as follows. We first need a finite field F of order 9 to represent the automobiles. For F, we will use the field we constructed in Example 1.10. For the cyclic generator a of F^*, we will use the element $x \in F$. Since $9 = 4t + 1$ gives $t = 2$, Proposition 2.9 yields two initial blocks in a $(9, 4, 3)$ generalized difference set in F. These initial blocks are $D_0 = \{1, x^2, x^4, x^6\} = \{1, 2x+1, 2, x+2\}$ and $D_1 = \{x, x^3, x^5, x^7\} = \{x, 2x+2, 2x, x+1\}$. Theorem 2.7 states that if we take the elements in F one at a time, and add these elements to each of the elements in D_0 and D_1, the 18 resulting sets will be the blocks in a $(9, 18, 8, 4, 3)$ block design. The blocks in this design are listed at the end of Sections 2.6 and 2.7. It can easily be verified that in this block design, each vehicle is evaluated 8 times, each test-driver evaluates 4 vehicles, and every possible pair of vehicles is evaluated by the same test-driver exactly 3 times.   □

2.6 Difference Sets with Maple

In this section, we will show how Maple can be used to construct initial blocks in generalized difference sets and corresponding block designs. We will consider the design resulting from the initial blocks in Example 2.8.

We will begin by defining the primitive polynomial $f(x) = x^2 + x + 2$ in $\mathbb{Z}_3[x]$ used to construct the elements in the finite field F.

```
>  f := x -> x^2+x+2:
```

```
>  Primitive(f(x)) mod 3;
```
$$true$$

Next, recall that since $v = 4t + 1 = 9$ implies $t = 2$, there will be 2 initial blocks. We assign the value of this parameter next.

```
>  t := 2:
```

Because the field elements are the objects that will fill the blocks, we need to store these elements in a way so that they can be recalled. We will do this by storing these elements in a vector. We first create the following vector with the same number of positions as the number of field elements.

```
>  field := Vector[row](4*t+1);
```
$$field := \begin{bmatrix} 0 & 0 & 0 & 0 & 0 & 0 & 0 & 0 & 0 \end{bmatrix}$$

We can then use the following commands to generate and store the field elements in the vector *field*. (Note the bracket [ ] syntax for accessing the positions in *field*.)

```
>  for i from 1 to 4*t do

>        field[i] := Powmod(x, i, f(x), x) mod 3:

>  od:

>  field[4*t+1] := 0:
```

We can view the entries in the vector *field* by entering the following command.

```
>  field;
```
$$\begin{bmatrix} x & 2x+1 & 2x+2 & 2 & 2x & x+2 & x+1 & 1 & 0 \end{bmatrix}$$

Next, we assign the parameter $k = 4$, which represents the number of objects contained in each initial block, and create a vector in which to store the initial blocks.

```
>  k := 4:
>  initblock := Vector[row](k);
```
$$initblock := \begin{bmatrix} 0 & 0 & 0 & 0 \end{bmatrix}$$

We can then generate and display the initial blocks by entering the following nested loop. In these commands, the outer loop spans the initial blocks, while the inner loop constructs the entries in each one.

```
>  for i from 0 to t-1 do
>      for j from 1 to k do
>          initblock[j] := Powmod(x, (j-1)*t+i, f(x), x)
           mod 3;
>      od:
>      print('Initial Block ', i, ' is ', initblock);
>  od:
```
$$Initial\ Block\ , 0,\ is\ , \begin{bmatrix} 1 & 2x+1 & 2 & x+2 \end{bmatrix}$$
$$Initial\ Block\ , 1,\ is\ , \begin{bmatrix} x & 2x+2 & 2x & x+1 \end{bmatrix}$$

In order to construct all of the blocks in the design, we first create the following vector in which to store the blocks, and initialize a counter bct that we will use to number the blocks.

```
>  block := Vector[row](k):
>  bct := 0:
```

We can then generate and display all of the blocks in the design by entering the following commands. In these commands, the outer loop spans the initial blocks, while the first inner loop constructs the entries in each one. The second inner (nested) loop adds the field elements one at a time to each of the elements in the initial blocks, yielding the blocks in the design.

```
>  for i from 0 to t-1 do
>      for j from 1 to k do
>          initblock[j] := Powmod(x, (j-1)*t+i, f(x), x)
           mod 3;
```

```
>        od:
>        for j from 1 to 4*t+1 do
>            for h from 1 to k do
>                block[h] := (field[j]+initblock[h]) mod 3;
>            od:
>            bct := bct+1;
>            print('Block ', bct, ' is ', block);
>        od:
>  od:
```

$$Block\ ,\ 1,\ is\ ,\ \left[\ x+1\ \ 1\ \ x+2\ \ 2\,x+2\ \right]$$

$$Block\ ,\ 2,\ is\ ,\ \left[\ 2\,x+2\ \ x+2\ \ 2\,x\ \ 0\ \right]$$

$$Block\ ,\ 3,\ is\ ,\ \left[\ 2\,x\ \ x\ \ 2\,x+1\ \ 1\ \right]$$

$$Block\ ,\ 4,\ is\ ,\ \left[\ 0\ \ 2\,x\ \ 1\ \ x+1\ \right]$$

$$Block\ ,\ 5,\ is\ ,\ \left[\ 2\,x+1\ \ x+1\ \ 2\,x+2\ \ 2\ \right]$$

$$Block\ ,\ 6,\ is\ ,\ \left[\ x\ \ 0\ \ x+1\ \ 2\,x+1\ \right]$$

$$Block\ ,\ 7,\ is\ ,\ \left[\ x+2\ \ 2\ \ x\ \ 2\,x\ \right]$$

$$Block\ ,\ 8,\ is\ ,\ \left[\ 2\ \ 2\,x+2\ \ 0\ \ x\ \right]$$

$$Block\ ,\ 9,\ is\ ,\ \left[\ 1\ \ 2\,x+1\ \ 2\ \ x+2\ \right]$$

$$Block\ ,\ 10,\ is\ ,\ \left[\ 2\,x\ \ 2\ \ 0\ \ 2\,x+1\ \right]$$

$$Block\ ,\ 11,\ is\ ,\ \left[\ 1\ \ x\ \ x+1\ \ 2\ \right]$$

$$Block\ ,\ 12,\ is\ ,\ \left[\ 2\ \ x+1\ \ x+2\ \ 0\ \right]$$

$$Block\ ,\ 13,\ is\ ,\ \left[\ x+2\ \ 2\,x+1\ \ 2\,x+2\ \ x\ \right]$$

$$Block\ ,\ 14,\ is\ ,\ \left[\ 0\ \ x+2\ \ x\ \ 1\ \right]$$

$$Block\ ,\ 15,\ is\ ,\ \left[\ 2\,x+2\ \ 1\ \ 2\ \ 2\,x\ \right]$$

$$Block\ ,\ 16,\ is\ ,\ \left[\ 2\,x+1\ \ 0\ \ 1\ \ 2\,x+2\ \right]$$

$$Block\ ,\ 17,\ is\ ,\ \left[\ x+1\ \ 2\,x\ \ 2\,x+1\ \ x+2\ \right]$$

$$Block\ ,\ 18,\ is\ ,\ \left[\ x\ \ 2\,x+2\ \ 2\,x\ \ x+1\ \right]$$

2.7   Difference Sets with MATLAB

In this section, we will show how MATLAB can be used to construct initial blocks in generalized difference sets and corresponding block designs. We will consider the design resulting from the initial blocks in Example 2.8.

We will begin by declaring the variable x as symbolic, and defining the primitive polynomial $f(x) = x^2 + x + 2 \in \mathbb{Z}_3[x]$ used to construct the elements in the finite field F.

```
>> syms x
>> f = @(x) x^2+x+2

f =

    @(x) x^2 + x + 2
```

Next, as in Section 1.6, to verify that $f(x)$ is primitive in $\mathbb{Z}_3[x]$, we use the user-written function **Primitive**, which we have written separately from this MATLAB session and saved as the M-file Primitive.m.

```
>> Primitive(f(x), 3)

ans =

TRUE
```

Recall that since $v = 4t + 1 = 9$ implies $t = 2$, there will be 2 initial blocks. We assign the value of this parameter next.

```
>> t = 2;
```

Because the field elements are the objects that will fill the blocks, we need to store these elements in a way so that they can be recalled. We will do this by storing these elements in a vector. In the following commands, we generate and store the field elements in the vector *field*. As in Section 1.6, to construct the field elements, we use the user-written function **Powmod**, which we have written separately from this MATLAB session and saved as the M-file Powmod.m.

```
>> for i = 1:4*t
        field(i) = Powmod(x, i, f(x), x, 3);
    end
```

```
>> field(4*t+1) = 0;
```

We can view the entries in the vector *field* by entering the following command.

```
>> field
```

```
field =
```

```
[   x,   2*x+1,   2*x+2,    2,   2*x,   x+2,   x+1,   1,   0]
```

Next, we assign the parameter $k = 4$, which represents the number of objects contained in each initial block.

```
>> k = 4;
```

We can then generate and display the initial blocks by entering the following nested loop. In these commands, the outer loop spans the initial blocks, while the inner loop constructs the entries in each one.

```
>> for i = 0:t-1
        for j = 1:k
            initblock(j) = Powmod(x, (j-1)*t+i, f(x), x, 3);
        end
        pretty(['Initial_Block' i 'is' initblock])
   end
```

```
+-                                          -+
| Initial_Block, 0, is, 1, 2 x + 1, 2, x + 2 |
+-                                          -+

+-                                          -+
| Initial_Block, 1, is, x, 2 x + 2, 2 x, x + 1 |
+-                                          -+
```

In order to construct all of the blocks in the design, we first initialize a counter *bct* that we will use to number the blocks.

```
>> bct = 0;
```

We can then generate and display all of the blocks in the design by entering the following commands. In these commands, the outer loop spans the initial blocks, while the first inner loop constructs the entries in each one.

The second inner (nested) loop adds the field elements one at a time to
each of the elements in the initial blocks, yielding the blocks in the design.

```
>> for i = 0:t-1
        for j = 1:k
            initblock(j) = Powmod(x, (j-1)*t+i, f(x), x, 3);
        end
        for j = 1:4*t+1
            for h = 1:k
                block(h) = mod(field(j)+initblock(h), 3);
            end
            bct = bct+1;
            pretty(['Block' bct 'is' block])
        end
   end
```

```
+-                                          -+
| Block, 1, is, x + 1, 1, x + 2, 2 x + 2 |
+-                                          -+

+-                                     -+
| Block, 2, is, 2 x + 2, x + 2, 2 x, 0 |
+-                                     -+

+-                                 -+
| Block, 3, is, 2 x, x, 2 x + 1, 1 |
+-                                 -+

+-                               -+
| Block, 4, is, 0, 2 x, 1, x + 1 |
+-                               -+

+-                                         -+
| Block, 5, is, 2 x + 1, x + 1, 2 x + 2, 2 |
+-                                         -+

+-                                   -+
| Block, 6, is, x, 0, x + 1, 2 x + 1 |
+-                                   -+

+-                               -+
| Block, 7, is, x + 2, 2, x, 2 x |
+-                               -+
```

```
+-                        -+
| Block, 8, is, 2, 2 x + 2, 0, x |
+-                        -+


+-                            -+
| Block, 9, is, 1, 2 x + 1, 2, x + 2 |
+-                            -+


+-                            -+
| Block, 10, is, 2 x, 2, 0, 2 x + 1 |
+-                            -+


+-                        -+
| Block, 11, is, 1, x, x + 1, 2 |
+-                        -+


+-                            -+
| Block, 12, is, 2, x + 1, x + 2, 0 |
+-                            -+


+-                                -+
| Block, 13, is, x + 2, 2 x + 1, 2 x + 2, x |
+-                                -+


+-                        -+
| Block, 14, is, 0, x + 2, x, 1 |
+-                        -+


+-                            -+
| Block, 15, is, 2 x + 2, 1, 2, 2 x |
+-                            -+


+-                                -+
| Block, 16, is, 2 x + 1, 0, 1, 2 x + 2 |
+-                                -+


+-                                    -+
| Block, 17, is, x + 1, 2 x, 2 x + 1, x + 2 |
+-                                    -+


+-                                -+
| Block, 18, is, x, 2 x + 2, 2 x, x + 1 |
+-                                -+
```

Exercises

1. Make a complete list of the blocks in the block design that results from the incidence matrix in Example 2.3. Then verify directly (i.e., without using Theorem 2.5) that these blocks are the blocks in a (7,7,3,3,1) block design.

2. Suppose a magazine editor wishes to obtain a comparison of 15 automobiles by assessing the opinions of 15 test-drivers. Construct a block design for this comparison. List the values of the parameters (v, b, r, k, λ) for the design, and state what each parameter represents.

3. Verify that the set D in Example 2.4 is a $(15, 7, 3)$ difference set in $\mathbb{Z}_{15}$.

4. Verify directly (i.e., without using Proposition 2.8) that the sets D_0, D_1, and D_2 in Example 2.6 are the initial blocks in a $(19, 3, 1)$ generalized difference set in $\mathbb{Z}_{19}$.

5. Make a complete list of the blocks in the block design in Example 2.6.

6. Verify directly (i.e., without using Proposition 2.9) that the sets D_0 and D_1 in Example 2.8 are the initial blocks in a $(9, 4, 3)$ generalized difference set in the finite field F in Example 1.10.

7. Construct a block design with $v = 17$ objects, and list the values of the parameters (v, b, r, k, λ) for the design.

8. Construct two different block designs with $v = 13$ objects, and list the values of the parameters (v, b, r, k, λ) for each design.

9. Suppose a magazine editor wishes to obtain a comparison of 25 automobiles by assessing the opinions of a certain number of test-drivers after each of the test-drivers evaluates three of the vehicles. Construct a block design for this comparison. (Construct initial blocks only if you use Proposition 2.8 or 2.9.) List the values of the parameters (v, b, r, k, λ) for the design, and state what each parameter represents.

10. Repeat Exercise 9, but this time assume that the magazine editor decides to have each of the test-drivers evaluate four of the vehicles instead of just three.

11. Repeat Exercise 9, but this time assume that the magazine editor decides to obtain a comparison of only 7 automobiles instead of 25. (Assume each test-driver evaluates three of the vehicles.)

12. A softball league is formed with 13 teams. The league has access to a field in the evenings, with enough time to play three games before the lights must be turned off. It is decided that on any given evening, three teams will show up at the field and play a total of three games, with each pair of teams playing exactly once. Is it possible for the league to schedule a full season so that each team plays every other team the same number of times? If so, how many times will each team play every other team, and how many evenings should the league reserve the field?

13. Repeat Exercise 12, but this time assume that the league has access to two adjacent fields in the evenings, with enough time to play three games on each field before the lights must be turned off, and that on any given evening, four teams will show up at the fields and play a total of six games, with each pair of teams playing exactly once.

14. Is it possible to use any of the techniques we presented in this chapter to construct a block design with $v = 529$ objects? What about with $v = 729$ objects? $v = 929$? Explain how you know for each one.

15. Prove that if H is a normalized Hadamard matrix, then so is the matrix $\begin{bmatrix} H & H \\ H & -H \end{bmatrix}$.

16. Prove Theorem 2.7.

17. Prove Proposition 2.8.

Computer Exercises[1]

18. Suppose a magazine editor wishes to obtain a comparison of 31 automobiles by assessing the opinions of 31 test-drivers. Construct a block design for this comparison. List the values of the parameters (v, b, r, k, λ) for the design, and state what each parameter represents.

19. Repeat Exercise 18, but this time assume that the magazine editor decides to use 155 test-drivers instead of just 31.

20. Suppose a magazine editor wishes to obtain a comparison of 49 automobiles by assessing the opinions of a certain number of test-drivers after each of the test-drivers evaluates four of the vehicles. Construct

---

[1]Maple by default suppresses output matrices that have more than 10 rows or columns. To display such matrices in Maple, the **interface** function with the **rtablesize** option may be used. We demonstrate the Maple **interface** function with the **rtablesize** option on page 61.

　　　　a block design for this comparison. List the values of the parameters (v, b, r, k, λ) for the design, and state what each parameter represents.

21. Repeat Exercise 20, but this time assume that the magazine editor decides to have each of the test-drivers evaluate only three of the vehicles instead of four.

22. Construct a block design with $v = 63$ objects. List the values of the parameters (v, b, r, k, λ) for the design.

23. Construct a block design with $v = 81$ objects. List the values of the parameters (v, b, r, k, λ) for the design.

24. Construct two different block designs with $v = 121$ objects. List the values of the parameters (v, b, r, k, λ) for each design.

25. Construct two different block designs with $v = 127$ objects. List the values of the parameters (v, b, r, k, λ) for each design.

Research Exercises

26. Investigate the life and career of the person for whom Hadamard matrices are named, and write a summary of your findings. Include in your summary Hadamard's political views and experiences, his most important mathematical contributions, some of his honors and awards, and some information about his family.

27. Investigate some other methods for constructing block designs different from those presented in this chapter, and write a summary of your findings.

28. Investigate *Kirkman's schoolgirl problem*, and write a summary of your findings. Include in your summary a statement of the problem, when and where the problem was first presented, some history of early solutions to the problem, and whether a block design provides a solution.

29. Investigate how the *Fano plane* models a particular block design, and write a summary of your findings.

30. Investigate how *Steiner systems* model block designs, and write a summary of your findings.

31. Investigate or create a mathematical use of difference sets or generalized difference sets outside the area of block designs, and write a summary of your findings or results.

Chapter 3

Error-Correcting Codes

In the next three chapters, we will present several types of error-correcting codes. A *code* is a set of messages, called *codewords*, which can be transmitted electronically between two parties. An *error-correcting* code is a code for which it is sometimes possible to detect and correct errors that occur during the transmission of codewords. Some applications of error-correcting codes include correction of errors that occur in information transmitted via the Internet, data stored in a computer, and music or video encoded on a compact disc or DVD. Error-correcting codes can also be used to correct errors that occur in information transmitted through space. For example, in Section 3.3, we will briefly discuss a specific error-correcting code that was used in the Mariner 9 space probe when it transmitted photographs back to Earth after entering into an orbit around Mars.

3.1 General Properties

In this chapter, we will look at some types of codes in which the codewords are vectors of a fixed length over $\mathbb{Z}_2$, with entries that we will call *bits*. We will denote the space of vectors of length n over $\mathbb{Z}_2$ as $\mathbb{Z}_2^n$. Thus, the codes that we will consider in this chapter will be subsets of $\mathbb{Z}_2^n$ for some n. A code in $\mathbb{Z}_2^n$ is not required to be a subspace of $\mathbb{Z}_2^n$. If a code is a subspace of $\mathbb{Z}_2^n$, then the code is called a *linear* code. We will present linear codes beginning in Section 3.6, and then continuing in Chapters 4 and 5.

The way that we will identify in general whether an error has occurred during the transmission of a codeword is to simply check whether the received vector is in the code. As a result, because our goal is to be able to detect and correct errors in received vectors, not all vectors in $\mathbb{Z}_2^n$ can be codewords in a particular code. If a received vector is in the code, then we will assume that it was the codeword that was sent. If not, then obviously an error occurred in at least one of the bits during transmission, and provided the received vector is of the correct length (i.e., provided no bits were dropped during transmission), we will attempt to correct the received vector to the codeword that was actually sent.

In general, we will use the "nearest neighbor" policy to correct received vectors, meaning we will assume that the fewest possible number of bit errors occurred, and correct a received vector to the codeword from which it differs in the fewest positions. This method of error correction is limited, of course, for there is not always a unique codeword that differs from a received vector in the fewest positions. Also, even when there is a unique codeword that differs from a received vector in the fewest positions, this codeword might not be the vector that was actually sent. However, since the nearest neighbor policy is usually the best approach to take when correcting errors in actual practice, it is the basis from which we will perform our error correction.

Example 3.1 Consider the code $C = \{(1010), (1110), (0011)\}$ in $\mathbb{Z}_2^4$. Suppose a codeword in C is transmitted, and we receive the vector $\mathbf{r}_1 = (0110)$. A quick search of C reveals that $\mathbf{c} = (1110)$ is the codeword from which $\mathbf{r}_1$ differs in the fewest positions. Thus, we would correct $\mathbf{r}_1$ to $\mathbf{c}$, and assume that the error in $\mathbf{r}_1$ was $\mathbf{r}_1 - \mathbf{c} = (1000)$. Now, suppose that a codeword in C is transmitted, and we receive the vector $\mathbf{r}_2 = (0010)$. Since two of the codewords in C differ from $\mathbf{r}_2$ in only one position, we cannot uniquely correct $\mathbf{r}_2$ using the nearest neighbor policy. So in C, we are not guaranteed to be able to uniquely correct a received vector even if it contains only a single bit error. □

To make the nearest neighbor policy error correction method more precise, we make the following definition. Let C be a code in $\mathbb{Z}_2^n$. For codewords $\mathbf{x}, \mathbf{y} \in C$, we define the *Hamming distance* $d(\mathbf{x}, \mathbf{y})$ from $\mathbf{x}$ to $\mathbf{y}$ to be the number of positions in which $\mathbf{x}$ and $\mathbf{y}$ differ. Thus, if $\mathbf{x} = (x_1, x_2, \ldots, x_n)$ and $\mathbf{y} = (y_1, y_2, \ldots, y_n)$, then $d(\mathbf{x}, \mathbf{y}) = \sum_{i=1}^{n} |x_i - y_i|$. We will call the smallest Hamming distance between all possible pairs of codewords in a code C the *minimum distance* of C, and denote this quantity by $d(C)$, or just d if there is no confusion regarding the code to which we are referring. For example, for the code C in Example 3.1, since the codewords (1010) and (1110) differ in only a single position, then $d = 1$.

Determining the number of bit errors that are guaranteed to be uniquely correctable in a given code is an important part of coding theory. To do this in general, we consider the following. For $\mathbf{x} \in \mathbb{Z}_2^n$ and positive integer r, let $S_r(\mathbf{x}) = \{\mathbf{y} \in \mathbb{Z}_2^n \mid d(\mathbf{x}, \mathbf{y}) \leq r\}$, called the *ball of radius r* around

x. Suppose C is a code with minimum distance d, and let t be the largest integer that satisfies $t < \frac{d}{2}$. Then $S_t(\mathbf{x}) \cap S_t(\mathbf{y})$ is empty for every pair $\mathbf{x}, \mathbf{y}$ of distinct codewords in C. Thus, if $\mathbf{z}$ is a received vector in $\mathbb{Z}_2^n$ with $d(\mathbf{u}, \mathbf{z}) \leq t$ for some codeword $\mathbf{u} \in C$, then $\mathbf{z} \in S_t(\mathbf{u})$, but $\mathbf{z} \notin S_t(\mathbf{v})$ for any other codeword $\mathbf{v} \in C$. That is, if a received vector $\mathbf{z}$ differs from a codeword $\mathbf{u}$ in t or fewer positions, then every other codeword will differ from $\mathbf{z}$ in more than t positions. So the nearest neighbor policy will always allow t or fewer bit errors to be uniquely corrected in the code, and the code is said to be *t-error correcting*.

Example 3.2 Consider the code $C = \{(00000000), (11100011), (00011111),$ $(11111100)\}$ in $\mathbb{Z}_2^8$. It can easily be verified that the minimum distance of C is $d = 5$. Since $t = 2$ is the largest integer that satisfies $t < \frac{d}{2}$, then C is two-error correcting. □

We will now address the problem of determining the number of vectors that are guaranteed to be uniquely correctable to a codeword in a given code. Suppose that C is a t-error correcting code in $\mathbb{Z}_2^n$. Note first that for any $\mathbf{x} \in \mathbb{Z}_2^n$, there will be $\binom{n}{k}$[1] vectors in $\mathbb{Z}_2^n$ that differ from $\mathbf{x}$ in exactly k positions. Also, any vector in $\mathbb{Z}_2^n$ that differs from $\mathbf{x}$ in exactly k positions will be in $S_t(\mathbf{x})$ provided $k \leq t$. Thus, the number of vectors in $S_t(\mathbf{x})$ will be $\binom{n}{0} + \binom{n}{1} + \cdots + \binom{n}{t}$. To determine the number of vectors in $\mathbb{Z}_2^n$ that are guaranteed to be uniquely correctable to a codeword in C, we must only count the number of vectors in $S_t(\mathbf{x})$ as $\mathbf{x}$ ranges through the codewords in C. Since the sets $S_t(\mathbf{x})$ are pairwise disjoint, the number of vectors in $\mathbb{Z}_2^n$ that differ from one of the codewords in C in t or fewer positions, and which are consequently guaranteed to be uniquely correctable to a codeword in C, is $|C| \cdot \left(\binom{n}{0} + \binom{n}{1} + \cdots + \binom{n}{t}\right)$. The fact that $|\mathbb{Z}_2^n| = 2^n$ then yields the following theorem, which gives a bound called the *Hamming bound* on the number of vectors that are guaranteed to be uniquely correctable to a codeword in a t-error correcting code.

Theorem 3.1 *Suppose that C is a t-error correcting code in $\mathbb{Z}_2^n$. Then* $|C| \cdot \left(\binom{n}{0} + \binom{n}{1} + \cdots + \binom{n}{t}\right) \leq 2^n.$

A code C in $\mathbb{Z}_2^n$ is said to be *perfect* if every vector in $\mathbb{Z}_2^n$ is guaranteed to be uniquely correctable to a codeword in C. More precisely, a code C in $\mathbb{Z}_2^n$ is perfect if the inequality in Theorem 3.1 is an equality. For the code C in Example 3.2, the factors in this inequality are $|C| = 4$, $\binom{8}{0} + \binom{8}{1} + \binom{8}{2} = 37$, and $2^8 = 256$. Thus, 108 of the vectors in $\mathbb{Z}_2^8$ are not guaranteed to be uniquely correctable to a codeword in this code (although some of these

---

[1]This notation $\binom{n}{k}$ represents the usual combinatorial value of n *choose* k, the number $\frac{n!}{k!(n-k)!}$ of different ways to choose k objects from a collection of n objects.

vectors may still be closest to a unique codeword). So we see that this code is far from perfect. In Section 3.6, we will present a type of code called a *Hamming code* that is perfect.

In practice, it is usually desirable to construct codes that have a large number of codewords and which are guaranteed to uniquely correct a large number of bit errors. However, the number of bit errors that are guaranteed to be uniquely correctable in a code is obviously related to the number of codewords in the code. Indeed, the problem of constructing a t-error correcting code C in $\mathbb{Z}_2^n$ with $|C|$ maximized for fixed values of n and t has been an important problem of recent mathematical interest. An equivalent problem is to find the maximum number of vectors in $\mathbb{Z}_2^n$ such that the balls of a fixed radius around the vectors can be arranged in the space without intersecting. This type of problem is called a *sphere packing* problem.

For the remainder of this chapter and the two subsequent chapters, we will present several methods for constructing various types of codes and correcting errors in these codes. To facilitate this, we will establish a set of parameters that we will use to describe codes. We will describe a code using the parameters (n, d) if the codewords in the code have length n positions and the code has minimum distance d.

3.2 Hadamard Codes

For our first method for constructing codes, we will refer back to block designs, the subject of Chapter 2. The following theorem states that the rows in an incidence matrix for a symmetric block design form the codewords in an error-correcting code.

Theorem 3.2 *Suppose A is an incidence matrix for a (v, v, r, r, λ) block design. Then the rows of A form a $(v, 2(r - \lambda))$ code with v codewords.*

Proof. There are v positions in each of the v rows of A. Thus, the rows of A form a code with v codewords, each of length v positions. It remains to be shown only that the minimum distance of this code is $2(r - \lambda)$. Consider any pair of rows R_1 and R_2 in A. Since every row of A contains ones in r positions, and each pair of rows of A contains ones in common in λ positions, there will be $r - \lambda$ positions in which R_1 contains a one and R_2 contains a zero, and $r - \lambda$ positions in which R_1 contains a zero and R_2 contains a one. This yields $2(r - \lambda)$ positions in which R_1 and R_2 differ. $\square$

Example 3.3 Theorem 3.2 states that the rows of the incidence matrix A in Example 2.3 form a $(7, 4)$ code with 7 codewords. $\square$

In Theorem 2.5, we showed that a normalized Hadamard matrix of order $4k \geq 8$ can be used to construct an incidence matrix for a

$(4k-1, 4k-1, 2k-1, 2k-1, k-1)$ symmetric block design. Theorem 3.2 states that the rows of such an incidence matrix A form codewords of length $4k-1$ positions in a code with minimum distance $2((2k-1)-(k-1)) = 2k$ and which contains $4k-1$ codewords. Also, recall that each of the rows of A will contain $2k$ zeros and $2k-1$ ones. Thus, there will be $2k$ positions in which the vector $(1\ 1\ \cdots\ 1)$ of length $4k-1$ positions differs from each of the rows of A. So by including the vector $(1\ 1\ \cdots\ 1)$ of length $4k-1$ positions with the rows of A, we obtain a $(4k-1, 2k)$ code with $4k$ codewords. In addition, as a result of Corollary 3.4 to the following theorem, no additional vectors can be included in this code without decreasing the minimum distance of the code. Because these $(4k-1, 2k)$ codes with $4k$ codewords are constructed from Hadamard matrices, we will call them *Hadamard codes*.

We will close this section with the following theorem and corollary, which verify the fact that no vectors can be joined to the codewords in a Hadamard code without decreasing the minimum distance of the code.

Theorem 3.3 *Suppose r is the number of codewords in an (n, d) code for some n, d with $d > \frac{n}{2}$. Then $r \le \frac{2d}{2d-n}$.*

Proof. Let $A = (a_{ij})$ be an $r \times n$ matrix with the codewords as rows, and let $S = \sum_{\mathbf{u},\mathbf{v}} d(\mathbf{u}, \mathbf{v})$ for all distinct pairs $\mathbf{u}, \mathbf{v}$ of codewords. Now, $d(\mathbf{u}, \mathbf{v}) \ge d$ for all pairs $\mathbf{u}, \mathbf{v}$ of codewords. Thus, $S \ge \binom{r}{2} d = \frac{r(r-1)}{2} d$. Let $t_0^{(i)}$ and $t_1^{(i)}$ be the number of times that 0 and 1 appear in the ith column of A, respectively. Then $t_1^{(i)} + t_0^{(i)} = r$ for all i. Also, for the set Ω consisting of all distinct pairs of rows of A, we have the following.

$$S = \sum_{\Omega} \sum_{j} |a_{ij} - a_{kj}| = \sum_{j} \sum_{\Omega} |a_{ij} - a_{kj}|$$

For each j, $\sum_{\Omega} |a_{ij} - a_{kj}|$ is equal to the number of times that any pair of distinct rows of A contain differing entries in the jth position. This number is $t_0^{(j)} t_1^{(j)}$, and so $S = \sum_{j} t_0^{(j)} \left(r - t_0^{(j)} \right)$. To find an upper bound on $t_0^{(j)} t_1^{(j)}$, consider the function $f(x) = x(r-x)$ for $0 \le x \le r$. Note that $f(x)$ is maximized at the point $(x, f(x)) = \left(\frac{r}{2}, \frac{r^2}{4} \right)$. Thus, $t_0^{(j)} t_1^{(j)} \le \frac{r^2}{4}$, and $S \le \frac{nr^2}{4}$. So $\frac{r(r-1)d}{2} \le \frac{nr^2}{4}$, and $r \left(d - \frac{n}{2} \right) \le d$. Therefore, $r \le \frac{d}{d - \frac{n}{2}} = \frac{2d}{2d-n}$. $\square$

Corollary 3.4 *Suppose r is the number of codewords in a $(4k-1, 2k)$ code for some k. Then $r \le 4k$.*

Proof. Exercise. $\square$

3.3 Reed-Muller Codes

In Section 3.2, we showed that a normalized Hadamard matrix can be used to construct an error-correcting code called a Hadamard code. We also showed that the number of codewords in a Hadamard code is maximal in the sense that no additional vectors can be included without decreasing the minimum distance of the code. However, as a consequence of the following theorem, by increasing the length of the codewords in a Hadamard code by only a single position, we can double the number of codewords without decreasing the minimum distance of the code.

Theorem 3.5 *Suppose A is the incidence matrix that results from a normalized Hadamard matrix of order $4k$, and suppose B is the matrix obtained by interchanging all zeros and ones in A. Let $\mathcal{A}$ be the matrix obtained by placing a one in front of all of the rows of A, and let $\mathcal{B}$ be the matrix obtained by placing a zero in front of all of the rows of B. Then the rows of $\mathcal{A}$ and $\mathcal{B}$ taken together form a $(4k, 2k)$ code with $8k - 2$ codewords.*

Proof. Exercise. □

Just as we included the vector $(1\ 1\ \cdots\ 1)$ with the rows of an incidence matrix when we formed Hadamard codes, we can do the same for the codes defined by Theorem 3.5. In fact, we can even do a little better. Note that each of the rows in the matrices $\mathcal{A}$ and $\mathcal{B}$ in Theorem 3.5 will contain $2k$ zeros and $2k$ ones. Thus, there will be $2k$ positions in which both of the vectors $(0\ 0\ \cdots\ 0)$ and $(1\ 1\ \cdots\ 1)$ of length $4k$ positions differ from each of the rows of $\mathcal{A}$ and $\mathcal{B}$. So by including the vectors $(0\ 0\ \cdots\ 0)$ and $(1\ 1\ \cdots\ 1)$ of length $4k$ positions with the rows of $\mathcal{A}$ and $\mathcal{B}$, we obtain a $(4k, 2k)$ code with $8k$ codewords. It is also true that no additional vectors can be included in this code without decreasing the minimum distance of the code, although we will not prove this fact. These $(4k, 2k)$ codes with $8k$ codewords are called *Reed-Muller* codes, and they have a storied history.

One notable use of a Reed-Muller code was in the Mariner 9 space probe, which transmitted photographs back to Earth after entering into an orbit around Mars in which it remains to this day. The specific code used in the probe was the $(32, 16)$ Reed-Muller code with 64 codewords constructed using the normalized Hadamard matrix H_{32}. Before being transmitted from the Mariner probe, photographs were broken down into a collection of pixels. Each pixel was assigned one of 64 levels of grayness, and then encoded into one of the 64 codewords in the code. Mariner 9 used this code to transmit a total of 7,329 images of Mars, covering the entire planet. Since transmissions from the probe had to cover such a long distance and were fairly weak, there was a marked potential for bit errors, making the high error correction capability of the code necessary.

3.4   Reed-Muller Codes with Maple

In this section, we will show how Maple can be used to construct and correct errors in Reed-Muller codes. We will consider the $(16, 8)$ Reed-Muller code.

We will begin by generating the normalized Hadamard matrix H_{16} used to construct the code.

```
>  with(LinearAlgebra):
>  H1 := Matrix([[1]]):
>  H2 := Matrix([[H1, H1], [H1, -H1]]):
>  H4 := Matrix([[H2, H2], [H2, -H2]]):
>  H8 := Matrix([[H4, H4], [H4, -H4]]):
>  H16 := Matrix([[H8, H8], [H8, -H8]]):
```

We can then obtain the incidence matrix A that results from H_{16} by entering the following sequence of commands.

```
>  A := DeleteRow(H16, 1):
>  A := DeleteColumn(A, 1):
>  f := x -> if x = -1 then 0 else 1 fi:
>  A := map(f, A);
```

$$A := \begin{bmatrix} \textit{15 x 15 Matrix} \\ \textit{Data Type: anything} \\ \textit{Storage: rectangular} \\ \textit{Order: Fortran\_order} \end{bmatrix}$$

By default, Maple only displays matrices of size 10×10 and smaller. As demonstrated in the preceding command, matrices that are larger than 10×10 are displayed with a placeholder. The Maple **interface** function with the **rtablesize** option can be used to display matrices that are larger than 10×10. Once a matrix dimension is specified with this function, all matrices of this size and smaller will be displayed until the Maple session is closed or until the function is entered again with a different dimension. By entering the following command, we cause Maple to display all matrices of size 50×50 and smaller throughout the remainder of this Maple session.

```
>  interface(rtablesize=50):
```

We can now see the incidence matrix that is stored as the variable A by entering the following command.

```
>  A;
```

$$
\begin{bmatrix}
0 & 1 & 0 & 1 & 0 & 1 & 0 & 1 & 0 & 1 & 0 & 1 & 0 & 1 & 0 \\
1 & 0 & 0 & 1 & 1 & 0 & 0 & 1 & 1 & 0 & 0 & 1 & 1 & 0 & 0 \\
0 & 0 & 1 & 1 & 0 & 0 & 1 & 1 & 0 & 0 & 1 & 1 & 0 & 0 & 1 \\
1 & 1 & 1 & 0 & 0 & 0 & 0 & 1 & 1 & 1 & 1 & 0 & 0 & 0 & 0 \\
0 & 1 & 0 & 0 & 1 & 0 & 1 & 1 & 0 & 1 & 0 & 0 & 1 & 0 & 1 \\
1 & 0 & 0 & 0 & 0 & 1 & 1 & 1 & 1 & 0 & 0 & 0 & 0 & 1 & 1 \\
0 & 0 & 1 & 0 & 1 & 1 & 0 & 1 & 0 & 0 & 1 & 0 & 1 & 1 & 0 \\
1 & 1 & 1 & 1 & 1 & 1 & 1 & 0 & 0 & 0 & 0 & 0 & 0 & 0 & 0 \\
0 & 1 & 0 & 1 & 0 & 1 & 0 & 0 & 1 & 0 & 1 & 0 & 1 & 0 & 1 \\
1 & 0 & 0 & 1 & 1 & 0 & 0 & 0 & 0 & 1 & 1 & 0 & 0 & 1 & 1 \\
0 & 0 & 1 & 1 & 0 & 0 & 1 & 0 & 1 & 1 & 0 & 0 & 1 & 1 & 0 \\
1 & 1 & 1 & 0 & 0 & 0 & 0 & 0 & 0 & 0 & 0 & 1 & 1 & 1 & 1 \\
0 & 1 & 0 & 0 & 1 & 0 & 1 & 0 & 1 & 0 & 1 & 1 & 0 & 1 & 0 \\
1 & 0 & 0 & 0 & 0 & 1 & 1 & 0 & 0 & 1 & 1 & 1 & 1 & 0 & 0 \\
0 & 0 & 1 & 0 & 1 & 1 & 0 & 0 & 1 & 1 & 0 & 1 & 0 & 0 & 1
\end{bmatrix}
$$

Next, to form the matrix B in Theorem 3.5, we need to interchange all zeros and ones in A. To do this, we will define and apply the following function g to each of the entries in A.

```
>  g := x -> if x = 0 then 1 else 0 fi:
>  B := map(g, A);
```

$$
B :=
\begin{bmatrix}
1 & 0 & 1 & 0 & 1 & 0 & 1 & 0 & 1 & 0 & 1 & 0 & 1 & 0 & 1 \\
0 & 1 & 1 & 0 & 0 & 1 & 1 & 0 & 0 & 1 & 1 & 0 & 0 & 1 & 1 \\
1 & 1 & 0 & 0 & 1 & 1 & 0 & 0 & 1 & 1 & 0 & 0 & 1 & 1 & 0 \\
0 & 0 & 0 & 1 & 1 & 1 & 1 & 0 & 0 & 0 & 0 & 1 & 1 & 1 & 1 \\
1 & 0 & 1 & 1 & 0 & 1 & 0 & 0 & 1 & 0 & 1 & 1 & 0 & 1 & 0 \\
0 & 1 & 1 & 1 & 1 & 0 & 0 & 0 & 0 & 1 & 1 & 1 & 1 & 0 & 0 \\
1 & 1 & 0 & 1 & 0 & 0 & 1 & 0 & 1 & 1 & 0 & 1 & 0 & 0 & 1 \\
0 & 0 & 0 & 0 & 0 & 0 & 0 & 1 & 1 & 1 & 1 & 1 & 1 & 1 & 1 \\
1 & 0 & 1 & 0 & 1 & 0 & 1 & 1 & 0 & 1 & 0 & 1 & 0 & 1 & 0 \\
0 & 1 & 1 & 0 & 0 & 1 & 1 & 1 & 1 & 0 & 0 & 1 & 1 & 0 & 0 \\
1 & 1 & 0 & 0 & 1 & 1 & 0 & 1 & 0 & 0 & 1 & 1 & 0 & 0 & 1 \\
0 & 0 & 0 & 1 & 1 & 1 & 1 & 1 & 1 & 1 & 1 & 0 & 0 & 0 & 0 \\
1 & 0 & 1 & 1 & 0 & 1 & 0 & 1 & 0 & 1 & 0 & 0 & 1 & 0 & 1 \\
0 & 1 & 1 & 1 & 1 & 0 & 0 & 1 & 1 & 0 & 0 & 0 & 0 & 1 & 1 \\
1 & 1 & 0 & 1 & 0 & 0 & 1 & 1 & 0 & 0 & 1 & 0 & 1 & 1 & 0
\end{bmatrix}
$$

Now, recall that to form the matrices $\mathcal{A}$ and $\mathcal{B}$ in Theorem 3.5, we must only place a one in front of all of the rows of A and a zero in front of all of the rows of B. To do this, we will define the following column vectors *colA* and *colB*. (For formatting purposes, we display the transpose of these vectors.)

```
> colA := Vector(RowDimension(A), 1):
```

```
> Transpose(colA);
```

$$\begin{bmatrix} 1 & 1 & 1 & 1 & 1 & 1 & 1 & 1 & 1 & 1 & 1 & 1 & 1 & 1 & 1 \end{bmatrix}$$

```
> colB := Vector(RowDimension(B), 0):
```

```
> Transpose(colB);
```

$$\begin{bmatrix} 0 & 0 & 0 & 0 & 0 & 0 & 0 & 0 & 0 & 0 & 0 & 0 & 0 & 0 & 0 \end{bmatrix}$$

By construction, the vectors *colA* and *colB* have the same number of positions as the number of rows in A and B, respectively. Thus, by placing these vectors as columns in front of A and B, respectively, we will obtain the matrices $\mathcal{A}$ and $\mathcal{B}$. We can do this by using the Maple **Matrix** function as follows.

```
> scriptA := Matrix([colA, A]):
```

```
> scriptB := Matrix([colB, B]):
```

The rows of the matrices $\mathcal{A}$ and $\mathcal{B}$ taken together form all but two of the codewords in the $(16, 8)$ Reed-Muller code. The two codewords not included in the rows of these matrices are the vectors $(0\ 0\ \cdots\ 0)$ and $(1\ 1\ \cdots\ 1)$ of length 16 positions. We will create these two vectors next.

```
> v_zero := Vector[row](ColumnDimension(scriptB), 0);
```

$$v\_zero := \begin{bmatrix} 0 & 0 & 0 & 0 & 0 & 0 & 0 & 0 & 0 & 0 & 0 & 0 & 0 & 0 & 0 & 0 \end{bmatrix}$$

```
> v_one := Vector[row](ColumnDimension(scriptB), 1);
```

$$v\_one := \begin{bmatrix} 1 & 1 & 1 & 1 & 1 & 1 & 1 & 1 & 1 & 1 & 1 & 1 & 1 & 1 & 1 & 1 \end{bmatrix}$$

We can then view the codewords in the $(16, 8)$ Reed-Muller code by using the Maple **Matrix** function as follows to stack together the matrices $\mathcal{A}$ and $\mathcal{B}$ and the vectors *v_zero* and *v_one*.

```
> cw := Matrix([[scriptA], [scriptB], [v_zero], [v_one]]);
```

$$cw := \begin{bmatrix}
1 & 0 & 1 & 0 & 1 & 0 & 1 & 0 & 1 & 0 & 1 & 0 & 1 & 0 & 1 & 0 \\
1 & 1 & 0 & 0 & 1 & 1 & 0 & 0 & 1 & 1 & 0 & 0 & 1 & 1 & 0 & 0 \\
1 & 0 & 0 & 1 & 1 & 0 & 0 & 1 & 1 & 0 & 0 & 1 & 1 & 0 & 0 & 1 \\
1 & 1 & 1 & 1 & 0 & 0 & 0 & 0 & 1 & 1 & 1 & 1 & 0 & 0 & 0 & 0 \\
1 & 0 & 1 & 0 & 0 & 1 & 0 & 1 & 1 & 0 & 1 & 0 & 0 & 1 & 0 & 1 \\
1 & 1 & 0 & 0 & 0 & 0 & 1 & 1 & 1 & 1 & 0 & 0 & 0 & 0 & 1 & 1 \\
1 & 0 & 0 & 1 & 0 & 1 & 1 & 0 & 1 & 0 & 0 & 1 & 0 & 1 & 1 & 0 \\
1 & 1 & 1 & 1 & 1 & 1 & 1 & 1 & 0 & 0 & 0 & 0 & 0 & 0 & 0 & 0 \\
1 & 0 & 1 & 0 & 1 & 0 & 1 & 0 & 0 & 1 & 0 & 1 & 0 & 1 & 0 & 1 \\
1 & 1 & 0 & 0 & 1 & 1 & 0 & 0 & 0 & 0 & 1 & 1 & 0 & 0 & 1 & 1 \\
1 & 0 & 0 & 1 & 1 & 0 & 0 & 1 & 0 & 1 & 1 & 0 & 0 & 1 & 1 & 0 \\
1 & 1 & 1 & 1 & 0 & 0 & 0 & 0 & 0 & 0 & 0 & 0 & 1 & 1 & 1 & 1 \\
1 & 0 & 1 & 0 & 0 & 1 & 0 & 1 & 0 & 1 & 0 & 1 & 1 & 0 & 1 & 0 \\
1 & 1 & 0 & 0 & 0 & 0 & 1 & 1 & 0 & 0 & 1 & 1 & 1 & 1 & 0 & 0 \\
1 & 0 & 0 & 1 & 0 & 1 & 1 & 0 & 0 & 1 & 1 & 0 & 1 & 0 & 0 & 1 \\
0 & 1 & 0 & 1 & 0 & 1 & 0 & 1 & 0 & 1 & 0 & 1 & 0 & 1 & 0 & 1 \\
0 & 0 & 1 & 1 & 0 & 0 & 1 & 1 & 0 & 0 & 1 & 1 & 0 & 0 & 1 & 1 \\
0 & 1 & 1 & 0 & 0 & 1 & 1 & 0 & 0 & 1 & 1 & 0 & 0 & 1 & 1 & 0 \\
0 & 0 & 0 & 0 & 1 & 1 & 1 & 1 & 0 & 0 & 0 & 0 & 1 & 1 & 1 & 1 \\
0 & 1 & 0 & 1 & 1 & 0 & 1 & 0 & 0 & 1 & 0 & 1 & 1 & 0 & 1 & 0 \\
0 & 0 & 1 & 1 & 1 & 1 & 0 & 0 & 0 & 0 & 1 & 1 & 1 & 1 & 0 & 0 \\
0 & 1 & 1 & 0 & 1 & 0 & 0 & 1 & 0 & 1 & 1 & 0 & 1 & 0 & 0 & 1 \\
0 & 0 & 0 & 0 & 0 & 0 & 0 & 0 & 1 & 1 & 1 & 1 & 1 & 1 & 1 & 1 \\
0 & 1 & 0 & 1 & 0 & 1 & 0 & 1 & 1 & 0 & 1 & 0 & 1 & 0 & 1 & 0 \\
0 & 0 & 1 & 1 & 0 & 0 & 1 & 1 & 1 & 1 & 0 & 0 & 1 & 1 & 0 & 0 \\
0 & 1 & 1 & 0 & 0 & 1 & 1 & 0 & 1 & 0 & 0 & 1 & 1 & 0 & 0 & 1 \\
0 & 0 & 0 & 0 & 1 & 1 & 1 & 1 & 1 & 1 & 1 & 1 & 0 & 0 & 0 & 0 \\
0 & 1 & 0 & 1 & 1 & 0 & 1 & 0 & 1 & 0 & 1 & 0 & 0 & 1 & 0 & 1 \\
0 & 0 & 1 & 1 & 1 & 1 & 0 & 0 & 1 & 1 & 0 & 0 & 0 & 0 & 1 & 1 \\
0 & 1 & 1 & 0 & 1 & 0 & 0 & 1 & 1 & 0 & 0 & 1 & 0 & 1 & 1 & 0 \\
0 & 0 & 0 & 0 & 0 & 0 & 0 & 0 & 0 & 0 & 0 & 0 & 0 & 0 & 0 & 0 \\
1 & 1 & 1 & 1 & 1 & 1 & 1 & 1 & 1 & 1 & 1 & 1 & 1 & 1 & 1 & 1
\end{bmatrix}$$

Recall that the $(4k, 2k)$ Reed-Muller code contains $8k$ codewords. Thus, there should be 32 rows in the matrix cw. We will verify this by entering the following command.

```
>   RowDimension(cw);
```
                                    32

We will now show how Maple can be used to correct a vector in $\mathbb{Z}_2^{16}$ to a codeword in the $(16, 8)$ Reed-Muller code. Note first that since the code has a minimum distance of 8, it will be three-error correcting. So now suppose that a codeword in the $(16, 8)$ Reed-Muller code is transmitted, and we receive the following vector.

```
>  r := Vector[row]([1, 0, 1, 0, 1, 0, 0, 1, 0, 1, 1, 0, 1,

   0, 0, 0]):
```

To determine if this vector contains a correctable number of bit errors, we can use the following commands. In these commands, we first assign the value of the Reed-Muller parameter $k = 4$, and then assign a value of 0 to an additional parameter rn which we will use in the subsequent **while** loop. This loop compares each of the rows in the matrix cw (i.e., each of the codewords in the code) with the vector r. The **Norm** function that appears in the loop counts the number of positions in which each row of cw differs from r. If a row is found in cw that differs from r in fewer than k positions, the loop is terminated, leaving k as the number of bit errors in r, and rn as the number of the row in cw that differs from r in fewer than k positions. If no row is found in cw that differs from r in fewer than k positions, the loop ends when it has run through each row of cw, leaving k with its initial value of 4.

```
>  k := 4:

>  rn := 0:

>  while rn < RowDimension(cw) do

>          rn := rn+1;

>          if Norm(Row(cw, rn)-r, 1) < k then

>              k := Norm(Row(cw, rn)-r, 1):

>              break:

>          fi:

>  od:
```

We can now enter the following command to see if r contains a correctable number of bit errors.

```
>  k;
```

3

This value for k indicates that r contains only three bit errors. The following value for rn reveals that the codeword that differs from r in three positions is the 22nd row of cw.

```
>   rn;
```

$$22$$

We can view this codeword by entering the following command.

```
>   Row(cw, rn);
```

$$\begin{bmatrix} 0 & 1 & 1 & 0 & 1 & 0 & 0 & 1 & 0 & 1 & 1 & 0 & 1 & 0 & 0 & 1 \end{bmatrix}$$

Also, we can see the positions in r that contained errors by entering the following command.

```
>   map(x -> x mod 2, Row(cw, rn)-r);
```

$$\begin{bmatrix} 1 & 1 & 0 & 0 & 0 & 0 & 0 & 0 & 0 & 0 & 0 & 0 & 0 & 0 & 0 & 1 \end{bmatrix}$$

3.5   Reed-Muller Codes with MATLAB

In this section, we will show how MATLAB can be used to construct and correct errors in Reed-Muller codes. We will consider the $(16, 8)$ Reed-Muller code.

We will begin by generating the normalized Hadamard matrix H_{16} used to construct the code.

```
>> H1 = [1];
>> H2 = [H1 H1; H1 -H1];
>> H4 = [H2 H2; H2 -H2];
>> H8 = [H4 H4; H4 -H4];
>> H16 = [H8 H8; H8 -H8];
```

We can then obtain the incidence matrix A that results from H_{16} by entering the following sequence of commands.

```
>> A = H16;
>> A(:, 1) = [];
>> A(1, :) = [];
>> f = find(A < 0);
>> A(f) = 0

A =
```

```
0  1  0  1  0  1  0  1  0  1  0  1  0  1  0
1  0  0  1  1  0  0  1  1  0  0  1  1  0  0
0  0  1  1  0  0  1  1  0  0  1  1  0  0  1
1  1  1  0  0  0  0  1  1  1  1  0  0  0  0
0  1  0  0  1  0  1  1  0  1  0  0  1  0  1
1  0  0  0  0  1  1  1  1  0  0  0  0  1  1
0  0  1  0  1  1  0  1  0  0  1  0  1  1  0
1  1  1  1  1  1  1  0  0  0  0  0  0  0  0
0  1  0  1  0  1  0  0  1  0  1  0  1  0  1
1  0  0  1  1  0  0  0  0  1  1  0  0  1  1
0  0  1  1  0  0  1  0  1  1  0  0  1  1  0
1  1  1  0  0  0  0  0  0  0  0  1  1  1  1
0  1  0  0  1  0  1  0  1  0  1  1  0  1  0
1  0  0  0  0  1  1  0  0  1  1  1  1  0  0
0  0  1  0  1  1  0  0  1  1  0  1  0  0  1
```

Next, to form the matrix B in Theorem 3.5, we need to interchange all zeros and ones in A. To do this, we will first create a matrix with the same size as A but containing a zero in every position. Since the MATLAB **size** function returns the number of rows and columns in a matrix, the following command assigns the number of rows and columns in A as the variables *rowdimA* and *coldimA*, respectively.

```
>> [rowdimA, coldimA] = size(A);
```

To see the number of rows in A, we can enter the following command.

```
>> rowdimA

rowdimA =

    15
```

The MATLAB **zeros** function is designed to create a matrix of a specified size containing a zero in every position. Thus, in the next command we create and store as the variable B a square matrix of the same size as A but containing a zero in every position.

```
>> B = zeros(rowdimA);
```

In the next command, we use the MATLAB **find** function to create and store an array g that represents the indices where A contains a zero.

```
>> g = find(A == 0);
```

We can now form the matrix B by assigning a value of one to the entries in B that are in positions corresponding to where A contains a zero.

```
>> B(g) = 1
```

B =

```
1  0  1  0  1  0  1  0  1  0  1  0  1  0  1
0  1  1  0  0  1  1  0  0  1  1  0  0  1  1
1  1  0  0  1  1  0  0  1  1  0  0  1  1  0
0  0  0  1  1  1  1  0  0  0  0  1  1  1  1
1  0  1  1  0  1  0  0  1  0  1  1  0  1  0
0  1  1  1  1  0  0  0  0  1  1  1  1  0  0
1  1  0  1  0  0  1  0  1  1  0  1  0  0  1
0  0  0  0  0  0  0  1  1  1  1  1  1  1  1
1  0  1  0  1  0  1  1  0  1  0  1  0  1  0
0  1  1  0  0  1  1  1  1  0  0  1  1  0  0
1  1  0  0  1  1  0  1  0  0  1  1  0  0  1
0  0  0  1  1  1  1  1  1  1  1  0  0  0  0
1  0  1  1  0  1  0  1  0  1  0  0  1  0  1
0  1  1  1  1  0  0  1  1  0  0  0  0  1  1
1  1  0  1  0  0  1  1  0  0  1  0  1  1  0
```

Now, recall that to form the matrices $\mathcal{A}$ and $\mathcal{B}$ in Theorem 3.5, we must only place a one in front of all of the rows of A and a zero in front of all of the rows of B. To do this, we will define the following column vectors *colA* and *colB*. (For formatting purposes, we display the transpose of these vectors.)

```
>> colA = ones(rowdimA, 1);
>> colA'
```

ans =

```
        1  1  1  1  1  1  1  1  1  1  1  1  1  1  1
```

```
>> [rowdimB, coldimB] = size(B);
>> colB = zeros(rowdimB, 1);
>> colB'
```

ans =

```
        0  0  0  0  0  0  0  0  0  0  0  0  0  0  0
```

By construction, the vectors *colA* and *colB* have the same number of positions as the number of rows in A and B, respectively. Thus, by placing these vectors as columns in front of A and B, respectively, we will obtain the matrices $\mathcal{A}$ and $\mathcal{B}$. We can do this by entering the following two commands.

```
>> scriptA = [colA A];
>> scriptB = [colB B];
```

The rows of the matrices $\mathcal{A}$ and $\mathcal{B}$ taken together form all but two of the codewords in the $(16, 8)$ Reed-Muller code. The two codewords not included in the rows of these matrices are the vectors $(0\ 0\ \cdots\ 0)$ and $(1\ 1\ \cdots\ 1)$ of length 16 positions. We will create these two vectors next.

```
>> [rowdimscriptB, coldimscriptB] = size(scriptB);
>> v_zero = zeros(1, coldimscriptB)

v_zero =

    0  0  0  0  0  0  0  0  0  0  0  0  0  0  0  0

>> v_one = ones(1, coldimscriptB)

v_one =

    1  1  1  1  1  1  1  1  1  1  1  1  1  1  1  1
```

We can then view the codewords in the $(16, 8)$ Reed-Muller code as follows by stacking together the matrices $\mathcal{A}$ and $\mathcal{B}$ and the vectors *v_zero* and *v_one*.

```
>> cw = [scriptA; scriptB; v_zero; v_one]

cw =

    1  0  1  0  1  0  1  0  1  0  1  0  1  0  1  0
    1  1  0  0  1  1  0  0  1  1  0  0  1  1  0  0
    1  0  0  1  1  0  0  1  1  0  0  1  1  0  0  1
    1  1  1  1  0  0  0  0  1  1  1  1  0  0  0  0
    1  0  1  0  0  1  0  1  1  0  1  0  0  1  0  1
    1  1  0  0  0  0  1  1  1  1  0  0  0  0  1  1
    1  0  0  1  0  1  1  0  1  0  0  1  0  1  1  0
    1  1  1  1  1  1  1  1  0  0  0  0  0  0  0  0
    1  0  1  0  1  0  1  0  0  1  0  1  0  1  0  1
    1  1  0  0  1  1  0  0  0  0  1  1  0  0  1  1
    1  0  0  1  1  0  0  1  0  1  1  0  0  1  1  0
```

```
1 1 1 1 0 0 0 0 0 0 0 0 1 1 1 1
1 0 1 0 0 1 0 1 0 1 0 1 1 0 1 0
1 1 0 0 0 0 1 1 0 0 1 1 1 1 0 0
1 0 0 1 0 1 1 0 0 1 1 0 1 0 0 1
0 1 0 1 0 1 0 1 0 1 0 1 0 1 0 1
0 0 1 1 0 0 1 1 0 0 1 1 0 0 1 1
0 1 1 0 0 1 1 0 0 1 1 0 0 1 1 0
0 0 0 0 1 1 1 1 0 0 0 0 1 1 1 1
0 1 0 1 1 0 1 0 0 1 0 1 1 0 1 0
0 0 1 1 1 1 0 0 0 0 1 1 1 1 0 0
0 1 1 0 1 0 0 1 0 1 1 0 1 0 0 1
0 0 0 0 0 0 0 0 1 1 1 1 1 1 1 1
0 1 0 1 0 1 0 1 1 0 1 0 1 0 1 0
0 0 1 1 0 0 1 1 1 1 0 0 1 1 0 0
0 1 1 0 0 1 1 0 1 0 0 1 1 0 0 1
0 0 0 0 1 1 1 1 1 1 1 1 0 0 0 0
0 1 0 1 1 0 1 0 1 0 1 0 0 1 0 1
0 0 1 1 1 1 0 0 1 1 0 0 0 0 1 1
0 1 1 0 1 0 0 1 1 0 1 0 1 1 0
0 0 0 0 0 0 0 0 0 0 0 0 0 0 0 0
1 1 1 1 1 1 1 1 1 1 1 1 1 1 1 1
```

Recall that the $(4k, 2k)$ Reed-Muller code contains $8k$ codewords. Thus, there should be 32 rows in the matrix cw. We will verify this by entering the following two commands.

```
>> [rowdimcw, coldimcw] = size(cw);
>> rowdimcw
```

```
rowdimcw =

   32
```

We will now show how MATLAB can be used to correct a vector in $\mathbb{Z}_2^{16}$ to a codeword in the $(16, 8)$ Reed-Muller code. Note first that since the code has a minimum distance of 8, it will be three-error correcting. So now suppose that a codeword in the $(16, 8)$ Reed-Muller code is transmitted, and we receive the following vector.

```
>> r = [1 0 1 0 1 0 0 1 0 1 1 0 1 0 0 0];
```

To determine if this vector contains a correctable number of bit errors, we can use the following commands. In these commands, we first assign the value of the Reed-Muller parameter $k = 4$, and then assign a value of 0

to an additional parameter *rn* which we will use in the subsequent **while**
loop. This loop compares each of the rows in the matrix *cw* (i.e., each of
the codewords in the code) with the vector *r*. The **norm** function that
appears in the loop counts the number of positions in which each row of *cw*
differs from *r*. If a row is found in *cw* that differs from *r* in fewer than *k*
positions, the loop is terminated, leaving *k* as the number of bit errors in
r, and *rn* as the number of the row in *cw* that differs from *r* in fewer than
k positions. If no row is found in *cw* that differs from *r* in fewer than *k*
positions, the loop ends when it has run through each row of *cw*, leaving *k*
with its initial value of 4.

```
>> k = 4;
>> rn = 0;
>> while rn < rowdimcw
        rn = rn+1;
        if norm(cw(rn, :)-r, 1) < k
            k = norm(cw(rn, :)-r, 1);
            break;
        end
    end
```

We can now enter the following command to see if *r* contains a correctable
number of bit errors.

```
>> k

k =

        3
```

This value for *k* indicates that *r* contains only three bit errors. The following
value for *rn* reveals that the codeword that differs from *r* in three positions
is the 22nd row of *cw*.

```
>> rn

rn =

        22
```

We can view this codeword by entering the following command.

```
>> cw(22, :)
```

```
ans =

    0  1  1  0  1  0  0  1  0  1  1  0  1  0  0  1
```

Also, we can see the positions in r that contained errors by entering the following command.

```
>> mod(cw(rn, :)-r, 2)

ans =

    1  1  0  0  0  0  0  0  0  0  0  0  0  0  0  1
```

3.6 Linear Codes

As we have shown, Hadamard and Reed-Muller codes are easy to construct and can have significant error correction capabilities. However, because Hadamard and Reed-Muller codes do not form vector spaces, they are not ideal for situations in which a very large number of codewords are needed. Because Hadamard and Reed-Muller codes do not form vector spaces, error correction in these types of codes must be done by comparing received vectors with each of the codewords one by one. While this error correction scheme does not pose any problems in relatively small codes, it would not be an efficient way to correct errors in a code with a very large number of codewords. In this section, we will present a method for constructing codes that do form vector spaces. We will then present some more efficient schemes for correcting errors in these codes.

Recall that a code that forms a vector space is called a *linear* code. We will describe a linear code using the parameters $[n, k]$ if the codewords in the code have length n positions and the code forms a vector space of dimension k. In this section, we will present linear codes constructed using generator matrices. Specifically, let $W = \mathbb{Z}_2^k$ and $V = \mathbb{Z}_2^n$ with $k < n$, and let G be a $k \times n$ matrix over $\mathbb{Z}_2$ of full row rank. Then $C = \{\mathbf{v} \in V \mid \mathbf{v} = \mathbf{w} \cdot G \text{ for some } \mathbf{w} \in W\}$ is a subspace of V of dimension k. Thus, the vectors in C are the codewords in an $[n, k]$ linear code in V with 2^k codewords. The matrix G is called a *generator* matrix for C.

Example 3.4 Let $W = \mathbb{Z}_2^2 = \{(00), (10), (01), (11)\}$, and choose the following generator matrix G.

$$G = \begin{bmatrix} 1 & 1 & 1 & 1 & 0 & 0 & 0 & 0 & 1 & 1 & 1 \\ 0 & 0 & 0 & 0 & 1 & 1 & 1 & 1 & 1 & 1 & 1 \end{bmatrix}$$

The resulting code is $C = \{(00000000000), (11110000111), (00001111111), (11111111000)\}$, which is an $[11, 2]$ linear code. $\square$

Note that the code C in Example 3.4 has a minimum distance of 7. Thus, C will be three-error correcting, whereas bit errors could not be corrected in $W = \mathbb{Z}_2^2$. Of course, the vectors in C are longer than the vectors in W, and consequently it would take more effort to transmit the vectors in C. However, the ability to correct up to three bit errors in C should be much more valuable than the extra effort required to transmit the vectors. Furthermore, W could still be used in the encoding and decoding of the actual information being transmitted. Specifically, information could be encoded as vectors in W, and then converted to vectors in C before being transmitted. Received vectors could then be corrected to codewords in C if necessary, and converted back into vectors in W to be decoded. In order for this process to be valid, we must be able to convert between W and C uniquely. This is precisely why we required G to have full row rank. With this requirement, G must have a right inverse, say B, and then $\mathbf{w} \in W$ can be retrieved uniquely from $\mathbf{w}G \in C$ by $\mathbf{w} = \mathbf{w}GB$.

We will now consider the problem of detecting errors in received vectors that occur from codewords in linear codes constructed using generator matrices. Because linear codes form vector spaces, there are techniques for identifying received vectors as codewords that are much more efficient than comparing the vectors with the codewords one by one. For a linear code C constructed from W using a generator matrix G of size $k \times n$, consider an $(n - k) \times n$ matrix H of full row rank over $\mathbb{Z}_2$ such that HG^T is equal to the zero matrix O. Since $HG^T = O$, then $HG^T\mathbf{w}^T$ will be equal to the zero vector $\mathbf{0}$ for all $\mathbf{w} \in W$. Thus, $H(\mathbf{w}G)^T = \mathbf{0}$ for all $\mathbf{w} \in W$, or, equivalently, $H\mathbf{c}^T = \mathbf{0}$ for all $\mathbf{c} \in C$. Since H has full row rank, it can be shown that $H\mathbf{c}^T = \mathbf{0}$ if and only if $\mathbf{c} \in C$, and so H can be used to identify codewords in C. The matrix H is called a *parity check* matrix for C.

To determine a parity check matrix H from a generator matrix G, note that $HG^T = O$ implies $GH^T = O$. Thus, the columns of H^T, which are the rows of H, are in the null space of G. As a result, to determine H from G, we must only find a basis for the null space of G, and place these basis vectors as rows in H. In practice, when constructing a linear code, it is often convenient to begin with a parity check matrix rather than a generator matrix. Since $HG^T = O$, then G can be determined from H in the same way that H can be determined from G. That is, G can be determined from H by finding a basis for the null space of H, and placing these basis vectors as rows in G.

Example 3.5 Consider a linear code C with the following parity check matrix H.

$$H = \begin{bmatrix} 0 & 0 & 0 & 1 & 1 & 1 & 1 \\ 0 & 1 & 1 & 0 & 0 & 1 & 1 \\ 1 & 0 & 1 & 0 & 1 & 0 & 1 \end{bmatrix}$$

To construct a generator matrix G for C, we must find a basis for the null space of H. To do this, we will consider H as the coefficient matrix for the following system of three homogeneous equations in seven unknowns.

$$
\begin{aligned}
x_4 + x_5 + x_6 + x_7 &= 0 \\
x_2 + x_3 + x_6 + x_7 &= 0 \\
x_1 + x_3 + x_5 + x_7 &= 0
\end{aligned}
$$

By solving these equations for x_1, x_2, and x_4 in terms of the others, we can find a basis for the null space of H by separately setting each of the variables x_3, x_5, x_6, and x_7 equal to 1 while setting the other three equal to 0. For example, setting $x_5 = 1$ and $x_3 = x_6 = x_7 = 0$, we obtain $x_1 = x_4 = 1$ and $x_2 = 0$. This yields the basis vector (1001100). This vector and the other three basis vectors constructed similarly form the rows in the following generator matrix G for C.

$$
G = \begin{bmatrix}
1 & 1 & 1 & 0 & 0 & 0 & 0 \\
1 & 0 & 0 & 1 & 1 & 0 & 0 \\
0 & 1 & 0 & 1 & 0 & 1 & 0 \\
1 & 1 & 0 & 1 & 0 & 0 & 1
\end{bmatrix}
$$

To construct the codewords in C, we would take $W = \mathbb{Z}_2^4$, and form $\mathbf{w}G$ for all $\mathbf{w} \in W$. The resulting code is a $[7, 4]$ linear code with 16 codewords. $\qquad\square$

The code in Example 3.5 is a special kind of code called a *Hamming code*. Hamming codes were originally used by AT&T to control errors in long-distance telephone calls. A distinguishing feature of Hamming codes is the form of the parity check matrices for the codes. Note that the columns of the parity check matrix H in Example 3.5 are an ordered list of the binary expressions of the integers $1, 2, \ldots, 7$. For example, the sixth column of H is $[1, 1, 0]^T$, whose entries are the coefficients in the expression $1 \cdot (2^2) + 1 \cdot (2^1) + 0 \cdot (2^0)$ of the integer 6. In general, to construct a Hamming code, we begin by placing the binary expressions of the integers $1, 2, \ldots, 2^m - 1$ for some integer $m > 1$ in order as the columns in a parity check matrix H of size $m \times (2^m - 1)$. Note that by stopping at an integer of the form $2^m - 1$, the columns of H will form all nonzero vectors of length m over $\mathbb{Z}_2$. The importance of this is for error correction in Hamming codes, which we will present later. From H, we determine a generator matrix G of size $(2^m - 1 - m) \times (2^m - 1)$ over $\mathbb{Z}_2$ by finding a basis for the null space of H over $\mathbb{Z}_2$. We then construct the codewords in the code by forming $\mathbf{w}G$ for all vectors $\mathbf{w}$ of length $2^m - 1 - m$ over $\mathbb{Z}_2$.

Example 3.6 The following matrix H is the parity check matrix for the $[15, 11]$ Hamming code.

$$H = \begin{bmatrix} 0 & 0 & 0 & 0 & 0 & 0 & 0 & 1 & 1 & 1 & 1 & 1 & 1 & 1 & 1 \\ 0 & 0 & 0 & 1 & 1 & 1 & 1 & 0 & 0 & 0 & 0 & 1 & 1 & 1 & 1 \\ 0 & 1 & 1 & 0 & 0 & 1 & 1 & 0 & 0 & 1 & 1 & 0 & 0 & 1 & 1 \\ 1 & 0 & 1 & 0 & 1 & 0 & 1 & 0 & 1 & 0 & 1 & 0 & 1 & 0 & 1 \end{bmatrix}$$

$\square$

All Hamming codes are one-error correcting (see Corollary 3.8) and perfect. Recall that a code in $\mathbb{Z}_2^n$ is perfect if every vector in $\mathbb{Z}_2^n$ is guaranteed to be uniquely correctable to a codeword in the code. The fact that Hamming codes are perfect is a consequence of the discussion immediately preceding Theorem 3.1 regarding the number of vectors that are guaranteed to be uniquely correctable to a codeword in a t-error correcting code. For example, because the $[7, 4]$ Hamming code is one-error correcting, the number of vectors in $\mathbb{Z}_2^7$ that are guaranteed to be uniquely correctable to a codeword in the code is $16 \cdot \left(\binom{7}{0} + \binom{7}{1} \right) = 128$. However, there are only $2^7 = 128$ vectors in $\mathbb{Z}_2^7$. Thus, every vector in $\mathbb{Z}_2^7$ is guaranteed to be uniquely correctable to a codeword in the $[7, 4]$ Hamming code, and the code is perfect. We will leave the general result as an exercise.

We have now shown an effective method for detecting errors in received vectors that occur from codewords in linear codes constructed using generator matrices. Specifically, if C is a linear code with parity check matrix H, then $Hc^T = 0$ if and only if $c \in C$. We will now consider the problem of correcting errors in received vectors that occur from codewords in these codes. Let C be a linear code in $\mathbb{Z}_2^n$ with parity check matrix H. Suppose $c \in C$ is transmitted, and we receive the vector $r \in \mathbb{Z}_2^n$. Then $r = c + e$ for some error vector $e \in \mathbb{Z}_2^n$ that contains ones in the positions where r and c differ and zeros elsewhere. Note that $Hr^T = Hc^T + He^T = He^T$. Thus, we can determine He^T by computing Hr^T. If we could then find e from He^T, we could form the corrected codeword as $c = r + e$.

Consider again the Hamming codes. Because they are one-error correcting and perfect, the only error vectors that we must consider are the vectors e_i that contain zeros in every position except for a single one in the ith position. Suppose a codeword in the $[2^m - 1, 2^m - 1 - m]$ Hamming code C is transmitted, and we receive the vector $r \in \mathbb{Z}_2^{2^m-1}$. If $r \notin C$, then since the columns in the parity check matrix H for C form all nonzero vectors of length m over $\mathbb{Z}_2$, it follows that Hr^T would have to be one of the columns of H, say the jth. Since the jth column of H is also He_j^T, then $Hr^T = He_j^T$, and the error in r would be e_j. Also, recall that the jth column of H will be the binary expression of the number j. So this is the remarkable error correction scheme that makes the perfect Hamming codes so mathematically beautiful and practical: if $H\mathbf{r}^T$ is the binary expression of the integer j, then the single bit error in $\mathbf{r}$ is in the jth position.

Example 3.7 Suppose a codeword in the $[7,4]$ Hamming code is transmitted, and we receive the vector $\mathbf{r} = (1011001)$. Using the parity check matrix H in Example 3.5, we find that $H\mathbf{r}^T = (001)^T$, which is the first column of H, and the binary expression of the number 1. Thus, we would assume that the error in $\mathbf{r}$ is $\mathbf{e}_1 = (1000000)$, and correct $\mathbf{r}$ to the codeword $\mathbf{c} = \mathbf{r} + \mathbf{e}_1 = (0011001)$. $\qquad\qquad\square$

Example 3.8 Suppose a codeword in the $[15,11]$ Hamming code is transmitted, and we receive the vector $\mathbf{r} = (101011100111000)$. Using the parity check matrix H in Example 3.6, we find that $H\mathbf{r}^T = (1011)^T$, which is the 11th column of H, and the binary expression of the number 11. Thus, we would assume that the error in $\mathbf{r}$ is $\mathbf{e}_{11} = (000000000010000)$, and correct $\mathbf{r}$ to the codeword $\mathbf{c} = \mathbf{r} + \mathbf{e}_{11} = (101011100101000)$. $\qquad\square$

We have now shown that all Hamming codes are one-error correcting. Next, we will consider the problem of determining the number of bit errors that are guaranteed to be uniquely correctable in other linear codes constructed using generator matrices. As we presented in Section 3.1, we can determine the number of bit errors that are guaranteed to be uniquely correctable in a particular code by finding the minimum distance of the code. Specifically, in a code with minimum distance d, we are guaranteed to be able to uniquely correct t bit errors for any $t < \frac{d}{2}$. In a code with a very large number of codewords, it might not be efficient to find the minimum distance of the code by actually computing the Hamming distance between every possible pair of codewords. However, because linear codes form vector spaces, there generally are techniques for determining the minimum distance of a code that are much more efficient than actually computing the Hamming distance between every possible pair of codewords. The following Theorems 3.6 and 3.7 provide such techniques.

For a codeword $\mathbf{c}$ in a linear code constructed using a generator matrix, we define the *Hamming weight* $w(\mathbf{c})$ of $\mathbf{c}$ to be the number of ones that appear in $\mathbf{c}$. That is, $w(\mathbf{c}) = d(\mathbf{c}, \mathbf{0})$, the Hamming distance between $\mathbf{c}$ and the zero vector.

Theorem 3.6 *Suppose C is a linear code constructed using a generator matrix, and let $w = min\{w(\mathbf{x}) \mid \mathbf{x} \in C,\ \mathbf{x} \neq \mathbf{0}\}$. Then $w = d(C)$.*

Proof. Since $w = w(\mathbf{c}) = d(\mathbf{c}, \mathbf{0})$ for some codeword $\mathbf{c} \in C$, it must be the case that $d(C) \leq w$. However, $d(C) = d(\mathbf{x}, \mathbf{y}) = w(\mathbf{x} - \mathbf{y})$ for some codewords $\mathbf{x}, \mathbf{y} \in C$. Since C is a vector space, then $\mathbf{x} - \mathbf{y} \in C$. Thus, $w \leq d(C)$. $\qquad\qquad\square$

Theorem 3.7 *Suppose C is a linear code with parity check matrix H, and let s be the minimum number of linearly dependent columns in H. Then $s = d(C)$.*

Proof. Let $w = \min\{w(\mathbf{x}) \mid \mathbf{x} \in C, \mathbf{x} \neq \mathbf{0}\}$, and suppose $C_{i_1}, C_{i_2}, \ldots, C_{i_s}$ are linearly dependent columns in H. Then $a_1 C_{i_1} + a_2 C_{i_2} + \cdots + a_s C_{i_s} = \mathbf{0}$ for some nonzero $a_1, a_2, \ldots, a_s$. Let $\mathbf{x}$ be a vector with length equal to the number of columns in H, and whose entries are a_j in positions i_j for $j = 1, 2, \ldots, s$ and zeros elsewhere. Then $H\mathbf{x}^T = \mathbf{0}$, and so $\mathbf{x} \in C$. Thus, $s \geq w = d(C)$. Conversely, let $\mathbf{y} \in C$ with $w(\mathbf{y}) = d(C)$, and let $i_1, i_2, \ldots, i_d$ be the nonzero positions in $\mathbf{y}$. Then $\mathbf{0} = H\mathbf{y}^T = C_{i_1} + C_{i_2} + \cdots + C_{i_d}$. Thus, columns $C_{i_1}, C_{i_2}, \ldots, C_{i_d}$ are linearly dependent, and $s \leq d(C)$. $\square$

The fact that all Hamming codes are one-error correcting can be shown as the following corollary to Theorem 3.7.

Corollary 3.8 *All Hamming codes are one-error correcting.*

Proof. Let C be a Hamming code with parity check matrix H. Note that the first three columns of H will be linearly dependent. Also, note that no two columns of H will be linearly dependent, since either they would have to be identical, or one would have to be the zero vector. By Theorem 3.7, it follows that $d(C) = 3$, and so C is one-error correcting. $\square$

We have now shown how errors can be corrected in Hamming codes. We will now consider the problem of correcting errors in received vectors that occur from codewords in more general linear codes constructed using generator matrices.

Let C be a t-error correcting linear code in $\mathbb{Z}_2^n$. A subset S of $\mathbb{Z}_2^n$ is called a *coset* of C if any two vectors in S differ by an element in C. Suppose $\mathbf{c} \in C$ is transmitted, and we receive the vector $\mathbf{r} \in \mathbb{Z}_2^n$ with $\mathbf{r} = \mathbf{c} + \mathbf{e}$ for some nonzero error vector $\mathbf{e}$. Since $\mathbf{r}$ and $\mathbf{e}$ differ by an element in C, then $\mathbf{r}$ and $\mathbf{e}$ will be in the same coset of C. Thus, if $\mathbf{r}$ contains t or fewer bit errors, we can find the error vector $\mathbf{e}$ that corresponds to $\mathbf{r}$ by finding the unique vector with the fewest ones in the coset that contains $\mathbf{r}$. In a code with a very large number of codewords, it might not be practical to construct all of the elements in the cosets. For such codes, the following theorem yields an equivalence on vectors in the same coset.

Theorem 3.9 *Suppose C is a linear code in $\mathbb{Z}_2^n$ with parity check matrix H. Then $\mathbf{u}, \mathbf{v} \in \mathbb{Z}_2^n$ are in the same coset of C if and only if $H\mathbf{u}^T = H\mathbf{v}^T$.*

Proof. Exercise. $\square$

Theorem 3.9 states that each coset of a linear code with parity check matrix H can be uniquely identified by $H\mathbf{u}^T$ for any vector $\mathbf{u}$ in the coset. We will call $H\mathbf{u}^T$ the *syndrome* of $\mathbf{u}$. Suppose a codeword $\mathbf{c}$ in a t-error correcting linear code in $\mathbb{Z}_2^n$ is transmitted, and we receive the vector $\mathbf{r} \in \mathbb{Z}_2^n$ with $\mathbf{r} = \mathbf{c} + \mathbf{e}$ for some nonzero error vector $\mathbf{e}$. If $\mathbf{r}$ contains t or fewer bit errors, then we can find $\mathbf{e}$ by finding the unique vector that contains t or fewer ones and which has the same syndrome as $\mathbf{r}$. If $\mathbf{r}$ contains more than t bit errors, then the syndrome of $\mathbf{r}$ will not match the syndromes of any of the vectors in $\mathbb{Z}_2^n$ that contain t or fewer ones. When a coset contains a unique vector with the fewest ones, we will call this vector the coset *leader*. Thus, for a t-error correcting linear code, each vector that contains t or fewer ones must be a coset leader.

Example 3.9 Let $W = \mathbb{Z}_2^2 = \{(00), (10), (01), (11)\}$, and choose the following generator matrix G.

$$G = \begin{bmatrix} 1 & 1 & 1 & 0 & 0 \\ 0 & 0 & 1 & 1 & 1 \end{bmatrix}$$

The resulting code is $C = \{(00000), (11100), (00111), (11011)\}$, which is a $[5, 2]$ linear code. It can easily be verified that the following matrix H is a parity check matrix for C.

$$H = \begin{bmatrix} 1 & 1 & 0 & 0 & 0 \\ 1 & 0 & 1 & 1 & 0 \\ 1 & 0 & 1 & 0 & 1 \end{bmatrix}$$

It can also easily be verified that C is one-error correcting. Thus, the only cosets leaders for C will be (00000) and the five vectors in $\mathbb{Z}_2^5$ that contain a single one. The following table shows these coset leaders and their corresponding syndromes.

Coset Leader	Syndrome
(00000)	$(000)^T$
(10000)	$(111)^T$
(01000)	$(100)^T$
(00100)	$(011)^T$
(00010)	$(010)^T$
(00001)	$(001)^T$

Suppose a codeword $\mathbf{c} \in C$ is transmitted, and we receive the vector $\mathbf{r}_1 = (00011)$. To correct this vector, we compute $H\mathbf{r}_1^T = (011)^T$. Since the coset leader (00100) also has this syndrome, we would assume that the error in $\mathbf{r}_1$ is $\mathbf{e} = (00100)$, and correct $\mathbf{r}_1$ to the codeword $\mathbf{c} = \mathbf{r}_1 + \mathbf{e} = (00111)$. Note that because each coset for C will contain 4 vectors, only 24 of the 32 vectors in $\mathbb{Z}_2^5$ will be in cosets that have coset leaders. For example, suppose a codeword in C is transmitted, and we receive the vector $\mathbf{r}_2 = (01001)$. To attempt to correct this vector, we compute $H\mathbf{r}_2^T = (101)^T$. None of the coset leaders for C also have this syndrome. Thus, $\mathbf{r}_2$ is not in a coset with a coset leader, and $\mathbf{r}_2$ cannot be corrected. $\qquad\square$

3.7 Hamming Codes with Maple

In this section, we will show how Maple can be used to construct and correct errors in Hamming codes. We will consider the $[15, 11]$ Hamming code.

Because some of the functions that we will use are in the Maple **List-Tools**, **LinearAlgebra**, and **Modular** packages, we will begin by including these packages. In addition, we will enter the following **interface** command to cause Maple to display all matrices of size 50×50 and smaller throughout the remainder of this Maple session.

```
>   with(ListTools):
>   with(LinearAlgebra):
>   with(Modular):
>   interface(rtablesize=50):
```

Next we will construct the parity check matrix H for the code. We first enter the length $m = 4$ of the vectors that form the columns of H.

```
>   m := 4:
```

Recall that the columns of H are binary expressions of the integers $1, 2, \ldots, 2^m - 1$. We can obtain the binary expression of an integer in Maple by using the Maple **convert** function. For example, we can obtain the binary expression of the integer 4 by entering the following command.

```
>   cb := convert(4, base, 2);
```
$$cb := [0,\, 0,\, 1]$$

The entries in the preceding result for cb are the coefficients in the expression $0 \cdot (2^0) + 0 \cdot (2^1) + 1 \cdot (2^2)$ of the integer 4. Note that cb contains only three positions, whereas for the columns of H we want binary vectors of length $m = 4$ positions. That is, to be placed as the fourth column of H, we would want the number 4 to be converted to the binary vector $[0, 0, 1, 0]$. Furthermore, the binary digits in cb are the reverse of how they should be expressed in the fourth column of H. To be directly placed as the fourth column of H, the number 4 should be converted to the binary vector $[0, 1, 0, 0]$. We can use the following command to take care of these two problems.

```
>  bv := Vector([seq(0, i=1..m-nops(cb)),

   op(Reverse(cb))]);
```

$$bv := \begin{bmatrix} 0 \\ 1 \\ 0 \\ 0 \end{bmatrix}$$

In the preceding command, the **Reverse** function, which is contained in the **ListTools** package, reverses the order of the entries in the list that is contained in the variable cb. The **seq** function pads the resulting reversed list with zeros so that the list will be of length $m = 4$ positions. Finally, the **Vector** function converts the list into a column vector.

We will now construct the parity check matrix H for the $[15, 11]$ Hamming code by placing binary expressions of length $m = 4$ positions for the integers $1, 2, \ldots, 2^m - 1$ in order as the columns of H. To do this, we first create an empty list H. We then use a **for** loop to build the parity check matrix column by column in H. The second **op** command that appears in the loop allows new columns to be attached to the list H. After the loop, we use **Matrix** to convert H from a list of column vectors into a matrix.

```
>  H := []:

>  for j from 1 to 2^m-1 do

>        cb := convert(j, base, 2):

>        bv := Vector([seq(0, i=1..m-nops(cb)),

         op(Reverse(cb))]);

>        H := [op(H), bv]:

>  od:

>  H := Matrix([H]);
```

$$H := \begin{bmatrix} 0 & 0 & 0 & 0 & 0 & 0 & 0 & 1 & 1 & 1 & 1 & 1 & 1 & 1 & 1 \\ 0 & 0 & 0 & 1 & 1 & 1 & 1 & 0 & 0 & 0 & 0 & 1 & 1 & 1 & 1 \\ 0 & 1 & 1 & 0 & 0 & 1 & 1 & 0 & 0 & 1 & 1 & 0 & 0 & 1 & 1 \\ 1 & 0 & 1 & 0 & 1 & 0 & 1 & 0 & 1 & 0 & 1 & 0 & 1 & 0 & 1 \end{bmatrix}$$

Next, we will construct a generator matrix G for the $[15, 11]$ Hamming code by finding a basis for the null space of H over $\mathbb{Z}_2$ and placing these basis vectors as rows in G. To do this, we will use the Maple **Basis** function, which is contained in the **Modular** package. In the following command we construct the generator matrix.

```
>  G := Basis(2, H, row, false, Matrix[row]);
```

$$G := \begin{bmatrix}
1 & 1 & 1 & 0 & 0 & 0 & 0 & 0 & 0 & 0 & 0 & 0 & 0 & 0 & 0 \\
1 & 0 & 0 & 1 & 1 & 0 & 0 & 0 & 0 & 0 & 0 & 0 & 0 & 0 & 0 \\
0 & 1 & 0 & 1 & 0 & 1 & 0 & 0 & 0 & 0 & 0 & 0 & 0 & 0 & 0 \\
1 & 1 & 0 & 1 & 0 & 0 & 1 & 0 & 0 & 0 & 0 & 0 & 0 & 0 & 0 \\
1 & 0 & 0 & 0 & 0 & 0 & 0 & 1 & 1 & 0 & 0 & 0 & 0 & 0 & 0 \\
0 & 1 & 0 & 0 & 0 & 0 & 0 & 1 & 0 & 1 & 0 & 0 & 0 & 0 & 0 \\
1 & 1 & 0 & 0 & 0 & 0 & 0 & 1 & 0 & 0 & 1 & 0 & 0 & 0 & 0 \\
0 & 0 & 0 & 1 & 0 & 0 & 0 & 1 & 0 & 0 & 0 & 1 & 0 & 0 & 0 \\
1 & 0 & 0 & 1 & 0 & 0 & 0 & 1 & 0 & 0 & 0 & 0 & 1 & 0 & 0 \\
0 & 1 & 0 & 1 & 0 & 0 & 0 & 1 & 0 & 0 & 0 & 0 & 0 & 1 & 0 \\
1 & 1 & 0 & 1 & 0 & 0 & 0 & 1 & 0 & 0 & 0 & 0 & 0 & 0 & 1
\end{bmatrix}$$

The first parameter in the preceding command is the modulus that is to be used in the calculations, and the second parameter is an input matrix. The third parameter indicates whether the vectors of interest in the input matrix are the rows or the columns. Since we want to find the null space of the vectors in the rows of H, we enter the third parameter as *row*. The fourth parameter allows the option of specifying a format for finding a basis for the rows (columns if the third parameter is specified as *column*) spanned by the input matrix. Since we do not need this output, we specify the fourth parameter as *false*. The fifth parameter allows the option of specifying a format for finding a basis for the null space of the rows (columns if the third parameter is specified as *column*) of the input matrix. Since this basis is to be stored as the rows in the generator matrix G, we specify this parameter as *Matrix[row]*.

Recall that the codewords in a linear code constructed using a generator matrix G are all vectors of the form $\mathbf{w}G$ where $\mathbf{w}$ is a vector over $\mathbb{Z}_2$ with the same number of positions as the number of rows in G. To see the number of positions in the vectors $\mathbf{w}$ for the $[15, 11]$ Hamming code, we can use the following command, which returns the number of rows in G.

```
>  RowDimension(G);
```

$$11$$

Thus, the vectors **w** for the $[15, 11]$ Hamming code should have 11 positions. For example, consider the following vector w.

```
>  w := Vector[row]([1, 0, 1, 1, 1, 0, 1, 1, 1, 1, 0]):
```

In the next command, we form the codeword wG in the $[15, 11]$ Hamming code that results from w. Note that in this command, we use the Maple . function for matrix multiplication. Note also that in this same command, we use the **map** function to reduce the entries in the product modulo 2.

```
>  c := map(x -> x mod 2, w.G);
```

$$c := \begin{bmatrix} 1 & 1 & 1 & 1 & 0 & 1 & 1 & 1 & 1 & 0 & 1 & 1 & 1 & 1 & 0 \end{bmatrix}$$

We will now show how Maple can be used to correct a vector in $\mathbb{Z}_2^{15}$ to a codeword in the $[15, 11]$ Hamming code. Suppose that a codeword in the $[15, 11]$ Hamming code is transmitted, and we receive the following vector.

```
>  r := Vector[row]([0, 1, 1, 1, 0, 0, 0, 1, 0, 1, 1, 1, 1,
   1, 0]):
```

To determine if this vector contains an error, we must compute the syndrome of r. Recall that the syndrome of a received vector **r** is Hr^T, which we calculate for r in the next command. Note that in this command, we must use the **Transpose** function, which is contained in the **Modular** package, in order for the product to be defined. Note also that the first parameter in the **Transpose** function is the modulus that is to be used, and the second parameter is an input vector or matrix.

```
>  syn := map(x -> x mod 2, H.Transpose(2, r));
```

$$syn := \begin{bmatrix} 0 \\ 0 \\ 1 \\ 1 \end{bmatrix}$$

Because this syndrome is nonzero, then we know that r is not a codeword in the $[15, 11]$ Hamming code. Recall that to correct the error in r, we must only convert this syndrome from the binary expression of an integer into the normal base 10 expression of the integer. This will reveal the number of the position in r that contains the error. To convert this syndrome from the binary expression of an integer into base 10, we can again use the Maple **convert** function, which we used earlier in this section.

```
>  ep := convert(cat(seq(syn[i], i=1..Dimension(syn))),
   decimal, binary);
```

$$ep := 3$$

In the preceding command, the **seq** function collects the entries in *syn* in a sequence, which is then joined together as the symbolic representation of a binary integer using the **cat** function. The **convert** function then converts the result to base 10. This output indicates that the error in r is in the third position. To correct this error, we first define the following vector e of length $2^m - 1$ positions with a zero in every position.

```
>  e := Vector[row](2^m-1, 0);
```

$$e := \begin{bmatrix} 0 & 0 & 0 & 0 & 0 & 0 & 0 & 0 & 0 & 0 & 0 & 0 & 0 & 0 & 0 \end{bmatrix}$$

Next, we replace the entry in the third position of e with a one.

```
>  e[ep] := 1:
```

We can then see the error vector that corresponds to r as follows.

```
>  e;
```

$$\begin{bmatrix} 0 & 0 & 1 & 0 & 0 & 0 & 0 & 0 & 0 & 0 & 0 & 0 & 0 & 0 & 0 \end{bmatrix}$$

Also, we can see the corrected codeword as follows.

```
>  c2 := map(x -> x mod 2, r+e);
```

$$c2 := \begin{bmatrix} 0 & 1 & 0 & 1 & 0 & 0 & 0 & 1 & 0 & 1 & 1 & 1 & 1 & 1 & 0 \end{bmatrix}$$

Finally, to verify that this vector $c2$ really is a codeword in the $(15, 11)$ Hamming code, we compute the following.

```
>  map(x -> x mod 2, H.Transpose(2, c2));
```

$$\begin{bmatrix} 0 \\ 0 \\ 0 \\ 0 \end{bmatrix}$$

3.8   Hamming Codes with MATLAB

In this section, we will show how MATLAB can be used to construct and correct errors in Hamming codes. We will consider the [15, 11] Hamming code.

We will begin by constructing the parity check matrix H for the code. We first enter the length $m = 4$ of the vectors that form the columns of H.

```
>> m = 4;
```

Recall that the columns of H are binary expressions of the integers $1, 2, \ldots, 2^m - 1$. We can obtain the binary expression of an integer in MATLAB by using the MATLAB **dec2bin** function. For example, we can obtain the binary expression of the integer 4 by entering the following command.

```
>> cb = dec2bin(4, m)

cb =

0100
```

The first parameter in the preceding command is the normal base 10 expression of an integer, while the second parameter is the number of bits that are to be used in the binary expression of the integer. The output displayed for cb is a string containing the coefficients in the expression $4 = 0 \cdot (2^3) + 1 \cdot (2^2) + 0 \cdot (2^1) + 0 \cdot (2^0)$ as characters. However, rather than having these coefficients stored as characters in a string, we want these coefficients to be stored as positions in a vector. We can take care of this by entering the following **for** loop.

```
>> for i = 1:length(cb)
       bv(i) = str2num(cb(i));
   end
```

The MATLAB **length** function is designed to count the number of characters in a string. Thus, the preceding **for** loop takes each of the binary digits in the string cb, converts them from characters into numbers using the MATLAB **str2num** function, and stores the resulting integers as positions in the vector bv. To see the contents of bv, we can enter the following command.

```
>> bv

bv =

    0    1    0    0
```

Also, we can convert bv from a row vector into a column vector by taking its transpose.

```
>> bv'

ans =

     0
     1
     0
     0
```

We will now construct the parity check matrix H for the $[15, 11]$ Hamming code by placing binary expressions of length $m = 4$ positions for the integers $1, 2, \ldots, 2^m - 1$ in order as the columns of H. To do this, we first create an empty array H. We then use a nested **for** loop to build the parity check matrix column by column in H.

```
>> H = [];
>> for j = 1:2^m-1
        cb = dec2bin(j, m);
        for i = 1:length(cb)
            bv(i) = str2num(cb(i));
        end
        H = [H bv'];
    end
>> H

H =

     0  0  0  0  0  0  0  1  1  1  1  1  1  1  1
     0  0  0  1  1  1  1  0  0  0  0  1  1  1  1
     0  1  1  0  0  1  1  0  0  1  1  0  0  1  1
     1  0  1  0  1  0  1  0  1  0  1  0  1  0  1
```

We will convert H into a symbolic variable by entering the following command.

```
>> H = sym(H)

H =

[ 0, 0, 0, 0, 0, 0, 0, 1, 1, 1, 1, 1, 1, 1, 1]
[ 0, 0, 0, 1, 1, 1, 1, 0, 0, 0, 0, 1, 1, 1, 1]
```

```
[ 0, 1, 1, 0, 0, 1, 1, 0, 0, 1, 1, 0, 0, 1, 1]
[ 1, 0, 1, 0, 1, 0, 1, 0, 1, 0, 1, 0, 1, 0, 1]
```

Next, we will construct a generator matrix G for the $[15, 11]$ Hamming code by finding a basis for the null space of H over $\mathbb{Z}_2$ and placing these basis vectors as rows in G. To do this, we will first find a basis for the null space of H by using the MATLAB **null** function as follows.

```
>> nH = mod(null(H), 2)

nH =

[ 1, 1, 0, 1, 1, 0, 1, 0, 1, 0, 1]
[ 1, 0, 1, 1, 0, 1, 1, 0, 0, 1, 1]
[ 1, 0, 0, 0, 0, 0, 0, 0, 0, 0, 0]
[ 0, 1, 1, 1, 0, 0, 0, 1, 1, 1, 1]
[ 0, 1, 0, 0, 0, 0, 0, 0, 0, 0, 0]
[ 0, 0, 1, 0, 0, 0, 0, 0, 0, 0, 0]
[ 0, 0, 0, 1, 0, 0, 0, 0, 0, 0, 0]
[ 0, 0, 0, 0, 1, 1, 1, 1, 1, 1, 1]
[ 0, 0, 0, 0, 1, 0, 0, 0, 0, 0, 0]
[ 0, 0, 0, 0, 0, 1, 0, 0, 0, 0, 0]
[ 0, 0, 0, 0, 0, 0, 1, 0, 0, 0, 0]
[ 0, 0, 0, 0, 0, 0, 0, 1, 0, 0, 0]
[ 0, 0, 0, 0, 0, 0, 0, 0, 1, 0, 0]
[ 0, 0, 0, 0, 0, 0, 0, 0, 0, 1, 0]
[ 0, 0, 0, 0, 0, 0, 0, 0, 0, 0, 1]
```

This output for nH is a matrix whose columns are basis vectors for the null space of H. Thus, we can construct a generator matrix G for the $[15, 11]$ Hamming code by simply defining G to be the transpose of nH.

```
>> G = nH'

G =

[ 1, 1, 1, 0, 0, 0, 0, 0, 0, 0, 0, 0, 0, 0, 0]
[ 1, 0, 0, 1, 1, 0, 0, 0, 0, 0, 0, 0, 0, 0, 0]
[ 0, 1, 0, 1, 0, 1, 0, 0, 0, 0, 0, 0, 0, 0, 0]
[ 1, 1, 0, 1, 0, 0, 1, 0, 0, 0, 0, 0, 0, 0, 0]
[ 1, 0, 0, 0, 0, 0, 0, 1, 1, 0, 0, 0, 0, 0, 0]
[ 0, 1, 0, 0, 0, 0, 0, 1, 0, 1, 0, 0, 0, 0, 0]
[ 1, 1, 0, 0, 0, 0, 0, 1, 0, 0, 1, 0, 0, 0, 0]
[ 0, 0, 0, 1, 0, 0, 0, 1, 0, 0, 0, 1, 0, 0, 0]
```

```
[ 1, 0, 0, 1, 0, 0, 0, 1, 0, 0, 0, 0, 1, 0, 0]
[ 0, 1, 0, 1, 0, 0, 0, 1, 0, 0, 0, 0, 0, 1, 0]
[ 1, 1, 0, 1, 0, 0, 0, 1, 0, 0, 0, 0, 0, 0, 1]
```

Recall that the codewords in a linear code constructed using a generator matrix G are all vectors of the form $\mathbf{w}G$ where $\mathbf{w}$ is a vector over $\mathbb{Z}_2$ with the same number of positions as the number of rows in G. To see the number of positions in the vectors $\mathbf{w}$ for the $[15, 11]$ Hamming code, we can use the following two commands, which show the number of rows in G.

```
>> [rowdimG, coldimG] = size(G);
>> rowdimG

rowdimG =

    11
```

Thus, the vectors $\mathbf{w}$ for the $[15, 11]$ Hamming code should have 11 positions. For example, consider the following vector w.

```
>> w = [1 0 1 1 1 0 1 1 1 1 0];
```

In the next command, we form the codeword wG in the $[15, 11]$ Hamming code that results from w. Note that in this command, we use the MATLAB **mod** function to reduce the entries in the product modulo 2.

```
>> c = mod(w*G, 2)

c =

[ 1, 1, 1, 1, 0, 1, 1, 1, 1, 0, 1, 1, 1, 1, 0]
```

We will now show how MATLAB can be used to correct a vector in $\mathbb{Z}_2^{15}$ to a codeword in the $[15, 11]$ Hamming code. Suppose that a codeword in the $[15, 11]$ Hamming code is transmitted, and we receive the following vector.

```
>> r = [0 1 1 1 0 0 0 1 0 1 1 1 1 1 0];
```

To determine if this vector contains an error, we must compute the syndrome of r. Recall that the syndrome of a received vector $\mathbf{r}$ is $H\mathbf{r}^T$, which we calculate for r in the next command.

```
>> syn = mod(H*r', 2)
```

```
syn =

   0
   0
   1
   1
```

Because this syndrome is nonzero, then we know that r is not a codeword in the $[15, 11]$ Hamming code. Recall that to correct the error in r, we must only convert this syndrome from the binary expression of an integer into the normal base 10 expression of the integer. This will reveal the number of the position in r that contains the error. To convert this syndrome from the binary expression of an integer into base 10, we enter the following command.

```
>> bin2dec(int2str(int8(syn')))
```

```
ans =

   3
```

The preceding command included several MATLAB functions. The reason for this is because the variable *syn* that is being converted from the binary expression of an integer into the base 10 expression of the integer was stored by MATLAB as a symbolic variable. However, the MATLAB **int2str** function, which is designed to convert an integer into a string, requires that its input be an integer. So in the preceding command, after we used the transpose function to convert *syn* from a column vector into a row vector, we used the MATLAB **int8** function to convert each entry of *syn* from symbolic form into an integer. We then used the MATLAB **int2str** function to convert the resulting integers into a string, and finally we used the MATLAB **bin2dec** function to convert the resulting binary string into the base 10 expression of an integer. The output indicates that the error in r is in the third position. To correct this error, we first define the following vector e of length $2^m - 1$ positions with a zero in every position.

```
>> e = zeros(1, 2^m-1)
```

```
e =

   0  0  0  0  0  0  0  0  0  0  0  0  0  0  0
```

Next, we replace the entry in the third position of e with a one.

```
>> e(3) = 1;
```

We can then see the error vector that corresponds to r as follows.

```
>> e

e =

    0  0  1  0  0  0  0  0  0  0  0  0  0  0  0
```

Also, we can see the corrected codeword as follows.

```
>> c2 = mod(r+e, 2)

c2 =

    0  1  0  1  0  0  0  1  0  1  1  1  1  1  0
```

Finally, to verify that this vector $c2$ really is a codeword in the $(15, 11)$ Hamming code, we compute the following.

```
>> mod(H*c2', 2)

ans =

    0
    0
    0
    0
```

Exercises

1. Let C be the two-error correcting code in Example 3.2.

   (a) Find two vectors in $\mathbb{Z}_2^8$ that are not codewords in C but which are guaranteed to be uniquely correctable to a codeword in C (i.e., which are one or two bit errors from a codeword in C).

   (b) Find two vectors in $\mathbb{Z}_2^8$ that are at least three bit errors from each of the codewords in C, but which are also uniquely correctable to a codeword in C.

   (c) Find two vectors in $\mathbb{Z}_2^8$ that are not uniquely correctable to a codeword in C.

   (d) Given that 108 of the vectors in $\mathbb{Z}_2^8$ are at least three bit errors from each of the codewords in C, how many non-codewords in $\mathbb{Z}_2^8$ are either one or two bit errors from a codeword in C?

2. Using Theorem 3.1, find the maximum number of bit errors that are guaranteed to be uniquely correctable in a code of length 7 with 4 codewords. Then using Theorem 3.3, show that it is actually not possible to construct a code of length 7 with 4 codewords that is guaranteed to uniquely correct this number of bit errors.

3. Let C be the $(7, 4)$ code with 7 codewords in Example 3.3.

   (a) Find the number of non-codewords in $\mathbb{Z}_2^7$ that are guaranteed to be uniquely correctable to a codeword in C.

   (b) Correct the vector (0101101) to a codeword in C.

   (c) Find a vector in $\mathbb{Z}_2^7$ that is not uniquely correctable to a codeword in C.

4. Use a Hadamard matrix to construct the codewords in a $(15, 8)$ code with 16 codewords. What is the maximum number of bit errors that are guaranteed to be uniquely correctable in this code? Correct the vector (001011010001010) to a codeword in this code.

5. Use a Hadamard matrix to construct the codewords in an $(8, 4)$ code with 16 codewords. What is the maximum number of bit errors that are guaranteed to be uniquely correctable in this code? Correct the vector (01000110) to a codeword in this code.

6. Is it possible to construct a $[6, 2]$ linear code that is two-error correcting? Clearly explain how you know. (Hint: See Theorem 3.1.)

7. Construct a generator matrix for a one-error correcting linear code with 8 codewords. What would the values of the parameters $[n, k]$ for this code be? What would the size of a parity check matrix for this code be?

8. Construct generator and parity check matrices for a two-error correcting linear code with 4 codewords.

9. Make a list of each of the codewords in the $[7, 4]$ Hamming code.

10. Correct the vectors (0011101) and (0100101) to codewords in the $[7, 4]$ Hamming code. Then verify that the corrected vectors really are codewords in the $[7, 4]$ Hamming code.

11. Correct the vectors (011111011111110) and (101100111000111) to codewords in the $[15, 11]$ Hamming code. Then verify that the corrected vectors really are codewords in the $[15, 11]$ Hamming code.

12. Find the number of codewords in the $[15, 11]$ Hamming code.

13. Let C be the $[31, 26]$ Hamming code.

    (a) Find the size of the parity check matrix for C.

    (b) Find the size of a generator matrix for C.

    (c) Find the number of codewords in C.

14. Find the number of codewords in the $[2^m - 1, 2^m - 1 - m]$ Hamming code.

15. For the code C in Example 3.9, which of the vectors (11101), (01011), and (10101) can be corrected to codewords in C using the coset method? Correct the ones that can be corrected. For any that cannot be corrected, clearly explain how you know they cannot be corrected.

16. Let $W = \mathbb{Z}_2^2 = \{(00), (01), (10), (11)\}$, and choose the following generator matrix G.

$$G = \begin{bmatrix} 1 & 1 & 0 & 0 & 1 & 1 \\ 0 & 0 & 1 & 1 & 1 & 1 \end{bmatrix}$$

    (a) Construct the codewords in the linear code C that results from G. What is the maximum number of bit errors that are guaranteed to be uniquely correctable in C?

    (b) Construct a parity check matrix for C.

    (c) Make a list of each of the coset leaders for C and their corresponding syndromes.

    (d) Which of the vectors (100011), (001100), and (111100) can be corrected to codewords in C using the coset method? Correct the ones that can be corrected. For any that cannot be corrected, clearly explain how you know they cannot be corrected.

    (e) How many vectors in $\mathbb{Z}_2^6$ are not in a coset for C that has a coset leader?

17. Make a list of each of the coset leaders for the $[7, 4]$ Hamming code and their corresponding syndromes.

18. Prove Corollary 3.4.

19. Prove Theorem 3.5.

20. Prove that all of the $[2^m - 1, 2^m - 1 - m]$ Hamming codes are perfect.

21. Prove Theorem 3.9.

22. Without using Theorem 3.7 or Corollary 3.8, show that all of the $[2^m - 1, 2^m - 1 - m]$ Hamming codes are one-error correcting. (Hint: Consider the number of codewords and the number of vectors in the cosets for the Hamming codes.)

23. A *metric space* is a set M with a real-valued function $d(\cdot, \cdot)$ on $M \times M$ that satisfies each of the following properties for every $x, y, z \in M$.

    - $d(x, y) \geq 0$, and $d(x, y) = 0$ if and only if $x = y$
    - $d(x, y) = d(y, x)$
    - $d(x, z) \leq d(x, y) + d(y, z)$

    Prove or disprove that a code C in $\mathbb{Z}_2^n$ with the Hamming distance function $d(\cdot, \cdot)$ is a metric space.

24. An *equivalence relation* is a relation $\sim$ on a set A that satisfies each of the following properties for every $a, b, c \in A$.

    - $a \sim a$
    - $a \sim b \Rightarrow b \sim a$
    - $a \sim b$ and $b \sim c \Rightarrow a \sim c$

    Let C be a linear code in $V = \mathbb{Z}_2^n$. Define a relation $\sim$ on V by $\mathbf{x} \sim \mathbf{y}$ if $\mathbf{x}$ and $\mathbf{y}$ are in the same coset of C. Prove or disprove that $\sim$ is an equivalence relation on V.

Computer Exercises

25. Use a Hadamard matrix to construct the codewords in a $(31, 16)$ code with 32 codewords. What is the maximum number of bit errors that are guaranteed to be uniquely correctable in this code? Correct the vector $(0011110000101100001011000011110)$ to a codeword in this code.

26. As we mentioned in Section 3.3, the $(32, 16)$ Reed-Muller code was used in the Mariner 9 space probe when it transmitted photographs of Mars back to Earth.

    (a) Construct the codewords in the $(32, 16)$ Reed-Muller code. What is the maximum number of bit errors that are guaranteed to be uniquely correctable in this code?

    (b) Correct the vector $(11100101011010011110101101101001)$ to a codeword in the $(32, 16)$ Reed-Muller code.

27. Find a parity check matrix for the code in Example 3.4.

28. Find a parity check matrix for the code for which you constructed a generator matrix in Exercise 7.

29. Let C be the $[31, 26]$ Hamming code.

   (a) Construct the parity check matrix H and a generator matrix G for C.

   (b) Find the number of codewords in C.

   (c) Construct the codeword $\mathbf{w}G$ in C that results from the vector $\mathbf{w} = (1011010111011011111110111000)$.

   (d) Correct the vector $(110101110011011011010101111110111)$ to a codeword in C.

   (e) Correct the vector $(110101011011101011001100001101)$ to a codeword in C.

30. Let C be the $[63, 57]$ Hamming code.

   (a) Construct the parity check matrix and a generator matrix for C.

   (b) Find the number of codewords in C.

   (c) Construct two of the codewords in C.

31. Consider the Maple command on page 83 in which we used the **convert** function to convert the syndrome of r from the binary expression of an integer into the normal base 10 expression of the integer, thereby revealing the position in r that contained an error. Recall that we could have also identified the position in r that contained an error by finding the number of the column in H that matched the syndrome of r. Write a routine or sequence of commands in Maple to replace the **convert** command on page 83 in which you find the position in r that contains an error by finding the number of the column in H that matches the syndrome of r.

32. Consider the MATLAB command on page 88 in which we used the **bin2dec** function to convert the syndrome of r from the binary expression of an integer into the normal base 10 expression of the integer, thereby revealing the position in r that contained an error. Recall that we could have also identified the position in r that contained an error by finding the number of the column in H that matched the syndrome of r. Write a routine or sequence of commands in MATLAB to replace the **bin2dec** command on page 88 in which you find the position in r that contains an error by finding the number of the column in H that matches the syndrome of r.

Research Exercises

33. Investigate or create a mathematical use of Hadamard matrices outside the areas of block designs and error-correcting codes, and write a summary of your findings or results.

34. Investigate *Walsh* codes and their use in *code division multiple access*, or *CDMA*, communication, and write a summary of your findings.

35. Investigate the lives and careers of the people for whom Reed-Muller codes are named (Irving Reed and David Muller), and write a summary of your findings.

36. Investigate some specific details about the use of a Reed-Muller code in the Mariner 9 space probe, and write a summary of your findings.

37. Investigate an actual real-life use of a Hadamard or Reed-Muller code outside of the Mariner 9 space probe or code division multiple access communication, and write a summary of your findings.

38. Investigate the life and career of the person for whom the Hamming distance, bound, codes, and weight are named, and write a summary of your findings. Include in your summary some highlights of Hamming's work during and after World War II, his most important scientific contributions, and some of his honors and awards.

39. Investigate some specific details about the use of Hamming codes to control errors in long-distance telephone calls, and write a summary of your findings.

40. Investigate an actual real-life use of a Hamming code outside of controlling errors in long-distance telephone calls, and write a summary of your findings.

41. Investigate Hamming codes extended by an extra parity bit, also known as *SECDED* (an acronym for *single error correction, double error detection*), and write a summary of your findings.

Chapter 4

BCH Codes

The most useful codes we presented in Chapter 3 were Hamming codes because they are linear and perfect. However, Hamming codes are not ideal if the occurrence of more than one bit error in a single codeword is likely. Since Hamming codes are only one-error correcting, if more than one bit error occurs during transmission of a Hamming codeword, the received vector will not be correctable to the codeword that was sent. Moreover, since Hamming codes are perfect, if more than one bit error occurs, the received vector will be uniquely correctable, but to the wrong codeword. In this chapter, we will present a type of code called a *BCH* code that is linear and can be constructed to be multiple-error correcting. BCH codes are named for their creators, Bose, Chaudhuri, and Hocquenghem.

4.1   Construction

One way BCH codes differ from the codes we presented in Chapter 3 is that BCH codewords are polynomials rather than vectors. To construct a BCH code, we begin with the polynomial $f(x) = x^m - 1 \in \mathbb{Z}_2[x]$ for some positive integer m. Then $R = \mathbb{Z}_2[x]/(f(x))$ is a ring that can be represented by all polynomials in $\mathbb{Z}_2[x]$ of degree less than m. Suppose $g(x) \in \mathbb{Z}_2[x]$ divides $f(x)$. Then the set C of all multiples of $g(x)$ in $\mathbb{Z}_2[x]$ of degree less than m is a vector space in R with dimension $m - \deg(g(x))$. Thus, the polynomials in C are the codewords in an $[m, m - \deg(g(x))]$ linear code in R with $2^{m-\deg(g(x))}$ codewords. The polynomial $g(x)$ is called a *generator polynomial* for the code, and we consider the codewords in the code to have length m positions because we view each term in a polynomial codeword as a codeword position. A codeword $c(x) \in \mathbb{Z}_2[x]$ with m terms can then naturally be expressed as a unique vector in $\mathbb{Z}_2^m$ by listing the coefficients of $c(x)$ in order (including coefficients of zero) for increasing powers of x. In this chapter, we will assume BCH codewords are transmitted in this form.

Example 4.1 Let $f(x) = x^7 - 1$, and choose $g(x) = x^3 + x + 1$. It can easily be verified that $g(x)$ is a divisor of $f(x)$ in $\mathbb{Z}_2[x]$. Then the code consisting of all multiples of $g(x)$ in $\mathbb{Z}_2[x]$ of degree less than 7 is a $[7,4]$ linear code with basis $\{g(x), xg(x), x^2g(x), x^3g(x)\}$ and 16 codewords. One of the codewords in this code is $(x^2 + x)g(x) = x^5 + x^4 + x^3 + x$, which we will assume would be transmitted as the vector (0101110) in $\mathbb{Z}_2^7$ since $x^5 + x^4 + x^3 + x = 0 + 1x + 0x^2 + 1x^3 + 1x^4 + 1x^5 + 0x^6$.   □

In order for a code constructed as we have described (using a generator polynomial $g(x)$ that divides $f(x) = x^m - 1$ for some positive integer m) to be a BCH code, the generator polynomial must be chosen in the following way. Let $a_1, a_2, \ldots, a_s$ for some $s < m$ be roots of $f(x)$ with minimum polynomials[1] $m_1(x), m_2(x), \ldots, m_s(x)$ in $\mathbb{Z}_2[x]$, respectively. Then choose $g(x)$ to be the least common multiple in $\mathbb{Z}_2[x]$ of the $m_i(x)$ for $i = 1, 2, \ldots, s$. Choosing $g(x)$ this way is useful because of how it allows errors to be corrected in the resulting code, as we will present in Section 4.2. However, choosing a generator polynomial this way does not alone necessarily yield a BCH code. For the resulting code to be a BCH code, m and the roots a_i must also be chosen in a special way, which we will describe next.

Let $m = 2^n - 1$ for some positive integer n, and let $p(x)$ be a primitive polynomial of degree n in $\mathbb{Z}_2[x]$. Then $\mathbb{Z}_2[x]/(p(x))$ is a finite field of order 2^n, whose nonzero elements can be generated by the field element x. For the remainder of this chapter and throughout the following chapter, in which we will look at another type of code with polynomial codewords, we will denote the element x in our finite fields by a. Then for the roots $a_1, a_2, \ldots, a_s$ described in the preceding paragraph, choose $a_i = a^i$ for $i = 1, 2, \ldots, s$. Choosing $a_1, a_2, \ldots, a_s$ in this way is useful because of how it allows the generator polynomial $g(x)$ for the resulting code to be determined. The polynomials $m_1(x), m_2(x), \ldots, m_s(x)$ described in the preceding paragraph are then the minimum polynomials of a^i for $i = 1, 2, \ldots, s$, respectively. Thus, we can determine $g(x)$ by forming a product that includes a single factor of each unique $m_i(x)$ for $i = 1, 2, \ldots, s$. As a consequence of Lagrange's Theorem (Theorem 1.4), each a^i will be a root of $f(x) = x^m - 1$ for $i = 1, 2, \ldots, s$. Therefore, $g(x)$ will divide $f(x)$.

Before we look at some examples, we should remark that the computations required with BCH codes are obviously more complex than for the types of codes we presented previously. However, because BCH codewords are polynomials over $\mathbb{Z}_2$, some of the computations required follow

---

[1] A polynomial $m(x)$ is called the *minimum polynomial* of a if $m(x)$ is the unique monic polynomial of smallest degree of which a is a root.

immediately from other computations. For instance, note that over $\mathbb{Z}_2$, $(x_1 + x_2 + \cdots + x_r)^2 = x_1^2 + x_2^2 + \cdots + x_r^2$, since all cross terms will contain a factor of 2. Thus, for any $h(x) \in \mathbb{Z}_2[x]$ and positive integer k, $h(a^{2k}) = (h(a^k))^2$. For example, $h(a^{12}) = (h(a^6))^2 = (h(a^3))^4$. The utility of this will be apparent in the following examples.

Example 4.2 Let $f(x) = x^7 - 1$, and choose $p(x) = x^3 + x + 1 \in \mathbb{Z}_2[x]$. Then for $a = x$ in the finite field $\mathbb{Z}_2[x]/(p(x))$ of order 8, we list the field elements that correspond to the first seven powers of a as follows.

Power	Field Element
a^1	a
a^2	a^2
a^3	$a + 1$
a^4	$a^2 + a$
a^5	$a^2 + a + 1$
a^6	$a^2 + 1$
a^7	1

Let C be the BCH code that results from the first four powers of a. To determine the generator polynomial $g(x)$ for C, we must find the minimum polynomials $m_1(x)$, $m_2(x)$, $m_3(x)$, and $m_4(x)$. However, since $p(x)$ is primitive and $a = x$, it follows that $p(a) = 0$. Furthermore, since $p(x)$ is a polynomial over $\mathbb{Z}_2$, it follows that $p(a^2) = p(a)^2 = 0$ and $p(a^4) = p(a)^4 = 0$. Therefore, $m_1(x) = m_2(x) = m_4(x) = p(x)$. Also, since a^3 is a root of $f(x)$, the minimum polynomial $m_3(x)$ of a^3 must be one of the irreducible factors of $x^7 - 1 = (x^3 + x + 1)(x^3 + x^2 + 1)(x + 1)$. By substituting a^3 into each of these factors, we can find that a^3 is a root of $x^3 + x^2 + 1$, and so $m_3(x) = x^3 + x^2 + 1$. Thus, $g(x) = m_1(x) \cdot m_3(x) = x^6 + x^5 + x^4 + x^3 + x^2 + x + 1$. The code that results from this polynomial is a $[7, 1]$ BCH code with two codewords. $\square$

Example 4.3 Let $f(x) = x^{15} - 1$, and choose $p(x) = x^4 + x + 1 \in \mathbb{Z}_2[x]$. Then for $a = x$ in the finite field $\mathbb{Z}_2[x]/(p(x))$ of order 16, we list the field elements that correspond to the first 15 powers of a as follows.

Power	Field Element	Power	Field Element
a^1	a	a^9	$a^3 + a$
a^2	a^2	a^{10}	$a^2 + a + 1$
a^3	a^3	a^{11}	$a^3 + a^2 + a$
a^4	$a + 1$	a^{12}	$a^3 + a^2 + a + 1$
a^5	$a^2 + a$	a^{13}	$a^3 + a^2 + 1$
a^6	$a^3 + a^2$	a^{14}	$a^3 + 1$
a^7	$a^3 + a + 1$	a^{15}	1
a^8	$a^2 + 1$		

Let C be the BCH code that results from the first six powers of a. To determine the generator polynomial $g(x)$ for C, we must find the minimum polynomials $m_1(x), m_2(x), \ldots, m_6(x)$. Note that since $p(a) = 0$, then $p(a^2) = p(a^4) = 0$. Thus, $m_1(x) = m_2(x) = m_4(x) = p(x)$. Also, since a^3 and a^5 are roots of $f(x)$, then $m_3(x)$ and $m_5(x)$ must be irreducible factors of $x^{15} - 1 = (x+1)(x^2 + x + 1)(x^4 + x + 1)(x^4 + x^3 + 1)(x^4 + x^3 + x^2 + x + 1)$. By substituting a^3 and a^5 into each of these irreducible factors, we can find that $m_3(x) = x^4 + x^3 + x^2 + x + 1$ and $m_5(x) = x^2 + x + 1$. Furthermore, $m_3(a^6) = m_3(a^3)^2 = 0$, and so $m_6(x) = m_3(x)$. Therefore, $g(x) = m_1(x) \cdot m_3(x) \cdot m_5(x) = x^{10} + x^8 + x^5 + x^4 + x^2 + x + 1$. The code that results from this generator polynomial is a $[15, 5]$ BCH code with 32 codewords. $\square$

Although BCH codes are clearly not as easy to construct as Hadamard and Reed-Muller codes, BCH codes are linear while Hadamard and Reed-Muller codes are not. Also, unlike Hamming codes, which are only one-error correcting, BCH codes can be constructed to be multiple-error correcting. Specifically, in Section 4.2, we will show that a BCH code that results from the first $2t$ powers of a will be t-error correcting. For example, since the BCH code in Example 4.3 resulted from the first six powers of a, the code will be three-error correcting.

4.2   Error Correction

As we mentioned in Section 4.1, the generator polynomial for a BCH code is chosen in a special way because of how it allows errors to be corrected in the resulting code. In this section, we will present the BCH code error correction method. Before doing so, we first note the following theorem.

Theorem 4.1 *Suppose $p(x) \in \mathbb{Z}_2[x]$ is a primitive polynomial of degree n, and let C be the BCH code that results from the first s powers of $a = x$ in the finite field $\mathbb{Z}_2[x]/(p(x))$. Then $c(x) \in \mathbb{Z}_2[x]$ of degree less than $2^n - 1$ is in C if and only if $c(a^i) = 0$ for $i = 1, 2, \ldots, s$.*

Proof. Let $m_i(x)$ be the minimum polynomial of a^i in $\mathbb{Z}_2[x]$ for every $i = 1, 2, \ldots, s$, and let $g(x)$ be the least common multiple in $\mathbb{Z}_2[x]$ of the $m_i(x)$ for $i = 1, 2, \ldots, s$. If $c(x) \in C$, then $c(x) = g(x) \cdot h(x)$ for some $h(x) \in \mathbb{Z}_2[x]$. Thus, $c(a^i) = g(a^i) \cdot h(a^i) = 0 \cdot h(a^i) = 0$ for $i = 1, 2, \ldots, s$. Conversely, if $c(a^i) = 0$ for $i = 1, 2, \ldots, s$, then $m_i(x)$ divides $c(x)$ for $i = 1, 2, \ldots, s$. Thus, $g(x)$ divides $c(x)$, and $c(x) \in C$. $\square$

We will now outline the BCH error correction method. Let $p(x) \in \mathbb{Z}_2[x]$ be a primitive polynomial of degree n, and let C be the BCH code that results from the first $2t$ powers of $a = x$ in the finite field $\mathbb{Z}_2[x]/(p(x))$. We will show in Theorem 4.2 that C is then t-error correcting. Suppose $c(x) \in C$ is transmitted, and we receive the polynomial $r(x) \in \mathbb{Z}_2[x]$ of degree less than $2^n - 1$. Then $r(x) = c(x) + e(x)$ for some error polynomial $e(x)$ in $\mathbb{Z}_2[x]$ of degree less than $2^n - 1$ that contains exactly and only the terms in which $r(x)$ and $c(x)$ differ. To correct $r(x)$, we must only determine $e(x)$, for we could then compute $c(x) = r(x) + e(x)$. However, Theorem 4.1 implies $r(a^i) = e(a^i)$ for $i = 1, 2, \ldots, 2t$. Thus, by knowing $r(x)$, we also know some information about $e(x)$. We will call the values of $r(a^i)$ for $i = 1, 2, \ldots, 2t$ the *syndromes* of $r(x)$.

Suppose $e(x) = x^{m_1} + x^{m_2} + \cdots + x^{m_p}$ for some integer error positions $m_1 < m_2 < \cdots < m_p$ with $p \le t$ and $m_p < 2^n - 1$. To correct $r(x)$, we must only find the error positions $m_1, m_2, \ldots, m_p$. To do this, we begin by computing the syndromes of $r(x)$, which we will denote by $s_1 = r(a)$, $s_2 = r(a^2)$, $\ldots$, $s_{2t} = r(a^{2t})$. Next, we introduce the following *error locator* polynomial $E(z)$, called so because its roots (unknown at this point) reveal the error positions in $r(x)$.

$$
\begin{aligned}
E(z) &= (z - a^{m_1})(z - a^{m_2}) \cdots (z - a^{m_p}) \\
&= z^p + \sigma_1 z^{p-1} + \sigma_2 z^{p-2} + \cdots + \sigma_p
\end{aligned}
$$

The coefficients $\sigma_1, \sigma_2, \ldots, \sigma_p$ of the preceding polynomial $E(z)$ are the elementary symmetric functions in $a^{m_1}, a^{m_2}, \ldots, a^{m_p}$. That is, $\sigma_1, \sigma_2, \ldots, \sigma_p$ are given by the following.

$$
\sigma_1 = \sum_{1 \le i \le p} a^{m_i}
$$

$$
\sigma_2 = \sum_{1 \le i < j \le p} a^{m_i} a^{m_j}
$$

$$
\vdots
$$

$$
\sigma_p = a^{m_1} a^{m_2} \cdots a^{m_p}
$$

Next, note that if we evaluate $E(a^{m_j})$ for all $1 \le j \le p$, and then multiply each result by $(a^{m_j})^k$ for some $1 \le k \le p$, since a^{m_j} is a root of $E(z)$ for all $1 \le j \le p$, we obtain the following system of equations.

$$
(a^{m_1})^k \cdot ((a^{m_1})^p + \sigma_1 (a^{m_1})^{(p-1)} + \cdots + \sigma_p) = 0
$$
$$
(a^{m_2})^k \cdot ((a^{m_2})^p + \sigma_1 (a^{m_2})^{(p-1)} + \cdots + \sigma_p) = 0
$$
$$
\vdots
$$
$$
(a^{m_p})^k \cdot ((a^{m_p})^p + \sigma_1 (a^{m_p})^{(p-1)} + \cdots + \sigma_p) = 0
$$

By distributing the $(a^{m_j})^k$ in each of the preceding equations and then adding the results together, we obtain the following single equation.

$$e(a^{k+p}) + \sigma_1 e(a^{k+p-1}) + \sigma_2 e(a^{k+p-2}) + \cdots + \sigma_p e(a^k) \;=\; 0$$

Since $r(a^i) = e(a^i)$ for all $i = 1, 2, \ldots, 2t$, then this equation is equivalent to the following equation.

$$s_{k+p} + \sigma_1 s_{k+p-1} + \sigma_2 s_{k+p-2} + \cdots + \sigma_p s_k \;=\; 0$$

Because this equation holds for all $1 \le k \le p$, this yields a system of p linear equations in the p unknowns $\sigma_1, \sigma_2, \ldots, \sigma_p$, which can easily be verified to be equivalent to the following single matrix equation.

$$\begin{bmatrix} s_1 & s_2 & \cdots & s_p \\ s_2 & s_3 & \cdots & s_{p+1} \\ \vdots & \vdots & & \vdots \\ s_p & s_{p+1} & \cdots & s_{2p-1} \end{bmatrix} \begin{bmatrix} \sigma_p \\ \sigma_{p-1} \\ \vdots \\ \sigma_1 \end{bmatrix} = \begin{bmatrix} s_{p+1} \\ s_{p+2} \\ \vdots \\ s_{2p} \end{bmatrix} \qquad (4.1)$$

If the $p \times p$ coefficient matrix in (4.1) is nonsingular, then we would be able to solve this equation uniquely for $\sigma_1, \sigma_2, \ldots, \sigma_p$. After finding $\sigma_1, \sigma_2, \ldots, \sigma_p$, we could then form the error locator polynomial $E(z)$, and determine $a^{m_1}, a^{m_2}, \ldots, a^{m_p}$ by trial and error as the roots of $E(z)$. This would reveal the error positions $m_1, m_2, \ldots, m_p$ in $r(x)$.

Since we would generally not know the number of errors in a received polynomial before attempting to correct it, we will begin the BCH error correction process on a received polynomial $r(x)$ by assuming that $r(x)$ contains the maximum number of correctable errors in the code. That is, in a t-error correcting code, we will begin the BCH error correction process on a received polynomial $r(x)$ by assuming that $r(x)$ contains t errors. If $r(x)$ does not contain exactly t errors, then the $t \times t$ coefficient matrix in (4.1) will be singular. In this case, we will simply reduce the number of assumed errors to $t-1$, and repeat the error correction procedure using only the first $2(t-1)$ syndromes of $r(x)$. If $r(x)$ also does not contain exactly $t-1$ errors (i.e., if the $(t-1) \times (t-1)$ coefficient matrix in (4.1) is also singular), we will repeat the procedure again assuming $t-2$ errors. We will continue to repeat this procedure as many times as necessary, each time reducing the number of assumed errors by one and using twice as many syndromes as the number of assumed errors, until the coefficient matrix in (4.1) is nonsingular.

We will now look at an example of the BCH error correction method in the code that results from the generator polynomial in Example 4.3.

Example 4.4 Let C be the BCH code that results from the generator polynomial $g(x) = x^{10} + x^8 + x^5 + x^4 + x^2 + x + 1$ in Example 4.3. Suppose that a codeword in C is transmitted as a vector in $\mathbb{Z}_2^{15}$, and we receive the vector $\mathbf{r} = (101111110010000) \in \mathbb{Z}_2^{15}$. Note first that this vector converts to the polynomial $r(x) = 1 + x^2 + x^3 + x^4 + x^5 + x^6 + x^7 + x^{10} \in \mathbb{Z}_2[x]$. Since C is three-error correcting, then to correct $r(x)$, we begin by computing the first six syndromes of $r(x)$. Using the table of powers of a and corresponding field elements in Example 4.3, we can compute these syndromes as follows.

$$
\begin{aligned}
s_1 &= r(a) \\
&= 1 + a^2 + a^3 + a^4 + a^5 + a^6 + a^7 + a^{10} \\
&= 1 + a^2 + a^3 + a + 1 + a^2 + a + a^3 + a^2 + a^3 + a + 1 + a^2 + a + 1 \\
&= a^3
\end{aligned}
$$

$$
s_2 = r(a^2) = (r(a))^2 = (a^3)^2 = a^6
$$

$$
\begin{aligned}
s_3 &= r(a^3) \\
&= 1 + a^6 + a^9 + a^{12} + a^{15} + a^{18} + a^{21} + a^{30} \\
&= 1 + a^6 + a^9 + a^{12} + 1 + a^3 + a^6 + 1 \\
&= a^3 + a + a^3 + a^2 + a + 1 + a^3 + 1 \\
&= a^6
\end{aligned}
$$

$$
s_4 = r(a^4) = (r(a))^4 = (a^3)^4 = a^{12}
$$

$$
\begin{aligned}
s_5 &= r(a^5) \\
&= 1 + a^{10} + a^{15} + a^{20} + a^{25} + a^{30} + a^{35} + a^{50} \\
&= 1 + a^{10} + 1 + a^5 + a^{10} + 1 + a^5 + a^5 \\
&= a^2 + a + 1 \\
&= a^{10}
\end{aligned}
$$

$$
s_6 = r(a^6) = (r(a^3))^2 = (a^6)^2 = a^{12}
$$

Now, assuming that $r(x)$ contains three errors, we must find σ_1, σ_2, and σ_3 that satisfy the following equation.

$$
\begin{bmatrix} a^3 & a^6 & a^6 \\ a^6 & a^6 & a^{12} \\ a^6 & a^{12} & a^{10} \end{bmatrix} \begin{bmatrix} \sigma_3 \\ \sigma_2 \\ \sigma_1 \end{bmatrix} = \begin{bmatrix} a^{12} \\ a^{10} \\ a^{12} \end{bmatrix} \tag{4.2}
$$

Next, we will find the determinant of the 3×3 coefficient matrix in (4.2).

$$
\begin{aligned}
\begin{vmatrix} a^3 & a^6 & a^6 \\ a^6 & a^6 & a^{12} \\ a^6 & a^{12} & a^{10} \end{vmatrix} &= a^{19} + a^{24} + a^{24} + a^{18} + a^{27} + a^{22} \\
&= a^4 + a^3 + a^{12} + a^7 \\
&= a + 1 + a^3 + a^3 + a^2 + a + 1 + a^3 + a + 1 \\
&= a^{12}
\end{aligned}
$$

Thus, the coefficient matrix in (4.2) is nonsingular, and $r(x)$ contains three errors. We can then use Cramer's Rule[2] to determine σ_1, σ_2, and σ_3 as follows. First, we calculate the following determinant.

$$\begin{vmatrix} a^{12} & a^6 & a^6 \\ a^{10} & a^6 & a^{12} \\ a^{12} & a^{12} & a^{10} \end{vmatrix} \;=\; a^{28} + a^{30} + a^{28} + a^{24} + a^{36} + a^{26}$$

$$= \; 1 + a^9 + a^6 + a^{11}$$

$$= \; 1 + a^3 + a + a^3 + a^2 + a^3 + a^2 + a$$

$$= \; a^{14}$$

Cramer's Rule then yields the following.

$$\sigma_3 \;=\; \frac{a^{14}}{a^{12}} \;=\; a^2$$

Next, we calculate the following determinant.

$$\begin{vmatrix} a^3 & a^{12} & a^6 \\ a^6 & a^{10} & a^{12} \\ a^6 & a^{12} & a^{10} \end{vmatrix} \;=\; a^{23} + a^{30} + a^{24} + a^{22} + a^{27} + a^{28}$$

$$= \; a^8 + 1 + a^9 + a^7 + a^{12} + a^{13}$$

$$= \; a^2 + 1 + 1 + a^3 + a + a^3 + a + 1 + a^3 + a^2$$

$$\quad + a + 1 + a^3 + a^2 + 1$$

$$= \; a^{10}$$

Cramer's Rule then yields the following.

$$\sigma_2 \;=\; \frac{a^{10}}{a^{12}} \;=\; a^{-2} \;=\; a^{13}$$

Finally, we calculate the following determinant.

$$\begin{vmatrix} a^3 & a^6 & a^{12} \\ a^6 & a^6 & a^{10} \\ a^6 & a^{12} & a^{12} \end{vmatrix} \;=\; a^{21} + a^{22} + a^{30} + a^{24} + a^{25} + a^{24}$$

$$= \; a^6 + a^7 + 1 + a^{10}$$

$$= \; a^3 + a^2 + a^3 + a + 1 + 1 + a^2 + a + 1$$

$$= \; 1$$

---

[2]Cramer's Rule states that for a matrix equation $A\mathbf{x} = \mathbf{b}$ with square nonsingular coefficient matrix A and column vector $\mathbf{x}$ of unknowns, the ith unknown from the top in $\mathbf{x}$ will equal $\frac{\det(A_i)}{\det(A)}$, where A_i is obtained from A by replacing the ith column of A with $\mathbf{b}$.

Cramer's Rule then yields the following.

$$\sigma_1 = \frac{1}{a^{12}} = a^{-12} = a^3$$

The resulting error locator polynomial is $E(z) = z^3 + a^3 z^2 + a^{13} z + a^2$. By evaluating $E(z)$ at successive powers of a, we can easily find that the roots of $E(z)$ are 1, a^5, and a^{12}. Thus, the error in $r(x)$ is $e(x) = 1 + x^5 + x^{12}$, and we can correct $r(x)$ to the following codeword $c(x)$.

$$c(x) = r(x) + e(x) = x^2 + x^3 + x^4 + x^6 + x^7 + x^{10} + x^{12}$$

It can easily be verified that this polynomial $c(x)$ is a multiple of $g(x)$.

Now, suppose that another codeword in C is transmitted, and we receive the vector $\mathbf{r} = (100100010011010) \in \mathbb{Z}_2^{15}$. This vector converts to the polynomial $r(x) = 1 + x^3 + x^7 + x^{10} + x^{11} + x^{13} \in \mathbb{Z}_2[x]$. To correct $r(x)$, we begin by computing the first six syndromes of $r(x)$, which turn out as follows.

$$
\begin{aligned}
s_1 &= a^5 \\
s_2 &= a^{10} \\
s_3 &= a^2 \\
s_4 &= a^5 \\
s_5 &= 1 \\
s_6 &= a^4
\end{aligned}
$$

Assuming that $r(x)$ contains three errors, we must find σ_1, σ_2, and σ_3 that satisfy the following equation.

$$
\begin{bmatrix} a^5 & a^{10} & a^2 \\ a^{10} & a^2 & a^5 \\ a^2 & a^5 & 1 \end{bmatrix}
\begin{bmatrix} \sigma_3 \\ \sigma_2 \\ \sigma_1 \end{bmatrix}
=
\begin{bmatrix} a^5 \\ 1 \\ a^4 \end{bmatrix}
\tag{4.3}
$$

However, the determinant of the 3×3 coefficient matrix in (4.3) turns out to be 0. Thus, this coefficient matrix is singular, and $r(x)$ does not contain exactly three errors. Next, we assume that $r(x)$ contains two errors, and use only the first four syndromes of $r(x)$. Assuming that $r(x)$ contains only two errors, we must find σ_1 and σ_2 that satisfy the following equation.

$$
\begin{bmatrix} a^5 & a^{10} \\ a^{10} & a^2 \end{bmatrix}
\begin{bmatrix} \sigma_2 \\ \sigma_1 \end{bmatrix}
=
\begin{bmatrix} a^2 \\ a^5 \end{bmatrix}
\tag{4.4}
$$

The determinant of the 2×2 coefficient matrix in (4.4) is nonzero. Thus, this coefficient matrix is nonsingular, and $r(x)$ contains two errors. We can then use Cramer's Rule to determine σ_1 and σ_2, which turn out to be $\sigma_1 = a^5$ and $\sigma_2 = a^3$. The resulting error locator polynomial is $E(z) = z^2 + a^5 z + a^3$. The roots of $E(z)$ can be determined to be a and a^2. Thus, the error in

$r(x)$ is $e(x) = x + x^2$, and we can correct $r(x)$ to the following codeword $c(x)$.

$$c(x) = r(x) + e(x) = 1 + x + x^2 + x^3 + x^7 + x^{10} + x^{11} + x^{13}$$

It can easily be verified that this polynomial $c(x)$ is a multiple of $g(x)$.   □

We will close this section by proving the following fundamental result regarding BCH codes that we have already mentioned and used.

Theorem 4.2 *Suppose $p(x) \in \mathbb{Z}_2[x]$ is a primitive polynomial of degree n, and let C be the BCH code that results from the first $2t$ powers of $a = x$ in the finite field $\mathbb{Z}_2[x]/(p(x))$. Then C is t-error correcting.*

Proof. Let $m = 2^n - 1$, and consider the following matrix H.

$$H = \begin{bmatrix} 1 & a & a^2 & \cdots & a^{m-1} \\ 1 & a^2 & (a^2)^2 & \cdots & (a^2)^{m-1} \\ 1 & a^3 & (a^3)^2 & \cdots & (a^3)^{m-1} \\ \vdots & \vdots & \vdots & & \vdots \\ 1 & a^{2t} & (a^{2t})^2 & \cdots & (a^{2t})^{m-1} \end{bmatrix}$$

Note that for a polynomial $r(x) = b_0 + b_1 x + \cdots + b_{m-1} x^{m-1} \in \mathbb{Z}_2[x]$, if we let $\mathbf{b} = (b_0, b_1, \ldots, b_{m-1})$, then $H\mathbf{b}^T = (s_1, s_2, \ldots, s_{2t})^T$. Thus, $r(x) \in C$ if and only if $H\mathbf{b}^T$ is the zero vector, and H serves as a parity check matrix for C. We will now show that the minimum number of linearly dependent columns in H is $2t+1$. First, we will show that any $2t$ columns in H must be linearly independent. Choose any nonnegative integers $j_1 < j_2 < \cdots < j_{2t}$ less than m. Then the columns in H in these positions form the following $2t \times 2t$ matrix.

$$\begin{bmatrix} a^{j_1} & a^{j_2} & \cdots & a^{j_{2t}} \\ (a^2)^{j_1} & (a^2)^{j_2} & \cdots & (a^2)^{j_{2t}} \\ \vdots & \vdots & & \vdots \\ (a^{2t})^{j_1} & (a^{2t})^{j_2} & \cdots & (a^{2t})^{j_{2t}} \end{bmatrix}$$

The determinant of this matrix can be expressed as follows.

$$\begin{vmatrix} 1 & 1 & \cdots & 1 \\ a^{j_1} & a^{j_2} & & a^{j_{2t}} \\ \vdots & \vdots & & \vdots \\ (a^{j_1})^{2t-1} & (a^{j_2})^{2t-1} & \cdots & (a^{j_{2t}})^{2t-1} \end{vmatrix} \cdot a^{j_1} a^{j_2} \cdots a^{j_{2t}}$$

This quantity is nonzero because it is the determinant of a Vandermonde matrix with distinct columns. Thus, any $2t$ columns in H must be linearly independent. Also, since H has $2t$ rows, we know that any $2t+1$ columns in H must be linearly dependent. Therefore, the minimum number of linearly dependent columns in H is $2t + 1$. By Theorem 3.7, we know that the minimum distance of C is $2t + 1$, and so C is t-error correcting. $\qquad\square$

4.3 BCH Codes with Maple

In this section, we will show how Maple can be used to construct and correct errors in BCH codes. We will consider the code in Examples 4.3 and 4.4.

4.3.1 Construction

Because some of the functions that we will use are in the Maple **Linear-Algebra** package, we will begin by including this package. In addition, we will enter the following **interface** command to cause Maple to display all matrices of size 200×200 and smaller throughout the remainder of this Maple session.

```
> with(LinearAlgebra):
```

```
> interface(rtablesize=200):
```

We will now define the primitive polynomial $p(x) = x^4 + x + 1 \in \mathbb{Z}_2[x]$ used to construct the code.

```
> p := x -> x^4+x+1:
```

```
> Primitive(p(x)) mod 2;
```
$$true$$

Next, we will use the Maple **degree** function to assign the number of elements in the underlying finite field as the variable *fs*, and use the Maple **Vector** function to create a vector in which to store the field elements.

```
> fs := 2^(degree(p(x)));
```
$$fs := 16$$

```
> field := Vector[row](fs):
```

We can then use the following commands to generate and store the field elements in the vector *field*. Since for BCH codes we denote the field element x by a, we use the parameters a and $p(a)$ in the following **Powmod** command.

```
> for i from 1 to fs-1 do
>        field[i] := Powmod(a, i, p(a), a) mod 2:
> od:
```

```
>  field[fs] := 0:
```

We can view the entries in the vector *field* by entering the following command.

```
>  field;
```

$$\left[a,\, a^2,\, a^3,\, a+1,\, a^2+a,\, a^3+a^2,\, a^3+a+1,\, a^2+1,\, a^3+a,\, a^2+a\right.$$
$$\left.+\,1,\, a^3+a^2+a,\, a^3+a^2+a+1,\, a^3+a^2+1,\, a^3+1,\, 1,\, 0\right]$$

Because working with BCH codes requires frequent conversions between polynomial field elements and powers of a, it will be useful for us to establish an association between the polynomial field elements and corresponding powers of a for this field. We will establish this association in a table. We first use the Maple **table** function to create the following table.

```
>  ftable := table():
```

Then by entering the following commands, we establish an association between the polynomial field elements and corresponding powers of a for this field in *ftable*. (Note the bracket [ ] syntax for accessing the positions in *field* and the entries in *ftable*.)

```
>  for i from 1 to fs-2 do
>      ftable[field[i]] := a^i:
>  od:
>  ftable[field[fs-1]] := 1:
>  ftable[field[fs]] := 0:
```

We can view the entries in *ftable* by entering the following **print** command.

```
>  print(ftable);
```

$$table\left(\left[0=0,\, 1=1,\, a^3+a^2=a^6,\, a^3+a^2+a=a^{11},\, a^2+a+1=a^{10},\right.\right.$$
$$a^3+a+1=a^7,\, a^3+a=a^9,\, a^3=a^3,\, a^2=a^2,\, a^3+1=a^{14},\, a+1$$
$$=a^4,\, a^2+1=a^8,\, a=a,\, a^3+a^2+1=a^{13},\, a^2+a=a^5,\, a^3+a^2$$
$$\left.\left.+\,a+1=a^{12}\right]\right)$$

The following command illustrates how *ftable* can be used. Specifically, by entering the following command, we retrieve the power of a that corresponds to the polynomial field element a^3+a^2.

```
>  ftable[a^3+a^2];
```

$$a^6$$

We will now construct the generator polynomial for the BCH code in Example 4.3. To do this, we first define the polynomial $f(x) = x^{15} - 1$ of which each power of a is a root.

```
>  f := x -> x^(fs-1)-1:
```

Next, we use the Maple **Factor** function to find the irreducible factors of $f(x)$ in $\mathbb{Z}_2[x]$.

```
>  factf := Factor(f(x)) mod 2;
```

$$factf := \left(x^4 + x^3 + 1\right)\left(x^4 + x + 1\right)(x + 1)\left(x^2 + x + 1\right)\left(x^4 + x^3 + x^2 + x + 1\right)$$

To construct the generator polynomial in Example 4.3, we will need to access these factors of $f(x)$ separately. The following **nops** command returns the number of factors in *factf*.

```
>  nops(factf);
```

$$5$$

We can access these factors individually by using the Maple **op** function. For example, we can use the following **op** command to assign the third factor in the expression *factf* as the variable *f3*.

```
>  f3 := op(3, factf);
```

$$f3 := x + 1$$

We can then use the Maple **unapply** function as follows to convert the expression *f3* into a function.

```
>  f3 := unapply(f3, x);
```

$$f3 := x \rightarrow x + 1$$

We can now evaluate *f3* as a function. For example, we can evaluate *f3* at a^6 by entering the following command.

```
>  f3(a^6);
```

$$a^6 + 1$$

Also, we can use the Maple **Rem** function to retrieve the polynomial field element that corresponds to the preceding output.

```
>  Rem(f3(a^6), p(a), a) mod 2;
```

$$a^3 + a^2 + 1$$

Since the preceding output is not 0, then we know that a^6 is not a root of *f3*, or, equivalently, that *f3* is not the minimum polynomial of a^6.

We will now find the minimum polynomials that are the factors in the generator polynomial. We first assign the number $t = 3$ of errors that are to be correctable in the code. Then in the subsequent nested loop, we find and display the minimum polynomials of $a, a^2, \ldots, a^{2t}$. In these commands, the outer loop spans the powers of a, while the inner loop evaluates the factors of *factf* at these powers. Since each a^i is a root of *factf*, each will be the root of an irreducible factor of *factf*. The factor of which a^i is a root will be the minimum polynomial of a^i. The **if** and **print** statements that appear in these commands cause the correct minimum polynomial of each a^i to be displayed. The **break** statement terminates the inner loop when the correct minimum polynomial is found.

```
>   t := 3:

>   for i from 1 to 2*t do

>        for j from 1 to nops(factf) do

>            fj := unapply(op(j, factf), x):

>            if Rem(fj(a^i), p(a), a) mod 2 = 0 then

>                print(a^i, '  is a root of  ', fj(x)):

>                break:

>            fi:

>        od:

>   od:
```

$$a, \ \textit{is a root of } , x^4 + x + 1$$

$$a^2, \ \textit{is a root of } , x^4 + x + 1$$

$$a^3, \ \textit{is a root of } , x^4 + x^3 + x^2 + x + 1$$

$$a^4, \ \textit{is a root of } , x^4 + x + 1$$

$$a^5, \ \textit{is a root of } , x^2 + x + 1$$

$$a^6, \ \textit{is a root of } , x^4 + x^3 + x^2 + x + 1$$

Thus, the minimum polynomial of a is x^4+x+1, which is also the minimum polynomial of a^2 and a^4. Also, the minimum polynomial of a^3 and a^6 is $x^4 + x^3 + x^2 + x + 1$, and the minimum polynomial of a^5 is $x^2 + x + 1$, just as we noted in Example 4.3. Next, we assign as variables one copy of each of these three unique minimum polynomials.

```
>  m1 := x -> x^4+x+1:

>  m3 := x -> x^4+x^3+x^2+x+1:

>  m5 := x -> x^2+x+1:
```

We can then define the generator polynomial $g(x)$ as the product of these three minimum polynomials.

```
>  g := m1(x)*m3(x)*m5(x);
```
$$g := \left(x^4 + x + 1\right)\left(x^4 + x^3 + x^2 + x + 1\right)\left(x^2 + x + 1\right)$$

We can expand this expression for $g(x)$ by using the Maple **Expand** function as follows.

```
>  g := Expand(g) mod 2;
```
$$g := 1 + x + x^2 + x^4 + x^5 + x^8 + x^{10}$$

Also, we can sort the terms in this expression for $g(x)$ by using the Maple **sort** function as follows.

```
>  g := sort(g);
```
$$g := x^{10} + x^8 + x^5 + x^4 + x^2 + x + 1$$

Finally, we can convert this expression for $g(x)$ into a function by entering the following **unapply** command.

```
>  g := unapply(g, x);
```
$$g := x \rightarrow x^{10} + x^8 + x^5 + x^4 + x^2 + x + 1$$

4.3.2   Error Correction

We will now show how Maple can be used to correct both of the received polynomials in Example 4.4. First we will consider the following received polynomial $r(x)$.

```
>  r := x -> 1+x^2+x^3+x^4+x^5+x^6+x^7+x^10:
```

Recall that to correct $r(x)$, we begin by computing the first $2t$ syndromes of $r(x)$. Before doing this, we will create the following vector syn of length $2t$ in which to store the syndromes.

```
>  syn := Vector[row](2*t):
```

We can then use the following loop to compute and store the first $2t$ syndromes of $r(x)$ in the vector *syn*.

```
>   for i from 1 to 2*t do

>         syn[i] := ftable[Rem(r(a^i), p(a), a) mod 2];

>   od:

>   syn;
```

$$\begin{bmatrix} a^3 & a^6 & a^6 & a^{12} & a^{10} & a^{12} \end{bmatrix}$$

We can now retrieve particular syndromes from the vector *syn*. For example, we can enter the following command to retrieve the fifth syndrome of $r(x)$.

```
>   syn[5];
```

$$a^{10}$$

Next, we will construct the 3×3 coefficient matrix in (4.2), and assign this matrix as the variable A.

```
>   A := Matrix([[syn[1], syn[2], syn[3]], [syn[2], syn[3],

    syn[4]], [syn[3], syn[4], syn[5]]]);
```

$$A := \begin{bmatrix} a^3 & a^6 & a^6 \\ a^6 & a^6 & a^{12} \\ a^6 & a^{12} & a^{10} \end{bmatrix}$$

We can find the determinant of A by using the Maple **Determinant** function as follows.

```
>   Determinant(A);
```

$$a^{19} - a^{27} + 2\,a^{24} - a^{22} - a^{18}$$

To find the field element that corresponds to this determinant of A, we can use the Maple **Rem** function as follows.

```
>    d := Rem(Determinant(A), p(a), a) mod 2;
```

$$d := a^3 + a^2 + a + 1$$

We can then use *ftable* as follows to find the power of a that corresponds to the determinant of A.

```
>   d := ftable[d];
```

$$d := a^{12}$$

Since this determinant is nonzero, then we know that $r(x)$ contains three errors. As in Example 4.4, we will use Cramer's Rule to determine σ_1, σ_2, and σ_3 in (4.2). First, we will construct the column vector on the right side of (4.2), and assign this vector as the variable B.

> B := Vector([syn[4], syn[5], syn[6]]);

$$B := \begin{bmatrix} a^{12} \\ a^{10} \\ a^{12} \end{bmatrix}$$

We can use the Maple **Column** function to choose particular columns from the matrix A. For example, the following command returns the second column of A.

> Column(A, 2);

$$\begin{bmatrix} a^6 \\ a^6 \\ a^{12} \end{bmatrix}$$

Also, we can use the Maple **Column** and **Matrix** functions to form the matrices required for Cramer's Rule. For example, in the following command, we form a new matrix $A1$ from A by replacing the first column of A with the vector B.

> A1 := Matrix([B, Column(A, 2), Column(A, 3)]);

$$A1 := \begin{bmatrix} a^{12} & a^6 & a^6 \\ a^{10} & a^6 & a^{12} \\ a^{12} & a^{12} & a^{10} \end{bmatrix}$$

Next, we will find the determinant of $A1$ expressed as a power of a.

> dA1 := ftable[Rem(Determinant(A1), p(a), a) mod 2];

$$dA1 := a^{14}$$

Thus, by Cramer's Rule, we can find σ_3 in (4.2) by entering the following command.

> sigma3 := dA1/d;

$$\sigma 3 := a^2$$

Similarly, by Cramer's Rule, we can find σ_2 in (4.2) as follows.

```
>  A2 := Matrix([Column(A, 1), B, Column(A, 3)]);
```

$$A2 := \begin{bmatrix} a^3 & a^{12} & a^6 \\ a^6 & a^{10} & a^{12} \\ a^6 & a^{12} & a^{10} \end{bmatrix}$$

```
>  dA2 := ftable[Rem(Determinant(A2), p(a), a) mod 2];
```

$$dA2 := a^{10}$$

```
>  sigma2 := dA2/d;
```

$$\sigma 2 := \frac{1}{a^2}$$

We will use the following command to express σ_2 as a positive power of a.

```
>  sigma2 := sigma2*a^(fs-1);
```

$$\sigma 2 := a^{13}$$

Finally, by Cramer's Rule, we can find σ_1 in (4.2) as follows.

```
>  A3 := Matrix([Column(A, 1), Column(A, 2), B]);
```

$$A3 := \begin{bmatrix} a^3 & a^6 & a^{12} \\ a^6 & a^6 & a^{10} \\ a^6 & a^{12} & a^{12} \end{bmatrix}$$

```
>  dA3 := ftable[Rem(Determinant(A3), p(a), a) mod 2];
```

$$dA3 := 1$$

```
>  sigma1 := dA3/d;
```

$$\sigma 1 := \frac{1}{a^{12}}$$

```
>  sigma1 := sigma1*a^(fs-1);
```

$$\sigma 1 := a^3$$

Next, we will construct the resulting error locator polynomial, and assign this polynomial as the variable *EL*.

```
>  EL := z^3+sigma1*z^2+sigma2*z+sigma3;
```

$$EL := z^3 + a^3 z^2 + a^{13} z + a^2$$

```
>   EL := unapply(EL, z);
```
$$EL := z \to z^3 + a^3 z^2 + a^{13} z + a^2$$

Then in the following loop, we find the roots of *EL* by trial and error.

```
>   for i from 0 to fs-2 do
>           if Rem(EL(a^i), p(a), a) mod 2 = 0 then
>               print(a^i, '  is a root of EL(z) =  ', EL(z));
>           fi:
>   od:
```

$$1, \ \textit{is a root of } EL(z) = \ , \ z^3 + a^3 z^2 + a^{13} z + a^2$$
$$a^5, \ \textit{is a root of } EL(z) = \ , \ z^3 + a^3 z^2 + a^{13} z + a^2$$
$$a^{12}, \ \textit{is a root of } EL(z) = \ , \ z^3 + a^3 z^2 + a^{13} z + a^2$$

Thus, as we noted in Example 4.4, the roots of *EL* are 1, a^5, and a^{12}, and the error in $r(x)$ is the following polynomial $e(x)$.

```
>   e := x -> 1+x^5+x^12:
```

In the next command we form the corrected codeword $c(x)$.

```
>   c := (r(x)+e(x)) mod 2;
```
$$c := x^2 + x^3 + x^4 + x^6 + x^7 + x^{10} + x^{12}$$

Finally, to verify that $c(x)$ is a multiple of $g(x)$, we enter the following.

```
>   (Factor(c) mod 2)/(Factor(g(x)) mod 2);
```
$$x^2$$

We will now consider the second received polynomial in Example 4.4, which is the following polynomial $r(x)$.

```
>   r := x -> 1+x^3+x^7+x^10+x^11+x^13:
```

To correct this received polynomial, we begin by computing and storing the first $2t$ syndromes of $r(x)$ in the vector *syn*.

```
>   for i from 1 to 2*t do
>           syn[i] := ftable[Rem(r(a^i), p(a), a) mod 2];
>   od:
>   syn;
```
$$\begin{bmatrix} a^5 & a^{10} & a^2 & a^5 & 1 & a^4 \end{bmatrix}$$

Next, we will construct the 3×3 coefficient matrix in (4.3).

```
>  A := Matrix([[syn[1], syn[2], syn[3]], [syn[2], syn[3],
   syn[4]], [syn[3], syn[4], syn[5]]]);
```

$$A := \begin{bmatrix} a^5 & a^{10} & a^2 \\ a^{10} & a^2 & a^5 \\ a^2 & a^5 & 1 \end{bmatrix}$$

To find the determinant of this coefficient matrix A, we enter the following command.

```
>  d := ftable[Rem(Determinant(A), p(a), a) mod 2];
```

$$d := 0$$

Since this determinant is zero, then we know that $r(x)$ does not contain exactly three errors. So next we will assume that $r(x)$ contains only two errors, and construct the 2×2 coefficient matrix in (4.4). To do this, we will first use the Maple **DeleteRow** and **DeleteColumn** functions to delete the last row and last column from A.

```
>  A := DeleteRow(A, 3):
>  A := DeleteColumn(A, 3);
```

$$A := \begin{bmatrix} a^5 & a^{10} \\ a^{10} & a^2 \end{bmatrix}$$

Next, we will find the determinant of this new coefficient matrix A.

```
>  d := ftable[Rem(Determinant(A), p(a), a) mod 2];
```

$$d := a^{13}$$

Since this determinant is nonzero, then we know that $r(x)$ contains two errors. We can then again use Cramer's Rule to determine σ_1 and σ_2 in (4.4). First, we will construct the column vector on the right side of (4.4), and assign this vector as the variable B.

```
>  B := Vector([syn[3], syn[4]]);
```

$$B := \begin{bmatrix} a^2 \\ a^5 \end{bmatrix}$$

We can then use Cramer's Rule to determine σ_1 and σ_2 as follows.

```
> A1 := Matrix([B, Column(A, 2)]);
```

$$A1 := \begin{bmatrix} a^2 & a^{10} \\ a^5 & a^2 \end{bmatrix}$$

```
> dA1 := ftable[Rem(Determinant(A1), p(a), a) mod 2];
```

$$dA1 := a$$

```
> sigma2 := dA1/d;
```

$$\sigma2 := \frac{1}{a^{12}}$$

```
> sigma2 := sigma2*a^(fs-1);
```

$$\sigma2 := a^3$$

```
> A2 := Matrix([Column(A, 1), B]);
```

$$A2 := \begin{bmatrix} a^5 & a^2 \\ a^{10} & a^5 \end{bmatrix}$$

```
> dA2 := ftable[Rem(Determinant(A2), p(a), a) mod 2];
```

$$dA2 := a^3$$

```
> sigma1 := dA2/d;
```

$$\sigma1 := \frac{1}{a^{10}}$$

```
> sigma1 := sigma1*a^(fs-1);
```

$$\sigma1 := a^5$$

Next, we will construct the resulting error locator polynomial, and assign this polynomial as the variable *EL*.

```
> EL := z^2+sigma1*z+sigma2;
```

$$EL := z^2 + a^5 z + a^3$$

```
> EL := unapply(EL, z);
```

$$EL := z \rightarrow z^2 + a^5 z + a^3$$

Then in the following loop, we find the roots of EL by trial and error.

```
>   for i from 0 to fs-2 do
>        if Rem(EL(a^i), p(a), a) mod 2 = 0 then
>            print(a^i, '  is a root of EL(z) =  ', EL(z));
>        fi:
>   od:
```

$$a, \ is \ a \ root \ of \ EL(z) \ = \ , \ z^2 + a^5 z + a^3$$
$$a^2, \ is \ a \ root \ of \ EL(z) \ = \ , \ z^2 + a^5 z + a^3$$

Thus, the error in $r(x)$ is the following polynomial $e(x)$.

```
>   e := x -> x+x^2:
```

In the next command we form the corrected codeword $c(x)$.

```
>   c := (r(x)+e(x)) mod 2;
```
$$c := 1 + x^3 + x^7 + x^{10} + x^{11} + x^{13} + x^2 + x$$

```
>   sort(c);
```
$$x^{13} + x^{11} + x^{10} + x^7 + x^3 + x^2 + x + 1$$

Finally, to verify that $c(x)$ is a multiple of $g(x)$, we enter the following.

```
>   (Factor(c) mod 2)/(Factor(g(x)) mod 2);
```
$$(x+1)\left(x^2 + x + 1\right)$$

4.4   BCH Codes with MATLAB

In this section, we will show how MATLAB can be used to construct and correct errors in BCH codes. We will consider the code in Examples 4.3 and 4.4.

4.4.1   Construction

We will begin by declaring the variables a and x as symbolic variables, and defining the primitive polynomial $p(x) = x^4 + x + 1 \in \mathbb{Z}_2[x]$ used to construct the code. As we did in Section 1.6, we will establish that the polynomial is primitive using the user-written function **Primitive**, which we have written separately from this MATLAB session and saved as the M-file Primitive.m.

```
>> syms a x
>> p = @(x) x^4+x+1;
>> Primitive(p(x), 2)

ans =

TRUE
```

Next, we will use the MuPAD **degree** function within MATLAB to assign the number of elements in the underlying finite field as the variable *fs*.

```
>> fs = 2^(feval(symengine, 'degree', p(x)))

fs =

16
```

The MATLAB **feval** function with the first parameter *symengine* given in the previous command is designed to allow MuPAD functions to be used in the MATLAB workspace.

Variable types in MATLAB can be displayed using the MATLAB **class** function. For example, the variable type for *fs* is displayed by the following command.

```
>> class(fs)

ans =

sym
```

The preceding output indicates that *fs* is a symbolic variable. However, in the commands that will follow in this section, we will use *fs* as an array index, and array indices in MATLAB must be positive integers. To take care of this, we will use the following MATLAB **double** command to convert *fs* from a symbolic variable into a numeric variable.

```
>> fs = double(fs)

fs =

16

>> class(fs)
```

```
ans =

double
```

We are now ready to generate the elements in the underlying finite field. As in Section 1.6, to construct the field elements, we use the user-written function **Powmod**, which we have written separately from this MATLAB session and saved as the M-file Powmod.m. We will store the field elements in the vector *field*. Since for BCH codes we denote the field element x by a, we use the parameters a and $p(a)$ in the following **Powmod** command.

```
>> for i = 1:fs-1
        field(i) = Powmod(a, i, p(a), a, 2);
   end
>> field(fs) = 0;
```

We can view the entries in the vector *field* by entering the following command.

```
>> field

field =

[ a, a^2, a^3, a + 1, a^2 + a, a^3 + a^2, a^3 + a + 1,
  a^2 + 1, a^3 + a, a^2 + a + 1, a^3 + a^2 + a,
  a^3 + a^2 + a + 1, a^3 + a^2 + 1, a^3 + 1, 1, 0]
```

Because working with BCH codes requires frequent conversions between polynomial field elements and powers of a, it will be very useful for us to establish an association between the polynomial field elements and corresponding powers of a for this field. We will establish this association using the user-written function **ftable**, which we have written separately from this MATLAB session and saved as the M-file ftable.m. The following command illustrates how the function **ftable** can be used. Specifically, by entering the following command, we retrieve the power of a that corresponds to the polynomial field element $a^3 + a^2$.

```
>> ftable(a^3+a^2, field)

ans =

a^6
```

The second parameter in the preceding command is the array in which we have stored the elements in the underlying finite field.

We will now construct the generator polynomial for the BCH code in Example 4.3. To do this, we first define the polynomial $f(x) = x^{15} - 1$ of which each power of a is a root.

```
>> f = @(x) x^(fs-1)-1

f =


    @(x) x^(fs - 1) - 1
```

Next, to find the irreducible factors of $f(x)$ in $\mathbb{Z}_2[x]$, we will use the user-written function **Factor**, which we have written separately from this MATLAB session and saved as the M-file Factor.m.

```
>> factf = Factor(f(x), 2)

factf =

(x + 1)*(x^4 + x^3 + 1)*(x^2 + x + 1)*(x^4 + x + 1)*
(x^4 + x^3 + x^2 + x + 1)
```

To construct the generator polynomial in Example 4.3, we will need to access these factors of $f(x)$ separately. The following MuPAD **nops** command returns the number of factors in *factf*.

```
>> feval(symengine, 'nops', factf)

ans =

5
```

We can access these factors individually by using the MuPAD **op** function. For example, we can use the following command to assign the fifth factor in the expression *factf* as the variable *f5*.

```
>> f5 = feval(symengine, 'op', factf, 5)

f5 =

x^4 + x^3 + x^2 + x + 1
```

We can then use the MATLAB **subs** function as follows to convert the expression *f5* into a function.

```
>> f5 = @(x) subs(f5, x);
```

We can now evaluate *f5* as a function. For example, we can evaluate *f5* at a^6 by entering the following command.

```
>> f5(a^6)

ans =

a^24 + a^18 + a^12 + a^6 + 1
```

To retrieve the polynomial field element that corresponds to the preceding output, we will use the user-written function **Rem**, which we have written separately from this MATLAB session and saved as the M-file Rem.m.

```
>> Rem(f5(a^6), p(a), a, 2)

ans =

0
```

Since the preceding output is 0, then we know that a^6 is a root of *f5*, or, equivalently, that *f5* is the minimum polynomial of a^6.

We will now find the minimum polynomials that are the factors in the generator polynomial. We first assign the number $t = 3$ of errors that are to be correctable in the code. Then in the subsequent nested loop, we find and display the minimum polynomials of $a, a^2, \ldots, a^{2t}$. In these commands, the outer loop spans the powers of a, while the inner loop evaluates the factors of *factf* at these powers. Since each a^i is a root of *factf*, each will be the root of an irreducible factor of *factf*. The factor of which a^i is a root will be the minimum polynomial of a^i. The **if** and **pretty** statements that appear in these commands cause the correct minimum polynomial of each a^i to be displayed. The **break** statement terminates the inner loop when the correct minimum polynomial is found.

```
>> t = 3;
>> for i = 1:2*t
        for j = 1:feval(symengine, 'nops', factf)
            fj = @(x) subs(feval(symengine, 'op', factf, ...
            j), x);
            if Rem(fj(a^i), p(a), a, 2) == 0
```

```
            pretty([a^i 'is_a_root_of' fj(x)])
            break;
        end
    end
end
```

```
+-                   4          -+
| a, is_a_root_of, x   + x + 1 |
+-                              -+
```

```
+- 2                 4          -+
| a , is_a_root_of, x   + x + 1 |
+-                              -+
```

```
+- 3                 4    3    2          -+
| a , is_a_root_of, x   + x  + x  + x + 1 |
+-                                        -+
```

```
+- 4                 4          -+
| a , is_a_root_of, x   + x + 1 |
+-                              -+
```

```
+- 5                 2          -+
| a , is_a_root_of, x   + x + 1 |
+-                              -+
```

```
+- 6                 4    3    2          -+
| a , is_a_root_of, x   + x  + x  + x + 1 |
+-                                        -+
```

Thus, the minimum polynomial of a is x^4+x+1, which is also the minimum polynomial of a^2 and a^4. Also, the minimum polynomial of a^3 and a^6 is $x^4 + x^3 + x^2 + x + 1$, and the minimum polynomial of a^5 is $x^2 + x + 1$, just as we noted in Example 4.3. Next, we assign as variables one copy of each of these three unique minimum polynomials.

```
>> m1 = @(x) x^4+x+1;
>> m3 = @(x) x^4+x^3+x^2+x+1;
>> m5 = @(x) x^2+x+1;
```

We can then define the generator polynomial $g(x)$ as the product of these three minimum polynomials.

```
>> g = m1(x)*m3(x)*m5(x)
```

```
g =
```

$$(x^2 + x + 1)*(x^4 + x + 1)*(x^4 + x^3 + x^2 + x + 1)$$

We can expand this expression for $g(x)$ by using the user-written function **Expand**, which we have written separately from this MATLAB session and saved as the M-file Expand.m.

```
>> g = Expand(g, 2)

g =

x^10 + x^8 + x^5 + x^4 + x^2 + x + 1
```

Finally, we can convert this expression for $g(x)$ into a function by entering the following **subs** command.

```
>> g = @(x) subs(g, x);
>> g(x)

ans =

x^10 + x^8 + x^5 + x^4 + x^2 + x + 1
```

4.4.2   Error Correction

We will now show how MATLAB can be used to correct both of the received polynomials in Example 4.4. First we will consider the following received polynomial $r(x)$.

```
>> r = @(x) 1+x^2+x^3+x^4+x^5+x^6+x^7+x^10;
```

Recall that to correct $r(x)$, we begin by computing the first $2t$ syndromes of $r(x)$. We can use the following loop to compute and store these syndromes in the vector *syn*.

```
>> for i = 1:2*t
        syn(i) = ftable(Rem(r(a^i), p(a), a, 2), field);
   end
>> syn

syn =

[ a^3,   a^6,   a^6,   a^12,   a^10,   a^12]
```

We can now retrieve particular syndromes from the vector *syn*. For example, we can enter the following command to retrieve the fifth syndrome of $r(x)$.

```
>> syn(5)

ans =

a^10
```

Next, we will construct the 3×3 coefficient matrix in (4.2), and assign this matrix as the variable A.

```
>> A = [syn(1) syn(2) syn(3); syn(2) syn(3) syn(4); ...
   syn(3) syn(4) syn(5)]

A =

[  a^3,   a^6,   a^6]
[  a^6,   a^6,  a^12]
[  a^6,  a^12,  a^10]
```

We can find the determinant of A by using the MATLAB **det** function as follows.

```
>> det(A)

ans =

- a^27 + 2*a^24 - a^22 + a^19 - a^18
```

To find the field element that corresponds to this determinant of A, we can use the **Rem** function as follows.

```
>> d = Rem(det(A), p(a), a, 2)

d =

a^3 + a^2 + a + 1
```

We can then use the function **ftable** as follows to find the power of a that corresponds to the determinant of A.

```
>> d = ftable(d, field)
```

```
d =

a^12
```

Since this determinant is nonzero, then we know that $r(x)$ contains three errors. As in Example 4.4, we will use Cramer's Rule to determine σ_1, σ_2, and σ_3 in (4.2). First, we will construct the column vector on the right side of (4.2), and assign this vector as the variable B.

```
>> B = [syn(4); syn(5); syn(6)]

B =

  a^12
  a^10
  a^12
```

Recall that we can use a colon in MATLAB to choose particular columns from a matrix. For example, the following command returns the second column of A.

```
>> A(:, 2)

ans =

  a^6
  a^6
  a^12
```

Also, we can very easily form the matrices required for Cramer's Rule using MATLAB. For example, in the following command, we form a new matrix $A1$ from A by replacing the first column of A with the vector B.

```
>> A1 = [B A(:, 2) A(:, 3)]

A1 =

[ a^12,   a^6,   a^6]
[ a^10,   a^6,  a^12]
[ a^12,  a^12,  a^10]
```

Next, we will find the determinant of $A1$ expressed as a power of a.

```
>> dA1 = ftable(Rem(det(A1), p(a), a, 2), field)
```

```
dA1 =

a^14
```

Thus, by Cramer's Rule, we can find σ_3 in (4.2) by entering the following command.

```
>> sigma3 = dA1/d

sigma3 =

a^2
```

Similarly, by Cramer's Rule, we can find σ_2 in (4.2) as follows.

```
>> A2 = [A(:, 1) B A(:, 3)]

A2 =

[  a^3, a^12,  a^6]
[  a^6, a^10, a^12]
[  a^6, a^12, a^10]

>> dA2 = ftable(Rem(det(A2), p(a), a, 2), field)

dA2 =

a^10

>> sigma2 = dA2/d

sigma2 =

1/a^2
```

We will use the following command to express σ_2 as a positive power of a.

```
>> sigma2 = sigma2*a^(fs-1)

sigma2 =

a^13
```

Finally, by Cramer's Rule, we can find σ_1 in (4.2) as follows.

```
>> A3 = [A(:, 1) A(:, 2) B]

A3 =

[  a^3,   a^6, a^12]
[  a^6,   a^6, a^10]
[  a^6, a^12, a^12]

>> dA3 = ftable(Rem(det(A3), p(a), a, 2), field)

dA3 =

     1

>> sigma1 = dA3/d

sigma1 =

1/a^12

>> sigma1 = sigma1*a^(fs-1)

sigma1 =

a^3
```

Next, after declaring the variable z as a symbolic variable, we will construct the resulting error locator polynomial, and assign this polynomial as the variable *EL*.

```
>> syms z
>> EL = z^3+sigma1*z^2+sigma2*z+sigma3

EL =

a^13*z + a^3*z^2 + a^2 + z^3

>> EL = @(z) subs(EL, z);
>> EL(z)

ans =

a^13*z + a^3*z^2 + a^2 + z^3
```

Then in the following loop, we find the roots of *EL* by trial and error.

```
>> for i = 0:fs-2
       if Rem(EL(a^i), p(a), a, 2) == 0
           pretty([a^i, 'is_a_root_of' EL(z)])
       end
   end
```

```
+-                     13     3 2   2    3-+
| 1, is_a_root_of, a   z + a  z + a + z |
+-                                     -+


+- 5                   13     3 2   2    3-+
| a , is_a_root_of, a   z + a  z + a + z |
+-                                     -+


+- 12                  13     3 2   2    3-+
| a , is_a_root_of, a   z + a  z + a + z |
+-                                     -+
```

Thus, as we noted in Example 4.4, the roots of *EL* are 1, a^5, and a^{12}, and the error in $r(x)$ is the following polynomial $e(x)$.

```
>> e = @(x) 1+x^5+x^12;
```

In the next command we form the corrected codeword $c(x)$.

```
>> c = mod(r(x)+e(x), 2)

c =

x^12 + x^10 + x^7 + x^6 + x^4 + x^3 + x^2
```

Finally, to verify that $c(x)$ is a multiple of $g(x)$, we enter the following.

```
>> Factor(c, 2)/Factor(g(x), 2)

ans =

x^2
```

We will now consider the second received polynomial in Example 4.4, which is the following polynomial $r(x)$.

```
>> r = @(x) 1+x^3+x^7+x^10+x^11+x^13;
```

To correct this received polynomial, we begin by computing and storing the first $2t$ syndromes of $r(x)$ in the vector *syn*.

```
>> for i = 1:2*t
       syn(i) = ftable(Rem(r(a^i), p(a), a, 2), field);
   end
>> syn

syn =

[ a^5, a^10, a^2, a^5, 1, a^4]
```

Next, we will construct the 3×3 coefficient matrix in (4.3).

```
>> A = [syn(1) syn(2) syn(3); syn(2) syn(3) syn(4); ...
   syn(3) syn(4) syn(5)]

A =

[  a^5, a^10,  a^2]
[ a^10,  a^2,  a^5]
[  a^2,  a^5,   1]
```

To find the determinant of this coefficient matrix A, we enter the following command.

```
>> d = ftable(Rem(det(A), p(a), a, 2), field)

d =

        0
```

Since this determinant is zero, then we know that $r(x)$ does not contain exactly three errors. So next we will assume that $r(x)$ contains only two errors, and construct the 2×2 coefficient matrix in (4.4). To do this, we will first delete the last row and last column from A.

```
>> A(3, :) = [];
>> A(:, 3) = []

A =

[  a^5, a^10]
[ a^10,  a^2]
```

Next, we will find the determinant of this new coefficient matrix A.

```
>> d = ftable(Rem(det(A), p(a), a, 2), field)

d =

a^13
```

Since this determinant is nonzero, then we know that $r(x)$ contains two errors. We can then again use Cramer's Rule to determine σ_1 and σ_2 in (4.4). First, we will construct the column vector on the right side of (4.4), and assign this vector as the variable B.

```
>> B = [syn(3); syn(4)]

B =

  a^2
  a^5
```

We can then use Cramer's Rule to determine σ_1 and σ_2 as follows.

```
>> A1 = [B A(:, 2)]

A1 =

[  a^2, a^10]
[  a^5,  a^2]

>> dA1 = ftable(Rem(det(A1), p(a), a, 2), field)

dA1 =

a

>> sigma2 = dA1/d

sigma2 =

1/a^12

>> sigma2 = sigma2*a^(fs-1)

sigma2 =
```

```
a^3

>> A2 = [A(:, 1) B]

A2 =

[  a^5,   a^2]
[ a^10,   a^5]

>> dA2 = ftable(Rem(det(A2), p(a), a, 2), field)

dA2 =

a^3

>> sigma1 = dA2/d

sigma1 =

1/a^10

>> sigma1 = sigma1*a^(fs-1)

sigma1 =

a^5
```

Next, we will construct the resulting error locator polynomial, and assign this polynomial as the variable *EL*.

```
>> EL = z^2+sigma1*z+sigma2

EL =

a^5*z + a^3 + z^2

>> EL = @(z) subs(EL, z);
>> EL(z)

ans =

a^5*z + a^3 + z^2
```

Then in the following loop, we find the roots of *EL* by trial and error.

```
>> for i = 0:fs-2
       if Rem(EL(a^i), p(a), a, 2) == 0
           pretty([a^i, 'is_a_root_of' EL(z)])
       end
   end
```

```
+-                       5     3    2-+
| a, is_a_root_of, a   z + a   + z  |
+-                                  -+
```

```
+- 2                     5     3    2-+
| a , is_a_root_of, a   z + a   + z  |
+-                                  -+
```

Thus, the error in $r(x)$ is the following polynomial $e(x)$.

```
>> e = @(x) x+x^2;
```

In the next command we form the corrected codeword $c(x)$.

```
>> c = mod(r(x)+e(x), 2)
```

```
c =
```

```
x^13 + x^11 + x^10 + x^7 + x^3 + x^2 + x + 1
```

Finally, to verify that $c(x)$ is a multiple of $g(x)$, we enter the following.

```
>> Factor(c, 2)/Factor(g(x), 2)
```

```
ans =
```

```
(x^2+x+1)*(x+1)
```

Exercises

1. Use the primitive polynomial $p(x) = x^3 + x^2 + 1 \in \mathbb{Z}_2[x]$ to construct the generator polynomial for a one-error correcting BCH code C. State the linear code parameters $[\cdot, \cdot]$ for C and the number of codewords in C, and construct a codeword of maximum possible degree.

2. Use the primitive polynomial $p(x) = x^3 + x^2 + 1 \in \mathbb{Z}_2[x]$ to construct the generator polynomial for a two-error correcting BCH code. How will the code that results from this generator polynomial compare to the code C in Example 4.2?

3. (a) Use the primitive polynomial $p(x) = x^3 + x^2 + 1 \in \mathbb{Z}_2[x]$ to construct the generator polynomial for a three-error correcting BCH code. How will the code that results from this generator polynomial compare to the code that results from the generator polynomial in Exercise 2?

   (b) Show that the code that results from the generator polynomial in part (a) is perfect.

4. Use the primitive polynomial $p(x) = x^4 + x + 1 \in \mathbb{Z}_2[x]$ to construct the generator polynomial for a $[15, 7]$ BCH code C. State the number of errors that can be corrected and the number of codewords in C, and construct a codeword of maximum possible degree.

5. Use the primitive polynomial $p(x) = x^4 + x^3 + 1 \in \mathbb{Z}_2[x]$ to construct the generator polynomial for a two-error correcting BCH code C. State the linear code parameters $[\cdot, \cdot]$ for C and the number of codewords in C, and construct a codeword of maximum possible degree.

6. Use the primitive polynomial $p(x) = x^4 + x^3 + 1 \in \mathbb{Z}_2[x]$ to construct the generator polynomial for a three-error correcting BCH code C. State the linear code parameters $[\cdot, \cdot]$ for C and the number of codewords in C, and construct a codeword of maximum possible degree.

7. Correct the following received vectors in the code C in Exercise 1.

   (a) $\mathbf{r} = (1101110)$

   (b) $\mathbf{r} = (1100010)$

8. Correct the following received vectors in the code C in Exercise 4.

   (a) $\mathbf{r} = (110011001100011)$

   (b) $\mathbf{r} = (001101111100110)$

   (c) $\mathbf{r} = (100001111110010)$

9. Correct the following received vectors in the code C in Exercise 5.

   (a) $\mathbf{r} = (000010000010011)$

   (b) $\mathbf{r} = (110000101001101)$

   (c) $\mathbf{r} = (010011100110010)$

10. Correct the following received vectors in the code C in Exercise 6.

  (a) $\mathbf{r} = (110111111001000)$
  (b) $\mathbf{r} = (111011001000111)$
  (c) $\mathbf{r} = (111111111111111)$
  (d) $\mathbf{r} = (001110110110010)$

11. Correct the following received vectors in the code C in Example 4.3.

  (a) $\mathbf{r} = (100011011001010)$
  (b) $\mathbf{r} = (101000011101100)$
  (c) $\mathbf{r} = (111011001010100)$
  (d) $\mathbf{r} = (011111010011010)$

12. Recall from the proof of Theorem 4.2 the description of parity check matrices for BCH codes.

  (a) Construct a parity check matrix for the code C in Exercise 1. Then verify that this parity check matrix "works" for the received vector in Exercise 7(a) and the codeword to which this vector can be corrected.

  (b) Construct a parity check matrix for the code C in Exercise 4. Then verify that this parity check matrix "works" for the received vector in Exercise 8(a) and the codeword to which this vector can be corrected.

  (c) Construct a parity check matrix for the code C in Exercise 5. Then verify that this parity check matrix "works" for the received vector in Exercise 9(a) and the codeword to which this vector can be corrected.

  (d) Construct a parity check matrix for the code C in Example 4.3. Then verify that this parity check matrix "works" for the received vector in Exercise 11(a) and the codeword to which this vector can be corrected.

13. Suppose $f(x) = x^m - 1$ for some integer $m > 0$, and let $a_1, a_2, \ldots, a_s$ for some integer $s < m$ be roots of $f(x)$ with minimum polynomials $m_1(x), m_2(x), \ldots, m_s(x)$, respectively. Prove that if $g(x)$ is the least common multiple of the $m_i(x)$ for $i = 1, 2, \ldots, s$, then $g(x)$ divides $f(x)$.

14. Suppose $m = 2^n - 1$ for some positive integer n, and let $p(x)$ be a primitive polynomial of degree n in $\mathbb{Z}_2[x]$. Then for the element $a = x$ in the finite field $\mathbb{Z}_2[x]/(p(x))$, use Lagrange's Theorem (Theorem 1.4) to prove that a^k is a root of $f(x) = x^m - 1$ for all positive integers k.

15. Prove that if $h(x) \in \mathbb{Z}_2[x]$, then $h(a^{2k}) = (h(a^k))^2$ for all positive integers k.

16. Suppose $p(x) \in \mathbb{Z}_2[x]$ is a primitive polynomial, and let C be a BCH code that results from powers of $a = x$ in the finite field $\mathbb{Z}_2[x]/(p(x))$. Prove that if C is a one-error correcting code, then the generator polynomial for C will be $p(x)$.

17. The irreducible factors of $f(x) = x^{31} - 1$ in $\mathbb{Z}_2[x]$ are $(x+1)$ and six primitive polynomials of degree 5. To construct a t-error correcting BCH code C with codewords of length 31 positions, we would begin with a primitive polynomial $p(x) \in \mathbb{Z}_2[x]$ of degree 5, and a certain number of powers of $a = x$ in the finite field $\mathbb{Z}_2[x]/(p(x))$. This would determine the generator polynomial and linear code parameters $[\cdot, \cdot]$ for C, and the number of codewords in C. Complete the following table for a BCH code with codewords of length 31 positions.

Number of Correctable Errors	Number of Powers of a Needed	Degree of Generator Polynomial	Linear Code Parameters	Number of Codewords
3				
4				
5				
6				

Computer Exercises

18. Find a primitive polynomial $p(x)$ of degree 5 in $\mathbb{Z}_2[x]$, and use $p(x)$ to construct the generator polynomial for a four-error correcting BCH code C. State the linear code parameters $[\cdot, \cdot]$ for C and the number of codewords in C, and construct a codeword of maximum possible degree.

19. Use the primitive polynomial $p(x) = x^6 + x^5 + 1 \in \mathbb{Z}_2[x]$ to construct the generator polynomial for a three-error correcting BCH code C. State the linear code parameters $[\cdot, \cdot]$ for C and the number of codewords C, and construct a codeword of maximum possible degree.

20. Use the primitive polynomial $p(x) = x^7 + x + 1 \in \mathbb{Z}_2[x]$ to construct the generator polynomial for a five-error correcting BCH code C. State the linear code parameters $[\cdot, \cdot]$ for C and the number of codewords C, and construct a codeword of maximum possible degree.

21. Correct the following received polynomials in the BCH code C in Exercise 19.

(a) $r(x) \;=\; x^3 + x^4 + x^5 + x^{12} + x^{14} + x^{18} + x^{19} + x^{20} + x^{21}$

(b) $r(x) \;=\; 1 + x + x^3 + x^4 + x^5 + x^7 + x^9 + x^{15} + x^{18} + x^{19}$
$+\, x^{20} + x^{22} + x^{24}$

(c) $r(x) \;=\; 1 + x + x^2 + x^9 + x^{10} + x^{13} + x^{15} + x^{16} + x^{17} + x^{18}$
$+\, x^{21} + x^{22} + x^{25} + x^{26} + x^{28}$

(d) $r(x) \;=\; 1 + x + x^2 + x^3 + x^5 + x^6 + x^7 + x^8 + x^9 + x^{11}$
$+\, x^{12} + x^{14} + x^{15} + x^{18} + x^{20} + x^{21} + x^{22} + x^{23}$

22. Correct the following received polynomials in the BCH code C in Exercise 20.

(a) $r(x) \;=\; 1 + x + x^3 + x^9 + x^{11} + x^{12} + x^{13} + x^{14} + x^{15} + x^{21}$
$+\, x^{22} + x^{23} + x^{25} + x^{26} + x^{28} + x^{30} + x^{33} + x^{34}$
$+\, x^{36} + x^{37}$

(b) $r(x) \;=\; x + x^4 + x^5 + x^6 + x^7 + x^8 + x^9 + x^{12} + x^{14} + x^{16}$
$+\, x^{17} + x^{18} + x^{19} + x^{20} + x^{21} + x^{23} + x^{26} + x^{29}$
$+\, x^{30} + x^{34} + x^{37}$

(c) $r(x) \;=\; 1 + x + x^3 + x^8 + x^9 + x^{11} + x^{13} + x^{14} + x^{16} + x^{18}$
$+\, x^{19} + x^{20} + x^{21} + x^{22} + x^{23} + x^{25} + x^{28} + x^{31}$
$+\, x^{32} + x^{36} + x^{39}$

(d) $r(x) \;=\; 1 + x + x^2 + x^3 + x^5 + x^6 + x^7 + x^8 + x^{14} + x^{20}$
$+\, x^{24} + x^{25} + x^{26} + x^{27} + x^{28} + x^{31} + x^{33} + x^{34}$
$+\, x^{35} + x^{36} + x^{37} + x^{38}$

(e) $r(x) \;=\; 1 + x + x^2 + x^5 + x^9 + x^{10} + x^{11} + x^{13} + x^{16} + x^{17}$
$+\, x^{21} + x^{22} + x^{24} + x^{26} + x^{27} + x^{32} + x^{33} + x^{34}$
$+\, x^{35} + x^{37} + x^{38} + x^{39}$

(f) $r(x) \;=\; x^2 + x^5 + x^6 + x^7 + x^8 + x^9 + x^{10} + x^{15} + x^{17} + x^{18}$
$+\, x^{19} + x^{20} + x^{21} + x^{22} + x^{24} + x^{27} + x^{30} + x^{31}$
$+\, x^{35} + x^{38} + x^{39}$

23. Recall from the proof of Theorem 4.2 the description of parity check matrices for BCH codes.

    (a) Construct a parity check matrix for the code C in Exercise 19. Then verify that this parity check matrix "works" for the received polynomial in Exercise 21(a) and the codeword to which this polynomial can be corrected.

    (b) Construct a parity check matrix for the code C in Exercise 20. Then verify that this parity check matrix "works" for the received polynomial in Exercise 22(a) and the codeword to which this polynomial can be corrected.

Research Exercises

24. Investigate the lives and careers of the people for whom BCH codes are named (Raj Bose, Dijen Ray-Chaudhuri, and Alexis Hocquenghem), and write a summary of your findings.

25. Investigate an actual real-life use of a BCH code, and write a summary of your findings.

26. In Exercise 3, you were asked to show that a particular BCH code is perfect. Are all BCH codes perfect? Investigate this, and write a summary of your findings.

27. All BCH codes are examples of *cyclic* codes. Investigate what it means for a code to be a cyclic code, and write a summary of your findings. Then write a clear explanation for how you know that all BCH codes are cyclic.

28. Hamming codes can actually be viewed as a special case of BCH codes. Investigate why this is true, and write a summary of your findings.

29. The type of BCH codes that we presented in this chapter are sometimes called *primitive*, or *narrow-sense*, to distinguish them from *general* BCH codes. Investigate the differences between primitive and general BCH codes, and write a summary of your findings.

30. The method we presented in this chapter for correcting errors in BCH codes is not unique. Investigate some of the other methods for correcting errors in BCH codes, and write a summary of your findings. Include in your summary a brief description of at least one of the other methods, and a statement of which method is the fastest.

Chapter 5

Reed-Solomon Codes

In this chapter, we will present a type of code called a *Reed-Solomon* code. Reed-Solomon codes, like BCH codes, have polynomial codewords, are linear, and can be constructed to be multiple-error correcting. However, Reed-Solomon codes are significantly better than BCH codes in many situations because they are ideal for correcting error bursts. When a binary codeword is transmitted, the received vector is said to contain an *error burst* if it contains several bit errors very close together. In data transmitted through space, error bursts are frequently caused by very brief periods of intense solar energy. It was for this reason that a Reed-Solomon code was used in the Voyager 2 satellite when it transmitted photographs of several of the planets in our solar system back to Earth. We will briefly discuss the use of a Reed-Solomon code in the Voyager 2 satellite in Section 5.6. In addition, there are a variety of other reasons why errors in binary codewords often occur naturally in bursts, such as power surges in cable and telephone wires, various types of interference, and scratches on compact discs. As a result, Reed-Solomon codes have a rich assortment of applications, and are claimed to be the most frequently used digital error-correcting codes in the world. They are used extensively in the encoding of music and video on CDs, DVDs, and Blu-ray discs, have played an integral role in the development of high-speed supercomputers, and will be an important tool in the future for dealing with complex communication and information transfer systems.

5.1 Construction

To construct a Reed-Solomon code, we begin by choosing a primitive polynomial $p(x)$ of degree n in $\mathbb{Z}_2[x]$, and forming the field $F = \mathbb{Z}_2[x]/(p(x))$ of order 2^n. As we did in Chapter 4, throughout this chapter we will denote the element x in our finite fields by a. Like BCH codewords, Reed-Solomon codewords are then polynomials of degree less than $2^n - 1$. However, unlike BCH codewords, which are elements in $\mathbb{Z}_2[x]$, Reed-Solomon codewords are elements in $F[x]$. To construct a t-error correcting Reed-Solomon code C, we use the generator polynomial $g(x) = (x - a)(x - a^2) \cdots (x - a^{2t})$ in $F[x]$. The codewords in C are then all multiples of $g(x)$ in $F(x)$ of degree less than $2^n - 1$. Theorem 4.2 can easily be modified to show that C will be t-error correcting. The codewords in C have length $2^n - 1$ positions, and form a vector space of dimension $2^n - 1 - 2t$. We will describe a Reed-Solomon code using the notation and parameters $RS(2^n - 1, t)$ if the codewords in the code have length $2^n - 1$ positions and the code is t-error correcting.

Example 5.1 Choose primitive polynomial $p(x) = x^4 + x + 1 \in \mathbb{Z}_2[x]$, and let F be the field $F = \mathbb{Z}_2[x]/(p(x))$. The nonzero elements in F are listed in the table in Example 4.3. Using this field F, we obtain the following generator polynomial $g(x)$ for an $RS(15, 2)$ code.

$$
\begin{aligned}
g(x) &= (x - a)(x - a^2)(x - a^3)(x - a^4) \\
&= \left(x^2 + (a^2 + a)x + a^3\right)\left(x^2 + (a^4 + a^3)x + a^7\right) \\
&= \left(x^2 + a^5 x + a^3\right)\left(x^2 + a^7 x + a^7\right) \\
&= x^4 + (a^7 + a^5)x^3 + (a^7 + a^{12} + a^3)x^2 + (a^{12} + a^{10})x + a^{10} \\
&= x^4 + a^{13}x^3 + a^6 x^2 + a^3 x + a^{10}
\end{aligned}
$$

So, for example, the following polynomial $c(x)$ is one of the codewords in the code.

$$
\begin{aligned}
c(x) &= \left(a^{10}x^9 + a^3 x^5 + a^2 x^3\right) g(x) \\
&= a^{10}x^{13} + a^{23}x^{12} + a^{16}x^{11} + a^{13}x^{10} + (a^{20} + a^3)x^9 + a^{16}x^8 \\
&\quad + (a^9 + a^2)x^7 + (a^6 + a^{15})x^6 + (a^{13} + a^8)x^5 + a^5 x^4 + a^{12}x^3 \\
&= a^{10}x^{13} + a^8 x^{12} + ax^{11} + a^{13}x^{10} + a^{11}x^9 + ax^8 + a^{11}x^7 \\
&\quad + a^{13}x^6 + a^3 x^5 + a^5 x^4 + a^{12}x^3
\end{aligned}
$$

$\square$

The fact that Reed-Solomon codewords are in $F[x]$ instead of $\mathbb{Z}_2[x]$ gives rise to two issues that we must address. First, unlike BCH codewords, Reed-Solomon codewords cannot be converted into binary vectors by simply listing the coefficients, since the coefficients would be elements in F instead of $\mathbb{Z}_2$. However, there is still an obvious method for converting a Reed-Solomon codeword $c(x)$ into a binary vector. Since the coefficients of $c(x)$ would be elements in a finite field of order 2^n, these coefficients could be expressed as polynomials in a of degree less than n with coefficients in

$\mathbb{Z}_2$. Thus, to convert $c(x)$ into a binary vector, we could list these binary coefficients of a in order in blocks of n bits (including coefficients of zero), with each block representing one coefficient of $c(x)$. For example, consider the codeword $c(x)$ in the Reed-Solomon code in Example 5.1. To convert $c(x)$ into a binary vector, we could begin by listing the terms in $c(x)$ with increasing powers of x as follows.

$$\begin{aligned}
c(x) \;=\;& a^{12}x^3 + a^5x^4 + a^3x^5 + a^{13}x^6 + a^{11}x^7 + ax^8 + a^{11}x^9 \\
& + a^{13}x^{10} + ax^{11} + a^8x^{12} + a^{10}x^{13}
\end{aligned}$$

Next, we could write the coefficients of $c(x)$ as follows, using the table in Example 4.3 to express each coefficient as a polynomial in a of degree less than four with increasing powers of a.

$$\begin{aligned}
c(x) \;=\;& (1 + a + a^2 + a^3)x^3 + (a + a^2)x^4 + a^3x^5 + (1 + a^2 + a^3)x^6 \\
& + (a + a^2 + a^3)x^7 + ax^8 + (a + a^2 + a^3)x^9 \\
& + (1 + a^2 + a^3)x^{10} + ax^{11} + (1 + a^2)x^{12} + (1 + a + a^2)x^{13}
\end{aligned}$$

Finally, we could express each coefficient of $c(x)$ as a block of four bits by listing the binary coefficients of a in order. For example, we could express the coefficient $0 + 1a + 1a^2 + 0a^3$ of x^4 in $c(x)$ as the block (0110). Using this method, we could convert the entire codeword $c(x)$ into the following binary vector of length 60 positions by stringing together these blocks of four bits, including four zeros for each of the constant, x, x^2, and x^{14} terms that could be present in codewords in the code but which are not present in $c(x)$.

(000000000000111101100001101101110100011110110100101011100000)

This natural method for converting Reed-Solomon codewords into binary vectors makes it clear why Reed-Solomon codes are ideal for correcting error bursts. Specifically, four bit errors in the binary vector that we just constructed could represent only a single coefficient error in $c(x)$. Thus, although $c(x)$ is a codeword in a code that is only two-error correcting, it may be possible to correct a received vector to $c(x)$ even if eight bit errors occurred during the transmission of the binary equivalent of $c(x)$. More generally, regarding this code we would say that provided only one error burst occurred during the transmission of the binary equivalent of a codeword, we would be guaranteed to be able to correct the received vector as long as the error burst was not longer than five bits. This is because any error burst not longer than five bits in the binary equivalent of a codeword could not span more than two of the coefficients in the codeword, while an error burst of length six bits in the binary equivalent of a codeword could span three of the coefficients in the codeword. Generalizing this statement to apply to any $RS(2^n - 1, t)$ code is not difficult, and will be left as an exercise.

As we stated previously, the fact that Reed-Solomon codewords are in $F[x]$ instead of $\mathbb{Z}_2[x]$ gives rise to two issues. The first issue was how Reed-Solomon codewords are converted into binary vectors. We have now shown an effective method for doing this, which also revealed why Reed-Solomon codes are ideal for correcting error bursts. The second issue is how to actually correct errors in Reed-Solomon codewords.

5.2   Error Correction

We should begin by noting that the error correction method for BCH codes that we presented in Chapter 4 yields the same information when it is applied to a received Reed-Solomon polynomial as when it is applied to a received BCH polynomial. However, the BCH error correction method cannot generally be used to correct errors in a received Reed-Solomon polynomial. Recall that the last step in the BCH error correction method involves finding the roots of an error locator polynomial, which reveals the error positions in a received polynomial. Because there are only two possible coefficients for each term in a BCH polynomial, knowledge of the error positions alone is sufficient to correct the polynomial. The BCH error correction method can also be used to find the error positions in a received Reed-Solomon polynomial. However, because there is more than one possible coefficient for each term in a Reed-Solomon polynomial, knowledge of the error positions alone is not generally sufficient to correct the polynomial. The specific error present within each error position would also have to be determined.

Rather than combining the BCH error correction method for identifying error positions in received polynomials with a separate method for actually correcting errors, we will present an entirely new method for both identifying and correcting errors in Reed-Solomon polynomials. Before stating this new Reed-Solomon error correction method, we first note the following analogue to Theorem 4.1.

Theorem 5.1 *Suppose that F is a field of order 2^n, and let C be an $RS(2^n - 1, t)$ code in $F[x]$. Then $c(x) \in F[x]$ of degree less than $2^n - 1$ is in C if and only if $c(a^i) = 0$ for $i = 1, 2, \ldots, 2t$.*

Proof. Exercise.   □

Theorem 5.1 is useful for correcting errors in Reed-Solomon codes in the same way that Theorem 4.1 is useful for correcting errors in BCH codes. Specifically, suppose F is a field of order 2^n, and let C be an $RS(2^n - 1, t)$

code in $F[x]$. If $c(x) \in C$ is transmitted and we receive the polynomial $r(x) \in F[x]$ of degree less than $2^n - 1$, then $r(x) = c(x) + e(x)$ for some error polynomial $e(x)$ in $F[x]$ of degree less than $2^n - 1$. To correct $r(x)$, we must only determine $e(x)$, for we could then compute $c(x) = r(x) + e(x)$. However, Theorem 5.1 implies $r(a^i) = e(a^i)$ for $i = 1, 2, \ldots, 2t$. Thus, by knowing $r(x)$, we also know some information about $e(x)$. We will again call the values of $r(a^i)$ for $i = 1, 2, \ldots, 2t$ the *syndromes* of $r(x)$.

We will now summarize and illustrate the Reed-Solomon error correction method.[1] Suppose F is a field of order 2^n, and let C be an $RS(2^n - 1, t)$ code in $F[x]$. If a codeword $c(x) \in C$ is transmitted and we receive the polynomial $r(x) = c(x) + e(x)$ for some error polynomial $e(x)$ in $F[x]$ of degree less than $2^n - 1$, then we can use the following steps to determine $e(x)$.

1. We first compute the syndromes of $r(x)$, which we will again denote by $s_1 = r(a)$, $s_2 = r(a^2)$, ... , $s_{2t} = r(a^{2t})$, and form the *syndrome polynomial* $S(z) = s_1 + s_2 z + s_3 z^2 + \cdots + s_{2t} z^{2t-1}$. (Note: If $r(x)$ is not in $\mathbb{Z}_2[x]$, then the shortcut $r(a^{2k}) = (r(a^k))^2$ that we used for finding syndromes in Chapter 4 will not apply.)

2. Next, we construct the Euclidean algorithm table (see Section 1.7) for the polynomials z^{2t} and $S(z)$ in $F[z]$, stopping at the first row j for which $\deg(r_j) < t$. (The **U** column may be omitted from the table.) We then let $R(z) = r_j$ and $V(z) = v_j$.

3. We can then find the error positions in $r(x)$ by finding the roots of $V(z)$. Specifically, if $a^{i_1}, a^{i_2}, \ldots, a^{i_k}$ are the roots of $V(z)$, then $r(x)$ contains errors in positions $x^{-i_1}, x^{-i_2}, \ldots, x^{-i_k}$. To find the coefficients of $e(x)$ at these error positions, if we denote the coefficient of the x^{-i} term in $e(x)$ by e_{-i}, then $e_{-i} = \dfrac{R(a^i)}{V'(a^i)}$.

We should also note that although the BCH error correction method that we presented in Chapter 4 cannot generally be used to correct errors in received Reed-Solomon polynomials, the Reed-Solomon error correction method that we just summarized can always be used to correct errors in received BCH polynomials.

We will now look at an example of the Reed-Solomon error correction method in the $RS(15, 2)$ code in Example 5.1.

Example 5.2 Let C be the $RS(15, 2)$ code with generator polynomial $g(x) = x^4 + a^{13}x^3 + a^6x^2 + a^3x + a^{10}$ in Example 5.1. Suppose that a

---

[1]Because verifying that the Reed-Solomon error correction method is valid requires an extensive discussion, we will postpone this verification until Section 5.3.

codeword in C is transmitted as a binary vector of length 60 positions, and we receive the following vector.

(000000000000111101100001101101110100011100101001101011100000)

To correct this vector, we must first convert it into a polynomial $r(x)$ in $F[x]$. By separating this vector into blocks of four bits, and then using each of these blocks as the coefficients of a polynomial in a of degree three, we obtain the following polynomial $r(x)$.

$$
\begin{aligned}
r(x) \;=\; & (1 + a + a^2 + a^3)x^3 + (a + a^2)x^4 + a^3 x^5 + (1 + a^2 + a^3)x^6 \\
& + (a + a^2 + a^3)x^7 + a x^8 + (a + a^2 + a^3)x^9 + a^2 x^{10} \\
& + (1 + a^3)x^{11} + (1 + a^2)x^{12} + (1 + a + a^2)x^{13}
\end{aligned}
$$

Next, using the table in Example 4.3 (in which the nonzero elements in F and corresponding powers of a are listed) to express each of these polynomial coefficients of $r(x)$ as a single power of a, we obtain the following.

$$
\begin{aligned}
r(x) \;=\; & a^{12}x^3 + a^5 x^4 + a^3 x^5 + a^{13}x^6 + a^{11}x^7 + a x^8 + a^{11}x^9 + a^2 x^{10} \\
& + a^{14}x^{11} + a^8 x^{12} + a^{10}x^{13}
\end{aligned}
$$

Since C is two-error correcting, then to correct $r(x)$, we must first compute the first four syndromes of $r(x)$. Using the table in Example 4.3, we can compute these syndromes as follows.

$$
\begin{aligned}
s_1 \;=\; & r(a) \\
=\; & a^{15} + a^9 + a^8 + a^{19} + a^{18} + a^9 + a^{20} + a^{12} + a^{25} + a^{20} + a^{23} \\
=\; & 1 + a^9 + a^8 + a^4 + a^3 + a^9 + a^5 + a^{12} + a^{10} + a^5 + a^8 \\
=\; & 1 + a + 1 + a^3 + a^3 + a^2 + a + 1 + a^2 + a + 1 \\
=\; & a
\end{aligned}
$$

$$
\begin{aligned}
s_2 \;=\; & r(a^2) \\
=\; & a^{18} + a^{13} + a^{13} + a^{25} + a^{25} + a^{17} + a^{29} + a^{22} + a^{36} + a^{32} + a^{36} \\
=\; & a^3 + a^2 + a^{14} + a^7 + a^2 \\
=\; & a^3 + a^3 + 1 + a^3 + a + 1 \\
=\; & a^9
\end{aligned}
$$

$$
\begin{aligned}
s_3 \;=\; & r(a^3) \\
=\; & a^{21} + a^{17} + a^{18} + a^{31} + a^{32} + a^{25} + a^{38} + a^{32} + a^{47} + a^{44} + a^{49} \\
=\; & a^6 + a^2 + a^3 + a + a^{10} + a^8 + a^2 + a^{14} + a^4 \\
=\; & a^3 + a^2 + a^3 + a + a^2 + a + 1 + a^2 + 1 + a^3 + 1 + a + 1 \\
=\; & a^{11}
\end{aligned}
$$

$$s_4 = r(a^4)$$
$$= a^{24} + a^{21} + a^{23} + a^{37} + a^{39} + a^{33} + a^{47} + a^{42} + a^{58} + a^{56} + a^{62}$$
$$= a^9 + a^6 + a^8 + a^7 + a^9 + a^3 + a^2 + a^{12} + a^{13} + a^{11} + a^2$$
$$= a^3 + a^2 + a^2 + 1 + a^3 + a + 1 + a^3 + a^3 + a^2 + a + 1 + a^3 + a^2$$
$$\quad + 1 + a^3 + a^2 + a$$
$$= a^5$$

The resulting syndrome polynomial is $S(z) = a + a^9 z + a^{11} z^2 + a^5 z^3$. Next, we need to construct the Euclidean algorithm table for z^4 and $S(z)$. We begin by dividing $S(z)$ into z^4, simplifying using the table in Example 4.3, which yields $z^4 = S(z)q_1 + r_1 = S(z)(a^{10}z + a) + (a^6 z^2 + a^{14}z + a^2)$. As a consequence, for the resulting Euclidean algorithm table, we will have $v_1 = 0 - 1(a^{10}z + a) = a^{10}z + a$. Since $\deg(r_1) \not< 2$, we must divide again. Dividing r_1 into $S(z)$ yields $S(z) = r_1 q_2 + r_2 = r_1(a^{14}z + a^{13}) + (a^{10}z + a^4)$. Thus, for the Euclidean algorithm table, we will have the following.

$$v_2 = 1 - (a^{10}z + a)(a^{14}z + a^{13})$$
$$= 1 - \left(a^{24}z^2 + (a^{23} + a^{15})z + a^{14}\right)$$
$$= 1 - \left(a^9 z^2 + (a^8 + 1)z + a^{14}\right)$$
$$= a^9 z^2 + a^2 z + a^3$$

Since $\deg(r_2) < 2$, we do not need to divide again, and the resulting Euclidean algorithm table (with the **U** column omitted) follows.

Row	Q	R	V
-1	–	z^4	0
0	–	$S(z)$	1
1	$a^{10}z + a$	$a^6 z^2 + a^{14}z + a^2$	$a^{10}z + a$
2	$a^{14}z + a^{13}$	$a^{10}z + a^4$	$a^9 z^2 + a^2 z + a^3$

Thus, $R(z) = a^{10}z + a^4$ and $V(z) = a^9 z^2 + a^2 z + a^3$. By evaluating $V(z)$ at successive powers of a, it can easily be verified that the roots of $V(z)$ are a^4 and a^5. As a result, the positions in $r(x)$ that contain errors are $x^{-4} = x^{11}$ and $x^{-5} = x^{10}$. To find the coefficients of the error polynomial $e(x)$ at these error positions, we first note that $V'(z) = a^2$. We can then find the coefficients of $e(x)$ as follows.

$$e_{10} = \frac{R(a^5)}{V'(a^5)} = \frac{a^{15} + a^4}{a^2} = \frac{a}{a^2} = a^{14}$$

$$e_{11} = \frac{R(a^4)}{V'(a^4)} = \frac{a^{14} + a^4}{a^2} = \frac{a^9}{a^2} = a^7$$

Thus, $e(x) = a^{14}x^{10} + a^7 x^{11}$. It can easily be verified that $r(x) + e(x)$ is the codeword $c(x)$ that we constructed in Example 5.1. □

5.3   Error Correction Method Proof

In this section, we will verify the Reed-Solomon error correction method that we summarized and illustrated in Section 5.2.[2]

Suppose F is a field of order 2^n, and let C be an $RS(2^n - 1, t)$ code in $F[x]$. If $c(x) \in C$ is transmitted and we receive the polynomial $r(x) \in F[x]$ of degree less than $2^n - 1$, then $r(x) = c(x) + e(x)$ for some error polynomial $e(x)$ in $F[x]$ of degree less than $2^n - 1$. We will denote this error polynomial by $e(x) = \sum_{j=0}^{m-1} e_j x^j$, with $m = 2^n - 1$ and $e_j \in F$. To determine $e(x)$, we begin by computing the first $2t$ syndromes of $r(x)$, which we will denote as follows for $i = 1, 2, \ldots, 2t$.

$$s_i \;=\; r(a^i) \;=\; e(a^i) \;=\; \sum_{j=0}^{m-1} e_j a^{ij}$$

Next, we use the preceding syndromes to form the syndrome polynomial $S(z) = \sum_{i=0}^{2t-1} s_{i+1} z^i$. Note then that $S(z)$ can be expressed as follows.

$$S(z) \;=\; \sum_{i=0}^{2t-1} \sum_{j=0}^{m-1} e_j a^{(i+1)j} z^i \;=\; \sum_{j=0}^{m-1} e_j a^j \sum_{i=0}^{2t-1} a^{ij} z^i$$

Let M be the set of integers that correspond to the error positions in $r(x)$. That is, let $M = \{0 \leq j \leq m - 1 \mid e_j \neq 0\}$. Note also the following.

$$
\begin{aligned}
S(z) \;&=\; \sum_{j \in M} e_j a^j \sum_{i=0}^{2t-1} a^{ij} z^i \\[2mm]
&=\; \sum_{j \in M} e_j a^j \left(\frac{1 - a^{j(2t)} z^{2t}}{1 - a^j z} \right) \\[2mm]
&=\; \sum_{j \in M} \frac{e_j a^j}{1 - a^j z} \;-\; \sum_{j \in M} \frac{e_j a^{j(2t+1)} z^{2t}}{1 - a^j z}
\end{aligned}
$$

---

[2]Because verifying that the Reed-Solomon error correction method is valid requires an extensive discussion, the reader may wish to postpone this section until completing the remainder of this chapter, or skip this section altogether, as understanding why the Reed-Solomon error correction method is valid is not a prerequisite for the remainder of this chapter or any of the subsequent chapters in this book.

Now, consider the following polynomials $R(z)$, $U(z)$, and $V(z)$.

$$R(z) = \sum_{j \in M} e_j a^j \prod_{\substack{i \in M \\ i \neq j}} (1 - a^i z)$$

$$U(z) = \sum_{j \in M} e_j a^{j(2t+1)} \prod_{\substack{i \in M \\ i \neq j}} (1 - a^i z)$$

$$V(z) = \prod_{i \in M} (1 - a^i z)$$

For these polynomials, the following is true.

$$S(z) = \frac{R(z)}{V(z)} + \frac{U(z) z^{2t}}{V(z)}$$

Also, the preceding equation is equivalent to the following.

$$V(z)S(z) + U(z)z^{2t} = R(z)$$

This last equation is called the *fundamental equation*. In this equation, $V(z)$ is called the *error locator* polynomial,[3] $R(z)$ the *error evaluator* polynomial, and $U(z)$ the *error coevaluator* polynomial. Also, note the following.

$$\gcd(R(z), V(z)) = \gcd(U(z), V(z)) = 1$$

We will now consider how to determine error locator, error evaluator, and error coevaluator polynomials. As we will show, these polynomials are the entries in the Euclidean algorithm table for z^{2t} and $S(z)$ in the first row j for which $\deg(r_j) < t$. The following results verify this. For convenience, in these results we will suppress the variable z whenever possible.

Theorem 5.2 *Suppose $VS + Uz^{2t} = R$ for some syndrome polynomial S, and let V_0, U_0, and R_0 be polynomials that satisfy $V_0 S + U_0 z^{2t} = R_0$, with $\deg(V_0) \leq t$, $\deg(U_0) < t$, and $\deg(R_0) < t$. Then there exists a polynomial $h \in F[z]$ such that $V_0 = hV$, $U_0 = hU$, and $R_0 = hR$. If it is also true that $\gcd(V_0, U_0) = 1$, then h is a constant.*

Proof. Note first that since $VS + Uz^{2t} = R$ and $V_0 S + U_0 z^{2t} = R_0$, the following will be true.

$$V_0 V S + V_0 U z^{2t} = V_0 R$$

---

[3]This error locator polynomial is not the same as the error locator polynomial in the BCH error correction method that we presented in Chapter 4.

Also, note that the following will be true as well.

$$VV_0S + VU_0z^{2t} = VR_0$$

Thus, by subtraction, we have the following.

$$(V_0U - VU_0)z^{2t} = V_0R - VR_0$$

By a degree argument, it can be verified that both sides of the preceding equation must be 0. As a result, we know that $V_0U - VU_0 = V_0R - VR_0 = 0$, or, equivalently, $V_0U = VU_0$ and $V_0R = VR_0$. Also, since $\gcd(U, V) = 1$, then there must exist polynomials $\alpha, \beta \in F[z]$ for which $\alpha U + \beta V = 1$. Thus, the following is true.

$$V_0\alpha U + V_0\beta V = V_0$$

However, since $V_0U = VU_0$, this yields the following.

$$V\alpha U_0 + V_0\beta V = V_0$$

The preceding equation is equivalent to the following.

$$(\alpha U_0 + V_0\beta)V = V_0$$

Now, let $h = \alpha U_0 + V_0\beta$. Then $hV = V_0$. Also, since $hVU = V_0U = VU_0$, then $hU = U_0$, and since $hVR = V_0R = VR_0$, then $hR = R_0$. Finally, since h must divide both V_0 and U_0, if $\gcd(V_0, U_0) = 1$, then h would have to be constant. $\square$

Theorem 5.3 *In the Euclidean algorithm table for z^{2t} and a syndrome polynomial S, let j be the first row for which $\deg(r_j) < t$. Define $R_0 = r_j$, $U_0 = u_j$, and $V_0 = v_j$. Then R_0, U_0, and V_0 satisfy all of the conditions in Theorem 5.2.*

Proof. By (1.5), we know that $R_0 = z^{2t}U_0 + SV_0$, or, equivalently, that $V_0S + U_0z^{2t} = R_0$. Furthermore, because $\deg(r_j) < t$, we know $\deg(R_0) < t$. Since $\deg(v_{j-1}) < \deg(v_j) = \deg(V_0)$ and $\deg(r_{j-1}) > \deg(r_j) = \deg(R_0)$, it follows that $\deg(v_{j-1}R_0) < \deg(V_0r_{j-1})$. However, by (1.7), we know $v_{j-1}R_0 - r_{j-1}V_0 = z^{2t}$. Therefore, it follows that $\deg(V_0r_{j-1}) \leq 2t$, and since $\deg(r_{j-1}) \geq t$, then we know that $\deg(V_0) \leq t$. Also, since $\deg(u_{j-1}) < \deg(u_j) = \deg(U_0)$, it follows $\deg(u_{j-1}R_0) < \deg(U_0r_{j-1})$. However, by (1.6), we know $u_{j-1}R_0 - r_{j-1}U_0 = S$. So, $\deg(U_0r_{j-1}) < 2t$, and since $\deg(r_{j-1}) \geq t$, then we know that $\deg(U_0) < t$. It remains to be shown only that $\gcd(V_0, U_0) = 1$. However, by (1.8), we know that $u_{j-1}V_0 - U_0v_{j-1} = 1$, and thus $\gcd(V_0, U_0) = 1$. $\square$

In summary, for a syndrome polynomial $S(z)$, to determine the error locator polynomial $V(z)$, error evaluator polynomial $R(z)$, and error coevaluator polynomial $U(z)$, we construct the Euclidean algorithm table for z^{2t} and $S(z)$. At the first row j for which $\deg(r_j) < t$, then $r_j = R_0 = hR(z)$, $u_j = U_0 = hU(z)$, and $v_j = V_0 = hV(z)$ for some polynomial h. However, since $\gcd(V_0, U_0) = 1$, then $h = 1$. Thus, $r_j = R(z)$, $u_j = U(z)$, and $v_j = V(z)$. By the definition of $V(z)$, we can find the error positions in $r(x)$ by finding the roots of $V(z)$. To find the coefficients of $e(x)$ at these error positions, note that since $V(z) = \prod_{i \in M} (1 - a^i z)$, the following is true.

$$V'(z) = \sum_{j \in M} -a^j \prod_{\substack{i \in M \\ i \neq j}} (1 - a^i z)$$

Also, recall the following.

$$R(z) = \sum_{j \in M} e_j a^j \prod_{\substack{i \in M \\ i \neq j}} (1 - a^i z)$$

By evaluating $V'(z)$ and $R(z)$ at a^{-j}, we obtain the following.

$$V'(a^{-j}) = -a^j \prod_{\substack{i \in M \\ i \neq j}} \left(1 - a^{(i-j)}\right)$$

$$R(a^{-j}) = e_j a^j \prod_{\substack{i \in M \\ i \neq j}} \left(1 - a^{(i-j)}\right)$$

Thus, $\dfrac{R(a^{-j})}{V'(a^{-j})} = e_j$, which reveals the coefficient of x^j in $e(x)$.

5.4 Reed-Solomon Codes with Maple

In this section, we will show how Maple can be used to construct and correct errors in Reed-Solomon codes. We will consider the code in Examples 5.1 and 5.2.

5.4.1 Construction

Because some of the functions that we will use are in the Maple **Linear-Algebra** package, we will begin by including this package. In addition, we will enter the following **interface** command to cause Maple to display all matrices of size 200×200 and smaller throughout the remainder of this Maple session.

```
>   with(LinearAlgebra):
```

```
>   interface(rtablesize=200):
```

We will now define the primitive polynomial $p(x) = x^4 + x + 1 \in \mathbb{Z}_2[x]$ used to construct the code.

```
>   p := x -> x^4+x+1:
```

```
>   Primitive(p(x)) mod 2;
```
$$true$$

Next, we will use the Maple **degree** function to assign the number of elements in the underlying finite field as the variable *fs*, and use the Maple **Vector** function to create a vector in which to store the field elements.

```
>   fs := 2^(degree(p(x)));
```
$$fs := 16$$

```
>   field := Vector[row](fs):
```

We can then use the following commands to generate and store the field elements in the vector *field*. Since for Reed-Solomon codes we denote the field element x by a, we use the parameters a and $p(a)$ in the following **Powmod** command, just as we did in the corresponding **Powmod** command when we used this same field with a BCH code in Section 4.3.

```
>   for i from 1 to fs-1 do
```

```
>       field[i] := Powmod(a, i, p(a), a) mod 2:
```

```
>   od:
```

```
>   field[fs] := 0:
```

```
>   field;
```
$$[a, a^2, a^3, a+1, a^2+a, a^3+a^2, a^3+a+1, a^2+1, a^3+a, a^2+a$$
$$+1, a^3+a^2+a, a^3+a^2+a+1, a^3+a^2+1, a^3+1, 1, 0]$$

Because working with Reed-Solomon codes requires frequent conversions between polynomial field elements and powers of a, we will establish an association between the polynomial field elements and corresponding powers of a in this field in the table *ftable*, also just as we did for the BCH code that we considered using this same field in Section 4.3.

```
>   ftable := table():

>   for i from 1 to fs-2 do

>         ftable[field[i]] := a^i:

>   od:

>   ftable[field[fs-1]] := 1:

>   ftable[field[fs]] := 0:

>   print(ftable);
```

$$table \left(\left[0 = 0,\, 1 = 1,\, a^3 + a^2 = a^6,\, a^3 + a^2 + 1 = a^7,\, a^3 + a^2 + a = a^{11}, \right. \right.$$
$$a^2 + a + 1 = a^{10},\, a^3 + a = a^9,\, a^2 = a^2,\, a^3 = a^3,\, a^3 + 1 = a^{14},\, a$$
$$+ 1 = a^4,\, a = a,\, a^2 + 1 = a^8,\, a^3 + a^2 + 1 = a^{13},\, a^3 + a^2 + a + 1$$
$$\left. \left. = a^{12},\, a^2 + a = a^5 \right] \right)$$

We will now construct the generator polynomial for the Reed-Solomon code in Example 5.1. To do this, we first assign the number $t = 2$ of coefficient errors that are to be correctable in the code. We then use the Maple **product** function to construct the generator polynomial.

```
>   t := 2:

>   g := product(x-a^j, j=1..2*t);
```
$$g := (x - a)\left(x - a^2\right)\left(x - a^3\right)\left(x - a^4\right)$$

The process of simplifying this generator polynomial so that its coefficients are expressed as powers of a can be nontrivial, as it can require many conversions between the polynomial field elements and the corresponding powers of a. Thus, we will do this expansion and simplification using the user-written function **rscoeff**, which we have written separately from this Maple session and saved as the text file rscoeff.mpl. To use this function, we must first read it into our Maple session as follows.

```
>   read "rscoeff.mpl";
```

The following command then illustrates how the function **rscoeff** can be used. Specifically, by entering the following command, we can see the expanded and simplified form of the generator polynomial g.

```
>   g := rscoeff(g, x, p(a), a);
```
$$g := x^4 + a^{13}x^3 + a^6x^2 + a^3x + a^{10}$$

The second parameter in the preceding command is the variable used in the first parameter, and the final two parameters are the primitive polynomial

$p(x)$ expressed in terms of the field element $a = x$, followed by this generator a for the nonzero elements in the underlying field.

Recall that the code that results from this generator polynomial is the set of all multiples of $g(x)$ in $F[x]$ of degree less than 31. By entering the following command, we can see one of the codewords in this code, specifically the codeword $c(x) = (a^{10}x^9 + a^3x^5 + a^2x^3)g(x)$. Also, note that we use **rscoeff** in this command so that the coefficients of $c(x)$ will be expressed as powers of a in expanded and simplified form.

```
>   c := rscoeff((a^10*x^9+a^3*x^5+a^2*x^3)*g, x, p(a), a);
```

$$c := a^{10} x^{13} + a^8 x^{12} + a x^{11} + a^{13} x^{10} + a^{11} x^9 + a x^8 + a^{11} x^7 + a^{13} x^6$$
$$+ a^3 x^5 + a^5 x^4 + a^{12} x^3$$

To perform the process of converting $c(x)$ into a binary vector, we will use the user-written function **binword**, which we have written separately from this Maple session and saved as the text file binword.mpl. To use this function, we must first read it into our Maple session as follows.

```
>   read "binword.mpl";
```

The following command then illustrates how the function **binword** can be used. Specifically, by entering the following command, we convert $c(x)$ into the binary vector *cbin*.

```
>   cbin := binword(c, degree(p(x)), p(a), a, fs-2);
```

$$cbin := [0, 0, 0, 0, 0, 0, 0, 0, 0, 0, 0, 0, 1, 1, 1, 1, 0, 1, 1, 0, 0, 0, 0, 1,$$
$$1, 0, 1, 1, 0, 1, 1, 1, 0, 1, 0, 0, 0, 1, 1, 1, 1, 0, 1, 1, 0, 1, 0, 0, 1, 0,$$
$$1, 0, 1, 1, 1, 0, 0, 0, 0, 0]$$

The second parameter in the preceding command is the number of bits that are to be used within each block in the resulting binary vector, and the next two parameters are the primitive polynomial $p(x)$ expressed in terms of the field element $a = x$, followed by this generator a for the nonzero elements in the underlying field. The final parameter is the largest possible degree *fs*–2 of the codewords in the code. Although the codewords in this code can be of degree up to 14, the degree of $c(x)$ is only 13. Note that **binword** recognizes that the term in $c(x)$ of degree 14 has a coefficient of zero, and inserts four zeros to reflect this in the resulting binary vector.

5.4.2   Error Correction

We will now show how Maple can be used to correct the received vector in Example 5.2. Suppose that a codeword in the Reed-Solomon code in

Example 5.1 is transmitted as a binary vector, and we receive the following vector *rbin*.

```
>  rbin := Vector[row]([0, 0, 0, 0, 0, 0, 0, 0, 0, 0, 0, 0,
    1, 1, 1, 1, 0, 1, 1, 0, 0, 0, 0, 1, 1, 0, 1, 1, 0, 1, 1,
    1, 0, 1, 0, 0, 0, 1, 1, 1, 0, 0, 1, 0, 1, 0, 0, 1, 1, 0,
    1, 0, 1, 1, 1, 0, 0, 0, 0, 0]):
```

To correct this received vector *rbin*, we must first convert it into a polynomial $r(x)$ that is in $F[x]$. To help with this conversion, we will use the user-written function **bincoeff**, which we have written separately from this Maple session and saved as the text file bincoeff.mpl. To use this function, we must first read it into our Maple session as follows.

```
>  read "bincoeff.mpl";
```

The following command then illustrates how the function **bincoeff** can be used. Specifically, by entering the following command, we obtain an ordered list of the coefficients in the polynomial expression of *rbin*. The first parameter in this command is the number of bits in *rbin* that will be used to form each of these coefficients.

```
>  pcoeff := bincoeff(degree(p(x)), rbin);
```

$pcoeff := \left[0, 0, 0, a^3 + a^2 + a + 1, a^2 + a, a^3, a^3 + a^2 + 1, a^3 + a^2 + a, \right.$
$\left. a, a^3 + a^2 + a, a^2, a^3 + 1, a^2 + 1, a^2 + a + 1, 0 \right]$

We can then construct a polynomial expression for $r(x)$ with these coefficients by entering the following command.

```
>  r := add(pcoeff[i]*x^(i-1), i=1..Dimension(pcoeff));
```

$r := \left(a^3 + a^2 + a + 1 \right) x^3 + \left(a^2 + a \right) x^4 + a^3 x^5 + \left(a^3 + a^2 + 1 \right) x^6$
$+ \left(a^3 + a^2 + a \right) x^7 + a x^8 + \left(a^3 + a^2 + a \right) x^9 + a^2 x^{10} + \left(a^3 \right.$
$\left. + 1 \right) x^{11} + \left(a^2 + 1 \right) x^{12} + \left(a^2 + a + 1 \right) x^{13}$

Also, we can use **rscoeff** as follows to simplify the coefficients in the preceding polynomial expression for $r(x)$.

```
>  r := rscoeff(r, x, p(a), a);
```

$r := a^{10} x^{13} + a^8 x^{12} + a^{14} x^{11} + a^2 x^{10} + a^{11} x^9 + a x^8 + a^{11} x^7$
$+ a^{13} x^6 + a^3 x^5 + a^5 x^4 + a^{12} x^3$

Finally, we can use the Maple **unapply** function as follows to convert the preceding polynomial expression for $r(x)$ into a function.

```
>  r := unapply(r, x);
```

$$r := x \to a^{10} x^{13} + a^8 x^{12} + a^{14} x^{11} + a^2 x^{10} + a^{11} x^9 + a x^8 + a^{11} x^7$$
$$+ a^{13} x^6 + a^3 x^5 + a^5 x^4 + a^{12} x^3$$

Recall now that to correct $r(x)$, we begin by computing the first $2t$ syndromes of $r(x)$. Before doing this, we will create the following vector *syn* of length $2t$ in which to store the syndromes.

```
>  syn := Vector[row](2*t):
```

We can then use the following loop to compute and store the first $2t$ syndromes of $r(x)$ in the vector *syn*.

```
>  for i from 1 to 2*t do

>        syn[i] := ftable[Rem(r(a^i), p(a), a) mod 2];

>  od:

>  syn;
```

$$\begin{bmatrix} a & a^9 & a^{11} & a^5 \end{bmatrix}$$

Next, we will use the Maple **add** function to construct the resulting syndrome polynomial $S(z)$, and then use the Maple **unapply** function to convert $S(z)$ from an expression into a function.

```
>  S := add(syn[j+1]*z^j, j=0..2*t-1);
```

$$S := a + a^9 z + a^{11} z^2 + a^5 z^3$$

```
>  S := unapply(S, z);
```

$$S := z \to a + a^9 z + a^{11} z^2 + a^5 z^3$$

We now need to construct the Euclidean algorithm table for the following polynomial $f(z) = z^{2t}$ and the syndrome polynomial $S(z)$.

```
>  f := z^(2*t);
```

$$f := z^4$$

To do the calculations for constructing this Euclidean algorithm table, we will use the user-written function **rseuclid**,[4] which we have written separately from this Maple session and saved as the text file rseuclid.mpl.

---

[4]The function **rseuclid** uses the user-written function **rscoeff** that we included previously in this Maple session.

To use this function, we must first read it into our Maple session as follows.

```
>   read "rseuclid.mpl";
```

The following command then illustrates how the function **rseuclid** can be used. Specifically, by entering the following command, we can see the entries in each row of the Euclidean algorithm table for $f(z)$ and $S(z)$ through the first row j for which $\deg(r_j) < t$. The fourth parameter in this command is the variable used in the second and third parameters, and the final two parameters are the primitive polynomial $p(x)$ expressed in terms of the field element $a = x$, followed by this generator a for the nonzero elements in the underlying field.

```
>   res := rseuclid(t, f, S(z), z, p(a), a);
```

$$Q \;=\; ,\, a^{10}\, z + a, \quad R \;=\; ,\, a^6\, z^2 + a^{14}\, z + a^2, \quad V \;=\; ,\, a^{10}\, z + a,$$
$$U \;=\; ,\, 1$$

$$Q \;=\; ,\, a^{14}\, z + a^{13}, \quad R \;=\; ,\, a^{10}\, z + a^4, \quad V \;=\; ,\, a^9\, z^2 + a^2\, z + a^3,$$
$$U \;=\; ,\, a^{14}\, z + a^{13}$$

$$res := a^{14}\, z + a^{13},\, a^{10}\, z + a^4,\, a^9\, z^2 + a^2\, z + a^3,\, a^{14}\, z + a^{13}$$

Note that the preceding command leaves the entries in the last computed row of the Euclidean algorithm table as the entries in the vector *res*. Thus, the polynomials $R(z)$ and $V(z)$ that we need for the Reed-Solomon error correction method will be the second and third entries in *res*. In the following commands, we assign these entries as the variables R and V, using **unapply** to convert each from an expression into a function.

```
>   R := res[2];
```
$$R := a^{10}z + a^4$$

```
>   R := unapply(R, z);
```
$$R := z \to a^{10}z + a^4$$

```
>   V := res[3];
```
$$V := a^9 z^2 + a^2 z + a^3$$

```
>   V := unapply(V, z);
```
$$V := z \to a^9 z^2 + a^2 z + a^3$$

Then in the following loop, we find the roots of V by trial and error, and for each root find and display the corresponding error position in $r(x)$.

```
> for i from 1 to fs-1 do
>       if (Rem(V(a^i), p(a), a) mod 2) = 0 then
>           print(a^i, ' is a root of ', V(z), ' error',
            'position is ', degree(a^(fs-1)/a^i, a));
>       fi:
> od;
```

$$a^4, \quad \textit{is a root of } , a^9 z^2 + a^2 z + a^3, \quad \textit{error, position is } , 11$$

$$a^5, \quad \textit{is a root of } , a^9 z^2 + a^2 z + a^3, \quad \textit{error, position is } , 10$$

To find the coefficients of the error polynomial $e(x)$ at these error positions, we will need the derivative of V. So next we use the Maple **diff** function to find this derivative.

```
> Vp := diff(V(z), z) mod 2;
```
$$Vp := a^2$$

```
> Vp := unapply(Vp, z);
```
$$Vp := z \to a^2$$

We can then find the coefficients of $e(x)$ as follows.

```
> e10 := ftable[Rem(R(a^5), p(a), a) mod 2]/
    ftable[Rem(Vp(a^5), p(a), a) mod 2]:
> e11 := ftable[Rem(R(a^4), p(a), a) mod 2]/
    ftable[Rem(Vp(a^4), p(a), a) mod 2]:
```

We will now use these coefficients to define $e(x)$.

```
> e := e10*x^10+e11*x^11;
```
$$e := \frac{x^{10}}{a} + a^7 x^{11}$$

```
> e := unapply(rscoeff(e, x, p(a), a), x);
```
$$e := x \to a^7 x^{11} + a^{14} x^{10}$$

Next, we add $r(x)$ and $e(x)$ to form the corrected codeword $c(x)$.

```
> c := rscoeff(r(x)+e(x), x, p(a), a);
```
$$c := a^{10} x^{13} + a^8 x^{12} + a x^{11} + a^{13} x^{10} + a^{11} x^9 + a x^8 + a^{11} x^7 + a^{13} x^6$$
$$+ a^3 x^5 + a^5 x^4 + a^{12} x^3$$

Finally, we will verify that $c(x)$ is a codeword by checking that its first $2t$ syndromes are all zero.

```
>  seq(Rem(subs(x=a^i, c), p(a), a) mod 2, i=1..2*t);
                    0, 0, 0, 0
```

To see the positions in *rbin* that contained errors, we can use the user-written function **binword** that we included previously in this Maple session to find the binary representation of $e(x)$.

```
>  ebin := binword(e(x), degree(p(x)), p(a), a, fs-2);
```

$ebin := [0, 0, 0, 0, 0, 0, 0, 0, 0, 0, 0, 0, 0, 0, 0, 0, 0, 0, 0, 0, 0, 0, 0,$
$0, 0, 0, 0, 0, 0, 0, 0, 0, 0, 0, 0, 0, 0, 0, 0, 1, 0, 0, 1, 1, 1, 0, 1, 0, 0,$
$0, 0, 0, 0, 0, 0, 0, 0, 0, 0]$

We can then use the Maple **add** function as follows to see the number of binary errors in *rbin*.

```
>  berrors := add(ebin[i], i=1..Dimension(ebin));
                    berrors := 5
```

Note that although this code is only two-error correcting, we were able to correct the received vector *rbin* despite the fact that it contained five binary errors. This is because the binary errors in *rbin* occurred close together (i.e., as an error burst), and thus resulted in only two coefficient errors in the corresponding polynomial $r(x)$.

5.5 Reed-Solomon Codes with MATLAB

In this section, we will show how MATLAB can be used to construct and correct errors in Reed-Solomon codes. We will consider the code in Examples 5.1 and 5.2.

5.5.1 Construction

We will begin by declaring the variables a and x as symbolic variables, and defining the primitive polynomial $p(x) = x^4 + x + 1 \in \mathbb{Z}_2[x]$ used to construct the code. As we did in Section 1.6, we will establish that the polynomial is primitive using the user-written function **Primitive**, which we have written separately from this MATLAB session and saved as the M-file Primitive.m.

```
>> syms a x
>> p = @(x) x^4+x+1;
```

```
>> Primitive(p(x), 2)

ans =

TRUE
```

Next, we will use the MuPAD **degree** function within MATLAB to assign the number of elements in the underlying finite field as the variable *fs*. We will also use the MATLAB **double** function in this command to convert the result from a symbolic variable into a numeric variable.

```
>> fs = double(2^(feval(symengine, 'degree', p(x))))

fs =

    16
```

We can then use the following commands to generate and store the field elements in the vector *field*. As in Section 1.6, to construct the field elements, we use the user-written function **Powmod**, which we have written separately from this MATLAB session and saved as the M-file Powmod.m. Since for Reed-Solomon codes we denote the field element x by a, we use the parameters a and $p(a)$ in the following **Powmod** command, just as we did in the corresponding **Powmod** command when we used this same field with a BCH code in Section 4.4.

```
>> for i = 1:fs-1
       field(i) = Powmod(a, i, p(a), a, 2);
    end
>> field(fs) = 0;
>> field

field =

[ a, a^2, a^3, a + 1, a^2 + a, a^3 + a^2, a^3 + a + 1,
  a^2 + 1, a^3 + a, a^2 + a + 1, a^3 + a^2 + a,
  a^3 + a^2 + a + 1, a^3 + a^2 + 1, a^3 + 1, 1, 0]
```

Because working with Reed-Solomon codes requires frequent conversions between polynomial field elements and powers of a, we will establish an association between the polynomial field elements and corresponding powers of a in this field, also just as we did for the BCH code that we considered using this same field in Section 4.4. As we did then, we will establish this association using the user-written function **ftable**, which we have written separately from this MATLAB session and saved as the M-file ftable.m.

We will now construct the generator polynomial for the Reed-Solomon code in Example 5.1. To do this, we first assign the number $t = 2$ of coefficient errors that are to be correctable in the code. We then use a MATLAB **for** loop to construct the generator polynomial.

```
>> t = 2;
>> g = 1;
>> for j = 1:2*t
        g = g*(x-a^j);
   end
>> g

g =

-(a - x)*(- a^2 + x)*(- a^3 + x)*(- a^4 + x)
```

The process of expanding and simplifying this generator polynomial so that its coefficients are expressed as powers of a can be nontrivial, as it can require many conversions between the polynomial field elements and corresponding powers of a. Thus, we will do this expansion and simplification using the user-written function **rscoeff**,[5] which we have written separately from this MATLAB session and saved as the M-file rscoeff.m. The following command illustrates how the function **rscoeff** can be used. Specifically, by entering the following command, we can see the expanded and simplified form of the generator polynomial g.

```
>> g = rscoeff(g, x, p(a), a)

g =

a^13*x^3 + a^10 + a^6*x^2 + a^3*x + x^4
```

The second parameter in the preceding command is the variable used in the first parameter, and the final two parameters are the primitive polynomial $p(x)$ expressed in terms of the field element $a = x$, followed by this generator a for the nonzero elements in the underlying field.

---

[5]The function **rscoeff** uses the user-written function **ftable** that we noted previously in this section, and the user-written function **Rem** that we used in Section 4.4 and will use again later in this section.

MuPAD displays polynomials within MATLAB with decreasing powers of its variables, using alphabetical order to decide which variable should take precedence. Since we are representing codewords using powers of a as the coefficients and x as the variable, we would prefer our polynomials be displayed with decreasing powers of x. To do this, we use the user-written function **sortpoly**, which we have written separately from this MATLAB session and saved as the M-file sortpoly.m. The following command illustrates how **sortpoly** can be used to display the generator polynomial g output by the previous command with decreasing powers of the variable x.

```
>> sortpoly(g, x)

ans =

x^4 + a^13*x^3 + a^6*x^2 + a^3*x + a^10
```

The data type output by the **sortpoly** function is a string, which is represented in MATLAB by the designation *char*. This can be observed by applying the MATLAB **class** function to the previous result, as demonstrated in the following command.

```
>> class(ans)

ans =

char
```

However, polynomials in MATLAB are normally manipulated as symbolic objects. Thus, we will use the following MATLAB **sym** command to convert the result of **sortpoly** into a symbolic object, while simultaneously displaying the polynomial terms with decreasing powers of the variable x.

```
>> sym(sortpoly(g, x))

ans =

a^13*x^3 + a^10 + a^6*x^2 + a^3*x + x^4

>> class(ans)

ans =

sym
```

Recall that the code that results from this generator polynomial is the set of all multiples of $g(x)$ in $F[x]$ of degree less than 31. By entering the following commands, we can see one of the codewords in this code, specifically the codeword $c(x) = (a^{10}x^9 + a^3x^5 + a^2x^3)g(x)$. Note that we use **rscoeff** so that the coefficients of $c(x)$ will be expressed as powers of a in expanded and simplified form, and **sortpoly** to express the answer with decreasing powers of the variable x.

```
>> c = rscoeff((a^10*x^9+a^3*x^5+a^2*x^3)*g, x, p(a), a)

c =

a^13*x^10 + a^13*x^6 + a^12*x^3 + a^11*x^9 + a^11*x^7
+ a^10*x^13 + a^8*x^12 + a^5*x^4 + a^3*x^5 + a*x^11
+ a*x^8

>> sortpoly(c, x)

ans =

a^10*x^13 + a^8*x^12 + a*x^11 + a^13*x^10 + a^11*x^9
+ a*x^8 + a^11*x^7 + a^13*x^6 + a^3*x^5 + a^5*x^4
+ a^12*x^3
```

To perform the process of converting $c(x)$ into a binary vector, we will use the user-written function **binword**, which we have written separately from this MATLAB session and saved as the M-file binword.m. The following command illustrates how the function **binword** can be used. Specifically, by entering the following command, we convert $c(x)$ into the binary vector *cbin*.

```
>> cbin = binword(c, double(feval(symengine, 'degree', ...
     p(x)))), p(a), a, fs-2)

cbin =

[ 0, 0, 0, 0, 0, 0, 0, 0, 0, 0, 0, 0, 1, 1, 1, 1, 0, 1, 1,
  0, 0, 0, 0, 1, 1, 0, 1, 1, 0, 1, 1, 1, 0, 1, 0, 0, 0, 1,
  1, 1, 1, 0, 1, 1, 0, 1, 0, 0, 1, 0, 1, 0, 1, 1, 1, 0, 0,
  0, 0, 0]
```

The second parameter in the preceding command is the number of bits that are to be used within each block in the resulting binary vector, and the next two parameters are the primitive polynomial $p(x)$ expressed in terms of the field element $a = x$, followed by this generator a for the nonzero elements in the underlying field. The final parameter is the largest possible degree $fs-2$ of the codewords in the code. Although the codewords in this code can be of degree up to 14, the degree of $c(x)$ is only 13. Note that **binword** recognizes that the term in $c(x)$ of degree 14 has a coefficient of zero, and inserts four zeros to reflect this in the resulting binary vector.

5.5.2   Error Correction

We will now show how MATLAB can be used to correct the received vector in Example 5.2. Suppose that a codeword in the code in Example 5.1 is transmitted as a binary vector, and we receive the following vector *rbin*.

```
>> rbin = [0, 0, 0, 0, 0, 0, 0, 0, 0, 0, 0, 0, 1, 1, 1, ...
   1, 0, 1, 1, 0, 0, 0, 0, 1, 1, 0, 1, 1, 0, 1, 1, 1, ...
   0, 1, 0, 0, 0, 1, 1, 1, 0, 0, 1, 0, 1, 0, 0, 1, 1, ...
   0, 1, 0, 1, 1, 1, 0, 0, 0, 0, 0];
```

To correct this received vector *rbin*, we must first convert it into a polynomial $r(x)$ that is in $F[x]$. To help with this conversion, we will use the user-written function **bincoeff**, which we have written separately from this MATLAB session and saved as the M-file bincoeff.m. The following command shows how **bincoeff** can be used. Specifically, by entering the following, we obtain an ordered list of the coefficients in the polynomial expression of *rbin*. The first parameter in this command is the number of bits in *rbin* that will be used to form each of these coefficients.

```
>> pcoeff = bincoeff(double(feval(symengine, 'degree', ...
   p(x))), rbin)

pcoeff =

[ 0, 0, 0, a^3 + a^2 + a + 1, a^2 + a, a^3, a^3 + a^2 + 1,
  a^3 + a^2 + a, a, a^3 + a^2 + a, a^2, a^3 + 1, a^2 + 1,
  a^2 + a + 1, 0]
```

We can then construct a polynomial expression for $r(x)$ with these coefficients by entering the following commands.

```
>> r = 0;
>> for i = 1:numel(pcoeff)
       r = r+pcoeff(i)*x^(i-1);
   end
```

```
>> r

r =

x^6*(a^3 + a^2 + 1) + x^11*(a^3 + 1) + x^12*(a^2 + 1)
+ x^3*(a^3 + a^2 + a + 1) + x^7*(a^3 + a^2 + a) +
x^9*(a^3 + a^2 + a) + a*x^8 + x^4*(a^2 + a)
+ x^13*(a^2 + a + 1) + a^3*x^5 + a^2*x^10
```

Also, we can use **rscoeff** as follows to simplify the coefficients in the preceding polynomial expression for $r(x)$.

```
>> r = rscoeff(r, x, p(a), a)

r =

a^14*x^11 + a^13*x^6 + a^12*x^3 + a^11*x^9 + a^11*x^7
+ a^10*x^13 + a^8*x^12 + a^5*x^4 + a^3*x^5 + a^2*x^10
+ a*x^8
```

Finally, we can use the MATLAB **subs** function as follows to convert the preceding polynomial expression for $r(x)$ into a function.

```
>> r = @(x) subs(r, 'x', x);
```

The second parameter in the preceding command is the variable that is to be used as the input variable for the function. The final parameter is the variable that is to replace every occurrence of the variable specified by the second parameter. To see the resulting function, we can enter the following command.

```
>> sortpoly(r(x), x)

ans =

a^10*x^13 + a^8*x^12 + a^14*x^11 + a^2*x^10 + a^11*x^9
+ a*x^8 + a^11*x^7 + a^13*x^6 + a^3*x^5 + a^5*x^4
+ a^12*x^3
```

Recall now that to correct $r(x)$, we begin by computing the first $2t$ syndromes of $r(x)$. We can use the following loop to compute and store these syndromes in the vector *syn*. This loop uses the user-written function **Rem**, which we used previously in Section 4.4, and which we have written separately from this MATLAB session and saved as the M-file Rem.m.

```
>> for i = 1:2*t
       syn(i) = ftable(Rem(r(a^i), p(a), a, 2), field);
   end
>> syn

syn =

[ a, a^9, a^11, a^5]
```

Next, after declaring z as a symbolic variable, we use a MATLAB **for** loop to construct the syndrome polynomial $S(z)$. We then use the MATLAB **subs** function to convert the resulting polynomial expression for $S(z)$ into a function.

```
>> syms z
>> S = 0;
>> for j = 0:2*t-1
       S = S+syn(j+1)*z^j;
   end
>> S

S =

a^11*z^2 + a^9*z + a^5*z^3 + a

>> S = @(z) subs(S, 'z', z);
>> sortpoly(S(z), z)

ans =

a^5*z^3 + a^11*z^2 + a^9*z + a
```

We now need to construct the Euclidean algorithm table for the following polynomial $f(z) = z^{2t}$ and the syndrome polynomial $S(z)$.

```
>> f = z^(2*t)

f =

z^4
```

To do the calculations for constructing this Euclidean algorithm table, we will use the user-written function **rseuclid**,[6] which we have written sepa-

---

[6]The function **rseuclid** uses the user-written functions **rscoeff** and **sortpoly** that we used previously in this section.

rately from this MATLAB session and saved as the M-file rseuclid.m. The following command illustrates how **rseuclid** can be used. Specifically, by entering the following command, we can see the entries in each row of the Euclidean algorithm table for $f(z)$ and $S(z)$ through the first row j for which $\deg(r_j) < t$. The fourth parameter in this command is the variable used in the second and third parameters, and the final two parameters are the primitive polynomial $p(x)$ expressed in terms of the field element $a = x$, followed by this generator a for the nonzero elements in the underlying field.

```
>> res = rseuclid(t, f, S(z), z, p(a), a)

Q: a^10*z + a    R:  a^6*z^2 + a^14*z + a^2
V:  a^10*z + a   U:  1

Q: a^14*z + a^13   R:  a^10*z + a^4
V:  a^9*z^2 + a^2*z + a^3   U:  a^14*z + a^13


res =

[ z*a^14 + a^13, z*a^10 + a^4, a^9*z^2 + a^3 + a^2*z,
  z*a^14 + a^13]
```

Note that the preceding command leaves the entries in the last computed row of the Euclidean algorithm table as the entries in the vector *res*. Thus, the polynomials $R(z)$ and $V(z)$ that we need for the Reed-Solomon error correction method will be the second and third entries in *res*. In the following commands, we assign these entries as the variables R and V, using **subs** to convert each from an expression into a function.

```
>> R = res(2)

R =

z*a^10 + a^4

>> R = @(z) subs(R, 'z', z);
>> sortpoly(R(z), z)

ans =

a^10*z + a^4
```

```
>> V = res(3)

V =

a^9*z^2 + a^3 + a^2*z

>> V = @(z) subs(V, 'z', z);
>> sortpoly(V(z), z)

ans =

a^9*z^2 + a^2*z + a^3
```

Then in the following loop, we find the roots of V by trial and error, and for each find and display the corresponding error position in $r(x)$. The **disp** command causes the output to be displayed in string format, thus allowing **sortpoly** to be used to display the contents of V with decreasing powers of z. The **blanks** parameter causes blank spaces in numbers specified by the numerical input (in this case 1) to be inserted into the string output.

```
>> for i = 1:fs-1
        if sym2poly(Rem(V(a^i), p(a), a, 2)) == 0
            pos = feval(symengine, 'degree', a^(fs-1)/ ...
            a^i, a);
            disp([char(a^i) blanks(1) 'is a root of' ...
            blanks(1) sortpoly(V(z), z)])
            fprintf('Error Position is %3.0f \n \n', ...
            double(pos))
        end
    end

    a^4 is a root of a^9*z^2 + a^2*z + a^3
    Error Position is   11

    a^5 is a root of a^9*z^2 + a^2*z + a^3
    Error Position is   10
```

To find the coefficients of the error polynomial $e(x)$ at these error positions, we will need the derivative of V. So next we use the MATLAB **diff** function to find this derivative.

```
>> Vp = feval(symengine, 'expr', feval(symengine, ...
    'poly', diff(V(z), z), 'Dom::IntegerMod(2)'))
```

```
Vp =

a^2
```

In the previous command, the inner **feval** function with the **poly** and **Dom::IntegerMod(2)** parameters indicates that the derivative is to be determined over the integers modulo 2. The outer **feval** function with the **expr** parameter extracts the polynomial result so that it can be used in the MATLAB workspace.

The next sequence of commands converts the previous result to a function and outputs the result.

```
>> Vp = @(z) subs(Vp, 'z', z);
>> sortpoly(Vp(z), z)

ans =

a^2
```

We can then find the coefficients of $e(x)$ as follows.

```
>> e10 = ftable(Rem(R(a^5), p(a), a, 2), field)/ ...
   ftable(Rem(Vp(a^5), p(a), a, 2), field);
>> e11 = ftable(Rem(R(a^4), p(a), a, 2), field)/ ...
   ftable(Rem(Vp(a^4), p(a), a, 2), field);
```

We will now use these coefficients to define $e(x)$.

```
>> e = e10*x^10+e11*x^11

e =

x^10/a + a^7*x^11

>> e = rscoeff(e, x, p(a), a)

e =

a^14*x^10 + a^7*x^11

>> e = @(x) subs(e, 'x', x);
>> sortpoly(e(x), x)

ans =

a^7*x^11 + a^14*x^10
```

Next, we add $r(x)$ and $e(x)$ to form the corrected codeword $c(x)$.

```
>> c = rscoeff(r(x)+e(x), x, p(a), a)

c =

a^13*x^10 + a^13*x^6 + a^12*x^3 + a^11*x^9 + a^11*x^7
+ a^10*x^13 + a^8*x^12 + a^5*x^4 + a^3*x^5 + a*x^11
+ a*x^8

>> sortpoly(c, x)

ans =

a^10*x^13 + a^8*x^12 + a*x^11 + a^13*x^10 + a^11*x^9
+ a*x^8 + a^11*x^7 + a^13*x^6 + a^3*x^5 + a^5*x^4
+ a^12*x^3
```

Finally, we will verify that $c(x)$ is a codeword by checking that its first $2t$ syndromes are all zero.

```
>> for i = 1:2*t
       sync(i) = Rem(subs(c, x, a^i), p(a), a, 2);
   end
>> sync

sync =

[ 0, 0, 0, 0]
```

To see the positions in *rbin* that contained errors, we can use the user-written function **binword** that we used previously in this section to find the binary representation of $e(x)$.

```
>> ebin = binword(e(x), double(feval(symengine, ...
   'degree', p(x))), p(a), a, fs-2)

ebin =

[ 0, 0, 0, 0, 0, 0, 0, 0, 0, 0, 0, 0, 0, 0, 0, 0, 0, 0,
  0, 0, 0, 0, 0, 0, 0, 0, 0, 0, 0, 0, 0, 0, 0, 0, 0, 0,
  0, 0, 1, 0, 0, 1, 1, 1, 0, 1, 0, 0, 0, 0, 0, 0, 0, 0,
  0, 0, 0]
```

We can then use the MATLAB **find** and **numel** functions as follows to see the number of binary errors in *rbin*.

```
>> numel(find(ebin == 1))

ans =

     5
```

Note that although this code is only two-error correcting, we were able to correct the received vector *rbin* despite the fact that it contained five binary errors. This is because the binary errors in *rbin* occurred close together (i.e., as an error burst), and thus resulted in only two coefficient errors in the corresponding polynomial $r(x)$.

5.6 Reed-Solomon Codes in Voyager 2

In August and September 1977, NASA launched the Voyager 1 and Voyager 2 satellites from Cape Canaveral, Florida. Upon reaching their initial destinations of Jupiter and Saturn, the Voyager satellites provided NASA with the most detailed analyses and images of these planets and their moons that had ever been observed. After leaving Jupiter and Saturn, Voyager 2 continued farther into the outer reaches of our solar system, and successfully transmitted data and images from Uranus and Neptune back to Earth. Without the use of a Reed-Solomon code in transmitting these images, the success achieved by Voyager 2 would have been very unlikely.

Photographs transmitted from outer space back to Earth are usually digitized into binary strings and sent over a space channel. Voyager 2 digitized its full-color images into binary strings of length 15,360,000 positions. Using an uncompressed spacecraft telecommunication system, these bits were transmitted one by one back to Earth, where the images were then reconstructed. This uncompressed system was the most reliable one available when Voyager 2 was launched, and was satisfactory for transmitting images from Jupiter and Saturn back to Earth. However, when Voyager 2 arrived at Uranus in January 1986, it was about twice as far from Earth as when it had been at Saturn. Since the transmission of bits back to Earth had already been stretched to a very slow rate from Saturn (around 44,800 bits per second), a new transmission method was necessary in order for NASA to be able to receive a large number of images from Uranus.

The problem of image transmission from Uranus was solved through the work of Robert Rice at California Institute of Technology's Jet Propulsion Laboratory. Rice developed an algorithm for implementing a compressed spacecraft telecommunication system that reduced by a factor of 2.5 the amount of data needed to transmit a single image from Uranus without causing any loss in image quality. However, there was a problem with Rice's algorithm. During long transmissions through space, compressed binary strings experience errors much more frequently than uncompressed strings, and Rice's algorithm was very sensitive to bit errors. In fact, if a received compressed binary string from Uranus contained even just a single bit error, the entire resulting image would be completely ruined. After considerable study, it was discovered that the bit errors that occurred during the long transmissions through space usually occurred in bursts. To account for these error bursts, a new system was designed in Voyager 2 for converting images into binary strings that utilized a Reed-Solomon code. These binary strings were then compressed and transmitted back to Earth, uncompressed using Rice's algorithm, and corrected using the Reed-Solomon error correction method. The process was remarkably successful.

After leaving Uranus, Voyager 2 continued its journey through space. In August 1989, the satellite transmitted data and visual images back to Earth that provided NASA with most of the information currently known about Neptune. At present, Voyager 2 is still in operation, and is still providing NASA with invaluable information about our solar system.

Exercises

1. Find the total number of codewords in the $RS(15, 2)$ code in Example 5.1.

2. Construct a polynomial codeword of maximum possible degree in the $RS(15, 2)$ code in Example 5.1, and then convert this polynomial codeword into a binary vector.

3. Convert the binary vector (000000000000000000000001011101011011 01101000000000000000000000) into the polynomial codeword that it represents in the $RS(15, 2)$ code in Example 5.1, and then verify that this polynomial really is a codeword in the code.

4. Provided that only one error burst occurs during the transmission of the binary equivalent of a polynomial codeword in an $RS(2^n - 1, t)$ code, what is the maximum error burst length (in terms of n and t) that we would be guaranteed to be able to correct?

5. Let C be the two-error correcting Reed-Solomon code that results from the primitive polynomial $p(x) = x^3 + x + 1 \in \mathbb{Z}_2[x]$.

   (a) Construct and simplify the generator polynomial for C.

(b) Construct two of the polynomial codewords in C, and then convert each of these polynomial codewords into binary vectors.

(c) Provided that only one error burst occurs during the transmission of the binary equivalent of a polynomial codeword in C, what is the maximum error burst length that we would be guaranteed to be able to correct?

(d) Find the total number of codewords in C.

6. Let C be the three-error correcting Reed-Solomon code that results from the primitive polynomial $p(x) = x^4 + x^3 + 1 \in \mathbb{Z}_2[x]$.

(a) Construct and simplify the generator polynomial for C.

(b) Construct two of the polynomial codewords in C, and then convert each of these polynomial codewords into binary vectors.

(c) Provided that only one error burst occurs during the transmission of the binary equivalent of a polynomial codeword in C, what is the maximum error burst length that we would be guaranteed to be able to correct?

(d) Find the total number of codewords in C.

7. Correct the following received polynomials in the Reed-Solomon code C in Exercise 5.

(a) $r(x) \;=\; a^5 x^6 + a^5 x^5 + a^6 x^4 + a x^2 + a^6 x + a^5$

(b) $r(x) \;=\; a x^5 + a^4 x^4 + a^6 x^3 + a^3 x^2 + a^5 x + a^4$

(c) $r(x) \;=\; a^4 x^6 + a^3 x^5 + x^4 + a^4 x^3 + a^2 x^2 + a^2 x + a^3$

8. Correct the following received polynomials in the $RS(15, 2)$ code in Example 5.1.

(a) $r(x) \;=\; a^6 x^{14} + a^4 x^{13} + a x^{12} + a^9 x^{11} + a^{14} x^{10} + a^{10} x^5$
$\qquad + a^8 x^4 + a x^3 + a^{13} x^2 + a^5 x$

(b) $r(x) \;=\; a^9 x^{13} + a^7 x^{12} + x^{11} + a^{10} x^{10} + x^9 + a^9 x^8 + a^6 x^7$
$\qquad + a^{13} x^6 + a^{14} x^4 + a^{12} x^3 + a^5 x^2 + a^2 x + a^9$

(c) $r(x) \;=\; a^{13} x^{13} + a^{11} x^{12} + a^4 x^{11} + a x^{10} + a^8 x^9 + a^{13} x^7$
$\qquad + a^4 x^5 + a x^4 + a^8 x^3$

9. Correct the following received polynomials in the Reed-Solomon code C in Exercise 6.

(a) $r(x) \;=\; a^7 x^{12} + a^2 x^{11} + a^8 x^{10} + a^8 x^9 + a^8 x^8 + a x^7 + a^{10} x^6$
$\qquad + a^6 x^5 + a^4 x^4 + a^7 x^3 + a x^2 + a^6 x + a$

(b) $r(x)$ $=$ $a^5x^{14} + a^{14}x^{13} + a^3x^{12} + a^4x^{11} + a^{14}x^{10} + a^4x^9$
$+ a^{11}x^8 + a^7x^7 + a^{11}x^6 + a^9x^5 + a^{14}x^4 + a^7x^3$
$+ a^{10}x + a^5$

(c) $r(x)$ $=$ $a^8x^{14} + a^{12}x^{13} + a^3x^{12} + a^8x^{11} + a^{12}x^{10} + a^3x^9$
$+ a^9x^8 + a^5x^7 + a^{13}x^6 + a^4x^5 + a^4x^4 + a^8x^3$
$+ a^{12}x^2 + a^{11}x + a^{13}$

(d) $r(x)$ $=$ $x^6 + a^{12}x^5 + x^4 + a^2x^3 + a^7x^2 + a^{11}x$

10. Use the Reed-Solomon error correction method to correct the following received polynomials in the BCH code that results from the generator polynomial in Example 4.3.

(a) $r(x)$ $=$ $1 + x^2 + x^3 + x^4 + x^5 + x^6 + x^7 + x^{10}$

(b) $r(x)$ $=$ $1 + x^3 + x^7 + x^{10} + x^{11} + x^{13}$

(c) $r(x)$ $=$ $1 + x + x^2 + x^4 + x^5 + x^8 + x^{10} + x^{12}$

(d) $r(x)$ $=$ $1 + x^4 + x^5 + x^7 + x^8 + x^{11} + x^{13}$

11. Suppose $p(x) \in \mathbb{Z}_2[x]$ is a primitive polynomial of degree n, and let C be the Reed-Solomon code that results from the first $2t$ powers of $a = x$ in the finite field $\mathbb{Z}_2[x]/(p(x))$. Prove that C is t-error correcting. (Hint: See the proof of Theorem 4.2.)

12. Prove Theorem 5.1.

Computer Exercises

13. Let C be the four-error correcting Reed-Solomon code that results from the primitive polynomial $p(x) = x^5 + x^3 + 1 \in \mathbb{Z}_2[x]$.

(a) Construct and simplify the generator polynomial for C.

(b) Construct two of the polynomial codewords in C, and then convert each of these polynomial codewords into binary vectors.

(c) Provided that only one error burst occurs during the transmission of the binary equivalent of a polynomial codeword in C, what is the maximum error burst length that we would be guaranteed to be able to correct?

(d) Find the total number of codewords in C.

14. Let C be the four-error correcting Reed-Solomon code that results from the primitive polynomial $p(x) = x^7 + x + 1 \in \mathbb{Z}_2[x]$.

(a) Construct and simplify the generator polynomial for C.

(b) Construct two of the polynomial codewords in C.

(c) Provided that only one error burst occurs during the transmission of the binary equivalent of a polynomial codeword in C, what is the maximum error burst length that we would be guaranteed to be able to correct?

(d) Find the total number of codewords in C.

15. Correct the following received binary vectors in the Reed-Solomon code C in Exercise 13.

(a) (11111101001011100000111010110100111001010110110110011
001010110101111100001101101100100110000111001111101000
111001100000000000000000000000000000000000000000000)

(b) (00000010111001000011110001010001001001001000111101011101
10110100110111000010111011100010010011100101001001010100
0110000010111010001010101000100111111100101010111)

(c) (1010010001010011010000001010101001111011000101101001011
011111010001010011010000011101111100000000110000010001
1011001000000000000000000000000000000000000000000000)

16. Correct the following received polynomials in the Reed-Solomon code C in Exercise 14.

(a) $r(x) = a^{100}x^{22} + a^{10}x^{21} + a^{100}x^{20} + a^{28}x^{19} + a^{114}x^{18}$
$+ a^{35}x^{17} + a^{81}x^{16} + a^{95}x^{15} + a^{56}x^{14} + a^{59}x^{13}$
$+ a^{38}x^{12} + a^{83}x^{11} + a^{42}x^{10}$

(b) $r(x) = a^{10}x^{108} + a^{60}x^{107} + a^{49}x^{106} + a^{115}x^{105} + a^{18}x^{104}$
$+ a^{124}x^{103} + a^{67}x^{102} + a^{87}x^{101} + a^{46}x^{100}$

(c) $r(x) = a^{100}x^{81} + a^{23}x^{80} + a^{12}x^{79} + a^{78}x^{78} + a^{108}x^{77}$
$+ a^{80}x^{76} + a^{25}x^{75} + a^{50}x^{74} + a^{9}x^{73}$

17. The Reed-Solomon code used in the Voyager 2 satellite when it transmitted images from Uranus back to Earth was the 16-error correcting code resulting from $p(x) = x^8 + x^4 + x^3 + x^2 + 1 \in \mathbb{Z}_2[x]$. Construct several of the polynomial codewords in this code, and then illustrate the Reed-Solomon error correction method in this code by correcting several received polynomials that contain errors. Write a summary of your results. Include in your summary the total number of codewords in the code, the length of the binary vectors that correspond to the codewords in the code, and, provided that only one error burst occurs during the transmission of the binary equivalent of a polynomial codeword in the code, the maximum error burst length that we would be guaranteed to be able to correct.

Research Exercises

18. Investigate the lives and careers of the people for whom Reed-Solomon codes are named, and write a summary of your findings. Include in your summary some highlights of Reed and Solomon's academic careers and their work at California Institute of Technology's Jet Propulsion Laboratory, their most important scientific contributions outside the area of error-correcting codes, and some of their honors and awards.

19. Investigate some specific details about the use of Reed-Solomon codes in encoding music and video on CDs, DVDs, and Blu-ray discs, and write a summary of your findings.

20. Investigate an actual real-life use of a Reed-Solomon code outside the areas of space communication and the encoding of music and video on CDs, DVDs, and Blu-ray discs, and write a summary of your findings.

21. Investigate some specific details about the discoveries made by the Voyager satellites, and write a summary of your findings.

22. Reed-Solomon codes can actually be viewed as a special case of BCH codes. Investigate why this is true, and write a summary of your findings.

23. Investigate the Berlekamp-Massey algorithm for correcting errors in BCH and Reed-Solomon codes, and write a summary of your findings.

24. Investigate the Berlekamp-Welch algorithm for correcting errors in BCH and Reed-Solomon codes, and write a summary of your findings.

25. Prior to the use of a Reed-Solomon code by the Voyager 2 satellite to transmit images of Uranus back to Earth, both Voyager satellites had used a type of error-correcting code called a *Golay* code to transmit images of Jupiter and Saturn back to Earth. Investigate Golay codes and their use in the Voyager satellites, and write a summary of your findings.

26. Investigate the life and career of Claude Shannon, the so-called Father of Information Theory, and write a summary of your findings. Include in your summary a brief overview of the science of information theory, and some of Shannon's influence on modern telecommunications and technology.

Chapter 6

Algebraic Cryptography

Cryptography is the study of techniques for disguising messages so that ideally only the intended recipient of a message can remove the disguise and read the message. Such techniques are called *ciphers*. One very simple type of cipher involves replacing every occurrence of each specific character in a message with a different specific character, yielding what is called a *substitution cipher*. Since substitution ciphers frequently appear as puzzles in newspapers and books, they are obviously easy for unintended recipients, or *intruders*, to "break" (i.e., remove the disguise and read the message), and should not be used when sending sensitive or personal information that really needs to be kept secret from intruders. The most famous cryptographic device ever used was the German Enigma machine during World War II. However, while mathematics was used extensively by the Allies in their successful effort to break the Enigma cipher, the machine itself did not actually use a mathematical operation to disguise messages. In the next five chapters, we will present several types of ciphers that do involve the use of mathematical operations to disguise messages, and which are thus examples of *algebraic* cryptography.

6.1 Two Elementary Cryptosystems

We will use the term *plaintext* to refer to the text of an undisguised message, and the term *ciphertext* to refer to the text of a disguised message. Also, we will call the process of converting a plaintext into a ciphertext the *encryption* or *encipherment* of the message, and the reverse process the *decryption* or *decipherment* of the message.

Because we would like to encrypt messages by applying mathematical operations, our plaintext characters will have to be mathematical objects.

We will express our messages using ring elements so that we will be able to both add and multiply message characters.

Definition 6.1 *A cryptosystem is an alphabet L containing all characters that can be used in messages (letters, numerals, punctuation marks, blank spaces, etc.), a commutative ring R with identity and of the same order as L, and bijections $\alpha : L \to R$ and $f : R \to R$.*

The idea we will take is that to encrypt a message that is expressed as a list of elements in L, we will first use α to convert the plaintext into a list of elements in R. We can then form the ciphertext by applying f to the plaintext ring elements, and, if desired, use α^{-1} to convert the ciphertext back into a list of elements in L. To recover the plaintext from the ciphertext, we can repeat this same procedure using f^{-1} in place of f. In order for only the intended recipient of the message to be able to recover the plaintext, only the intended recipient can know f^{-1}. We will always assume that everything else in a cryptosystem, with the exception of f, is public knowledge.[1]

For simplicity, throughout this chapter we will assume that all messages are written using only the characters in the alphabet $L = \{A, B, C, \ldots, Z\}$. Also, we will take $R = \mathbb{Z}_{26}$, and let the bijection $\alpha : L \to R$ be given by $\alpha(A) = 0$, $\alpha(B) = 1$, $\ldots$, $\alpha(Z) = 25$. For reference, we list each of these correspondences for α below.

A	B	C	D	E	F	G	H	I	J	K	L	M
0	1	2	3	4	5	6	7	8	9	10	11	12

N	O	P	Q	R	S	T	U	V	W	X	Y	Z
13	14	15	16	17	18	19	20	21	22	23	24	25

We will now consider two elementary cryptosystems with different types of encryption methods. That is, we will consider two elementary cryptosystems with different types of bijections $f : R \to R$.

6.1.1   Shift Ciphers

One of the simplest possible encryption methods involves listing the alphabet characters in some fixed order, and then converting each plaintext character into the ciphertext character located some specific number of positions to the right in the order, wrapping around to the start whenever

---

[1] It is true in some cryptosystems that f can be public knowledge without revealing f^{-1}. Such cryptosystems are called *public-key* systems. We will present two public-key cryptosystems, the RSA and ElGamal systems, in Chapters 8 and 9.

necessary. For instance, Julius Caesar used an example of this by converting each plaintext character into the ciphertext character located three positions to the right. Thus, Caesar's cipher would be, for our natural ordering of $L = \{A, B, C, \ldots, Z\}$, converting A into D, B into E, C into F, ..., W into Z, X into A, Y into B, and Z into C. Because this general encryption method can be viewed as just shifting each plaintext character to the ciphertext character located some specific number of positions to the right, the resulting cryptosystem is called a *shift cipher*. We can represent shift ciphers mathematically as follows.

Shift Cipher Encryption: *Choose $f : R \to R$ as $f(x) = (x + b) \bmod |R|$ for some $b \in R$.*

Example 6.1 Let $f(x) = (x + 17) \bmod 26$. Then the message ATTACK AT DAWN enciphers as follows.

	A	T	T	A	C	K	A	T	D	A	W	N
$\alpha \Rightarrow$	0	19	19	0	2	10	0	19	3	0	22	13
$f \Rightarrow$	17	10	10	17	19	1	17	10	20	17	13	4
$\alpha^{-1} \Rightarrow$	R	K	K	R	T	B	R	K	U	R	N	E

Thus, the corresponding ciphertext is RKKRTBRKURNE. To decrypt this ciphertext, we can repeat this same procedure using the inverse function $f^{-1}(x) = (x - b) \bmod |R| = (x - 17) \bmod 26$ in place of f. This ciphertext deciphers, of course, as follows.

	R	K	K	R	T	B	R	K	U	R	N	E
$\alpha \Rightarrow$	17	10	10	17	19	1	17	10	20	17	13	4
$f^{-1} \Rightarrow$	0	19	19	0	2	10	0	19	3	0	22	13
$\alpha^{-1} \Rightarrow$	A	T	T	A	C	K	A	T	D	A	W	N

$\square$

Two people wishing to exchange a secret message would certainly want to use a cryptosystem that an intruder would find impossible or at least very difficult to break. However, breaking a cryptosystem, a process known as *cryptanalysis*, is not always as difficult as it may at first appear. Recall that everything in a cryptosystem is assumed to be public knowledge except f and f^{-1}, and in practice it is usually assumed that even the form of f is publicly known and that only the parameters in f are unknown to intruders. We will call the parameters in an encryption function f the *keys* of the cryptosystem because if an intruder is able to determine these parameters, the intruder should then be able to find f^{-1} (i.e., "unlock" the system).

The cryptosystem in Example 6.1 has $b = 17$ as its only key. Indeed, this system is not very secure. However, this is not because it has only a single key or because it is so easy to use for encrypting messages, but rather because an intruder could very quickly determine the key by trial and error. We would assume that intruders know $f(x) = (x+b) \bmod 26$ for some $b \in \mathbb{Z}_{26}$. With only 26 elements in $\mathbb{Z}_{26}$, an intruder could very quickly take each of these key candidates separately, form the corresponding inverse function f^{-1}, and then use f^{-1} to attempt to decrypt the ciphertext. It is almost certain that only one of these decipherments, the correct plaintext, would be sensible English. Thus, even if the calculations were done with a pencil and paper, an intruder could break this system in only a few minutes.

An intruder could also use basic facts about letter frequencies in the English language to break the system. In ordinary English, the letters that naturally occur with the highest frequencies are, in order, E, T, A, O, I, N, and S. Table 6.1 shows a standard tabulation of letter frequencies in ordinary English.

Letter	Frequency (%)	Letter	Frequency (%)
A	8.167	N	6.749
B	1.492	O	7.507
C	2.782	P	1.929
D	4.253	Q	0.095
E	12.702	R	5.987
F	2.228	S	6.327
G	2.015	T	9.056
H	6.094	U	2.758
I	6.966	V	0.978
J	0.153	W	2.360
K	0.772	X	0.150
L	4.025	Y	1.974
M	2.406	Z	0.074

Table 6.1   Letter frequencies in ordinary English.

Assuming that an intercepted English ciphertext has a sufficient number of letters, it is reasonable to suppose that the letter that occurs in the ciphertext with the highest frequency will have resulted from one of the letters that naturally occurs in ordinary English with the highest frequency. Once the correct letter is guessed, the key will be quickly determined. We illustrate this cryptanalysis technique for shift ciphers in the following example.

Example 6.2 Suppose that an intruder intercepts the ciphertext RKKRT-BRKURNE in Example 6.1, and knows that this ciphertext was formed using a shift cipher. Since the letter that occurs in this ciphertext with the highest frequency is R, the intruder could try to break this system by supposing that this ciphertext letter resulted from the plaintext letter E. This would yield an encryption function of the form $f(x) = (x+b) \bmod 26$ that satisfies $f(4) = 17$, or, equivalently, $(4+b) \bmod 26 = 17$, which implies $b = 13$. The resulting encryption function is $f(x) = (x+13) \bmod 26$, with corresponding decryption function $f^{-1}(x) = (x - 13) \bmod 26$. If the intruder used this decryption function to try to decrypt the ciphertext, the result would be as follows.

	R	K	K	R	T	B	R	K	U	R	N	E
$\alpha \Rightarrow$	17	10	10	17	19	1	17	10	20	17	13	4
$f^{-1} \Rightarrow$	4	23	23	4	6	14	4	23	7	4	0	17
$\alpha^{-1} \Rightarrow$	E	X	X	E	G	O	E	X	H	E	A	R

Since this decipherment is not sensible English, the intruder would know that the ciphertext letter R must have resulted from some plaintext letter different from E. If the intruder supposed that the ciphertext letter R resulted from the plaintext letter T, this would also yield a decipherment that is not sensible English. However, if the intruder supposed that R resulted from the plaintext letter A, this would yield the correct decipherment. □

6.1.2 Affine Ciphers

A shift cipher is a special case of a type of cryptosystem called an *affine cipher*. The only difference between the general shift and affine cipher encryption functions is that for affine ciphers we allow the x term to be multiplied by an element in R. Thus, we can represent affine ciphers mathematically as follows.

Affine Cipher Encryption: *Choose* $f : R \to R$ *as* $f(x) = (ax+b) \bmod |R|$ *for some* $a, b \in R$ *with* $\gcd(a, |R|) = 1$.

Example 6.3 Let $f(x) = (3x + 4) \bmod 26$. Then the message ATTACK AT DAWN enciphers as follows.

	A	T	T	A	C	K	A	T	D	A	W	N
$\alpha \Rightarrow$	0	19	19	0	2	10	0	19	3	0	22	13
$f \Rightarrow$	4	9	9	4	10	8	4	9	13	4	18	17
$\alpha^{-1} \Rightarrow$	E	J	J	E	K	I	E	J	N	E	S	R

Thus, the corresponding ciphertext is EJJEKIEJNESR. To decrypt this ciphertext, we can repeat this same procedure using the inverse function $f^{-1}(x) = a^{-1}(x - b) \bmod |R| = 9(x - 4) \bmod 26$ in place of f. Note that $3^{-1} = 9 \bmod 26$ because $3 \cdot 9 = 27 = 1 \bmod 26$. Note also that we are guaranteed a multiplicative inverse of $a = 3$ will exist modulo 26 because of the requirement that $\gcd(a, |R|) = 1$. □

The cryptosystem in Example 6.3 has two keys, $a = 3$ and $b = 4$. While this system is certainly more mathematically secure than the cryptosystem in Example 6.1, the system in Example 6.3 is still not very secure. We would assume that intruders know $f(x) = (ax+b) \bmod 26$ for some $a, b \in \mathbb{Z}_{26}$ with $\gcd(a, 26) = 1$. Thus, since a would have to be one of the twelve elements in $\{1, 3, 5, 7, 9, 11, 15, 17, 19, 21, 23, 25\}$, and b would have to be one of the 26 elements in $\mathbb{Z}_{26}$, there would only be a total of $12 \cdot 26 = 312$ possible pairs (a, b) of keys for the system. An intruder using only a handheld calculator could easily test each of these possible pairs of keys in a relatively short amount of time. More importantly, an intruder using a computer that can perform millions of operations per second could test each of these possible pairs of keys immediately. So although the cryptosystem in Example 6.3 does have more possible keys than the system in Example 6.1, the system in Example 6.3 can still be broken very easily.

As was the case with shift ciphers, affine ciphers can also be broken using letter frequency analysis. Frequency analysis on an affine cipher would require matching two ciphertext letters with their corresponding plaintext letters, and then solving a system of two linear equations over the ring R. We illustrate this cryptanalysis technique in the following example.

Example 6.4 Suppose that an intruder intercepts the ciphertext EJJEK-IEJNESR in Example 6.3, and knows that this ciphertext was formed using an affine cipher. Since the two letters that occur in this ciphertext with the highest frequencies are E and J, the intruder could try to break this system by supposing that these ciphertext letters resulted from the plaintext letters E and T, respectively, which are the two letters that naturally occur in ordinary English with the highest frequencies. This would yield an encryption function of the form $f(x) = (ax + b) \bmod 26$ that satisfies $f(4) = 4$ and $f(19) = 9$, or, equivalently, $(4a + b) \bmod 26 = 4$ and $(19a + b) \bmod 26 = 9$. To solve this system of linear equations, the intruder could begin by subtracting the first equation from the second, yielding $15a \bmod 26 = 5$. In addition, because $15 \cdot 7 = 105 = 1 \bmod 26$, then $15^{-1} = 7 \bmod 26$, and $a = 7 \cdot 5 \bmod 26 = 9$. Substituting this value of a into the original first equation gives $(36 + b) \bmod 26 = 4$, and then solving for b yields $b = -32 \bmod 26 = 20$. The resulting encryption function is $f(x) = (9x + 20) \bmod 26$, with corresponding decryption function $f^{-1}(x) = 3(x - 20) \bmod 26$. However, if the intruder used this decryption function to try to decrypt the ciphertext, the result would not be sensible English, and the intruder would know that the ciphertext letters E and J must have resulted from some plaintext letters different from E and T, respectively. Similarly, following in order the letters that naturally occur with the highest frequencies in ordinary English, if the intruder supposed that the ciphertext letters E and J resulted from the plaintext letters T and

E, E and A, A and E, or T and A, respectively, in each case either the resulting system of linear equations would not have a unique solution, or the resulting decipherment would not be sensible English. Either way, the intruder would know that the supposed plaintext correspondences for the ciphertext letters E and J were not correct. However, when the intruder supposed that E and J resulted from the plaintext letters A and T, respectively, this would yield the correct decipherment. □

6.2 Shift and Affine Ciphers with Maple

In this section, we will show how Maple can be used to encrypt and decrypt messages using shift and affine ciphers. We will also show how cryptanalysis can be done on ciphertexts formed using each of these types of cryptosystems.

Because some of the functions that we will use are in the Maple **String-Tools** package, we will begin by including this package.

```
>  with(StringTools):
```

Next, we will establish the alphabet correspondences between the elements in $L = \{A, B, C, \ldots, Z\}$ and $R = \mathbb{Z}_{26}$ for the bijection $\alpha : L \to R$. To do this, we will construct the following array *letters* containing the elements in L in order.

```
>  letters := array(0..25, ["A", "B", "C", "D", "E", "F",
      "G", "H", "I", "J", "K", "L", "M", "N", "O", "P", "Q",
      "R", "S", "T", "U", "V", "W", "X", "Y", "Z"]):
```

We can then access the elements in L from their positions in this array, starting with A in position 0. For example, because C is the third entry in this array, it is returned by the following command.

```
>  letters[2];
```

$$\text{``C''}$$

In addition, so that we will be able to retrieve each element in $\mathbb{Z}_{26}$ by entering its corresponding element in L, we will also establish the correspondences for α in the following table.

```
>  ltable := table():
>  for i from 0 to 25 do
>      ltable[letters[i]] := i:
>  od:
```

Using *ltable*, we can retrieve each element in $\mathbb{Z}_{26}$ by entering its corresponding element in *L*. For example, the element in $\mathbb{Z}_{26}$ that corresponds to C is returned by the following command.

```
>  ltable["C"];
```

$$2$$

6.2.1   Shift Ciphers

We will now show how Maple can be used to encrypt the plaintext message ATTACK AT DAWN using a shift cipher. First, we will store this message as the following string.

```
>  message := "ATTACK AT DAWN";
```
$$message := \text{``ATTACK AT DAWN''}$$

Since our alphabet *L* does not include a blank space as a character, we must remove the blank spaces from *message*. We can do this as follows by using the **Select** function with the **IsAlpha** option.

```
>  message := Select(IsAlpha, message);
```
$$message := \text{``ATTACKATDAWN''}$$

Using the **IsAlpha** option within the preceding **Select** command instructs Maple to select only the alphabetic characters in *message*, ignoring all non-alphabetic characters, including blank spaces.

Recall that with a shift cipher, each plaintext character is encrypted separately. To facilitate this, in the next command we convert our plaintext string into the list *ptext* of plaintext characters.

```
>  ptext := convert(message, list);
```
$$ptext := [\text{``A''}, \text{``T''}, \text{``T''}, \text{``A''}, \text{``C''}, \text{``K''}, \text{``A''}, \text{``T''}, \text{``D''}, \text{``A''}, \text{``W''},$$
$$\text{``N''}]$$

Next, we will use the Maple **map** function and *ltable* to apply α to each of these plaintext characters.

```
>  ptext := map(i -> ltable[i], ptext);
```
$$ptext := [0, 19, 19, 0, 2, 10, 0, 19, 3, 0, 22, 13]$$

We will now enter the following command to create the general shift cipher function $(x + b) \bmod 26$, where x represents the plaintext number that is to be encrypted, and b represents the key for the system.

```
>  f := (x, b) -> (x+b) mod 26:
```

Now, suppose that we would like to encrypt our message ATTACK AT DAWN using the shift cipher function $(x + 17) \bmod 26$. In the next command, we use **map** to apply this function to each of the plaintext numerical values for this message contained in *ptext*.

```
>  ctext := map(f, ptext, 17);
```
$$ctext := [17, 10, 10, 17, 19, 1, 17, 10, 20, 17, 13, 4]$$

Next, we apply the inverse of α to convert these ciphertext numerical values back into letters.

```
>  ctext := map(i -> letters[i], ctext);
```
$ctext :=$ ["R", "K", "K", "R", "T", "B", "R", "K", "U", "R", "N", "E"]

These ciphertext characters are stored in *ctext* as a list. Thus, we can extract these characters by entering the following Maple **op** command.

```
>  op(ctext);
```
"R", "K", "K", "R", "T", "B", "R", "K", "U", "R", "N", "E"

Also, we can adjoin these characters into a single ciphertext string as follows using the Maple **cat** function, which is designed to concatenate strings.

```
>  ctext := cat(op(ctext));
```
$$ctext := \text{"RKKRTBRKURNE"}$$

In order to decrypt this ciphertext, we will first convert *ctext* back into a list of numerical values by entering the following two commands.

```
>  ctext := convert(ctext, list):
```

```
>  ctext := map(i -> ltable[i], ctext);
```
$$ctext := [17, 10, 10, 17, 19, 1, 17, 10, 20, 17, 13, 4]$$

Next, to each of these ciphertext numerical values, we will apply the inverse $(x - 17) \bmod 26$ of the function that we used to encrypt the message.

```
>  ptext := map(f, ctext, -17);
```
$$ptext := [0, 19, 19, 0, 2, 10, 0, 19, 3, 0, 22, 13]$$

We can now recover the plaintext characters by entering the following two commands.

```
>  ptext := map(i -> letters[i], ptext):

>  ptext := cat(op(ptext));
```
$$ptext := \text{“ATTACKATDAWN”}$$

We will now show how Maple can be used to break a shift cipher. Suppose we intercept the ciphertext LYHXYTPIOMUNGCXHCABN sent by our enemy, and we know that this ciphertext was formed using a shift cipher. In order to determine the plaintext, we will begin as follows by converting these ciphertext characters into their corresponding numerical values.

```
>  ctext := "LYHXYTPIOMUNGCXHCABN":

>  ctext := convert(ctext, list):

>  ctext := map(i -> ltable[i], ctext);
```
$$ctext := [11, 24, 7, 23, 24, 19, 15, 8, 14, 12, 20, 13, 6, 2, 23, 7, 2, 0, 1,$$
$$13]$$

Since a shift cipher with $R = \mathbb{Z}_{26}$ will only have 26 possible keys, we can easily take each of these key candidates separately, form the corresponding decryption function, and then use this function to attempt to decrypt the ciphertext. We will do this by entering the following **for** loop.

```
>  for b from 0 to 25 do

>      ptext := map(f, ctext, -b);

>      ptext := map(i -> letters[i], ptext);

>      ptext := cat(op(ptext));

>      lprint(cat("b = ", convert(b, string), " ", ptext));

>  od:
```

```
"b = 0 LYHXYTPIOMUNGCXHCABN"
"b = 1 KXGWXSOHNLTMFBWGBZAM"
"b = 2 JWFVWRNGMKSLEAVFAYZL"
"b = 3 IVEUVQMFLJRKDZUEZXYK"
"b = 4 HUDTUPLEKIQJCYTDYWXJ"
"b = 5 GTCSTOKDJHPIBXSCXVWI"
"b = 6 FSBRSNJCIGOHAWRBWUVH"
"b = 7 ERAQRMIBHFNGZVQAVTUG"
"b = 8 DQZPQLHAGEMFYUPZUSTF"
"b = 9 CPYOPKGZFDLEXTOYTRSE"
"b = 10 BOXNOJFYECKDWSNXSQRD"
"b = 11 ANWMNIEXDBJCVRMWRPQC"
"b = 12 ZMVLMHDWCAIBUQLVQOPB"
"b = 13 YLUKLGCVBZHATPKUPNOA"
"b = 14 XKTJKFBUAYGZSOJTOMNZ"
"b = 15 WJSIJEATZXFYRNISNLMY"
"b = 16 VIRHIDZSYWEXQMHRMKLX"
```

```
"b = 17 UHQGHCYRXVDWPLGQLJKW"
"b = 18 TGPFGBXQWUCVOKFPKIJV"
"b = 19 SFOEFAWPVTBUNJEOJHIU"
"b = 20 RENDEZVOUSATMIDNIGHT"
"b = 21 QDMCDYUNTRZSLHCMHFGS"
"b = 22 PCLBCXTMSQYRKGBLGEFR"
"b = 23 OBKABWSLRPXQJFAKFDEQ"
"b = 24 NAJZAVRKQOWPIEZJECDP"
"b = 25 MZIYZUQJPNVOHDYIDBCO"
```

Since the only decipherment that is sensible English is for $b = 20$, this value is the key for the system, and the resulting plaintext is RENDEZVOUS AT MIDNIGHT.

6.2.2 Affine Ciphers

We will now show how Maple can be used to encrypt the plaintext message ATTACK AT DAWN using an affine cipher. First, we will store this message as the following string.

```
>   message := Select(IsAlpha, "ATTACK AT DAWN");
```
$$message := \text{"ATTACKATDAWN"}$$

Next, will will convert this plaintext string into a list of plaintext characters, and then apply α to convert each of the plaintext characters into their corresponding numerical values.

```
>   ptext := convert(message, list):
```

```
>   ptext := map(i -> ltable[i], ptext);
```
$$ptext := [0, 19, 19, 0, 2, 10, 0, 19, 3, 0, 22, 13]$$

We will now enter the following command to create the general affine cipher function $(ax + b) \bmod 26$, where x represents the plaintext number that is to be encrypted, and a and b represent the keys for the system.

```
>   f := (x, a, b) -> (a*x+b) mod 26:
```

Now, suppose that we would like to encrypt our message ATTACK AT DAWN using the affine cipher function $(3x + 4) \bmod 26$. First, in the following command we verify that the value $a = 3$ satisfies the requirement that $\gcd(a, 26) = 1$.

```
>   gcd(3, 26);
```

In the next command, we use **map** to apply the function $(3x+4) \bmod 26$ to each of the plaintext numerical values for this message contained in *ptext*.

```
> ctext := map(f, ptext, 3, 4);
```
$$ctext := [4, 9, 9, 4, 10, 8, 4, 9, 13, 4, 18, 17]$$

Next, by entering the following two commands, we convert these ciphertext numerical values back into letters, and adjoin these letters into a single ciphertext string.

```
> ctext := map(i -> letters[i], ctext):
```
```
> ctext := cat(op(ctext));
```
$$ctext := \text{“EJJEKIEJNESR”}$$

In order to decrypt this ciphertext, we will first convert *ctext* back into a list of numerical values by entering the following two commands.

```
> ctext := convert(ctext, list):
```
```
> ctext := map(i -> ltable[i], ctext);
```
$$ctext := [4, 9, 9, 4, 10, 8, 4, 9, 13, 4, 18, 17]$$

Next, we will define the following function *invf*, which is the inverse function $a^{-1}(x - b) \bmod 26$ of the general affine cipher function $(ax + b) \bmod 26$.

```
> invf := (x, a, b) -> a^(-1)*(x-b) mod 26:
```

Then to each of the ciphertext numerical values in *ctext*, we will apply the inverse $3^{-1}(x - 4) \bmod 26$ of the function that we used to encrypt the message.

```
> ptext := map(invf, ctext, 3, 4);
```
$$ptext := [0, 19, 19, 0, 2, 10, 0, 19, 3, 0, 22, 13]$$

We can now recover the plaintext characters by entering the following two commands.

```
> ptext := map(i -> letters[i], ptext):
```
```
> ptext := cat(op(ptext));
```
$$ptext := \text{“ATTACKATDAWN”}$$

We will now show how Maple can be used to break an affine cipher. Suppose we intercept the ciphertext JDDYTANYSNAYNEQTALICRYRLQNICH sent by our enemy, and we know that this ciphertext was formed using an affine cipher.

```
> ctext := "JDDYTANYSNAYNEQTALICRYRLQNICH":
```

To determine the plaintext, we will first find the frequency with which each of the letters in this ciphertext occurs. To do this, we will use the user-written function **printletterfreq**, which we have written separately from this Maple session and saved as the text file printletterfreq.mpl.

```
> read "printletterfreq.mpl";
```

The following command then illustrates how the function **printletterfreq** can be used. Specifically, by entering the following command, we can see the frequency with which each of the letters in *ctext* occurs.

```
> printletterfreq(ctext);
```

```
"Letter Y occurs 4 times"
"Letter N occurs 4 times"
"Letter A occurs 3 times"
"Letter T occurs 2 times"
"Letter R occurs 2 times"
"Letter Q occurs 2 times"
"Letter L occurs 2 times"
"Letter I occurs 2 times"
"Letter D occurs 2 times"
"Letter C occurs 2 times"
"Letter S occurs 1 times"
"Letter J occurs 1 times"
"Letter H occurs 1 times"
"Letter E occurs 1 times"
```

Next, we will convert the ciphertext characters in *ctext* into their corresponding numerical values.

```
> ctext := convert(ctext, list):

> ctext := map(i -> ltable[i], ctext);
```

$$ctext := [9, 3, 3, 24, 19, 0, 13, 24, 18, 13, 0, 24, 13, 4, 16, 19, 0, 11, 8,$$
$$2, 17, 24, 17, 11, 16, 13, 8, 2, 7]$$

Since the two letters that occur in the ciphertext with the highest frequencies are Y and N, we will try to break this system by supposing these ciphertext letters resulted from the plaintext letters E and T, respectively. This would yield an encryption function of the form $f(x) = (ax+b) \bmod 26$ that satisfies $f(4) = 24$ and $f(19) = 13$, or, equivalently, $(4a + b) \bmod 26 = 24$ and $(19a+b) \bmod 26 = 13$. In order to solve this system of linear equations, we will first enter the following commands to make sure that any previously assigned values of a and b are cleared.

```
>   a := 'a':
>   b := 'b':
```

We can then solve the system of linear equations $(4a + b) \bmod 26 = 24$ and $(19a + b) \bmod 26 = 13$ by entering the following command. Note that in this command, we use the Maple **solve** function to solve the system, and the Maple **map** function to reduce the result modulo 26.

```
>   map(m -> m mod 26, solve({4*a+b=24, 19*a+b=13}));
```
$$\{a = 1,\, b = 20\}$$

The resulting encryption function is $(1x + 20) \bmod 26$, with corresponding decryption function $1^{-1}(x - 20) \bmod 26$. Next, we will use this decryption function to try to decrypt the ciphertext.

```
>   ptext := map(invf, ctext, 1, 20):
>   map(i -> letters[i], ptext);
```

["P", "J", "J", "E", "Z", "G", "T", "E", "Y", "T", "G", "E", "T",
    "K", "W", "Z", "G", "R", "O", "I", "X", "E", "X", "R", "W",
    "T", "O", "I", "N"]

Since the resulting decipherment is not sensible English, then we know that the ciphertext letters Y and N must have resulted from some plaintext letters different from E and T, respectively. Next, we will suppose that Y and N resulted from the plaintext letters T and E, respectively. This would yield an encryption function of the form $(ax+b) \bmod 26$ that satisfies $(19a + b) \bmod 26 = 24$ and $(4a + b) \bmod 26 = 13$. To solve this system of linear equations, we enter the following command.

```
>   map(m -> m mod 26, solve({19*a+b=24, 4*a+b=13}));
```
$$\{a = 25,\, b = 17\}$$

Next, we will use the corresponding decryption function to try to decrypt the ciphertext.

```
>   ptext := map(invf, ctext, 25, 17):
>   map(i -> letters[i], ptext);
```

["I", "O", "O", "T", "Y", "R", "E", "T", "Z", "E", "R", "T", "E",
    "N", "B", "Y", "R", "G", "J", "P", "A", "T", "A", "G", "B",
    "E", "J", "P", "K"]

Again, the resulting decipherment is not sensible English. Next, we suppose that the ciphertext letters Y and N resulted from the plaintext letters E and A, respectively. This would yield an encryption function of the form $(ax+b) \bmod 26$ that satisfies $(4a+b) \bmod 26 = 24$ and $(0a+b) \bmod 26 = 13$. To solve this system of linear equations, we enter the following command.

```
>  map(m -> m mod 26, solve({4*a+b=24, 0*a+b=13}));

Error, (in unknown) the modular inverse does not exist
```

This error message is returned because when Maple eliminates the b variable from the equations by, for example, subtracting the second equation from the first, the coefficient of a in the resulting equation $4a = 11$ does not have a multiplicative inverse in $\mathbb{Z}_{26}$. Thus, the system of equations does not have a unique solution, and the ciphertext letters Y and N must have resulted from some plaintext letters different from E and A, respectively. Furthermore, the same error message would be returned for the same reason if we supposed that Y and N resulted from the plaintext letters A and E, respectively. So next, we will suppose that Y and N resulted from the plaintext letters T and A, respectively. This would yield an encryption function of the form $(ax + b) \bmod 26$ that satisfies $(19a + b) \bmod 26 = 24$ and $(0a + b) \bmod 26 = 13$. To solve this system of linear equations, we enter the following command.

```
>  map(m -> m mod 26, solve({19*a+b=24, 0*a+b=13}));
```

$$\{a = 17, b = 13\}$$

Next, we will use the corresponding decryption function to try to decrypt the ciphertext.

```
>  ptext := map(invf, ctext, 17, 13):

>  map(i -> letters[i], ptext);
```

["M", "E", "E", "T", "I", "N", "A", "T", "L", "A", "N", "T", "A", "B", "R", "I", "N", "G", "P", "H", "O", "T", "O", "G", "R", "A", "P", "H", "S"]

This time the resulting decipherment is sensible English, and the plaintext message follows.

```
>  cat(op(%));
```

"MEETINATLANTABRINGPHOTOGRAPHS"

6.3 Shift and Affine Ciphers with MATLAB

In this section, we will show how MATLAB can be used to encrypt and decrypt messages using shift and affine ciphers. We will also show how cryptanalysis can be done on ciphertexts formed using each of these types of cryptosystems.

We will begin by establishing the correspondences between the elements in $L = \{A, B, C, \ldots, Z\}$ and $R = \mathbb{Z}_{26}$ for the bijection $\alpha : L \to R$. To do this, we will construct the following string *abet* containing the elements in L in order.

```
>> abet = 'ABCDEFGHIJKLMNOPQRSTUVWXYZ';
```

Strings in MATLAB are treated as arrays with their elements stored sequentially. The elements can then be accessed from their array positions, starting with the first element in position 1. For example, because C is the third entry in the array *abet* that we just defined, it is returned by the following command.

```
>> abet(3)

ans =

C
```

In the bijection $\alpha : L \to R$, recall that C maps to 2, not 3. More generally, for $\alpha : L \to R$, we want to be able to access the elements in L using position numbers 0 through 25, respectively, instead of 1 through 26. However, the starting index for an array in MATLAB cannot be smaller than 1. We will use the following function *letters* to take care of this problem.

```
>> letters = @(x) abet(x+1);
```

This function *letters* takes a single input parameter x, representing an element in $\mathbb{Z}_{26}$, adds 1 to x, and then locates the element in *abet* that corresponds to the result. For example, we can now retrieve the letter C by entering its corresponding element in $\mathbb{Z}_{26}$ as follows.

```
>> letters(2)

ans =

C
```

In addition, we will need to be able to retrieve each element in $\mathbb{Z}_{26}$ by entering its corresponding element in L. To do this, we will use the user-written function **ltable**, which we have written separately from this MATLAB session and saved as the M-file ltable.m. Using **ltable**, we can retrieve each element in $\mathbb{Z}_{26}$ by entering its corresponding element in L. For example, the element in $\mathbb{Z}_{26}$ that corresponds to C is returned by the following command.

```
>> ltable('C')

ans =

    2
```

6.3.1 Shift Ciphers

We will now show how MATLAB can be used to encrypt the plaintext message ATTACK AT DAWN using a shift cipher. First, we will store this message as the following string.

```
>> message = 'ATTACK AT DAWN'

message =

ATTACK AT DAWN
```

Since our alphabet L does not include a blank space as a character, we must remove the blank spaces from *message*. We can do this as follows by using the MATLAB **findstr** function.

```
>> message(findstr(message, ' ')) = []

message =

ATTACKATDAWN
```

Using **findstr** as in the preceding command instructs MATLAB to find the indices in which the array *message* contains a blank space, as indicated by the blank space enclosed by single quotes in the command. After these indices are located, the corresponding positions are deleted by being assigned an empty array command, as indicated by the empty brackets at the end of the command.

In the next command, we use **ltable** to apply α to each of the plaintext characters in our message, leaving the results in a list named *ptext*.

```
>> ptext = ltable(message)

ptext =
```

0	19	19	0	2	10	0	19	3	0	22	13

We will now enter the following command to create the general shift cipher function $(x + b) \bmod 26$, where x represents the plaintext number that is to be encrypted, and b represents the key for the system.

```
>> f = @(x, b) mod(x+b, 26);
```

Now, suppose that we would like to encrypt our message ATTACK AT DAWN using the shift cipher function $(x + 17) \bmod 26$. In the next command, we apply this function to each of the plaintext numerical values for this message contained in *ptext*.

```
>> ctext = f(ptext, 17)

ctext =
```

17	10	10	17	19	1	17	10	20	17	13	4

Next, we apply the inverse of α to convert these ciphertext numerical values back into letters.

```
>> ctext = letters(ctext)

ctext =

RKKRTBRKURNE
```

In order to decrypt this ciphertext, we will first convert *ctext* back into a list of numerical values by entering the following command.

```
>> ctext = ltable(ctext)

ctext =
```

17	10	10	17	19	1	17	10	20	17	13	4

Next, to each of these ciphertext numerical values, we will apply the inverse $(x - 17) \bmod 26$ of the function that we used to encrypt the message.

```
>> ptext = f(ctext, -17)

ptext =

    0   19   19    0    2   10    0   19    3    0   22   13
```

We can now recover the plaintext characters by entering the following.

```
>> ptext = letters(ptext)

ptext =

ATTACKATDAWN
```

We will now show how MATLAB can be used to break a shift cipher. Suppose we intercept the ciphertext **LYHXYTPIOMUNGCXHCABN** sent by our enemy, and we know that this ciphertext was formed using a shift cipher. In order to determine the plaintext, we will begin as follows by converting these ciphertext characters into their corresponding numerical values.

```
>> ctext = 'LYHXYTPIOMUNGCXHCABN';
>> ctext = ltable(ctext)

ctext =

  Columns 1 through 15

    11   24    7   23   24   19   15    8   14   12   20   13    6    2   23

  Columns 16 through 20

     7    2    0    1   13
```

Since a shift cipher with $R = \mathbb{Z}_{26}$ will only have 26 possible keys, we can easily take each of these key candidates separately, form the corresponding decryption function, and then use this function to attempt to decrypt the ciphertext. We will do this by entering the following **for** loop.

```
>> for b = 0:25
       ptext = f(ctext, -b);
       ptext = letters(ptext);
       fprintf('%s %2.0f %s \n', 'b = ', b, ptext)
   end
b =    0 LYHXYTPIOMUNGCXHCABN
```

```
b =    1  KXGWXSOHNLTMFBWGBZAM
b =    2  JWFVWRNGMKSLEAVFAYZL
b =    3  IVEUVQMFLJRKDZUEZXYK
b =    4  HUDTUPLEKIQJCYTDYWXJ
b =    5  GTCSTOKDJHPIBXSCXVWI
b =    6  FSBRSNJCIGOHAWRBWUVH
b =    7  ERAQRMIBHFNGZVQAVTUG
b =    8  DQZPQLHAGEMFYUPZUSTF
b =    9  CPYOPKGZFDLEXTOYTRSE
b =   10  BOXNOJFYECKDWSNXSQRD
b =   11  ANWMNIEXDBJCVRMWRPQC
b =   12  ZMVLMHDWCAIBUQLVQOPB
b =   13  YLUKLGCVBZHATPKUPNOA
b =   14  XKTJKFBUAYGZSOJTOMNZ
b =   15  WJSIJEATZXFYRNISNLMY
b =   16  VIRHIDZSYWEXQMHRMKLX
b =   17  UHQGHCYRXVDWPLGQLJKW
b =   18  TGPFGBXQWUCVOKFPKIJV
b =   19  SFOEFAWPVTBUNJEOJHIU
b =   20  RENDEZVOUSATMIDNIGHT
b =   21  QDMCDYUNTRZSLHCMHFGS
b =   22  PCLBCXTMSQYRKGBLGEFR
b =   23  OBKABWSLRPXQJFAKFDEQ
b =   24  NAJZAVRKQOWPIEZJECDP
b =   25  MZIYZUQJPNVOHDYIDBCO
```

Since the only decipherment that is sensible English is for $b = 20$, this value is the key for the system, and the resulting plaintext is RENDEZVOUS AT MIDNIGHT.

6.3.2   Affine Ciphers

We will now show how MATLAB can be used to encrypt the plaintext message ATTACK AT DAWN using an affine cipher. First, we will store this message as the following string.

```
>> message = 'ATTACK AT DAWN';
>> message(findstr(message, ' ')) = []

message =

ATTACKATDAWN
```

Next, will will use **ltable** to apply α to each of these plaintext characters, leaving the results in a list named *ptext*.

```
>> ptext = ltable(message)

ptext =
```

0	19	19	0	2	10	0	19	3	0	22	13

We will now enter the following command to create the general affine cipher function $(ax + b) \bmod 26$, where x represents the plaintext number that is to be encrypted, and a and b represent the keys for the system.

```
>> f = @(x, a, b) mod(a*x+b, 26);
```

Now, suppose that we would like to encrypt our message ATTACK AT DAWN using the affine cipher function $(3x + 4) \bmod 26$. First, in the following command we verify that the value $a = 3$ satisfies the requirement that $\gcd(a, 26) = 1$.

```
>> gcd(3, 26)

ans =

     1
```

Then in the next command, we apply the function $(3x + 4) \bmod 26$ to each of the plaintext numerical values for this message contained in *ptext*.

```
>> ctext = f(ptext, 3, 4)

ctext =
```

4	9	9	4	10	8	4	9	13	4	18	17

Next, we apply the inverse of α to convert these ciphertext numerical values back into letters.

```
>> ctext = letters(ctext)

ctext =

EJJEKIEJNESR
```

In order to decrypt this ciphertext, we will first convert *ctext* back into a list of numerical values by entering the following command.

```
>> ctext = ltable(ctext)
```

```
ctext =
```

    4   9   9   4   10   8   4   9   13   4   18   17

Next, we will define the following function *invf*, which is the inverse function $a^{-1}(x - b)$ mod 26 of the general affine cipher function $(ax + b)$ mod 26.

```
>> invf = @(x, a, b) double(mod(sym(a^(-1))*(x-b), 26));
```

In the preceding command, the **sym** function instructs MATLAB to convert the inverse of a into symbolic form. Without this command, MATLAB would use a decimal approximation for the inverse of a. The **double** function in this command then converts the final result from symbolic form back into numeric form.

In the next command, to each of the ciphertext numerical values contained in *ctext*, we apply the inverse $3^{-1}(x - 4)$ mod 26 of the function that we used to encrypt the message.

```
>> ptext = invf(ctext, 3, 4)
```

```
ptext =
```

    0   19   19   0   2   10   0   19   3   0   22   13

We can now recover the plaintext characters by entering the following command.

```
>> ptext = letters(ptext)
```

```
ptext =
```

```
ATTACKATDAWN
```

We now show how MATLAB can be used to break an affine cipher. Suppose we intercept the ciphertext JDDYTANYSNAYNEQTALICRYRLQNICH sent by our enemy, and we know that this ciphertext was formed using an affine cipher.

```
>> ctext = 'JDDYTANYSNAYNEQTALICRYRLQNICH';
```

To determine the plaintext, we will first find the frequency with which each of the letters in this ciphertext occurs. To do this, we will use the user-written function **printletterfreq**, which we have written separately from this MATLAB session and saved as the M-file printletterfreq.m. The following command illustrates how the function **printletterfreq** can be used. Specifically, by entering the following command, we can see the frequency with which each of the letters in *ctext* occurs.

```
>> printletterfreq(ctext)
```

```
Letter Y occurs  4 times
Letter N occurs  4 times
Letter A occurs  3 times
Letter T occurs  2 times
Letter R occurs  2 times
Letter Q occurs  2 times
Letter L occurs  2 times
Letter I occurs  2 times
Letter D occurs  2 times
Letter C occurs  2 times
Letter S occurs  1 times
Letter J occurs  1 times
Letter H occurs  1 times
Letter E occurs  1 times
```

Next, we will convert the ciphertext characters in *ctext* into their corresponding numerical values.

```
>> ctext = ltable(ctext)

ctext =

  Columns 1 through 15

    9   3   3  24  19   0  13  24  18  13   0  24  13   4  16

  Columns 16 through 29

   19   0  11   8   2  17  24  17  11  16  13   8   2   7
```

Since the two letters that occur in the ciphertext with the highest frequencies are Y and N, we will try to break this system by supposing these ciphertext letters resulted from the plaintext letters E and T, respectively. This would yield an encryption function of the form $f(x) = (ax+b) \bmod 26$ that satisfies $f(4) = 24$ and $f(19) = 13$, or, equivalently, $(4a + b) \bmod 26 = 24$ and $(19a+b) \bmod 26 = 13$. We can solve this system of linear equations by entering the following command. Note that in this command, we use the

solve function to solve the system, with the equations enclosed in single quotes, and the last two parameters containing the solution variables.

```
>> S = solve('4*a+b=24', '19*a+b=13', 'a', 'b');
```

The result of the preceding command is stored as the variable S, with the solution for a stored as the variable $S.a$ and the solution for b stored as the variable $S.b$. In the next command we use the MATLAB **mod** function to reduce the solutions modulo 26 and display the results.

```
>> mod([S.a, S.b], 26)

ans =

[  1, 20]
```

The resulting encryption function is $(1x + 20)$ mod 26, with corresponding decryption function $1^{-1}(x - 20)$ mod 26. Next, we will use this decryption function to try to decrypt the ciphertext.

```
>> ptext = invf(ctext, 1, 20);
>> letters(ptext)

ans =

PJJEZGTEYTGETKWZGROIXEXRWTOIN
```

Since the resulting decipherment is not sensible English, then we know that the ciphertext letters Y and N must have resulted from some plaintext letters different from E and T, respectively. Next, we will suppose that Y and N resulted from the plaintext letters T and E, respectively. This would yield an encryption function of the form $(ax+b)$ mod 26 that satisfies $(19a + b)$ mod 26 = 24 and $(4a + b)$ mod 26 = 13. To solve this system of linear equations, we enter the following two commands.

```
>> S = solve('19*a+b=24', '4*a+b=13', 'a', 'b');
>> mod([S.a, S.b], 26)

ans =

[ 25, 17]
```

Next, we will use the corresponding decryption function to try to decrypt the ciphertext.

```
>> ptext = invf(ctext, 25, 17);
>> letters(ptext)
```

ans =

IOOTYRETZERTENBYRGJPATAGBEJPK

Again, the resulting decipherment is not sensible English. Next, we suppose that the ciphertext letters Y and N resulted from the plaintext letters E and A, respectively. This would yield an encryption function $(ax + b) \bmod 26$ that satisfies $(4a + b) \bmod 26 = 24$ and $(0a + b) \bmod 26 = 13$. To solve this system of linear equations, we enter the following two commands.

```
>> S = solve('4*a+b=24', '0*a+b=13', 'a', 'b');
>> mod([S.a, S.b], 26)
```

```
Error using mupadmex
Error in MuPAD command: The modular inverse does not exist.
[modp]
   Evaluating: symobj::modp
Error in sym/privBinaryOp (line 839)
            Csym = mupadmex(op,args{1}.s, args{2}.s,
            varargin{:});
Error in sym/mod (line 21)
C = privBinaryOp(A, B, 'symobj::zip', 'symobj::modp');
```

This error message is returned because when MATLAB eliminates the b variable from the equations by, for example, subtracting the second equation from the first, the coefficient of a in the resulting equation $4a = 11$ does not have a multiplicative inverse in $\mathbb{Z}_{26}$. Thus, the system of equations does not have a unique solution, and the ciphertext letters Y and N must have resulted from some plaintext letters different from E and A, respectively. Furthermore, the same error message would be returned for the same reason if we supposed that Y and N resulted from the plaintext letters A and E, respectively. So next, we will suppose that Y and N resulted from the plaintext letters T and A, respectively. This would yield an encryption function of the form $(ax + b) \bmod 26$ that satisfies $(19a + b) \bmod 26 = 24$ and $(0a + b) \bmod 26 = 13$. To solve this system of linear equations, we enter the following two commands.

```
>> S = solve('19*a+b=24', '0*a+b=13', 'a', 'b');
>> mod([S.a, S.b], 26)
```

ans =

```
[ 17, 13]
```

Next, we will use the corresponding decryption function to try to decrypt the ciphertext.

```
>> ptext = invf(ctext, 17, 13);
>> letters(ptext)

ans =

MEETINATLANTABRINGPHOTOGRAPHS
```

This time the resulting decipherment is the correct plaintext.

6.4 Hill Ciphers

We have shown that neither shift nor affine ciphers are secure cryptosystems by presenting rather simple procedures for breaking them. We can also see that these systems are not secure because they both yield substitution ciphers. Shift and affine ciphers are actually even easier to break than non-mathematical substitution ciphers, for in each case our general procedure for breaking the system can easily be programmed on a computer, while breaking non-mathematical substitution ciphers can require frequency analysis that could be even more time-consuming due to the lack of mathematical structure. However, despite the fact that the advent of calculators and computers have rendered shift and affine ciphers obsolete, they are certainly still worth studying, and not just for their mathematical beauty. As it turns out, shift and affine ciphers can easily be generalized into cryptosystems that use matrices as keys instead of scalars, and which can be constructed with any desired level of mathematical security. For example, the generalization of affine ciphers with $b = 0$ and for which a is a matrix was first published by Lester Hill in 1929.

Hill Cipher Encryption: *Let A be an $n \times n$ invertible matrix over R (i.e., with $gcd(det(A), |R|) = 1$). Group the plaintext into row vectors $\mathbf{p}_i$ in R^n, and choose $f : R^n \to R^n$ as $f(\mathbf{p}_i) = \mathbf{p}_i A$ with each entry reduced modulo $|R|$. The resulting rows listed together form the ciphertext.*

Example 6.5 In this example, we will use a Hill cipher to encrypt the message MEET AT SEVEN. We will begin by converting this message into a list of elements in $\mathbb{Z}_{26}$.

$$
\begin{array}{cccccccccc}
\text{M} & \text{E} & \text{E} & \text{T} & \text{A} & \text{T} & \text{S} & \text{E} & \text{V} & \text{E} & \text{N} \\
\alpha \Rightarrow 12 & 4 & 4 & 19 & 0 & 19 & 18 & 4 & 21 & 4 & 13
\end{array}
$$

We will use the following 2×2 key matrix A to encrypt this message.

$$
A = \begin{bmatrix} 2 & 5 \\ 1 & 4 \end{bmatrix}
$$

Note that A is invertible over $\mathbb{Z}_{26}$ since $\gcd(\det(A), 26) = \gcd(3, 26) = 1$. To form the ciphertext, we will group the plaintext into row vectors $\mathbf{p}_i$ in $\mathbb{Z}_{26}^2$, and compute $\mathbf{p}_i A$ for all i with each entry reduced modulo 26. For example, the first ciphertext vector is determined as follows.

$$
\begin{aligned}
\mathbf{p}_1 A &= \begin{bmatrix} 12 & 4 \end{bmatrix} \begin{bmatrix} 2 & 5 \\ 1 & 4 \end{bmatrix} \\
&= \begin{bmatrix} 28 & 76 \end{bmatrix} \\
&= \begin{bmatrix} 2 & 24 \end{bmatrix}
\end{aligned}
$$

The remaining ciphertext vectors are determined similarly, and listed next. Note that since the original message does not completely fill the last plaintext vector $\mathbf{p}_6$, we fill this vector with an arbitrary element in $\mathbb{Z}_{26}$.

$$
\begin{aligned}
\mathbf{p}_2 A &= \begin{bmatrix} 4 & 19 \end{bmatrix} A &= \begin{bmatrix} 1 & 18 \end{bmatrix} \\
\mathbf{p}_3 A &= \begin{bmatrix} 0 & 19 \end{bmatrix} A &= \begin{bmatrix} 19 & 24 \end{bmatrix} \\
\mathbf{p}_4 A &= \begin{bmatrix} 18 & 4 \end{bmatrix} A &= \begin{bmatrix} 14 & 2 \end{bmatrix} \\
\mathbf{p}_5 A &= \begin{bmatrix} 21 & 4 \end{bmatrix} A &= \begin{bmatrix} 20 & 17 \end{bmatrix} \\
\mathbf{p}_6 A &= \begin{bmatrix} 13 & 25 \end{bmatrix} A &= \begin{bmatrix} 25 & 9 \end{bmatrix}
\end{aligned}
$$

The entire encipherment follows.

$$
\begin{array}{ccccccccccccc}
& \text{M} & \text{E} & \text{E} & \text{T} & \text{A} & \text{T} & \text{S} & \text{E} & \text{V} & \text{E} & \text{N} & \\
\alpha \Rightarrow & 12 & 4 & 4 & 19 & 0 & 19 & 18 & 4 & 21 & 4 & 13 & 25 \\
f \Rightarrow & 2 & 24 & 1 & 18 & 19 & 24 & 14 & 2 & 20 & 17 & 25 & 9 \\
\alpha^{-1} \Rightarrow & \text{C} & \text{Y} & \text{B} & \text{S} & \text{T} & \text{Y} & \text{O} & \text{C} & \text{U} & \text{R} & \text{Z} & \text{J}
\end{array}
$$

Thus, the ciphertext is CYBSTYOCURZJ. Although the last ciphertext character is in a position beyond the last plaintext character, it must be retained, as it would be necessary for decryption. $\square$

One thing that we can notice immediately from Example 6.5 is that Hill ciphers do not generally yield substitution ciphers. Also, encrypting messages using a Hill cipher requires nothing more than a little matrix multiplication with an invertible key matrix. A matrix A over R will be invertible if and only if the determinant of A has a multiplicative inverse in R. For $R = \mathbb{Z}_k$, as we have noted, this is equivalent to $\gcd(\det(A), k) = 1$.

To decrypt a ciphertext that has been formed using a Hill cipher with an $n \times n$ key matrix A, we would group the ciphertext into row vectors $\mathbf{c}_i$ in R^n, and then compute $f^{-1}(\mathbf{c}_i) = \mathbf{c}_i A^{-1}$ with each entry reduced modulo $|R|$. The inverse matrix A^{-1} can be determined over R using the following formula, where $\mathrm{adj}(A)$ represents the adjoint of A.

$$A^{-1} = \frac{1}{\det(A)} \cdot \mathrm{adj}(A) \tag{6.1}$$

To determine $\mathrm{adj}(A)$, we would first need to find the cofactors of A. These cofactors are defined as $C_{ij} = (-1)^{i+j} M_{ij}$ for $i, j = 1, 2 \ldots, n$, where M_{ij} is the determinant of the matrix obtained by deleting the ith row and jth column from A. Using these cofactors, the adjoint of A is defined as follows.

$$\mathrm{adj}(A) = \begin{bmatrix} C_{11} & C_{21} & \cdots & C_{n1} \\ C_{12} & C_{22} & \cdots & C_{n2} \\ \vdots & \vdots & & \vdots \\ C_{1n} & C_{2n} & \cdots & C_{nn} \end{bmatrix}$$

That is, $\mathrm{adj}(A)$ is defined as the transpose of the matrix of cofactors of A.

Example 6.6 To decrypt the ciphertext in Example 6.5, we would first need to find the inverse of the key matrix A. Using (6.1), we can determine this inverse as follows.

$$\begin{aligned} A^{-1} &= \frac{1}{3} \begin{bmatrix} 4 & -5 \\ -1 & 2 \end{bmatrix} \\ &= 9 \begin{bmatrix} 4 & 21 \\ 25 & 2 \end{bmatrix} \\ &= \begin{bmatrix} 36 & 189 \\ 225 & 18 \end{bmatrix} \\ &= \begin{bmatrix} 10 & 7 \\ 17 & 18 \end{bmatrix} \end{aligned}$$

We would then group the ciphertext into row vectors $\mathbf{c}_i$ in $\mathbb{Z}_{26}^2$, and compute $\mathbf{c}_i A^{-1}$ for all i with each entry reduced modulo 26. For example, the first plaintext vector is determined as follows.

$$\begin{aligned} \mathbf{c}_1 A^{-1} &= \begin{bmatrix} 2 & 24 \end{bmatrix} \begin{bmatrix} 10 & 7 \\ 17 & 18 \end{bmatrix} \\ &= \begin{bmatrix} 428 & 446 \end{bmatrix} \\ &= \begin{bmatrix} 12 & 4 \end{bmatrix} \end{aligned}$$

The remaining plaintext vectors are determined similarly, and listed next.

$$\mathbf{c}_2 A^{-1} = \begin{bmatrix} 1 & 18 \end{bmatrix} A^{-1} = \begin{bmatrix} 4 & 19 \end{bmatrix}$$
$$\mathbf{c}_3 A^{-1} = \begin{bmatrix} 19 & 24 \end{bmatrix} A^{-1} = \begin{bmatrix} 0 & 19 \end{bmatrix}$$
$$\mathbf{c}_4 A^{-1} = \begin{bmatrix} 14 & 2 \end{bmatrix} A^{-1} = \begin{bmatrix} 18 & 4 \end{bmatrix}$$
$$\mathbf{c}_5 A^{-1} = \begin{bmatrix} 20 & 17 \end{bmatrix} A^{-1} = \begin{bmatrix} 21 & 4 \end{bmatrix}$$
$$\mathbf{c}_6 A^{-1} = \begin{bmatrix} 25 & 9 \end{bmatrix} A^{-1} = \begin{bmatrix} 13 & 25 \end{bmatrix}$$

Applying α^{-1} to the entries in these plaintext vectors will reveal the original message. Note that because we chose an arbitrary element from $\mathbb{Z}_{26}$ to fill the last plaintext vector $\mathbf{p}_6$ in Example 6.5, there will be an extra character at the end of the message, which can be discarded. □

We will now consider how an intruder could break the Hill cipher in Example 6.5. We would assume that an intruder who intercepts the ciphertext in Example 6.6 would know that this ciphertext was formed by grouping the plaintext into row vectors $\mathbf{p}_i$ in $\mathbb{Z}_{26}^2$, and then multiplying each $\mathbf{p}_i$ by some 2×2 invertible key matrix A over $\mathbb{Z}_{26}$. Although the requirement that A must be invertible does not impose any specific restrictions on the entries in A (individually), an intruder would at least know that each entry in A would have to be an element in $\mathbb{Z}_{26}$. Thus, to find A by trial and error, an intruder would have to test a maximum of only $26^4 = 456{,}976$ matrices, many of which would fail to be invertible and not need to be tested any further. While it would not be realistic for an intruder to test each of these matrices by hand or even with a calculator, an intruder using a computer that can perform millions of operations per second could test each of these matrices very quickly and easily. Thus, a Hill cipher with a 2×2 key matrix A is also not a very secure cryptosystem. However, if A were chosen of size 3×3, then there would be a maximum of $26^9 = 5{,}429{,}503{,}678{,}976$ matrices for an intruder to test, and if A were chosen of size 5×5, then there would be a maximum of $26^{25} = 2.37 \times 10^{35}$ matrices for an intruder to test. So even for relatively small key matrices, Hill ciphers can be reasonably secure. More importantly, a Hill cipher can be constructed with any desired level of mathematical security by simply using a key matrix that is sufficiently large.

Hill ciphers do have one vulnerability that we should mention. It is not unreasonable to suppose that an intruder who intercepts a ciphertext formed using a Hill cipher might know or be able to correctly guess a small part of the plaintext. For example, the intruder may know where or from whom the message originated, and correctly guess that the first several characters in the plaintext are a time or location stamp, or that the last few characters in the plaintext are the originator's name. As it turns out, it may be possible for an intruder to break a Hill cipher relatively easily if the intruder knows or is somehow able to correctly guess a small part of the plaintext. More specifically, if a Hill cipher is used to encrypt a message with an $n \times n$ key matrix, it may be possible for an intruder to break the system relatively easily if the intruder knows or is somehow able to correctly guess n^2 characters from the plaintext. We illustrate this in the following example.

Example 6.7 Suppose that an intruder intercepts the ciphertext CYBST-YOCURZJ in Example 6.5, and knows that this ciphertext was formed using a Hill cipher with some 2×2 key matrix A over $\mathbb{Z}_{26}$, which we will denote as follows.

$$A = \begin{bmatrix} a & b \\ c & d \end{bmatrix}$$

Suppose also the intruder somehow knows that the last four ciphertext letters resulted from the plaintext letters VENZ. That is, suppose the intruder somehow knows the following from the encipherment in Example 6.5.

									V	E	N	Z
$\alpha \Rightarrow$									21	4	13	25
$f \Rightarrow$	2	24	1	18	19	24	14	2	20	17	25	9
$\alpha^{-1} \Rightarrow$	C	Y	B	S	T	Y	O	C	U	R	Z	J

The intruder would then know that the following matrix equations are true, with $a, b, c, d \in \mathbb{Z}_{26}$.

$$\begin{bmatrix} 21 & 4 \end{bmatrix} \begin{bmatrix} a & b \\ c & d \end{bmatrix} = \begin{bmatrix} 20 & 17 \end{bmatrix}$$

$$\begin{bmatrix} 13 & 25 \end{bmatrix} \begin{bmatrix} a & b \\ c & d \end{bmatrix} = \begin{bmatrix} 25 & 9 \end{bmatrix}$$

The preceding two matrix equations are equivalent to the following single matrix equation.

$$\begin{bmatrix} 21 & 4 \\ 13 & 25 \end{bmatrix} \begin{bmatrix} a & b \\ c & d \end{bmatrix} = \begin{bmatrix} 20 & 17 \\ 25 & 9 \end{bmatrix} \tag{6.2}$$

If this equation had unique solutions for a, b, c, and d in $\mathbb{Z}_{26}$, then these values would necessarily be the entries in the key matrix A for the system. Note that the following quantity has a multiplicative inverse in $\mathbb{Z}_{26}$.

$$\begin{vmatrix} 21 & 4 \\ 13 & 25 \end{vmatrix} = 473 = 5$$

Thus, the intruder would know that (6.2) does indeed have unique solutions for a, b, c, and d in $\mathbb{Z}_{26}$. Using (6.1), the intruder could find these values as follows.

$$\begin{bmatrix} a & b \\ c & d \end{bmatrix} = \begin{bmatrix} 21 & 4 \\ 13 & 25 \end{bmatrix}^{-1} \begin{bmatrix} 20 & 17 \\ 25 & 9 \end{bmatrix}$$

$$= \frac{1}{5} \begin{bmatrix} 25 & -4 \\ -13 & 21 \end{bmatrix} \begin{bmatrix} 20 & 17 \\ 25 & 9 \end{bmatrix}$$

$$= 21 \begin{bmatrix} 25 & 22 \\ 13 & 21 \end{bmatrix} \begin{bmatrix} 20 & 17 \\ 25 & 9 \end{bmatrix}$$

$$= \begin{bmatrix} 22050 & 13083 \\ 16485 & 8610 \end{bmatrix}$$

$$= \begin{bmatrix} 2 & 5 \\ 1 & 4 \end{bmatrix}$$

Note that $5^{-1} = 21 \bmod 26$ because $5 \cdot 21 = 105 = 1 \bmod 26$. Note also that this yields the key matrix A in Example 6.5. The intruder could then find A^{-1}, and use A^{-1} to decrypt the remainder of the ciphertext. □

For a message encrypted using a Hill cipher with an $n \times n$ key matrix, one obvious problem that an intruder could encounter when trying to break the system in the way illustrated in Example 6.7 is that even if the intruder knew n^2 characters from the plaintext, the analogue to (6.2) might not have unique solutions for the entries in the unknown key matrix. Also, even if (6.2) did have unique solutions for the entries in the unknown key matrix, it might not be possible to find these values in the way illustrated in Example 6.7. For example, if an intruder intercepted the ciphertext CYBSTYOCURZJ in Example 6.5 and knew that the first four ciphertext letters resulted from the plaintext letters MEET, the analogue to (6.2) would be the following.

$$\begin{bmatrix} 12 & 4 \\ 4 & 19 \end{bmatrix} \begin{bmatrix} a & b \\ c & d \end{bmatrix} = \begin{bmatrix} 2 & 24 \\ 1 & 18 \end{bmatrix} \tag{6.3}$$

Note that the following quantity does not have a multiplicative inverse in $\mathbb{Z}_{26}$.

$$\begin{vmatrix} 12 & 4 \\ 4 & 19 \end{vmatrix} = 212 = 4$$

Thus, even if (6.3) had unique solutions for a, b, c, and d in $\mathbb{Z}_{26}$, it would not be possible to find these values in the way illustrated in Example 6.7.

6.5 Hill Ciphers with Maple

In this section, we will show how Maple can be used to encrypt and decrypt messages using Hill ciphers. We will also show an example of successful cryptanalysis on a ciphertext formed using a Hill cipher given a number of plaintext characters equal to the number of entries in the key matrix.

Because some of the functions that we will use are in the Maple **String-Tools**, **LinearAlgebra**, and **Modular** packages, we will begin by including these packages. In addition, we will enter the following **interface** command to cause Maple to display all matrices of size 200×200 and smaller throughout the remainder of this Maple session.

```
>  with(StringTools):
```

```
>  with(LinearAlgebra):
```

```
>  with(Modular):
```

```
>  interface(rtablesize=200):
```

Next, we will establish the alphabet correspondences between the elements in $L = \{A, B, C, \ldots, Z\}$ and $R = \mathbb{Z}_{26}$ for the bijection $\alpha : L \to R$. To do this, we will construct the following array *letters* containing the elements in L in order.

```
>  letters := array(0..25, ["A", "B", "C", "D", "E", "F",

       "G", "H", "I", "J", "K", "L", "M", "N", "O", "P", "Q",

       "R", "S", "T", "U", "V", "W", "X", "Y", "Z"]):
```

In addition, so that we will be able to retrieve each element in $\mathbb{Z}_{26}$ by entering its corresponding element in L, we will also establish the correspondences for α in the following table.

```
>  ltable := table():
```

```
>  for i from 0 to 25 do
```

```
>        ltable[letters[i]] := i:
```

```
>  od:
```

Next, we will define the following 3×3 matrix A over $\mathbb{Z}_{26}$, which we will use as the key matrix for the system.

```
>  A := Matrix([[11, 6, 8], [0, 3, 14], [24, 0, 9]]);
```

$$A := \begin{bmatrix} 11 & 6 & 8 \\ 0 & 3 & 14 \\ 24 & 0 & 9 \end{bmatrix}$$

In order for A to be a valid key matrix for a Hill cipher, it must be invertible over R. We will verify this by entering the following two commands, in which we find the determinant of A in $\mathbb{Z}_{26}$, and check that this determinant satisfies the requirement that $\gcd(\det(A), 26) = 1$.

```
> Determinant(26, A);
```
$$21$$

```
> gcd(21, 26);
```
$$1$$

We will now show how Maple can be used to encrypt the message GO NAVY BEAT ARMY using a Hill cipher. First, we will store this message as the following string. Note that because our key matrix is of size 3×3, the number of letters in the message must be a multiple of three. Thus, we will append an extra arbitrary letter to the end of the message when we enter it in the following command.

```
> message := Select(IsAlpha, "GO NAVY BEAT ARMY A");
```
$$message := \text{“GONAVYBEATARMYA''}$$

Next, we will convert our plaintext string into a list of plaintext characters, and then apply α to convert each of these plaintext characters into their corresponding numerical values.

```
> ptext := convert(message, list):
> ptext := map(i -> ltable[i], ptext);
```
$$ptext := [6, 14, 13, 0, 21, 24, 1, 4, 0, 19, 0, 17, 12, 24, 0]$$

The following command returns the number of positions in the vector *ptext*. Note that this is, as required by the size of our key matrix, a multiple of three.

```
> nops(ptext);
```
$$15$$

Before encrypting our message, we must group the numerical values in *ptext* into row vectors with the same number of positions as the number of rows in A. To do this, we will first assign the number of rows in A as the following variable *rowsize*.

```
> rowsize := RowDimension(A);
```
$$rowsize := 3$$

Next, we will assign the number of row vectors into which *ptext* will be divided as the following variable *numrows*.

> numrows := nops(ptext)/rowsize;

$$numrows := 5$$

Note that the number of row vectors into which *ptext* will be divided is an integer. If the value returned for *numrows* here was not an integer, then we could simply go back to the command in which we entered our original plaintext and append another extra arbitrary letter or two to the end of the message.

We are now ready to divide *ptext* into row vectors of length *rowsize*. In the following command, we use the Maple **Matrix** function to do this, and to place these vectors in order as the rows in the matrix *pmat*.

> pmat := Matrix(numrows, rowsize, (i, j) ->

ptext[(i-1)*rowsize+j]);

$$pmat := \begin{bmatrix} 6 & 14 & 13 \\ 0 & 21 & 24 \\ 1 & 4 & 0 \\ 19 & 0 & 17 \\ 12 & 24 & 0 \end{bmatrix}$$

The first two parameters in the preceding command give the size of the resulting matrix. The final parameter is a function that fills the entries of this matrix with the corresponding entries in the array *ptext*.

Because *pmat* contains all of the plaintext row vectors for our message, we can find all of the corresponding ciphertext row vectors at the same time by multiplying *pmat* by A. In the following command, we compute this product and assign the result as *cmat*. Note that in this command, we use the Maple **Multiply** function to reduce the entries in *cmat* modulo 26.

> cmat := Multiply(26, pmat, A);

$$cmat := \begin{bmatrix} 14 & 0 & 23 \\ 4 & 11 & 16 \\ 11 & 18 & 12 \\ 19 & 10 & 19 \\ 2 & 14 & 16 \end{bmatrix}$$

We can then use the Maple **convert** function as follows to list the rows in *cmat* in order as the vector *ctext*.

```
>   ctext := convert(cmat, Vector[row]);
```
$$ctext := [14\ 0\ 23\ 4\ 11\ 16\ 11\ 18\ 12\ 19\ 10\ 19\ 2\ 14\ 16]$$

In the next command, we apply the inverse of α to convert the ciphertext numerical values in *ctext* back into letters.

```
>   ctext := map(i -> letters[i], ctext);
```
$ctext := [$ "O" "A" "X" "E" "L" "Q" "L" "S" "M" "T" "K" "T" "C" "O" "Q" $]$

The preceding output for *ctext* displays the ciphertext letters as entries in a vector. In order to use the Maple **cat** function to display these characters together as a string, we must first convert this vector representation for *ctext* into a list. We can do this by entering the following **convert** command.

```
>   ctext := convert(ctext, list);
```
$ctext := [$ "O", "A", "X", "E", "L", "Q", "L", "S", "M", "T", "K", "T", "C", "O", "Q" $]$

Although the output displayed for the preceding two commands is virtually identical, the conversion of *ctext* from a vector into a list must be done in order for us to be able to use the following **cat** command to adjoin the ciphertext letters into a single ciphertext string.

```
>   ctext := cat(op(ctext));
```
$$ctext := \text{``OAXELQLSMTKTCOQ''}$$

In order to decrypt this ciphertext, we will first convert *ctext* back into a list of numerical values by entering the following two commands.

```
>   ctext := convert(ctext, list):
```

```
>   ctext := map(i -> ltable[i], ctext);
```
$$ctext := [14, 0, 23, 4, 11, 16, 11, 18, 12, 19, 10, 19, 2, 14, 16]$$

Next, in the following command, we divide *ctext* into row vectors of length *rowsize*, and place these vectors in order as the rows in the matrix *cmat*.

```
>   cmat := Matrix(numrows, rowsize, (i, j) ->
    ctext[(i-1)*rowsize+j]);
```

$$cmat := \begin{bmatrix} 14 & 0 & 23 \\ 4 & 11 & 16 \\ 11 & 18 & 12 \\ 19 & 10 & 19 \\ 2 & 14 & 16 \end{bmatrix}$$

We can then recover all of the plaintext row vectors at the same time by multiplying *cmat* by the inverse of A. In the following command, we compute this product and assign the result as *pmat*. Note that in this command, to obtain the inverse of A we raise A to the power -1.

```
>  pmat := Multiply(26, cmat, A^(-1));
```

$$pmat := \begin{bmatrix} 6 & 14 & 13 \\ 0 & 21 & 24 \\ 1 & 4 & 0 \\ 19 & 0 & 17 \\ 12 & 24 & 0 \end{bmatrix}$$

Next, we will enter the following command to list the rows in *pmat* in order as the vector *ptext*.

```
>  ptext := convert(pmat, Vector[row]);
```
$$ptext := [6\ 14\ 13\ 0\ 21\ 24\ 1\ 4\ 0\ 19\ 0\ 17\ 12\ 24\ 0]$$

Finally, in the next three commands, we apply the inverse of α to convert the plaintext numerical values in *ptext* back into letters, convert the vector representation for *ptext* into a list, and adjoin the plaintext letters into a single plaintext string.

```
>  ptext := map(i -> letters[i], ptext):
```

```
>  ptext := convert(ptext, list):
```

```
>  ptext := cat(op(ptext));
```
$$ptext := \text{“GONAVYBEATARMYA”}$$

Now, suppose an intruder intercepts the ciphertext `OAXELQLSMTKTCOQ` in this section, and knows that this ciphertext was formed using a Hill cipher with a 3×3 key matrix over $\mathbb{Z}_{26}$. Suppose also that the intruder somehow knows that the first nine ciphertext letters resulted from the plaintext letters GONAVYBEA. In order to use this information to try to determine the key matrix, the intruder could begin as follows by entering the known part of the plaintext, and then converting this plaintext string into a list of numerical values.

```
>  ptext := "GONAVYBEA":

>  ptext := convert(ptext, list):

>  ptext := map(i -> ltable[i], ptext);
```
$$ptext := [6, 14, 13, 0, 21, 24, 1, 4, 0]$$

The intruder could then use these numerical values to form the following 3×3 matrix *pmat*.

```
>  pmat := Matrix(3, 3, (i, j) -> ptext[(i-1)*3+j]);
```
$$pmat := \begin{bmatrix} 6 & 14 & 13 \\ 0 & 21 & 24 \\ 1 & 4 & 0 \end{bmatrix}$$

The intruder could verify that *pmat* is invertible over $\mathbb{Z}_{26}$ by entering the following two commands.

```
>  Determinant(26, pmat);
```
$$7$$

```
>  gcd(7, 26);
```
$$1$$

Next, the intruder could enter the first nine ciphertext letters as follows, and then convert this ciphertext string into a list of numerical values.

```
>  ctext := "OAXELQLSM":

>  ctext := convert(ctext, list):

>  ctext := map(i -> ltable[i], ctext);
```
$$ctext := [14, 0, 23, 4, 11, 16, 11, 18, 12]$$

The intruder could then use these numerical values to form the following 3×3 matrix *cmat*.

```
> cmat := Matrix(3, 3, (i, j) -> ctext[(i-1)*3+j]);
```

$$cmat := \begin{bmatrix} 14 & 0 & 23 \\ 4 & 11 & 16 \\ 11 & 18 & 12 \end{bmatrix}$$

Now, the intruder knows that the ciphertext letters OAXELQLSM resulted from the plaintext letters GONAVYBEA, and that the ciphertext was formed using a Hill cipher with some 3×3 key matrix A over $\mathbb{Z}_{26}$, which we will denote as follows.

$$A = \begin{bmatrix} a & b & c \\ d & e & f \\ g & h & i \end{bmatrix}$$

Thus, the intruder would know that the following matrix equation involving A would have to be true.

$$\begin{bmatrix} 6 & 14 & 13 \\ 0 & 21 & 24 \\ 1 & 4 & 0 \end{bmatrix} \begin{bmatrix} a & b & c \\ d & e & f \\ g & h & i \end{bmatrix} = \begin{bmatrix} 14 & 0 & 23 \\ 4 & 11 & 16 \\ 11 & 18 & 12 \end{bmatrix} \tag{6.4}$$

Because the matrix *pmat* is invertible over $\mathbb{Z}_{26}$ (as we have verified), the intruder would then know that (6.4) has unique solutions for $a, b, c, \ldots, i$ in $\mathbb{Z}_{26}$. The intruder could find these values by entering the following command.

```
> Multiply(26, pmat^(-1), cmat);
```

$$\begin{bmatrix} 11 & 6 & 8 \\ 0 & 3 & 14 \\ 24 & 0 & 9 \end{bmatrix}$$

Note that the preceding result is the key matrix for the system. The intruder could then easily find the inverse of this key matrix, and use this inverse to decrypt the remainder of the ciphertext.

6.6   Hill Ciphers with MATLAB

In this section, we will show how MATLAB can be used to encrypt and decrypt messages using Hill ciphers. We will also show an example of successful cryptanalysis on a ciphertext formed using a Hill cipher given a number of plaintext characters equal to the number of entries in the key matrix.

We will begin by establishing the correspondences between the elements in $L = \{\text{A}, \text{B}, \text{C}, \ldots, \text{Z}\}$ and $R = \mathbb{Z}_{26}$ for the bijection $\alpha : L \to R$. To do this, just as we did in Section 6.3, we will first construct the following string *abet* containing the elements in L in order, and then define the subsequent function **letters** so that we will be able to access each element in L by entering its corresponding element in $\mathbb{Z}_{26}$.

```
>> abet = 'ABCDEFGHIJKLMNOPQRSTUVWXYZ';
>> letters = @(x) abet(x+1);
```

In addition, we will need to be able to retrieve each element in $\mathbb{Z}_{26}$ by entering its corresponding element in L. To do this, we will use the same user-written function **ltable** that we used in Section 6.3, which we have written separately from this MATLAB session and saved as the M-file ltable.m.

Next, we will define the following 3×3 matrix A over $\mathbb{Z}_{26}$, which we will use as the key matrix for the system.

```
>> A = [11 6 8; 0 3 14; 24 0 9]

A =

    11     6     8
     0     3    14
    24     0     9
```

In order for A to be a valid key matrix for a Hill cipher, it must be invertible over R. We will verify this by entering the following two commands, in which we find the determinant of A in $\mathbb{Z}_{26}$, and check that this determinant satisfies the requirement that $\gcd(\det(A), 26) = 1$.

```
>> mod(det(A), 26)

ans =

    21

>> gcd(21, 26)

ans =

     1
```

We will now show how MATLAB can be used to encrypt the message GO NAVY BEAT ARMY using a Hill cipher. First, we will store this message as the following string. Note that because our key matrix is of size 3×3, the number of letters in the message must be a multiple of three. Thus, we will append an extra arbitrary letter to the end of the message when we enter it in the following command.

```
>> message = 'GO NAVY BEAT ARMY A';
>> message(findstr(message, ' ')) = []
```

message =

GONAVYBEATARMYA

Next, we will apply α to convert each of these plaintext characters into their corresponding numerical values.

```
>> ptext = ltable(message)
```

ptext =

   6   14   13   0   21   24   1   4   0   19   0   17   12   24   0

The following command returns the number of positions in the vector *ptext*. Note that this is, as required by the size of our key matrix, a multiple of three.

```
>> numel(ptext)
```

ans =

   15

Before encrypting our message, we must group the numerical values in *ptext* into row vectors with the same number of positions as the number of rows in A. To do this, we will first assign the number of rows in A as the following variable *rowsize*.

```
>> rowsize = size(A, 1)
```

rowsize =

   3

Next, we will assign the number of row vectors into which *ptext* will be divided as the following variable *numrows*.

```
>> numrows = numel(ptext)/rowsize

numrows =

    5
```

Note that the number of row vectors into which *ptext* will be divided is an integer. If the value returned for *numrows* here was not an integer, then we could simply go back to the command in which we entered our original plaintext and append another extra arbitrary letter or two to the end of the message.

We are now ready to divide *ptext* into row vectors of length *rowsize*. In the following command, we use the MATLAB **reshape** function to do this, and to place these vectors in order as the rows in the matrix *pmat*. Because **reshape** by default places vector entries into a matrix as columns, we initially place the plaintext vectors of length *rowsize* from *ptext* as columns in *pmat*. We then convert these columns into rows by taking the transpose.

```
>> pmat = reshape(ptext, rowsize, numrows)'

pmat =

     6    14    13
     0    21    24
     1     4     0
    19     0    17
    12    24     0
```

Because *pmat* contains all of the plaintext row vectors for our message, we can find all of the corresponding ciphertext row vectors at the same time by multiplying *pmat* by *A*. In the following command, we compute this product and assign the result as *cmat*. Note that in this command, we use the MATLAB **mod** function to reduce the entries in *cmat* modulo 26.

```
>> cmat = mod(pmat*A, 26)

cmat =

    14     0    23
     4    11    16
    11    18    12
    19    10    19
     2    14    16
```

We can then use the MATLAB **reshape** function as follows to list the rows in *cmat* in order as the vector *ctext*.

```
>> ctext = reshape(cmat', 1, numel(cmat'))

ctext =

    14   0  23   4  11  16  11  18  12  19  10  19   2  14  16
```

In the next command, we apply the inverse of α to convert the ciphertext numerical values in *ctext* back into letters.

```
>> ctext = letters(ctext)

ctext =

OAXELQLSMTKTCOQ
```

In order to decrypt this ciphertext, we will first convert *ctext* back into a list of numerical values by entering the following command.

```
>> ctext = ltable(ctext)

ctext =

    14   0  23   4  11  16  11  18  12  19  10  19   2  14  16
```

Next, in the following command, we divide *ctext* into row vectors of length *rowsize*, and place these vectors in order as the rows in the matrix *cmat*.

```
>> cmat = reshape(ctext, rowsize, numrows)'

cmat =

    14    0   23
     4   11   16
    11   18   12
    19   10   19
     2   14   16
```

We can then recover all of the plaintext row vectors at the same time by multiplying *cmat* by the inverse of A. In the following command, we compute this product and assign the result as *pmat*. Note that in this command, to obtain the inverse of A we raise A to the power -1.

```
>> pmat = mod(cmat*double(mod(sym(A)^(-1), 26)), 26)

pmat =

     6    14    13
     0    21    24
     1     4     0
    19     0    17
    12    24     0
```

Next, we will enter the following command to list the rows in *pmat* in order as the vector *ptext*.

```
>> ptext = reshape(pmat', 1, numel(pmat'))

ptext =

    6   14   13   0   21   24   1   4   0   19   0   17   12   24   0
```

Finally, we apply the inverse of α to convert the plaintext numerical values in *ptext* back into letters.

```
>> ptext = letters(ptext)

ptext =

GONAVYBEATARMYA
```

Now, suppose an intruder intercepts the ciphertext OAXELQLSMTKTCOQ in this section, and knows that this ciphertext was formed using a Hill cipher with a 3×3 key matrix over $\mathbb{Z}_{26}$. Suppose also that the intruder somehow knows that the first nine ciphertext letters resulted from the plaintext letters GONAVYBEA. In order to use this information to try to determine the key matrix, the intruder could begin as follows by entering the known part of the plaintext, and then converting this plaintext string into a list of numerical values.

```
>> ptext = 'GONAVYBEA';
>> ptext = ltable(ptext)

ptext =

    6   14   13   0   21   24   1   4   0
```

The intruder could then use these numerical values to form the following 3×3 matrix *pmat*.

```
>> pmat = reshape(ptext, 3, 3)'

pmat =

    6    14    13
    0    21    24
    1     4     0
```

The intruder could verify that *pmat* is invertible over $\mathbb{Z}_{26}$ by entering the following two commands.

```
>> mod(det(pmat), 26)

ans =

    7

>> gcd(7, 26)

ans =

    1
```

Next, the intruder could enter the first nine ciphertext letters as follows, and then convert this ciphertext string into a list of numerical values.

```
>> ctext = 'OAXELQLSM';
>> ctext = ltable(ctext)

ctext =

    14    0    23    4    11    16    11    18    12
```

The intruder could then use these numerical values to form the following 3×3 matrix *cmat*.

```
>> cmat = reshape(ctext, 3, 3)'

cmat =

    14     0    23
     4    11    16
    11    18    12
```

Now, the intruder knows that the ciphertext letters OAXELQLSM resulted from the plaintext letters GONAVYBEA, and that the ciphertext was formed using a Hill cipher with some 3×3 key matrix A over $\mathbb{Z}_{26}$, which we will denote as follows.

$$A = \begin{bmatrix} a & b & c \\ d & e & f \\ g & h & i \end{bmatrix}$$

Thus, the intruder would know that the following matrix equation involving A would have to be true.

$$\begin{bmatrix} 6 & 14 & 13 \\ 0 & 21 & 24 \\ 1 & 4 & 0 \end{bmatrix} \begin{bmatrix} a & b & c \\ d & e & f \\ g & h & i \end{bmatrix} = \begin{bmatrix} 14 & 0 & 23 \\ 4 & 11 & 16 \\ 11 & 18 & 12 \end{bmatrix} \tag{6.5}$$

Because the matrix $pmat$ is invertible over $\mathbb{Z}_{26}$ (as we have verified), the intruder would then know that (6.5) has unique solutions for $a, b, c, \ldots, i$ in $\mathbb{Z}_{26}$. The intruder could find these values by entering the following command.

```
>> mod(double(mod(sym(pmat)^(-1), 26))*cmat, 26)

ans =

    11    6    8
     0    3   14
    24    0    9
```

Note that the preceding result is the key matrix for the system. The intruder could then easily find the inverse of this key matrix, and use this inverse to decrypt the remainder of the ciphertext.

6.7 The Two-Message Problem

Recall that the Hill cryptosystem can be used to obtain any desired level of security by simply choosing a key matrix that is sufficiently large. However, recall also that for the Hill system with an $n \times n$ key matrix, we discussed a technique in Section 6.4 by which an intruder may be able to break the system relatively easily if the intruder knows or is somehow able to correctly guess n^2 characters from the plaintext. This technique illustrates the general fact that it is sometimes possible for an intruder to break a cryptosystem in an unusual way provided the intruder knows some additional information about the system (such as, for example, n^2 characters from the plaintext). In this section, we discuss a technique by which an intruder may be able to break a slight modification of the Hill system in an unusual way. We first discuss the modification of the system, which consists only of using a key matrix that is involutory. A matrix K is said to be *involutory* if $K^2 = I$ (i.e., if $K = K^{-1}$).

Involutory Hill Cipher Encryption: *Let K be an $n \times n$ involutory matrix over R (i.e., with $K^2 = I$ over R). Group the plaintext into row vectors $\mathbf{p}_i$ in R^n, and choose $f : R^n \to R^n$ as $f(\mathbf{p}_i) = \mathbf{p}_i K$ with each entry reduced modulo $|R|$. The resulting rows listed together form the ciphertext.*

This involutory Hill cipher encryption method is in fact the method proposed by Hill when he first published his cryptosystem in 1929. The reason for Hill's suggestion of using an involutory key matrix is obvious, for then the same matrix could be used to decrypt a message that was used to encrypt the message. Although this simplification of the Hill system is not as significant now as in 1929 after the development of calculators and personal computers, the elimination of having to determine the inverse of the key matrix was very noteworthy in 1929. In fact, Hill even invented and patented a machine designed to perform the calculations in his cryptosystem, and argued that by using an involutory key matrix, one could both encrypt and decrypt a single message without changing the settings.

Note that in order for the involutory Hill cipher encryption method to be useful, there should be relatively many $n \times n$ involutory matrices for each $n > 1$. For any $n > 1$, if there were only relatively few $n \times n$ involutory matrices, then an involutory key matrix of size $n \times n$ would certainly not yield a secure cryptosystem. We would assume in general that an intruder to such a system would know the size of the key matrix and the fact that the key matrix was involutory. Thus, if there were only relatively few involutory matrices of the known size, an intruder could break the system very easily by simply testing each one. However, it is not the case that there are only relatively few $n \times n$ involutory matrices for any $n > 1$. It is well known (and was well known in 1929) that for any matrices A of size $r \times s$ and B of size $s \times r$, the block matrix in the following theorem is involutory.

Theorem 6.2 *Suppose A is an $r \times s$ matrix over a ring R, and B is an $s \times r$ matrix over R. Then the following matrix is involutory over R.*

$$\begin{bmatrix} BA - I & B \\ 2A - ABA & I - AB \end{bmatrix}$$

Proof. Exercise.　　　　　　　　　　　　　　　　　　　　　　　　　　$\square$

As a result of Theorem 6.2, it is not unreasonable to suppose that an intruder should find the involutory Hill system not significantly less difficult to break than the usual Hill system. However, as mentioned, it is sometimes possible for an intruder to break a cryptosystem in an unusual way provided the intruder knows some additional information about the system. As we discuss next, an intruder may be able to break the involutory Hill system in an unusual and relatively easy way provided the intruder intercepts two ciphertexts formed from the same plaintext using different key matrices of the same size. In this scenario, the problem of breaking the system is called the *two-message problem*.

Suppose we intercept ciphertexts C and C' formed from the same plaintext P using the involutory Hill cryptosystem with distinct $n \times n$ key matrices K and K', respectively. That is, suppose a plaintext P is grouped into row vectors $\mathbf{p}_i$ of length n, and we intercept ciphertext vectors $\mathbf{c}_i$ and $\mathbf{c}'_i$ formed as follows for all i, where K and K' are distinct $n \times n$ involutory matrices.

$$\mathbf{c}_i = \mathbf{p}_i K \tag{6.6}$$
$$\mathbf{c}'_i = \mathbf{p}_i K' \tag{6.7}$$

The two-message problem is to determine the plaintext vectors $\mathbf{p}_i$ for all i from knowledge of the ciphertext vectors $\mathbf{c}_i$ and $\mathbf{c}'_i$. Note that since K and K' are involutory, they are their own corresponding decryption matrices. Thus, if we could determine K or K', we would be done. To do this, note that because K is involutory, (6.6) is equivalent to the following equation for all i.

$$\mathbf{p}_i = \mathbf{c}_i K$$

By substituting this expression for $\mathbf{p}_i$ into (6.7), we obtain the following equation for all i.

$$\mathbf{c}'_i = \mathbf{c}_i K K' \tag{6.8}$$

Now suppose the plaintext P is at least n^2 characters in length, and that there are n values of i, say $i_1, i_2, \ldots, i_n$, for which the following $n \times n$ matrix S is invertible.

$$S = \begin{bmatrix} \mathbf{c}_{i_1} \\ \mathbf{c}_{i_2} \\ \vdots \\ \mathbf{c}_{i_n} \end{bmatrix}$$

Then we can determine KK' as follows. First define the following $n \times n$ matrix T.

$$T = \begin{bmatrix} \mathbf{c}'_{i_1} \\ \mathbf{c}'_{i_2} \\ \vdots \\ \mathbf{c}'_{i_n} \end{bmatrix}$$

Since (6.8) holds for every i, it follows that $T = SKK'$. Thus, we can determine KK' as $KK' = S^{-1}T$.

Recall, as we stated above, if we could determine K or K', then the two-message problem would be solved. However, as we have just shown, with a very mild assumption, we should not have any difficulty finding KK'. After finding KK', we consider for an unknown matrix X the equation $(KK')X = X(KK')^{-1}$, or, equivalently, the following equation.

$$(KK')X \; = \; X(K'K) \tag{6.9}$$

Note that both K and K' are involutory solutions to (6.9). Thus, if we found all involutory solutions to (6.9), the resulting collection of matrices would include both K and K'. To find the plaintext, we could then just decrypt one of the ciphertexts with each involutory solution to (6.9). Most likely only one of these decryptions, the correct plaintext, would make any sense. (To save time, we could decrypt only a portion of one of the ciphertexts with each involutory solution to (6.9). This would reveal which involutory solution to (6.9) was the correct key matrix for that ciphertext. We could then use this key matrix to decrypt the rest of the ciphertext.)

We now summarize the complete solution process for the two-message problem as follows.

1. Determine KK'.

2. Find all of the involutory solutions to (6.9).

3. Decrypt one of the ciphertexts with each of the involutory solutions to (6.9). The correct key matrix, which is one of these involutory solutions, will yield the correct plaintext.

The procedure for completing the first step in this process was described previously, and can generally be done in a straightforward manner. The calculations for the third step can also generally be done in a straightforward manner. Due to the potentially large number of involutory solutions to (6.9), these calculations can be long and tedious, but it is at least possible to do them. It is in the second step of this process that the essential difficulties of the two-message problem lie. This step can be considered in two parts. First, we can determine the general solution X to (6.9) by solving a system of n^2 linear equations for the unknown elements in X. After finding this general solution, we can find the involutory solutions to (6.9) by imposing the condition $X^2 = I$ on the general solution X. This requires solving a system of up to n^2 quadratic equations, thus providing many possible difficulties, especially for large n. However, the potential difficulties that can be incurred in solving a system of up to n^2 quadratic equations do not compare (time-wise) to those incurred in breaking the involutory Hill system by using trial and error to determine the key matrix.

Exercises

1. Use the Caesar shift cipher function $f(x) = (x+3)$ mod 26 to encrypt the message EXIT STRATEGY.

2. Decrypt the ciphertext CHAGOHGGXNFTGG, which has been formed using the shift cipher function $f(x) = (x + 19)$ mod 26.

3. Decrypt the ciphertext DOHKCDGYCDOZ, which has been formed using a shift cipher.

4. Use the affine cipher function $f(x) = (5x + 2)$ mod 26 to encrypt the message ABLE WAS I ERE I SAW ELBA.

5. Decrypt the ciphertext DVJWPTDQPOLA, which has been formed using the affine cipher function $f(x) = (21x + 3)$ mod 26.

6. Decrypt the ciphertext RALIAIVXABVGSV, which has been formed using an affine cipher.

7. Suppose that a plaintext P is encrypted, yielding ciphertext C_1, and then C_1 is encrypted, yielding a new ciphertext C_2. The encryption of C_1 is called a *superencryption* of P.

   (a) If a plaintext P was first encrypted with a shift cipher, would superencrypting P with another shift cipher increase the mathematical security in the final ciphertext? Completely explain your answer, and be as specific as possible.

   (b) If a plaintext P was first encrypted with an affine cipher, would superencrypting P with another affine cipher increase the mathematical security in the final ciphertext? Completely explain your answer, and be as specific as possible.

   (c) If a plaintext P was first encrypted with an affine cipher, would superencrypting P with a shift cipher increase the mathematical security in the final ciphertext? Completely explain your answer, and be as specific as possible.

   (d) If a plaintext P was first encrypted with a Hill cipher, would superencrypting P using another Hill cipher with a key matrix of the same size increase the mathematical security in the final ciphertext? Explain your answer, and be as specific as possible.

   (e) If a plaintext P was first encrypted with a Hill cipher, would superencrypting P using another Hill cipher with a key matrix of a different size increase the mathematical security in the final ciphertext? Explain your answer, and be as specific as possible.

8. In the ancient Hebrew Atbash cipher, each plaintext letter was encrypted into the ciphertext letter the same number of positions from the end of the alphabet as the plaintext letter was from the start of the alphabet. For example, this cipher with our alphabet would encrypt plaintext letter A as ciphertext letter Z, B as Y, C as X, ..., Y as B, and Z as A. Show that this cipher (with our alphabet) is an affine cipher.

9. For a ciphertext formed using an affine cipher $f(x) = (ax+b) \bmod 26$, suppose it takes you three minutes to try to decrypt the ciphertext for each possible pair (a, b) of keys. How long would it take you on average to recover the plaintext if you just kept trying possible pairs of keys until you found the correct ones?

10. For a ciphertext formed using a Hill cipher $f(\mathbf{p}_i) = \mathbf{p}_i A$ with a 2×2 key matrix A over $\mathbb{Z}_{26}$, suppose it takes you five minutes to try to decrypt the ciphertext with a potential key matrix (whether the matrix is invertible over $\mathbb{Z}_{26}$ or not). How long would it take you on average to recover the plaintext if you just kept trying potential key matrices until you found the correct one?

11. Repeat Exercise 10 assuming A is of size 3×3 instead of 2×2.

12. The ASCII alphabet is a complete alphabet for the English language, and contains 95 characters, including both capital and lowercase letters, the digits 0–9, punctuation, a blank space, and some others.

   (a) For an affine cipher with a 95-character alphabet (i.e., with function $f(x) = (ax + b) \bmod 95$, for $a, b \in \mathbb{Z}_{95}$ with $\gcd(a, 95) = 1$), how many possible pairs (a, b) of keys are there?

   (b) For a ciphertext formed using an affine cipher with a 95-character alphabet, suppose it takes you three minutes to try to decrypt the ciphertext for each possible pair (a, b) of keys. How long would it take you on average to recover the plaintext if you just kept trying possible pairs of keys until you found the correct ones?

   (c) For a ciphertext formed using a Hill cipher $f(\mathbf{p}_i) = \mathbf{p}_i A$ with a 2×2 key matrix A over $\mathbb{Z}_{95}$, suppose it takes you five minutes to try to decrypt the ciphertext with a potential key matrix (whether the matrix is invertible over $\mathbb{Z}_{95}$ or not). How long would it take you on average to recover the plaintext if you just kept trying potential key matrices until you found the correct one?

13. Let A be the following 2×2 matrix over $\mathbb{Z}_{26}$.

$$A = \begin{bmatrix} 13 & 7 \\ 8 & 21 \end{bmatrix}$$

(a) Use a Hill cipher with the key matrix A to encrypt the message SEND TARGET STATUS.

(b) Decrypt NDJLWLTBWFVXGSNV, a ciphertext which has been formed using a Hill cipher with the key matrix A.

14. Let A be the following 3×3 matrix over $\mathbb{Z}_{26}$.

$$A = \begin{bmatrix} 1 & 2 & 1 \\ 3 & 1 & 0 \\ 0 & 2 & 1 \end{bmatrix}$$

Use a Hill cipher with the key matrix A to encrypt the message BULL DURHAM.

15. Suppose you intercept the ciphertext FLBIPURCRGAO, and you know that this ciphertext was formed using a Hill cipher with some 2×2 key matrix over $\mathbb{Z}_{26}$. Suppose you also somehow know that the first four ciphertext letters resulted from the plaintext letters NCST. Decrypt the remainder of the ciphertext.

16. The Hill cipher function $f(\mathbf{p}_i) = \mathbf{p}_i A$, which we viewed as a generalization of the affine cipher function $f(x) = ax + b$ with $b = 0$, can be written to include the analog of a nonzero constant b. This more general Hill cipher function would be $f(\mathbf{p}_i) = \mathbf{p}_i A + \mathbf{b}$, where $\mathbf{b}$ is a fixed row vector of the same length as the plaintext vectors $\mathbf{p}_i$.

(a) Use the Hill cipher function $f(\mathbf{p}_i) = \mathbf{p}_i A + \mathbf{b}$ with the key matrix A in Exercise 13 and the key vector $\mathbf{b} = \begin{bmatrix} 1 & 2 \end{bmatrix}$ over $\mathbb{Z}_{26}$ to encrypt the message ABORT MISSION.

(b) Decrypt the ciphertext UXSJOEWNOJHE, which has been formed using the Hill cipher function $f(\mathbf{p}_i) = \mathbf{p}_i A + \mathbf{b}$ with the key matrix A in Exercise 13 and the key vector $\mathbf{b} = \begin{bmatrix} 1 & 2 \end{bmatrix}$ over $\mathbb{Z}_{26}$.

(c) Use the Hill cipher function $f(\mathbf{p}_i) = \mathbf{p}_i A + \mathbf{b}$ with the key matrix A in Exercise 14 and the key vector $\mathbf{b} = \begin{bmatrix} 1 & 2 & 3 \end{bmatrix}$ over $\mathbb{Z}_{26}$ to encrypt the message WOLFPACK.

(d) Does a Hill cipher function of the form $f(\mathbf{p}_i) = \mathbf{p}_i A + \mathbf{b}$ yield more mathematical security than a Hill cipher function of the form $f(\mathbf{p}_i) = \mathbf{p}_i A$? If so, how much more security? Completely explain your answer, and be as specific as possible.

17. The general Hill cipher function $f(\mathbf{p}_i) = \mathbf{p}_i A + \mathbf{b}$ described in Exercise 16 can be made more secure by allowing the row vector $\mathbf{b}$ to vary with the plaintext vectors $\mathbf{p}_i$. This stronger general Hill cipher function would be $f(\mathbf{p}_i) = \mathbf{p}_i A + \mathbf{b}_i$, where the $\mathbf{b}_i$ are varying row vectors of the same length as the plaintext vectors $\mathbf{p}_i$. To reduce the difficulty in keeping a record of the $\mathbf{b}_i$, these vectors can be chosen to depend uniquely on the plaintext vectors $\mathbf{p}_i$, the ciphertext vectors $\mathbf{c}_i$, or the previous $\mathbf{b}_i$. For example, with plaintext vectors $\mathbf{p}_i$ of length n, three simple methods for choosing the $\mathbf{b}_i$ are as follows.

- $\mathbf{b}_i = \mathbf{p}_{i-1}B$, where B is a fixed $n \times n$ matrix and $\mathbf{p}_0$ is given.
- $\mathbf{b}_i = \mathbf{c}_{i-1}B$, where B is a fixed $n \times n$ matrix and $\mathbf{c}_0$ is given.
- $\mathbf{b}_i = \begin{bmatrix} r_i & r_{i+1} & \cdots & r_{i+n-1} \end{bmatrix}$, where $\{r_j\}$ is a recursive sequence with necessary initial values given.

(a) Use the Hill cipher function $f(\mathbf{p}_i) = \mathbf{p}_i A + \mathbf{p}_{i-1}B$ with the matrix A in Exercise 14 and the following matrix B and vector $\mathbf{p}_0$ over $\mathbb{Z}_{26}$ to encrypt the message MEET AT SIX.

$$B = \begin{bmatrix} 1 & 0 & 0 \\ 1 & 1 & 1 \\ 0 & 0 & 1 \end{bmatrix} \qquad \mathbf{p}_0 = \begin{bmatrix} 1 & 2 & 3 \end{bmatrix}$$

(b) Use the Hill cipher function $f(\mathbf{p}_i) = \mathbf{p}_i A + \mathbf{c}_{i-1}B$ with the following matrices A and B and vector $\mathbf{c}_0$ over $\mathbb{Z}_{26}$ to encrypt the message NEED BACKUP.

$$A = \begin{bmatrix} 2 & 5 \\ 1 & 4 \end{bmatrix} \qquad B = \begin{bmatrix} 1 & 0 \\ 1 & 1 \end{bmatrix} \qquad \mathbf{c}_0 = \begin{bmatrix} 1 & 2 \end{bmatrix}$$

(c) Decrypt the ciphertext RTOWKRPTLS, which has been formed using the Hill cipher function $f(\mathbf{p}_i) = \mathbf{p}_i A + \mathbf{c}_{i-1}B$ with the matrices A and B and vector $\mathbf{c}_0$ in part (b).

(d) Use the Hill cipher function $f(\mathbf{p}_i) = \mathbf{p}_i A + \begin{bmatrix} r_i & r_{i+1} \end{bmatrix}$ with the matrix A in part (b) and recursive sequence $\{r_j\}$ over $\mathbb{Z}_{26}$ given by the following formula to encrypt the message GO PACK.

$$r_{j+2} = (r_j + r_{j+1}) \bmod 26, \text{ with } r_1 = 3 \text{ and } r_2 = 5$$

(e) Decrypt the ciphertext JAYGKI, which has been formed using the Hill cipher function $f(\mathbf{p}_i) = \mathbf{p}_i A + \begin{bmatrix} r_i & r_{i+1} \end{bmatrix}$ with the matrix A in part (b) and recursive sequence $\{r_j\}$ in part (d).

18. Prove that $a \in \mathbb{Z}_k$ will have a multiplicative inverse in $\mathbb{Z}_k$ if and only if $\gcd(a, k) = 1$.

19. Let A be an $n \times n$ matrix over $\mathbb{Z}_k$. Recall the following, where $\mathrm{adj}(A)$ represents the adjoint of A, and I is the $n \times n$ identity matrix.

$$A \cdot \mathrm{adj}(A) \ = \ \mathrm{adj}(A) \cdot A \ = \ \det(A) \cdot I$$

Use this fact and Exercise 18 to prove that A will be invertible over $\mathbb{Z}_k$ if and only if $\gcd(\det(A), k) = 1$.

20. For matrices A of size $r \times s$ and B of size $s \times r$ over a ring R, consider the following matrix.

$$K = \left[\begin{array}{cc} BA - I & B \\ 2A - ABA & I - AB \end{array} \right]$$

Find the size of K, and show that K is involutory over R.

21. Use Exercise 20 to construct a 3×3 involutory matrix over $\mathbb{Z}_{26}$. Then use your result as the key matrix K in an involutory Hill cipher to encrypt a plaintext of your choice with at least six characters. Also, show how to decrypt the resulting ciphertext.

Computer Exercises

22. Use a shift cipher to encrypt the plaintext message JULIUS CAESAR THOUGHT SHIFT CIPHERS WERE SECURE.

23. Decrypt the ciphertext AOMPSHVOHWGKVMHVSFCAOBSADWFSKOGQCBEIS-FSR, which has been formed using a shift cipher.

24. Use an affine cipher to encrypt the message THE GERMANS THOUGHT THE ENIGMA MACHINE WAS SECURE.

25. Decrypt the ciphertext YHYLXMTOLKQYIPYTXAVXZGKTAMPXZYXNTASGX-ZGGLOKWY, which has been formed using an affine cipher.

26. Repeat Exercise 10 assuming A is of size 5×5 instead of 2×2.

27. Write a single Maple or MATLAB procedure that will encrypt or decrypt a given message using a shift cipher. This procedure should have three input parameters. The first input parameter should be a string of characters representing a plaintext or ciphertext, the second input parameter should be an integer representing the key for the system, and the third input parameter should be an indicator for whether the message is to be encrypted or decrypted.

28. Write a single Maple or MATLAB procedure that will encrypt or decrypt a given message using an affine cipher. This procedure should have four input parameters. The first input parameter should be a string of characters representing a plaintext or ciphertext, the second and third input parameters should be integers representing the keys for the system, and the fourth input parameter should be an indicator for whether the message is to be encrypted or decrypted.

29. Write a single Maple or MATLAB procedure that will take each of the 312 possible pairs (a, b) of keys for an affine cipher with $R = \mathbb{Z}_{26}$ separately, form the corresponding decryption function, and then use this function to attempt to decrypt a ciphertext. Then use your procedure to decrypt the ciphertext in Exercise 25.

30. Let A be the following 5×5 matrix over $\mathbb{Z}_{26}$.

$$
A = \begin{bmatrix}
9 & 10 & 0 & 20 & 7 \\
4 & 3 & 14 & 23 & 16 \\
7 & 2 & 5 & 7 & 5 \\
21 & 1 & 25 & 3 & 1 \\
1 & 5 & 4 & 3 & 0
\end{bmatrix}
$$

   (a) Use a Hill cipher with the key matrix A to encrypt the message ABORT MISSION PROCEED WITH SECONDARY ORDERS.

   (b) Decrypt ZQBTDDBIGZZDCQFRXFBXPVJERZRSBA, a ciphertext that has been formed using a Hill cipher with the key matrix A.

31. Suppose you intercept the ciphertext TLIYLDFSURYFGIEFNIMSMLWQOS-QHFGVP, and you know that this ciphertext was formed using a Hill cipher with some 4×4 key matrix over $\mathbb{Z}_{26}$. Suppose you also somehow know that the first several words in the corresponding plaintext are THE BEST MOVIE OF ALL TIME IS. Decrypt the remainder of the ciphertext.

32. Suppose you intercept the ciphertext PZGCR RMELU ZDLPN APKUA FWEUV EMTQG GNNZM IQLGA JEOUQ NMDDB ZIMAR PASIZ CARVF DYOHS RLYIX AOBGE KXYBW NNVMO IGEHK IFNYO QKLVW DOXBL WYKJG YDSGT YICAG, and you know that this ciphertext was formed using a Hill cipher with some 5×5 key matrix over $\mathbb{Z}_{26}$. Suppose you also somehow know that the first several words in the corresponding plaintext are JACK LEVINE WAS A PROFESSOR, and that the last 15 letters in the plaintext that was encrypted are HILLCIPHERSAAAA. Decrypt the remainder of the ciphertext.

33. Write a single Maple or MATLAB procedure that will encrypt or decrypt a given message using a Hill cipher. This procedure should have three input parameters. The first input parameter should be a string of characters representing a plaintext or ciphertext, the second input parameter should be a matrix representing the key for the system, and the third input parameter should be an indicator for whether the message is to be encrypted or decrypted.

34. Recall from Exercise 16 general Hill ciphers with $f(\mathbf{p}_i) = \mathbf{p}_i A + \mathbf{b}$. Write a single Maple or MATLAB procedure that will encrypt or decrypt a given message using a general Hill cipher. This procedure should have four input parameters. The first should be a string of characters representing a plaintext or ciphertext, the second input parameter should be a matrix representing the key matrix A for the system, the third input parameter should be a vector representing the key vector $\mathbf{b}$ for the system, and the fourth input parameter should be an indicator for whether the message is to be encrypted or decrypted.

35. Recall from Exercise 17 the stronger general Hill cipher function of the form $f(\mathbf{p}_i) = \mathbf{p}_i A + \mathbf{b}_i$, and consider the following matrix B and vector $\mathbf{p}_0$ over $\mathbb{Z}_{26}$.

$$B = \begin{bmatrix} 13 & 22 & 4 & 4 & 3 \\ 2 & 0 & 4 & 6 & 8 \\ 1 & 25 & 17 & 23 & 9 \\ 3 & 2 & 6 & 3 & 12 \\ 7 & 4 & 5 & 3 & 12 \end{bmatrix} \qquad \mathbf{p}_0 = \begin{bmatrix} 1 & 1 & 1 & 1 & 1 \end{bmatrix}$$

(a) Use the Hill cipher function $f(\mathbf{p}_i) = \mathbf{p}_i A + \mathbf{p}_{i-1} B$ with the matrix A in Exercise 30, matrix B, and vector $\mathbf{p}_0$ to encrypt the message ATTACK FLANK AT SUNRISE.

(b) Decrypt the ciphertext ZRLGVCKZHWLMOSHXOGBU, which has been formed using the Hill cipher function $f(\mathbf{p}_i) = \mathbf{p}_i A + \mathbf{p}_{i-1} B$ with the matrix A in Exercise 30, matrix B, and vector $\mathbf{p}_0$.

36. Use Exercise 20 to construct a 5×5 involutory matrix over $\mathbb{Z}_{26}$. Then use your result as the key matrix K in an involutory Hill cipher to encrypt a plaintext of your choice with at least 15 characters. Also, show how to decrypt the resulting ciphertext.

Research Exercises

37. Investigate some specific details about the *ROT13* algorithm, including its origin and some other instances of its use in real life, and write a summary of your findings.

38. Find a copy of the spoof academic paper *On the 2ROT13 Encryption Algorithm*, and write a summary of the description of 2ROT13 and opinions about 2ROT13 and other ciphers given in this article.

39. Investigate some specific details about the *ROT47* algorithm, and write a complete summary of your findings.

40. Investigate some specific details about Julius Caesar's use of encryption during the Gallic Wars, and write a summary of your findings.

41. Investigate some specific details about *permutation* ciphers, and write a summary of your findings. Include in your summary at least one example of encryption and decryption with a permutation cipher.

42. Investigate some specific details about a mechanical device that Lester Hill claimed could be used to implement Hill ciphers and for which he and a partner were awarded a U.S. patent, and write a summary of your findings.

43. Find a copy of the article written by Lester Hill and published in *The American Mathematical Monthly* in which Hill ciphers were formally introduced. Write a summary of this article, including how the system as it is described in this article differs from and is similar to the Hill cipher system as it is described in this chapter.

44. Hill ciphers are *polygraphic* ciphers. Investigate what it means for a cipher to be polygraphic, and write a summary of your findings. Include in your summary a description and example of at least one other type of polygraphic cipher.

45. Hill ciphers are *block* ciphers. Investigate what it means for a cipher to be a block cipher, and write a summary of your findings. Include in your summary a description and example of at least one other type of block cipher.

46. Three years before Hill ciphers were published by Lester Hill, they appeared in a detective magazine in an article written by a young mathematician named Jack Levine. Find a copy of this article, and write a summary of how Hill ciphers are described in it. Then find copies of some of Levine's later writings on Hill ciphers and cryptography in general, and write a summary of your findings. Finally, investigate the life and career of Levine, and write a summary of your findings. Include in your final summary some highlights of Levine's work for the U.S. government during World War II and his academic career. (Note: You may find a tribute written by Joel Brawley in memory of Levine that appeared in *Cryptologia* very helpful.)

Chapter 7

Vigenère Ciphers

Shift and affine ciphers are examples of *monoalphabetic* ciphers, meaning that each occurrence of every possible plaintext character is replaced with the same ciphertext character. As we showed in Chapter 6, monoalphabetic ciphers are vulnerable to frequency analysis, since using the same cipher correspondences, or *cipher alphabet*, throughout the encryption causes character frequencies to be preserved in the ciphertext. A *polyalphabetic* cipher is a system in which the cipher alphabet can change from one plaintext character to the next. The German Enigma machine was a polyalphabetic cipher device, as was a cipher wheel constructed by Thomas Jefferson that was produced and used as a U.S. Army field cipher after World War I.

In this chapter, we will present a type of polyalphabetic cipher that was developed in the late 1500s by a Frenchman named Blaise de Vigenère. As we will show, Vigenère ciphers are more secure than monoalphabetic ciphers, since character frequencies are disguised in ciphertexts and thus basic frequency analysis can no longer be employed by intruders. However, as we will also show, Vigenère ciphers can still be broken through a little perseverance and ingenuity.

7.1 Encryption and Decryption

Two people wishing to exchange a secret message using a Vigenère cipher must first agree upon a keyword for the system. The originator of the message would then write this keyword below the plaintext, repeated as many times as necessary, with one keyword character beneath each plaintext character. Then for each plaintext character, the corresponding ciphertext character is found using the *Vigenère square*, which is shown in Table 7.1.

	(Plaintext Character)
	A B C D E F G H I J K L M N O P Q R S T U V W X Y Z
A	A B C D E F G H I J K L M N O P Q R S T U V W X Y Z
B	B C D E F G H I J K L M N O P Q R S T U V W X Y Z A
C	C D E F G H I J K L M N O P Q R S T U V W X Y Z A B
D	D E F G H I J K L M N O P Q R S T U V W X Y Z A B C
E	E F G H I J K L M N O P Q R S T U V W X Y Z A B C D
F	F G H I J K L M N O P Q R S T U V W X Y Z A B C D E
G	G H I J K L M N O P Q R S T U V W X Y Z A B C D E F
H	H I J K L M N O P Q R S T U V W X Y Z A B C D E F G
I	I J K L M N O P Q R S T U V W X Y Z A B C D E F G H
J	J K L M N O P Q R S T U V W X Y Z A B C D E F G H I
K	K L M N O P Q R S T U V W X Y Z A B C D E F G H I J
L	L M N O P Q R S T U V W X Y Z A B C D E F G H I J K
M	M N O P Q R S T U V W X Y Z A B C D E F G H I J K L
N	N O P Q R S T U V W X Y Z A B C D E F G H I J K L M
O	O P Q R S T U V W X Y Z A B C D E F G H I J K L M N
P	P Q R S T U V W X Y Z A B C D E F G H I J K L M N O
Q	Q R S T U V W X Y Z A B C D E F G H I J K L M N O P
R	R S T U V W X Y Z A B C D E F G H I J K L M N O P Q
S	S T U V W X Y Z A B C D E F G H I J K L M N O P Q R
T	T U V W X Y Z A B C D E F G H I J K L M N O P Q R S
U	U V W X Y Z A B C D E F G H I J K L M N O P Q R S T
V	V W X Y Z A B C D E F G H I J K L M N O P Q R S T U
W	W X Y Z A B C D E F G H I J K L M N O P Q R S T U V
X	X Y Z A B C D E F G H I J K L M N O P Q R S T U V W
Y	Y Z A B C D E F G H I J K L M N O P Q R S T U V W X
Z	Z A B C D E F G H I J K L M N O P Q R S T U V W X Y

Table 7.1   The Vigenère square.

With a Vigenère cipher, for each plaintext character, the corresponding ciphertext character is the character in the Vigenère square where the column labeled with the plaintext character intersects the row labeled with the keyword character. We illustrate this encryption technique in the following example.

Example 7.1 Using a Vigenère cipher with the keyword TRIXIE, the message HAVING A PET CAN MAKE YOU HAPPY enciphers as follows.

```
 Plaintext:  H A V I N G A P E T C A N M A K E Y O U H A P P Y
 Keyword:    T R I X I E T R I X I E T R I X I E T R I X I E T
Ciphertext:  A R D F V K T G M Q K E G D I H M C H L P X X T R
```

Thus, the corresponding ciphertext is ARDFVKTGMQKEGDIHMCHLPXXTR.   □

To decrypt a ciphertext that has been formed using a Vigenère cipher, we would first write the keyword above the ciphertext, repeated as many times as necessary, with one keyword character above each ciphertext character. Then for each ciphertext character, we would find the row in the Vigenère square labeled with the corresponding keyword character, and locate the ciphertext character in this row. The character labeling the column in which this ciphertext letter appears is the plaintext character.

Example 7.2 The ciphertext in Example 7.1 deciphers, of course, as follows.

> Keyword: T R I X I E T R I X I E T R I X I E T R I X I E T
> Ciphertext: A R D F V K T G M Q K E G D I H M C H L P X X T R
> Plaintext: H A V I N G A P E T C A N M A K E Y O U H A P P Y

□

To describe Vigenère ciphers mathematically, note that if we associate each character in the alphabet $L = \{A, B, C, \ldots, Z\}$ with its corresponding element in the ring $R = \mathbb{Z}_{26}$ under the bijection $\alpha : L \to R$ that we used in Chapter 6 (i.e., $A \mapsto 0$, $B \mapsto 1$, $C \mapsto 2, \ldots,$ $Z \mapsto 25$), then the Vigenère square can be viewed as an addition table for R. For example, to encrypt the plaintext character H using a Vigenère cipher with the keyword character T, we can take the numerical representations 7 of H and 19 of T, and compute $(7 + 19) \bmod 26 = 0$, which is the numerical representation of the letter A that would result from using the Vigenère square to do the encryption directly.

More precisely, for encrypting a plaintext of length m characters using a Vigenère cipher with a keyword of length n characters, if we convert the plaintext and keyword into lists of elements in R, then we can represent the Vigenère encryption process mathematically as follows.

Vigenère Cipher Encryption: *Let* $\mathbf{x} = (x_0, x_1, \ldots, x_{m-1})$ *be a vector in* R^m *representing the plaintext, and let* $\mathbf{k} = (k_0, k_1, \ldots, k_{n-1})$ *be a vector in* R^n *representing the keyword. Choose the mapping* $f : R \to R$ *as follows for* $i = 0, 1, \ldots, m - 1$.

$$f(x_i) = \left(x_i + k_{(i \bmod n)}\right) \bmod |R| \tag{7.1}$$

The vector $f(\mathbf{x}) = (f(x_0), f(x_1), \ldots, f(x_{m-1}))$ *forms the ciphertext.*

Example 7.3 In this example, we will use a Vigenère cipher with the keyword TRIXIE to encrypt the message HAVING A PET CAN mathematically. We will begin by converting this message and keyword into lists of elements in $\mathbb{Z}_{26}$.

		H	A	V	I	N	G	A	P	E	T	C	A	N
α	$\Rightarrow$	7	0	21	8	13	6	0	15	4	19	2	0	13

		T	R	I	X	I	E
α	$\Rightarrow$	19	17	8	23	8	4

Thus, we can see that the Vigenère square encryption will be completely equivalent to adding the vectors $(7, 0, 21, 8, 13, 6, 0, 15, 4, 19, 2, 0, 13)$ and $(19, 17, 8, 23, 8, 4, 19, 17, 8, 23, 8, 4, 19)$ in $\mathbb{Z}_{26}^{13}$ componentwise. The resulting encipherment follows.

		H	A	V	I	N	G	A	P	E	T	C	A	N
α	$\Rightarrow$	7	0	21	8	13	6	0	15	4	19	2	0	13
f	$\Rightarrow$	0	17	3	5	21	10	19	6	12	16	10	4	6
α^{-1}	$\Rightarrow$	A	R	D	F	V	K	T	G	M	Q	K	E	G

Thus, the resulting ciphertext is ARDFVKTGMQKEG. To decrypt this ciphertext, we can repeat this same procedure but using the inverse function $f^{-1}(x_i) = \left(x_i - k_{(i \bmod 6)}\right) \bmod 26$ in place of f. □

7.2 Cryptanalysis

The fact that Vigenère ciphers are polyalphabetic prevents the use of basic frequency analysis in breaking the system. Indeed, the main purpose of using a polyalphabetic cipher is to distribute the frequencies so that the characters in the ciphertext are all likely to occur around the same number of times. This can be seen in the ciphertext ARDFVKTGMQKEGDIHMCHLPXXTR in Example 7.1, in which the number of times that each character occurs is shown in the following list.

A	B	C	D	E	F	G	H	I	J	K	L	M
1	0	1	2	1	1	2	2	1	0	2	1	2

N	O	P	Q	R	S	T	U	V	W	X	Y	Z
0	0	1	1	2	0	2	0	1	0	2	0	0

As we can see, the character frequencies in this ciphertext are fairly evenly distributed, indicating that it is unlikely that the ciphertext was formed using a monoalphabetic cipher.

Through careful analysis, however, it is still possible to use frequency analysis to break a ciphertext that has been formed using a Vigenère cipher. To do so, an intruder must first find the length of the keyword that was used in the system. In order to describe how this can be done, we first need to present some additional background material on character probability distributions.

7.2.1 The Index of Coincidence

If the character frequencies in a ciphertext are highly variable, the cipher is more likely to be monoalphabetic, whereas if the frequencies are more evenly distributed, the cipher is more likely to be polyalphabetic. The *index of coincidence* is a number that measures variation in character frequencies. Specifically, the index of coincidence for a piece of text is the probability that two characters selected at random from the text will be identical. Using the frequency percentages shown in Table 6.1 for ordinary English, we can see that the probability of selecting the letter A at random twice from a piece of ordinary English text is $(0.08167)^2 = 0.00667$. Similarly, the probability of selecting the letter B at random twice from a piece of ordinary English text is $(0.01492)^2 = 0.00022$. Continuing in this manner, we obtain the following index of coincidence for ordinary English.

$$(0.08167)^2 + (0.01492)^2 + \cdots + (0.00074)^2 = 0.0655$$

In a mythical language in which the frequencies of the letters in the alphabet $L = \{A, B, \ldots, Z\}$ are distributed exactly evenly, for any first letter selected at random, the probability that a second letter selected at random will match the first is $\frac{1}{26} = 0.0385$. Thus, the index of coincidence for the mythical language is 0.0385.

Monoalphabetic ciphers preserve letter frequencies, and thus we would expect that a ciphertext produced by a monoalphabetic cipher would have an index of coincidence of 0.0655. Polyalphabetic ciphers, on the other hand, distribute letter frequencies evenly, and thus we would expect that a ciphertext produced by a polyalphabetic cipher would have an index of coincidence of 0.0385. Of course, a typical ciphertext would not be long enough to have an index of coincidence exactly equal to one of these values. However, we would at least expect that the index of coincidence for a typical ciphertext would be somewhere around 0.0655 if it was produced by a monoalphabetic cipher, and closer to 0.0385 if it was produced by a polyalphabetic cipher. Also, we should note that for a particular ciphertext, if we let $m_0, m_1, \ldots, m_{25}$ be the frequencies with which the letters $A, B, \ldots, Z$ appear in the ciphertext, and if we denote the total length of the ciphertext by $m = m_0 + m_1 + \cdots + m_{25}$, then it is not difficult to show that the index of coincidence for the ciphertext will be given by the following formula.

$$I = \frac{1}{m(m-1)} \sum_{i=0}^{25} m_i(m_i - 1) \tag{7.2}$$

Example 7.4 Consider the ciphertext PAPCP SRSIC RKILT GYFXG ETWAI JIUPG RLTGH ACMOQ RWXYT JIEDF NVEAC ZUUEJ TLOHA WHEET RFDCT JGSGZ LKRSC ZRVLU PCONM FPDTC XWJYI XIJHT TAMKA ZCCXW STNTE DTTGJ MFISE

GEKIP RPTGG EIQRG UEHGR GGEHE EJDWI PEHXP DOSFI CEIMG CCAFJ GGOUP
MNTCS KXQXD LQGSI PDKRJ POFQV VXYTJ IEDFN VEACZ UUEJT LOHWG JEHYI
KIPRP ZAGRI PMS, which was formed using a Vigenère cipher, and in which
the number of times that each character occurs is shown in the following
list.

A	B	C	D	E	F	G	H	I	J	K	L	M
10	0	15	9	20	9	20	9	18	13	7	7	7

N	O	P	Q	R	S	T	U	V	W	X	Y	Z
5	7	16	5	13	10	18	8	5	7	9	5	6

Using (7.2), we find that the index of coincidence for this ciphertext is as
follows.

$$I = \frac{1}{258 \cdot 257}\left((10 \cdot 9) + (0 \cdot -1) + (15 \cdot 14) + \cdots + (6 \cdot 5)\right) = 0.0449$$

This suggests that the ciphertext was formed using a polyalphabetic cipher.
In contrast, consider the ciphertext OJJOT OWDUO BIZOZ EQBUH ERIUZ GUITS
WOZEN UBFSS BEUBZ DEKSQ BZOUB SQIOR EOSVY EIZER BBSRZ DWORS TUBOB
SRZDW ORSTU BOIZO ZEQBU HERIU ZGUIT SWOZE NUBRO TEUMD UBZDE EOIZE
RBJSR ZUSBS VZDEB SRZDW ORSTU BOJUE NKSBZ REMUS BRONV SRNQB UHERI
UZGUI TSWOZ ENUBR ONVSR NUBZD EKSQB ZOUBS QIORE OSVIS QZDYE IZERB
HURMU BUO, which was formed from the same plaintext using an affine cipher,
and in which the number of times that each character occurs is shown in
the following list.

A	B	C	D	E	F	G	H	I	J	K	L	M
0	28	0	10	24	1	3	4	14	4	3	0	3

N	O	P	Q	R	S	T	U	V	W	X	Y	Z
8	26	0	8	22	25	8	28	5	7	0	2	25

The index of coincidence for this ciphertext is 0.0742, which suggests that
it was formed using a monoalphabetic cipher.   □

7.2.2 Determining the Keyword Length

The index of coincidence is a tool that we can use to determine the likely
keyword length for a Vigenère cipher. To see this, consider the following
example.

Example 7.5 Consider again the following encryption from Example 7.1.

    Plaintext: H A V I N G A P E T C A N M A K E Y O U H A P P Y
    Keyword: T R I X I E T R I X I E T R I X I E T R I X I E T
    Ciphertext: A R D F V K T G M Q K E G D I H M C H L P X X T R

Note that every sixth plaintext letter starting with the first is encrypted using the same keyword letter T. Thus, all of these plaintext letters are encrypted using a shift cipher with a key of 19. Similarly, every sixth plaintext letter starting with the second is encrypted using a shift cipher with a key of 17. □

Example 7.5 reveals an important fact about Vigenère ciphers; each specific keyword letter yields its own little shift cipher. In a Vigenère ciphertext, all of the letters that were encrypted using the same specific keyword letter form what we call a *coset*. In Example 7.5, the coset that corresponds to the keyword letter T consists of the letters A, T, G, H, and R. Similarly, the coset that corresponds to the first keyword letter I consists of the letters D, M, I, and P.

The number of cosets in a Vigenère ciphertext is the same as the length of the keyword. Also, since all of the letters in each coset would be formed using the same shift cipher, the index of coincidence for each coset should be similar to that of text encrypted using a monoalphabetic cipher. Suppose n is the number of cosets, and let I_j be the index of coincidence for the jth coset. To measure how indicative a collection of cosets is of having been produced by monoalphabetic ciphers, we use the average of the indices I_j for the cosets. We will denote this average index of coincidence by $\overline{I_{1:n}}$.

$$\overline{I_{1:n}} = \frac{1}{n} \sum_{j=1}^{n} I_j \tag{7.3}$$

Of course, when confronted with a Vigenère ciphertext, we would not know the value of n. So the idea we will take is that we will calculate $\overline{I_{1:n}}$ for several values of n, and look for the smallest value of n for which $\overline{I_{1:n}}$ is noticeably larger than for the surrounding values of n. Such a value of n would be the likely keyword length for the cipher. We illustrate this process in the following example.

Example 7.6 Consider again the ciphertext PAPCP SRSIC RKILT GYFXG ETWAI JIUPG RLTGH ACMOQ RWXYT JIEDF NVEAC ZUUEJ TLOHA WHEET RFDCT JGSGZ LKRSC ZRVLU PCONM FPDTC XWJYI XIJHT TAMKA ZCCXW STNTE DTTGJ MFISE GEKIP RPTGG EIQRG UEHGR GGEHE EJDWI PEHXP DOSFI CEIMG CCAFJ GGOUP MNTCS KXQXD LQGSI PDKRJ POFQV VXYTJ IEDFN VEACZ UUEJT LOHWG JEHYI KIPRP ZAGRI PMS in Example 7.4, which was formed using a Vigenère cipher. To try to find the length of the keyword for the cipher, we will calculate $\overline{I_{1:n}}$ for $n = 1, 2, \ldots, 9$. Thus, for each of these values of n, we must find I_j for $j = 1, 2, \ldots, n$. These values of I_j can be found using (7.2), and are listed along with their corresponding averages $\overline{I_{1:n}}$ in the following table.

n	I_1	I_2	I_3	I_4	I_5	I_6	I_7	I_8	I_9	$\overline{I_{1:n}}$
1	0.045									0.045
2	0.060	0.046								0.053
3	0.060	0.058	0.048							0.055
4	0.059	0.042	0.059	0.043						0.051
5	0.044	0.043	0.037	0.038	0.064					0.045
6	0.080	0.058	0.069	0.052	0.084	0.082				**0.071**
7	0.050	0.039	0.045	0.032	0.051	0.048	0.041			0.044
8	0.042	0.045	0.054	0.042	0.079	0.036	0.050	0.040		0.049
9	0.076	0.052	0.049	0.057	0.052	0.062	0.061	0.058	0.040	0.056

As we can see, $\overline{I_{1:n}}$ is noticeably larger at $n = 6$ than for the surrounding values of n. Thus, $n = 6$ is the likely keyword length for the cipher. □

Knowing the length of the keyword for a Vigenère cipher is the first major step in determining the keyword itself. Next, we will present how the keyword itself can be determined.

7.2.3   Determining the Keyword

After the length of the keyword for a Vigenère cipher is found, the keyword itself can be determined one character at a time. Before we present a method for doing this, we first need to review some basic concepts about vectors in $\mathbb{R}^n$.

Let $\mathbf{x} = (x_1, x_2, \ldots, x_n)$ and $\mathbf{y} = (y_1, y_2, \ldots, y_n)$ be any pair of vectors in $\mathbb{R}^n$. Then the *dot product* of $\mathbf{x}$ and $\mathbf{y}$, denoted $\mathbf{x} \cdot \mathbf{y}$, is the sum of the products of the respective entries in $\mathbf{x}$ and $\mathbf{y}$.

$$\mathbf{x} \cdot \mathbf{y} = x_1 y_1 + x_2 y_2 + \cdots + x_n y_n$$

Also, if $\mathbf{x}$ and $\mathbf{y}$ are both nonzero (i.e., with components that are not all zero), then $\mathbf{x}$ and $\mathbf{y}$ are said to be *parallel* if one is a scalar multiple of the other. That is, nonzero vectors $\mathbf{x}$ and $\mathbf{y}$ are parallel if there exists a scalar $c \in \mathbb{R}$ such that $\mathbf{y} = c\mathbf{x}$ (i.e., such that $y_i = cx_i$ for $i = 1, 2, \ldots, n$). The *Euclidean norm* of $\mathbf{x}$, denoted $||\mathbf{x}||$, is a measure of the magnitude of $\mathbf{x}$, and is defined as follows.

$$||\mathbf{x}|| = \sqrt{x_1^2 + x_2^2 + \cdots + x_n^2}$$

It is easy to see that $||\mathbf{x}||^2 = \mathbf{x} \cdot \mathbf{x}$. The following is a list of six additional properties concerning dot products and Euclidean norms of vectors in $\mathbb{R}^n$. In this list, $\mathbf{x}$, $\mathbf{y}$, and $\mathbf{z}$ represent vectors in $\mathbb{R}^n$, and c represents a scalar in $\mathbb{R}$.

1. $\mathbf{x} \cdot \mathbf{y} = \mathbf{y} \cdot \mathbf{x}$

2. $(c\mathbf{x}) \cdot \mathbf{y} = \mathbf{x} \cdot (c\mathbf{y}) = c(\mathbf{x} \cdot \mathbf{y})$

3. $\mathbf{x} \cdot (\mathbf{y} + \mathbf{z}) = \mathbf{x} \cdot \mathbf{y} + \mathbf{x} \cdot \mathbf{z}$

4. $||\mathbf{x}|| \geq 0$, and $||\mathbf{x}|| = 0$ if and only if $\mathbf{x} = \mathbf{0}$

5. $||c\mathbf{x}|| = |c| \, ||\mathbf{x}||$

6. $||\mathbf{x} + \mathbf{y}|| \leq ||\mathbf{x}|| + ||\mathbf{y}||$

A final property that we will state concerning dot products and Euclidean norms is given in the following theorem. This property is called the *Cauchy-Schwarz Inequality*, and provides an important connection between the dot product of a pair of vectors and the lengths of the vectors. It is this property that we will use to find Vigenère cipher keyword characters.

Theorem 7.1 (Cauchy-Schwarz Inequality) *For any pair of vectors* $\mathbf{x}$ *and* $\mathbf{y}$ *in* $\mathbb{R}^n$, *it follows that* $\mathbf{x} \cdot \mathbf{y} \leq ||\mathbf{x}|| \, ||\mathbf{y}||$. *Moreover, for nonzero* $\mathbf{x}$ *and* $\mathbf{y}$, *this inequality will be an equality if and only if* $\mathbf{x}$ *and* $\mathbf{y}$ *are parallel.*

Proof. Exercise. □

Recall that in a Vigenère cipher, each specific keyword letter yields its own little monoalphabetic shift cipher, and all of the ciphertext letters that are formed using the same keyword letter form a coset. Let $\mathbf{x}_0$ be a vector containing the relative frequencies shown in Table 6.1 for letters in ordinary English.

$$\mathbf{x}_0 = (0.0817, 0.0149, 0.0278, 0.0425, 0.1270, 0.0223, 0.0202, 0.0609,$$
$$0.0697, 0.0015, 0.0077, 0.0403, 0.0241, 0.0675, 0.0751, 0.0193,$$
$$0.0010, 0.0599, 0.0633, 0.0906, 0.0276, 0.0098, 0.0236, 0.0015,$$
$$0.0197, 0.0007)$$

In addition, let $\mathbf{x}_i$ for $i = 1, 2, \ldots, 25$ be the vector that results from shifting the entries in $\mathbf{x}_0$ to the right by i positions, with the entries at the end of $\mathbf{x}_0$ wrapping around to the start. For example, let $\mathbf{x}_2$ be the following vector.

$$\mathbf{x}_2 = (0.0197, 0.0007, 0.0817, 0.0149, \ldots, 0.0236, 0.0015)$$

Also, let $\mathbf{y} = (y_0, y_1, \ldots, y_{25})$ be a vector containing the relative frequencies for the letters in a particular Vigenère ciphertext coset. Since all of the letters in the coset would have been formed using the same monoalphabetic shift cipher, we would expect that the relative frequencies in $\mathbf{y}$ would be distributed similarly to how the relative frequencies for letters in ordinary English are distributed, except that the relative frequencies in **y** would be shifted to the right by some number of positions. To determine this number of positions, we can use Theorem 7.1. In particular, we can use the fact that the dot product $\mathbf{x} \cdot \mathbf{y}$ in the Cauchy-Schwarz Inequality will be as large as possible when the vectors **x** and **y** are as close to being parallel as possible. More specifically, given the vectors $\mathbf{x}_i$ for $i = 0, 1, \ldots, 25$ containing the relative frequencies for letters in ordinary English shifted to the right by i positions, and a vector **y** containing the relative frequencies for the letters in a particular Vigenère ciphertext coset, we will compute $\mathbf{x}_i \cdot \mathbf{y}$ for each i. The value of i for which $\mathbf{x}_i \cdot \mathbf{y}$ is the largest will represent the number of positions such that the vector that results from shifting the entries in $\mathbf{x}_0$ to the right by this number of positions most closely resembles the relative frequencies for the letters in the coset. This will give the likely shift cipher, and, consequently, keyword letter, that produced the coset. Performing this same procedure for each coset will recover the entire keyword. We illustrate this process in the following example.

Example 7.7 Consider again the ciphertext PAPCP SRSIC RKILT GYFXG ETWAI JIUPG RLTGH ACMOQ RWXYT JIEDF NVEAC ZUUEJ TLOHA WHEET RFDCT JGSGZ LKRSC ZRVLU PCONM FPDTC XWJYI XIJHT TAMKA ZCCXW STNTE DTTGJ MFISE GEKIP RPTGG EIQRG UEHGR GGEHE EJDWI PEHXP DOSFI CEIMG CCAFJ GGOUP MNTCS KXQXD LQGSI PDKRJ POFQV VXYTJ IEDFN VEACZ UUEJT LOHWG JEHYI KIPRP ZAGRI PMS in Example 7.6, which was formed using a Vigenère cipher. In Example 7.6, we found that 6 is the likely keyword length for the cipher. We will now show how the first letter in the keyword can be determined. We start with the first coset in the ciphertext, PRIXI RCXDC THDGC PPJHA STIIG UGDXI CGTXI PXDCT JIG, in which the number of times that each character occurs is shown in the following list.

A	B	C	D	E	F	G	H	I	J	K	L	M
1	0	5	4	0	0	5	2	7	2	0	0	0

N	O	P	Q	R	S	T	U	V	W	X	Y	Z
0	0	4	0	2	1	4	1	0	0	5	0	0

Next, we divide each of these frequencies by the total number of letters in the entire coset, resulting in the entries in the following vector **y** of relative frequencies for the letters in the coset.

$$\mathbf{y} = (0.0233, 0, 0.1163, 0.0930, 0, 0, 0.1163, 0.0465, 0.1628, 0.0465,$$
$$0, 0, 0, 0, 0, 0.0930, 0, 0.0465, 0.0233, 0.0930, 0.0233, 0, 0,$$
$$0.1163, 0, 0)$$

Then for each of the vectors $\mathbf{x}_i$ for $i = 0, 1, \ldots, 25$ that result from shifting the entries in $\mathbf{x}_0$ to the right by i positions, we form the dot product $\mathbf{x}_i \cdot \mathbf{y}$.

$$
\begin{aligned}
\mathbf{x}_0 \cdot \mathbf{y} &= 0.0409 \\
\mathbf{x}_1 \cdot \mathbf{y} &= 0.0402 \\
\mathbf{x}_2 \cdot \mathbf{y} &= 0.0486 \\
&\vdots \\
\mathbf{x}_{25} \cdot \mathbf{y} &= 0.0383
\end{aligned}
$$

The largest dot product in this list is $\mathbf{x}_{15} \cdot \mathbf{y} = 0.0720$. Thus, of all the possible shifts of the entries in $\mathbf{x}_0$, shifting the entries in $\mathbf{x}_0$ to the right by 15 positions yields the vector $\mathbf{x}_i$ that most closely resembles $\mathbf{y}$. Using our bijection $\alpha : L \to R$, we see that $\alpha^{-1}(15) = \mathrm{P}$. As a result, we can conclude that the likely first keyword letter for the cipher is P. These computations and similar computations required for the other five cosets are summarized in the following table.

Shift	Key	$\mathbf{x}_i \cdot \mathbf{y}$					
i	Letter	Coset 1	Coset 2	Coset 3	Coset 4	Coset 5	Coset 6
0	A	0.041	0.046	**0.068**	0.030	0.055	0.037
1	B	0.040	0.038	0.047	0.042	0.033	0.041
2	C	0.049	0.035	0.036	**0.066**	0.029	0.053
3	D	0.034	0.027	0.037	0.037	0.034	0.039
4	E	0.051	0.037	0.045	0.031	**0.068**	0.035
5	F	0.036	0.032	0.034	0.036	0.035	0.048
6	G	0.038	0.036	0.041	0.043	0.032	0.041
7	H	0.025	0.043	0.037	0.036	0.024	0.033
8	I	0.034	0.034	0.032	0.036	0.045	0.029
9	J	0.035	0.029	0.029	0.039	0.034	0.035
10	K	0.037	0.039	0.035	0.030	0.034	0.029
11	L	0.040	**0.067**	0.043	0.039	0.042	0.028
12	M	0.032	0.040	0.039	0.039	0.042	0.038
13	N	0.032	0.033	0.040	0.040	0.046	0.045
14	O	0.037	0.036	0.035	0.036	0.028	0.050
15	P	**0.072**	0.040	0.046	0.050	0.039	0.035
16	Q	0.042	0.030	0.044	0.044	0.039	0.033
17	R	0.034	0.041	0.036	0.041	0.046	0.040
18	S	0.026	0.040	0.024	0.032	0.030	**0.067**
19	T	0.043	0.031	0.038	0.033	0.048	0.036
20	U	0.036	0.035	0.038	0.034	0.032	0.025
21	V	0.043	0.042	0.036	0.038	0.034	0.028
22	W	0.033	0.041	0.035	0.040	0.035	0.050
23	X	0.033	0.039	0.033	0.037	0.037	0.030
24	Y	0.041	0.050	0.027	0.044	0.043	0.038
25	Z	0.038	0.038	0.036	0.030	0.036	0.039

From the preceding table, we can see that the likely keyword for the cipher is PLACES. If we use this potential keyword to try to decrypt the ciphertext, we do in fact obtain the correct plaintext: APPALACHIAN STATE UNIVERSITY IS LOCATED IN BOONE IN THE MOUNTAINOUS AREA OF WESTERN NORTH CAROLINA NORTH CAROLINA STATE UNIVERSITY IS LOCATED IN RALEIGH IN THE EASTERN PORTION OF THE NORTH CAROLINA PIEDMONT REGION RADFORD UNIVERSITY IS LOCATED IN RADFORD IN THE MOUNTAINOUS AREA OF SOUTHWESTERN VIRGINIA.   □

7.3 Vigenère Ciphers with Maple

In this section, we will show how Maple can be used to encrypt and decrypt messages using Vigenère ciphers. We will also show an example of successful cryptanalysis on a ciphertext formed using a Vigenère cipher.

Because some of the functions that we will use are in the Maple **String-Tools** and **ListTools** packages, we will begin by including these packages.

```
> with(StringTools):
```

```
> with(ListTools):
```

Next, we will establish the correspondences between the elements in the set $L = \{A, B, \ldots, Z\}$ and $R = \mathbb{Z}_{26}$ for the bijection $\alpha : L \to R$. To do this, we will construct the following array *letters* containing the elements in L in order.

```
> letters := array(0..25, ["A", "B", "C", "D", "E", "F",
      "G", "H", "I", "J", "K", "L", "M", "N", "O", "P", "Q",
      "R", "S", "T", "U", "V", "W", "X", "Y", "Z"]):
```

In addition, so that we will be able to retrieve each element in $\mathbb{Z}_{26}$ by entering its corresponding element in L, we will also establish the correspondences for α in the following table.

```
> ltable := table():
```

```
> for i from 0 to 25 do
```

```
>       ltable[letters[i]] := i:
```

```
> od:
```

7.3.1 Encryption and Decryption

We will now show how Maple can be used to encrypt the message HAVING A PET CAN MAKE YOU HAPPY using a Vigenère cipher with the keyword TRIXIE. First, we will store this message as the following string, using the **Select** function with the **IsAlpha** option to remove the blank spaces from the message.

```
> message := Select(IsAlpha, "HAVING A PET CAN MAKE YOU
  HAPPY");
```

$$message := \text{"HAVINGAPETCANMAKEYOUHAPPY"}$$

Next, we will convert our plaintext string into a list of plaintext characters, and then apply α to convert each of these plaintext characters into their corresponding numerical values.

```
> ptext := convert(message, list):

> ptext := map(i -> ltable[i], ptext);
```

$$ptext := [7, 0, 21, 8, 13, 6, 0, 15, 4, 19, 2, 0, 13, 12, 0, 10, 4, 24, 14,$$
$$20, 7, 0, 15, 15, 24]$$

In the next two commands, we convert the keyword for the system into a list of plaintext characters, and then apply α to convert each of these plaintext characters into their corresponding numerical values.

```
> keyword := convert("TRIXIE", list):

> keyword := map(i -> ltable[i], keyword);
```

$$keyword := [19, 17, 8, 23, 8, 4]$$

Next, we will enter the following command to create the general Vigenère cipher function. This function is defined using (7.1) modified slightly to reflect the fact that list elements in Maple are accessed starting with the first element in position 1 rather than 0.

```
> f := (x, k) -> [seq((x[i]+k[((i-1) mod nops(k))+1]) mod
  26, i=1..nops(x))]:
```

The function defined by the preceding command takes two input parameters. The first parameter is a vector x over $\mathbb{Z}_{26}$ representing the message, and the second parameter is a vector k over $\mathbb{Z}_{26}$ representing the keyword. For example, in the following command we use f to encrypt the plaintext *ptext* using the keyword that we defined previously.

```
> ctext := f(ptext, keyword);
```

$$ctext := [0, 17, 3, 5, 21, 10, 19, 6, 12, 16, 10, 4, 6, 3, 8, 7, 12, 2, 7, 11,$$
$$15, 23, 23, 19, 17]$$

Then by entering the following two commands, we convert these ciphertext numerical values back into letters, and adjoin these letters into a single ciphertext string.

```
>  ctext := map(i -> letters[i], ctext):
>  ctext := cat(op(ctext));
```
    ctext := "ARDFVKTGMQKEGDIHMCHLPXXTR"

In order to decrypt this ciphertext, we will first convert *ctext* back into a list of numerical values by entering the following two commands.

```
>  ctext := convert(ctext, list):
>  ctext := map(i -> ltable[i], ctext);
```
ctext := [0, 17, 3, 5, 21, 10, 19, 6, 12, 16, 10, 4, 6, 3, 8, 7, 12, 2, 7, 11, 15, 23, 23, 19, 17]

Then to decrypt the ciphertext, we must only apply the same function f that we used to encrypt the plaintext, but with the negation of the keyword that we used to encrypt the plaintext.

```
>  ptext := f(ctext, -keyword);
```
ptext := [7, 0, 21, 8, 13, 6, 0, 15, 4, 19, 2, 0, 13, 12, 0, 10, 4, 24, 14, 20, 7, 0, 15, 15, 24]

We can now recover the plaintext characters by entering the following two commands.

```
>  ptext := map(i -> letters[i], ptext):
>  ptext := cat(op(ptext));
```
    ptext := "HAVINGAPETCANMAKEYOUHAPPY"

7.3.2   Cryptanalysis

Suppose that we intercept the ciphertext PAPCP SRSIC RKILT GYFXG ETWAI JIUPG RLTGH ACMOQ RWXYT JIEDF NVEAC ZUUEJ TLOHA WHEET RFDCT JGSGZ LKRSC ZRVLU PCONM FPDTC XWJYI XIJHT TAMKA ZCCXW STNTE DTTGJ MFISE GEKIP RPTGG EIQRG UEHGR GGEHE EJDWI PEHXP DOSFI CEIMG CCAFJ GGOUP MNTCS KXQXD LQGSI PDKRJ POFQV VXYTJ IEDFN VEACZ UUEJT LOHWG JEHYI KIPRP ZAGRI PMS sent by our enemy, and we know that this ciphertext was formed using a Vigenère cipher.

```
>  ctext := Select(IsAlpha, "PAPCP SRSIC RKILT GYFXG ETWAI

   JIUPG RLTGH ACMOQ RWXYT JIEDF NVEAC ZUUEJ TLOHA WHEET
```

```
RFDCT JGSGZ LKRSC ZRVLU PCONM FPDTC XWJYI XIJHT TAMKA

ZCCXW STNTE DTTGJ MFISE GEKIP RPTGG EIQRG UEHGR GGEHE

EJDWI PEHXP DOSFI CEIMG CCAFJ GGOUP MNTCS KXQXD LQGSI

PDKRJ POFQV VXYTJ IEDFN VEACZ UUEJT LOHWG JEHYI KIPRP

ZAGRI PMS");
```

$ctext :=$
"PAPCPSRSICRKILTGYFXGETWAIJIUPGRLTGHACMOQR\
WXYTJIEDFNVEACZUUEJTLOHAWHEETRFDCTJGSGZL\
KRSCZRVLUPCONMFPDTCXWJYIXIJHTTAMKAZCCXWS\
TNTEDTTGJMFISEGEKIPRPTGGEIQRGUEHGRGGEHEEJ\
DWIPEHXPDOSFICEIMGCCAFJGGOUPMNTCSKXQXDLQ\
GSIPDKRJPOFQVVXYTJIEDFNVEACZUUEJTLOHWGJEH\
YIKIPRPZAGRIPMS"

Note first that we can use the Maple **length** function as follows to determine the number of characters in *ctext*.

```
>  length(ctext);
```

$$258$$

To find the frequency with which each of the letters in *ctext* occurs, we will use the user-written function **letterfreq**, which we have written separately from this Maple session and saved as the text file letterfreq.mpl. To use this function, we must first read it into our Maple session as follows.

```
>  read "letterfreq.mpl";
```

The following command then illustrates how the function **letterfreq** can be used. Specifically, by entering the following command, we obtain a vector containing the relative frequency with which each of the letters in *ctext* occurs, with the letters ordered from A through Z.

```
>  letterfreq(ctext);
```

$$\left[\frac{5}{129}, 0, \frac{5}{86}, \frac{3}{86}, \frac{10}{129}, \frac{3}{86}, \frac{10}{129}, \frac{3}{86}, \frac{3}{86}, \frac{13}{43}, \frac{7}{258}, \frac{7}{258}, \frac{7}{258}, \frac{5}{258}, \right.$$
$$\left. \frac{7}{258}, \frac{8}{129}, \frac{5}{258}, \frac{13}{258}, \frac{5}{129}, \frac{3}{43}, \frac{4}{129}, \frac{5}{258}, \frac{7}{258}, \frac{3}{86}, \frac{5}{258}, \frac{1}{43} \right]$$

We can convert these relative frequencies into decimal form by entering the following command. The parameter *4* in brackets in this command specifies that the results are to be displayed with four significant digits.

>   `evalf[4](letterfreq(ctext));`

[0.03876, 0., 0.05814, 0.03488, 0.07752, 0.03488, 0.07752, 0.03488,
    0.06977, 0.05039, 0.02713, 0.02713, 0.02713, 0.01938, 0.02713,
    0.06202, 0.01938, 0.05039, 0.03876, 0.06977, 0.03101, 0.01938,
    0.02713, 0.03488, 0.01938, 0.02326]

We can also convert these relative frequencies into actual frequencies by entering the following command.

>   `evalm(length(ctext)*letterfreq(ctext));`

[10, 0, 15, 9, 20, 9, 20, 9, 18, 13, 7, 7, 7, 5, 7, 16, 5, 13, 10, 18, 8, 5, 7,
    9, 5, 6]

To break the ciphertext *ctext*, we must find the keyword for the system. Recall that the first step in finding this keyword is to find its length. In order to do this, we will need to be able to calculate the index of coincidence for a sequence of letters. To do this, we will use the user-written function **ic**,[1] which we have written separately from this Maple session and saved as the text file ic.mpl. To use this function, we must first read it into our Maple session as follows.

>   `read "ic.mpl";`

The following command then illustrates how the function **ic** can be used. Specifically, by entering the following command, we obtain the index of coincidence for the ciphertext *ctext*.

>   `ic(ctext);`

0.04491297922

This index of coincidence suggests that *ctext* was indeed formed using a polyalphabetic cipher.

Recall that the number of cosets in a Vigenère ciphertext is the same as the length of the keyword. To extract cosets from a ciphertext, we will use the user-written function **vigencoset**, which we have written separately from this Maple session and saved as the text file vigencoset.mpl. To use this function, we must first read it into our Maple session as follows.

>   `read "vigencoset.mpl";`

---

[1]The function **ic** uses the user-written function **letterfreq**.

The following command then illustrates how the function **vigencoset** can be used. Specifically, by entering the following command, we obtain each of the cosets that result from dividing the ciphertext *ctext* into four cosets.

```
>   cosets := vigencoset(4, ctext);
```

cosets :=
[
"PPIIYEIPTCRTDEUTAEDGLCLOPXIHMCSEGIERGRHGEI\
XSECJUTXLIRFXINCEOJIRGM",
"ASCLFTJGGMWJFAULWTCSKZUNDWXTKCTDJSKPEGG\
EJPPFICGPCQQPJQYEVZJHEKPRS",
"PRRTXWIRHOXINCEOHRTGRRPMTJITAXNTMEITIURH\
DEDIMAGMSXGDPVTDEUTWHIZI",
"CSKGGAULAQYEVZJHEFJZSVCFCYJAZWTTFGPGQEGE\
WHOCGFONKDSKOVJFAULGYPAP"]

We can then access each of the cosets contained in *cosets* separately. For example, to access the second coset contained in *cosets*, we can enter the following command.

```
>   cosets[2];
```

"ASCLFTJGGMWJFAULWTCSKZUNDWXTKCTDJSKPEGGEJP\
PFICGPCQQPJQYEVZJHEKPRS"

Next, note that we can use the Maple **seq** and **ic** functions as follows to calculate the index of coincidence for each of the four cosets contained in *cosets*.

```
>   ics := [seq(evalf[4](ic(cosets[i])),
    i=1..nops(cosets))];
```
$$ics := [0.05865, 0.04231, 0.05853, 0.04266]$$

Also, we can use the Maple **add** function as follows to calculate the average index of coincidence for the four cosets contained in *cosets*.

```
>   a := add(i, i=ics)/nops(cosets);
```
$$a := 0.05053750000$$

Recall that to find the length of the keyword for the system, we need to divide the ciphertext into several different numbers of cosets, and calculate the average index of coincidence for the cosets in each division. We should then look for a value for the number of cosets for which this average index of coincidence is noticeably larger than for the surrounding numbers of cosets. In order to do this, we will enter the following **for** loop, which is designed to calculate the average index of coincidence for the cosets in each division of *ctext* ranging from a single coset through nine cosets.

```
>   for n from 1 to 9 do

>       cosets := vigencoset(n, ctext);

>       ics := [seq(evalf[4](ic(cosets[i])),

        i=1..nops(cosets))];

>       a := evalf[4](add(i, i=ics)/nops(cosets));

>       ics := convert(ics, Vector);

>       printf("%a [%5.3f] %5.3f\n", n, ics, a);

>   od:

1 [0.045] 0.045
2 [0.060 0.046] 0.053
3 [0.060 0.058 0.048] 0.055
4 [0.059 0.042 0.059 0.043] 0.051
5 [0.044 0.043 0.037 0.038 0.064] 0.045
6 [0.080 0.058 0.069 0.052 0.084 0.082] 0.071
7 [0.050 0.039 0.045 0.032 0.051 0.048 0.041] 0.044
8 [0.042 0.045 0.054 0.042 0.079 0.036 0.050 0.040] 0.049
9 [0.076 0.052 0.049 0.057 0.052 0.062 0.061 0.058 0.040] 0.056
```

In the preceding **for** loop, for each number n of cosets ranging from one through nine, the index of coincidence for each particular coset in each division is stored in the vector *ics*, and the average index of coincidence for the cosets in each division is stored as the variable a. The vector *ics* is also converted using the **convert** function with the **Vector** option so that it can be printed using the **printf** function. The **printf** function is designed to do formatted printing in Maple. In each row of the output, the number n of cosets is listed first, followed by the index of coincidence values in *ics*, and then the average index of coincidence a. The first parameter in the **printf** statement is a string enclosed in double quotes that gives formatting instructions for how each subsequent parameter is to be printed. The *%a* in the **printf** statement instructs Maple to print the number n of cosets in standard form. The first *%5.3f* in the **printf** statement instructs Maple to print the values in *ics* as fixed point numbers with widths of five and three digits shown to the right of the decimal. The second *%5.3f* instructs Maple to print the value of a also with a width of five and three digits shown to the right of the decimal. As we can see from the output, the average index of coincidence of 0.071 for six cosets is noticeably larger than the average index of coincidence for the surrounding numbers of cosets. Thus, the likely keyword length for the cipher is six.

In order to find the actual letters in the keyword for the system, we will begin as follows by using **vigencoset** to split *ctext* into six cosets.

```
>  cosets := vigencoset(6, ctext);
```

cosets :=

[

"PRIXIRCXDCTHDGCPPJHASTIIGUGDXICGTXIPXDCTJI\
G",

"ASLGJLMYFZLECZZCDYTZTTSPEEEWPCCOCDPOYFZL\
EPR",

"PITEITOTNUOETLROTITCNGERIHHIDEAUSLDFTNUOH\
RI",

"CCGTUGQJVUHTJKVNCXACTJGPQGEPOIFPKQKQJVU\
HYPP",

"PRYWPHRIEEARGRLMXIMXEMETRREESMJMXGRVIEE\
WIZM",

"SKFAGAWEAJWFSSUFWJKWDFKGGGJHFGGNQSJVEAJ\
GKAS"]

We will demonstrate how the first keyword letter can be found from the first coset. We start by finding and storing as the following vector y the relative frequency with which each letter occurs in the first coset.

```
>  y := evalf[4](letterfreq(cosets[1]));
```

$y := [0.02326, 0., 0.1163, 0.09302, 0., 0., 0.1163, 0.04651, 0.1628,$
$0.04651, 0., 0., 0., 0., 0., 0.09302, 0., 0.04651, 0.02326, 0.09302,$
$0.02326, 0., 0., 0.1163, 0., 0.]$

Then in the next command, we store as the following vector x the relative frequency with which each letter occurs in ordinary English. The values in this list were obtained from Table 6.1.

```
>  x := [0.08167, 0.01492, 0.02782, 0.04253, 0.12702,
        0.02228, 0.02015, 0.06094, 0.06966, 0.00153, 0.00772,
        0.04025, 0.02406, 0.06749, 0.07507, 0.01929, 0.00095,
        0.05987, 0.06327, 0.09056, 0.02758, 0.00978, 0.02360,
        0.00150, 0.01974, 0.00074]:
```

Next, we will use the Maple **Rotate** function, which is contained in the **ListTools** package and is designed to shift the values in a list either to the left or right. The **Rotate** function takes two input parameters. The first input parameter is the list that is to be shifted. The second input parameter is an integer, which, if nonnegative, indicates that the values in the list are to be shifted to the left this number of positions, with the entries at the start wrapping around to the end. If the second input parameter is negative, this indicates that the values in the list are to be shifted to the right the absolute value of this number of positions, with the entries at the end wrapping around to the start. For example, to shift the values stored in x two positions to the right, we can enter the following command.

```
>  Rotate(x, -2);
```

[0.01974, 0.00074, 0.08167, 0.01492, 0.02782, 0.04253, 0.12702,
      0.02228, 0.02015, 0.06094, 0.06966, 0.00153, 0.00772, 0.04025,
      0.02406, 0.06749, 0.07507, 0.01929, 0.00095, 0.05987, 0.06327,
      0.09056, 0.02758, 0.00978, 0.02360, 0.00150]

To find the first letter in the keyword, we need to find the number of times the entries in x should be shifted to the right so that the entries in the resulting vector most closely resemble the entries in y. To do this, we must shift the entries in x to the right by i positions for $i = 0, 1, \ldots, 25$, and compute the dot product of each of the resulting vectors with y. We will do this by entering the following loop, in which we use the Maple **DotProduct** function which is contained in the **ListTools** package to compute these dot products, and store the results in the list *dprodxy*.

```
>  dprodxy := [] :
>  for i from 0 to 25 do
>      dprodxy := [op(dprodxy),
         evalf[4](DotProduct(Rotate(x, -i), y))];
>  od:
```

To see the resulting dot products, we enter the following command.

```
>  dprodxy;
```

[0.04094, 0.04018, 0.04864, 0.03401, 0.05082, 0.03630, 0.03778,
      0.02474, 0.03437, 0.03503, 0.03654, 0.04019, 0.03243, 0.03161,
      0.03663, 0.07200, 0.04207, 0.03384, 0.02552, 0.04268, 0.03613,
      0.04314, 0.03256, 0.03277, 0.04076, 0.03827]

To find the number of times the entries in x should be shifted to the right so that the entries in the resulting vector most closely resemble the entries in y, we need to determine the position in *dprodxy* where the largest dot product is located. We can find the largest value in *dprodxy* by using the Maple **max** function as follows.

```
>  maxdprod := max(op(dprodxy));
```
$$maxdprod := 0.07200$$

To determine the position in *dprodxy* where this largest value is located, we can use the Maple **member** function as follows.

```
>  member(maxdprod, dprodxy, 'i'):
```

The preceding command assigns as the variable i the position number where the largest value in *dprodxy* is located. To see the resulting value of i, we can enter the following command.

```
>  i;
```

$$16$$

However, because lists in Maple are expressed starting with the first element in position 1 rather than 0, this value is actually one greater than the number of times the entries in x should be shifted to the right so that the entries in the resulting vector most closely resemble the entries in y. That is, the actual number of times the entries should be shifted is the following.

```
>  shift := i-1:
```

In the following nested **for** loop, we compute and store in a list k the number of times the entries in x should be shifted to the right so that the entries in the resulting vector most closely resemble the entries in y, where y is constructed to contain the relative frequency with which each letter occurs in each of the six cosets in *cosets* separately.

```
>  k := []:
>  for j from 1 to nops(cosets) do
>          dprodxy := []:
>          y := letterfreq(cosets[j]);
>          for i from 0 to 25 do
>              dprodxy := [op(dprodxy),
>                  evalf[4](DotProduct(Rotate(x, -i), y))];
>          od:
>          maxdprod := max(op(dprodxy));
>          member(maxdprod, dprodxy, 'shift'):
>          k := [op(k), shift-1];
>  od:
```

Thus, the following list k contains the numerical values that correspond to the letters in the keyword for the system.

```
>  k;
```
$$[15, 11, 0, 2, 4, 18]$$

In the following command, we convert these numerical values into their corresponding letters so that we can see the keyword for the system.

```
>  map(i -> letters[i], k);
```
$$[\text{``P''}, \text{``L''}, \text{``A''}, \text{``C''}, \text{``E''}, \text{``S''}]$$

Next, in order to decrypt *ctext*, we will convert the letters in *ctext* into their corresponding numerical values.

```
>  ctext := convert(ctext, list):
```

```
>  ctext := map(i -> ltable[i], ctext):
```

In the following command, we recover the numerical values in the corresponding plaintext.

```
>  ptext := f(ctext, -k):
```

Finally, we recover the letters in the corresponding plaintext by entering the following two commands.

```
>  ptext := map(i -> letters[i], ptext):
```

```
>  ptext := cat(op(ptext));
```

ptext :=

"APPALACHIANSTATEUNIVERSITYISLOCATEDINBOONEI\
NTHEMOUNTAINOUSAREAOFWESTERNNORTHCAROLIN\
ANORTHCAROLINASTATEUNIVERSITYISLOCATEDINRAL\
EIGHINTHEEASTERNPORTIONOFTHENORTHCAROLINA\
PIEDMONTREGIONRADFORDUNIVERSITYISLOCATEDIN\
RADFORDINTHEMOUNTAINOUSAREAOFSOUTHWESTER\
NVIRGINIA"

7.4　Vigenère Ciphers with MATLAB

In this section, we will show how MATLAB can be used to encrypt and decrypt messages using Vigenère ciphers. We will also show an example of successful cryptanalysis on a ciphertext formed using a Vigenère cipher.

We will begin by establishing the correspondences between the elements in the set $L = \{\text{A}, \text{B}, \ldots, \text{Z}\}$ and $R = \mathbb{Z}_{26}$ for the bijection $\alpha : L \to R$. To do this, just as we did in Section 6.3, we will first construct the following string *abet* containing the elements in L in order, and then define the subsequent function **letters** so that we will be able to access each element in L by entering its corresponding element in $\mathbb{Z}_{26}$.

```
>> abet = 'ABCDEFGHIJKLMNOPQRSTUVWXYZ';
>> letters = @(x) abet(x+1);
```

Also, we will need to be able to retrieve each element in $\mathbb{Z}_{26}$ by entering its corresponding element in L. To do this, we will use the same user-written function **ltable** that we used in Section 6.3, which we have written separately from this MATLAB session and saved as the M-file ltable.m.

7.4.1 Encryption and Decryption

We will now show how MATLAB can be used to encrypt the message HAVING A PET CAN MAKE YOU HAPPY using a Vigenère cipher with the keyword TRIXIE. First, we will store this message as the following string, using **findstr** to remove the blank spaces from the message.

```
>> message = 'HAVING A PET CAN MAKE YOU HAPPY';
>> message(findstr(message, ' ')) = []

message =

HAVINGAPETCANMAKEYOUHAPPY
```

Next, we will use **ltable** to apply α to each of these plaintext characters, leaving the results in a list named *ptext*.

```
>> ptext = ltable(message)

ptext =

  Columns 1 through 15

     7   0  21   8  13   6   0  15   4  19   2   0  13  12   0

  Columns 16 through 25

    10   4  24  14  20   7   0  15  15  24
```

We will now use **ltable** to apply α to each of the plaintext characters in the keyword.

```
>> keyword = ltable('TRIXIE')

keyword =

    19    17     8    23     8     4
```

In order to encrypt and decrypt messages using Vigenère ciphers, we will use the user-written function **vigenere**, which we have written separately from this MATLAB session and saved as the M-file vigenere.m. The following command shows how the function **vigenere** can be used. Specifically, by entering the following command, we encrypt the plaintext *ptext* using a Vigenère cipher with the keyword that we defined previously.

```
>> ctext = vigenere(ptext, keyword)

ctext =

  Columns 1 through 15

     0   17    3    5   21   10   19    6   12   16   10    4    6    3    8

  Columns 16 through 25

     7   12    2    7   11   15   23   23   19   17
```

The first parameter in the preceding command is a vector over $\mathbb{Z}_{26}$ representing the message, and the second parameter is a vector over $\mathbb{Z}_{26}$ representing the keyword. In the resulting vector *ctext*, we can convert the ciphertext numerical values back into letters by entering the following command.

```
>> ctext = letters(ctext)

ctext =

ARDFVKTGMQKEGDIHMCHLPXXTR
```

In order to decrypt this ciphertext, we will first convert *ctext* back into a list of numerical values by entering the following command.

```
>> ctext = ltable(ctext)

ctext =
```

```
Columns 1 through 15

   0  17   3   5  21  10  19   6  12  16  10   4   6   3   8

Columns 16 through 25

   7  12   2   7  11  15  23  23  19  17
```

Then to decrypt the ciphertext, we must only apply the same **vigenere** function that we used to encrypt the plaintext, but with the negation of the keyword that we used to encrypt the plaintext.

```
>> ptext = vigenere(ctext, -keyword)

ptext =

Columns 1 through 15

   7   0  21   8  13   6   0  15   4  19   2   0  13  12   0

Columns 16 through 25

  10   4  24  14  20   7   0  15  15  24
```

We can now recover the plaintext characters by entering the following command.

```
>> ptext = letters(ptext)

ptext =

HAVINGAPETCANMAKEYOUHAPPY
```

7.4.2 Cryptanalysis

Suppose that we intercept the ciphertext PAPCP SRSIC RKILT GYFXG ETWAI JIUPG RLTGH ACMOQ RWXYT JIEDF NVEAC ZUUEJ TLOHA WHEET RFDCT JGSGZ LKRSC ZRVLU PCONM FPDTC XWJYI XIJHT TAMKA ZCCXW STNTE DTTGJ MFISE GEKIP RPTGG EIQRG UEHGR GGEHE EJDWI PEHXP DOSFI CEIMG CCAFJ GGOUP MNTCS KXQXD LQGSI PDKRJ POFQV VXYTJ IEDFN VEACZ UUEJT LOHWG JEHYI KIPRP ZAGRI PMS sent by our enemy, and we know that this ciphertext was formed using a Vigenère cipher.

```
>> ctext = ['PAPCP SRSIC RKILT GYFXG ETWAI JIUPG RLTGH' ...
```

```
        'ACMOQ RWXYT JIEDF NVEAC ZUUEJ TLOHA WHEET RFDCT' ...
        'JGSGZ LKRSC ZRVLU PCONM FPDTC XWJYI XIJHT TAMKA' ...
        'ZCCXW STNTE DTTGJ MFISE GEKIP RPTGG EIQRG UEHGR' ...
        'GGEHE EJDWI PEHXP DOSFI CEIMG CCAFJ GGOUP MNTCS' ...
        'KXQXD LQGSI PDKRJ POFQV VXYTJ IEDFN VEACZ UUEJT' ...
        'LOHWG JEHYI KIPRP ZAGRI PMS'];
>> ctext(findstr(ctext, ' ')) = [];
```

When a string is displayed in a MATLAB window, the entire string is displayed as a single row. For longer strings, this can cause only a small part of the output to actually be shown. In order to cause a string that is to be displayed in a MATLAB window to appear in its entirety, we will use the user-written function **stringprint**, which we have written separately from this MATLAB session and saved as the M-file stringprint.m. The following command illustrates how the function **stringprint** can be used. Specifically, by entering the following command, we can see the contents of the string *ctext* displayed in its entirety in rows with 50 characters each.

```
>> stringprint(ctext, 50)


PAPCPSRSICRKILTGYFXGETWAIJIUPGRLTGHACMOQRWXYTJIEDF
NVEACZUUEJTLOHAWHEETRFDCTJGSGZLKRSCZRVLUPCONMFPDTC
XWJYIXIJHTTAMKAZCCXWSTNTEDTTGJMFISEGEKIPRPTGGEIQRG
UEHGRGGEHEEJDWIPEHXPDOSFICEIMGCCAFJGGOUPMNTCSKXQXD
LQGSIPDKRJPOFQVVXYTJIEDFNVEACZUUEJTLOHWGJEHYIKIPRP
ZAGRIPMS
```

Next, note that we can use the MATLAB **length** function as follows to determine the number of characters in *ctext*.

```
>> length(ctext)

ans =

   258
```

To find the frequency with which each of the letters in *ctext* occurs, we will use the user-written function **letterfreq**, which we have written separately from this MATLAB session and saved as the M-file letterfreq.m. The following command illustrates how the function **letterfreq** can be used. Specifically, by entering the following command, we obtain a vector containing the relative frequency with which each of the letters in *ctext* occurs, with the letters ordered from A through Z.

```
>> letterfreq(ctext)

ans =

[5/129,    0,    5/86,    3/86,    10/129,    3/86,    10/129,    3/86,
 3/43,    13/258,    7/258,    7/258,    7/258,    5/258,    7/258,
 8/129,    5/258,    13/258,    5/129,    3/43,    4/129,    5/258,
 7/258,    3/86,    5/258,    1/43]
```

We can convert these relative frequencies into decimal form by entering the following command.

```
>> double(letterfreq(ctext))

ans =

  Columns 1 through 7

    0.0388        0   0.0581   0.0349   0.0775   0.0349   0.0775

  Columns 8 through 14

    0.0349   0.0698   0.0504   0.0271   0.0271   0.0271   0.0194

  Columns 15 through 21

    0.0271   0.0620   0.0194   0.0504   0.0388   0.0698   0.0310

  Columns 22 through 26

    0.0194   0.0271   0.0349   0.0194   0.0233
```

We can also convert these relative frequencies into actual frequencies by entering the following command.

```
>> length(ctext)*letterfreq(ctext)

ans =

[10,    0,    15,    9,    20,    9,    20,    9,    18,    13,    7,    7,    7,
  5,    7,    16,    5,    13,    10,    18,    8,    5,    7,    9,    5,    6]
```

To break the ciphertext *ctext*, we must find the keyword for the system. Recall that the first step in the process for determining the keyword is to find the length of the keyword. In order to do this, we will need to be able to calculate the index of coincidence for a sequence of letters. To do this, we will use the user-written function **ic**, which we have written separately from this MATLAB session and saved as the M-file ic.m. This function takes a single input parameter, representing a sequence of letters enclosed in quotes, for which the index of coincidence is to be calculated. Thus, to calculate the index of coincidence for the ciphertext *ctext*, we can enter the following command.

```
>> ic(ctext)

ans =

    0.0449
```

This index of coincidence suggests that *ctext* was indeed formed using a polyalphabetic cipher.

Recall that the number of cosets in a Vigenère ciphertext is the same as the length of the keyword. To extract cosets from a ciphertext, we will use the user-written function **vigencoset**, which we have written separately from this MATLAB session and saved as the M-file vigencoset.m. The following command illustrates how the function **vigencoset** can be used. Specifically, by entering the following command, we cause the ciphertext *ctext* to be divided into five cosets.

```
>> cosets = vigencoset(5, ctext);
```

We can then use the MATLAB **char** function as follows to display each of the five resulting cosets. The output of this command displays the cosets separately by rows.

```
>> char(cosets)

ans =

PSRGEJRARJNZTWRJLZPFXXTZSDMGREUGEPDCCGMKLPPVIVULJKZP
ARKYTILCWIVULHFGKRCPWIACTTFEPIEGJEOECGNXQDOXEEUOEIAM
PSIFWUTMXEEUOEDSRVODJJMCNTIKTQHEDHSIAOTQGKFYDAEHHPGS
CILXAPGOYDAEHECGSLNTYHKXTGSIGRGHWXFMFUCXSRQTFCJWYRR
PCTGIGHQTFCJATTZCUMCITAWEJEPGGREIPIGJPSDIJVJNZTGIPI
```

We can also use the **char** function to access each of the cosets contained in *cosets* separately. For example, to access the second coset contained in *cosets*, we can enter the following command.

```
>> char(cosets(2))

ans =

ARKYTILCWIVULHFGKRCPWIACTTFEPIEGJEOECGNXQDOXEEUOEIAM
```

Next, note that we can use the **ic** function as follows to calculate the index of coincidence for each of the five cosets contained in *cosets*.

```
>> ics = ic(cosets)

ics =

    0.0437    0.0430    0.0370    0.0377    0.0643
```

Also, we can use the MATLAB **sum** function as follows to calculate the average index of coincidence for the five cosets contained in *cosets*.

```
>> a = sum(ics)/numel(ics)

a =

    0.0451
```

Recall that to find the length of the keyword for the system, we need to divide the ciphertext into several different numbers of cosets, and calculate the average index of coincidence for the cosets in each division. We should then look for a value for the number of cosets for which this average index of coincidence is noticeably larger than for the surrounding numbers of cosets. In order to do this, we will enter the following **for** loop, which is designed to calculate the average index of coincidence for the cosets in each division of *ctext* ranging from a single coset through nine cosets.

```
>> for n = 1:9
       cosets = vigencoset(n, ctext);
       ics = ic(cosets);
       a = sum(ics)/numel(ics);
       fprintf('%3.0f [', n)
       fprintf(' %5.3f ', ics)
       fprintf('] %5.3f\n', a)
   end

1 [0.045] 0.045
2 [0.060 0.046] 0.053
```

```
3 [0.060 0.058 0.048] 0.055
4 [0.059 0.042 0.059 0.043] 0.051
5 [0.044 0.043 0.037 0.038 0.064] 0.045
6 [0.080 0.058 0.069 0.052 0.084 0.082] 0.071
7 [0.050 0.039 0.045 0.032 0.051 0.048 0.041] 0.044
8 [0.042 0.045 0.054 0.042 0.079 0.036 0.050 0.040] 0.049
9 [0.076 0.052 0.049 0.057 0.052 0.062 0.061 0.058 0.040] 0.056
```

In the preceding **for** loop, for each number n of cosets ranging from one through nine, the index of coincidence for each particular coset in each division is stored in the vector *ics*, and the average index of coincidence for the cosets in each division is stored as the variable a. The **fprintf** function is designed to do formatted printing in MATLAB. In each row of the output, the number n of cosets is listed first, followed by the index of coincidence values in *ics*, and then the average index of coincidence a. The first parameter in each **fprintf** statement is a string enclosed in single quotes that gives formatting instructions for how each subsequent parameter is to be printed. The *%3.0f* in the first **fprintf** statement instructs MATLAB to print the number n of cosets as fixed point numbers with widths of three and zero digits shown to the right of the decimal. The *%5.3f* in the second **fprintf** statement instructs MATLAB to print the values in *ics* as fixed point numbers with widths of five and three digits shown to the right of the decimal. The *%5.3f* in the last **fprintf** statement instructs MATLAB to print the value of a also with a width of five and three digits shown to the right of the decimal. As we can see from the output, the average index of coincidence of 0.071 for six cosets is noticeably larger than the average index of coincidence for the surrounding numbers of cosets. Thus, the likely keyword length for the cipher is six.

In order to find the actual letters in the keyword for the system, we will begin as follows by using **vigencoset** to split *ctext* into six cosets.

```
>> cosets = vigencoset(6, ctext);
>> char(cosets)

ans =

PRIXIRCXDCTHDGCPPJHASTIIGUGDXICGTXIPXDCTJIG
ASLGJLMYFZLECZZCDYTZTTSPEEEWPCCOCDPOYFZLEPR
PITEITOTNUOETLROTITCNGERIHHIDEAUSLDFTNUOHRI
CCGTUGQJVUHTJKVNCXACTJGPQGEPOIFPKQKQJVUHYPP
PRYWPHRIEEARGRLMXIMXEMETRREESMJMXGRVIEEWIZM
SKFAGAWEAJWFSSUFWJKWDFKGGGJHFGGNQSJVEAJGKAS
```

We will demonstrate how the first keyword letter can be found from the first coset. We start by finding and storing as the following vector y the relative frequency with which each letter occurs in the first coset.

```
>> y = double(letterfreq(cosets(1)))

y =

  Columns 1 through 7

    0.0233         0   0.1163   0.0930         0         0   0.1163

  Columns 8 through 14

    0.0465   0.1628   0.0465         0         0         0         0

  Columns 15 through 21

         0   0.0930         0   0.0465   0.0233   0.0930   0.0233

  Columns 22 through 26

         0         0   0.1163         0         0
```

Then in the next command, we store as the following vector x the relative frequency with which each letter occurs in ordinary English. The values in this list were obtained from Table 6.1.

```
>> x = [0.08167, 0.01492, 0.02782, 0.04253, 0.12702, ...
        0.02228, 0.02015, 0.06094, 0.06966, 0.00153, ...
        0.00772, 0.04025, 0.02406, 0.06749, 0.07507, ...
        0.01929, 0.00095, 0.05987, 0.06327, 0.09056, ...
        0.02758, 0.00978, 0.02360, 0.00150, 0.01974, 0.00074];
```

Next, we will use the MATLAB **circshift** function, which is designed to shift the entries in a matrix. The **circshift** function takes two input parameters. The first input parameter is the matrix in which the entries are to be shifted. The second input parameter is a vector. The first entry in this vector is an integer, which, if nonnegative, indicates that the rows in the matrix are to be shifted down this number of positions, with the rows at the bottom wrapping around to the top. If the first entry in the vector is negative, this indicates that the rows in the matrix are to be shifted up the absolute value of this number of positions, with the rows at the top wrapping around to the bottom. The second entry in the vector is an integer, which, if nonnegative, indicates that the columns in the matrix are to be shifted to the right this number of positions, with the columns at the end wrapping around to the start. If the second entry in the vector is negative, this indicates that the columns in the matrix are to be shifted to the left the absolute value of this number of positions, with the columns at the start wrapping around to the end. For example, to shift the values stored in x two positions to the right, we can enter the following command.

```
>> circshift(x, [0, 2])

ans =

  Columns 1 through 7

    0.0197  0.0007  0.0817  0.0149  0.0278  0.0425  0.1270

  Columns 8 through 14

    0.0223  0.0202  0.0609  0.0697  0.0015  0.0077  0.0403

  Columns 15 through 21

    0.0241  0.0675  0.0751  0.0193  0.0010  0.0599  0.0633

  Columns 22 through 26

    0.0906  0.0276  0.0098  0.0236  0.0015
```

To find the first letter in the keyword, we need to find the number of times the entries in x should be shifted to the right so that the entries in the resulting vector most closely resemble the entries in y. To do this, we must shift the entries in x to the right by i positions for $i = 0, 1, \ldots, 25$, and compute the dot product of each of the resulting vectors with y. We will do this by entering the following loop, in which we compute these dot products, and store the results in the vector *dprodxy*.

```
>> dprodxy = [];
>> for i = 0:25
       dprodxy = [dprodxy, circshift(x, [1, i])*y'];
   end
```

To see the resulting dot products, we enter the following command.

```
>> dprodxy

dprodxy =
```

```
Columns 1 through 7

    0.0410   0.0402   0.0486   0.0340   0.0508   0.0363   0.0378

Columns 8 through 14

    0.0247   0.0344   0.0350   0.0366   0.0402   0.0324   0.0316

Columns 15 through 21

    0.0366   0.0720   0.0421   0.0338   0.0255   0.0427   0.0361

Columns 22 through 26

    0.0431   0.0326   0.0328   0.0408   0.0383
```

To find the number of times the entries in x should be shifted to the right so that the entries in the resulting vector most closely resemble the entries in y, we need to determine the position in *dprodxy* where the largest dot product is located. We can find the largest value in *dprodxy* by using the MATLAB **max** function as follows.

```
>> maxdprod = max(dprodxy)

maxdprod =

    0.0720
```

To determine the position in *dprodxy* where this largest value is located, we can use the MATLAB **find** function as follows.

```
>> i = find(dprodxy == maxdprod)

i =

    16
```

However, because vectors in MATLAB are expressed starting with the first entry in position 1 rather than 0, this value is actually one greater than the number of times the entries in x should be shifted to the right so that the entries in the resulting vector most closely resemble the entries in y. That is, the actual number of times the entries should be shifted is the following.

```
>> shift = i-1;
```

In the following nested **for** loop, we compute and store in a vector k the number of times the entries in x should be shifted to the right so that the entries in the resulting vector most closely resemble the entries in y, where y is constructed to contain the relative frequency with which each letter occurs in each of the six cosets in *cosets* separately.

```
>> k = [];
>> for j = 1:numel(cosets)
        dprodxy = [];
        y = double(letterfreq(cosets(j)));
        for i = 0:25
            dprodxy = [dprodxy, circshift(x, [1, i])*y'];
        end
        maxdprod = max(dprodxy);
        shift = find(dprodxy == maxdprod);
        k = [k, shift-1];
    end
```

Thus, the following vector k contains the numerical values that correspond to the letters in the keyword for the system.

```
>> k

k =

    15    11     0     2     4    18
```

In the following command, we convert these numerical values into their corresponding letters so that we can see the keyword for the system.

```
>> letters(k)

ans =

PLACES
```

Next, in order to decrypt *ctext*, we will convert the letters in *ctext* into their corresponding numerical values.

```
>> ctext = ltable(ctext);
```

In the following command, we recover the numerical values in the corresponding plaintext.

```
>> ptext = vigenere(ctext, -k);
```

Finally, we recover the letters in the corresponding plaintext by entering the following command.

```
>> stringprint(letters(ptext), 50)

APPALACHIANSTATEUNIVERSITYISLOCATEDINBOONEINTHEMOU
NTAINOUSAREAOFWESTERNNORTHCAROLINANORTHCAROLINASTA
TEUNIVERSITYISLOCATEDINRALEIGHINTHEEASTERNPORTIONO
FTHENORTHCAROLINAPIEDMONTREGIONRADFORDUNIVERSITYIS
LOCATEDINRADFORDINTHEMOUNTAINOUSAREAOFSOUTHWESTERN
VIRGINIA
```

Exercises

1. (a) Use the Vigenère square to encrypt the message THE EAGLE HAS LANDED with the keyword APOLLO.

    (b) Use the function f in (7.1) to encrypt the message THE EAGLE HAS LANDED with the keyword APOLLO.

2. The ciphertext SEFZTIHMCCUKOWH has been formed using a Vigenère cipher with the keyword GEORGE.

    (a) Use the Vigenère square to decrypt this ciphertext.

    (b) Use the inverse of the function f in (7.1) to decrypt this ciphertext.

3. Vigenère ciphers as we have described them (and as they are almost always described in literature) are actually not the ciphers Blaise de Vigenère himself described in his own writings. Vigenère did not use a keyword, but rather a single key letter, called the *priming* key. The first plaintext letter was encrypted with the priming key (still using the Vigenère square), and then each subsequent plaintext letter was encrypted with the previous plaintext letter as the key. The resulting type of cipher is sometimes called a Vigenère *autokey* cipher, since the key letters after the priming key are dictated by the message.

    (a) Use a Vigenère autokey cipher with priming key P to encrypt the message EBENEZER SCROOGE.

    (b) Decrypt the ciphertext RJHRCPWVLKMORF, which has been formed using a Vigenère autokey cipher with priming key P.

    (c) Create a general mathematical description of Vigenère autokey ciphers similar to our mathematical description of Vigenère ciphers immediately before Example 7.3.

(d) A second type of autokey cipher that Vigenère described used the ciphertext letters as the key letters after the priming key. That is, for this second type of Vigenère autokey cipher, the first plaintext letter was encrypted with the priming key (still using the Vigenère square), and then each subsequent plaintext letter was encrypted with the previous ciphertext letter as the key. Create a general mathematical description of this second type of Vigenère autokey cipher similar to our mathematical description of Vigenère ciphers immediately before Example 7.3.

4. Consider a ciphertext in which the number of times that each character occurs is shown in the following list.

A	B	C	D	E	F	G	H	I	J	K	L	M
2	14	9	8	8	12	7	5	19	14	8	4	9

N	O	P	Q	R	S	T	U	V	W	X	Y	Z
8	12	10	3	14	15	10	7	13	2	10	15	4

(a) Find the index of coincidence for the ciphertext.

(b) Based on your answer to part (a), which type of cipher is more likely to have produced the ciphertext, monoalphabetic or polyalphabetic? Why?

5. Consider a ciphertext in which the number of times that each character occurs is shown in the following list.

A	B	C	D	E	F	G	H	I	J	K	L	M
8	31	17	19	0	4	5	1	7	4	0	6	14

N	O	P	Q	R	S	T	U	V	W	X	Y	Z
10	22	5	3	6	0	18	3	19	11	2	8	19

(a) Find the index of coincidence for the ciphertext.

(b) Based on your answer to part (a), which type of cipher is more likely to have produced the ciphertext, monoalphabetic or polyalphabetic? Why?

6. Show that (7.2) is a correct formula for the index of coincidence for a sample of text.

7. For a ciphertext formed using a Vigenère cipher, the index of coincidence for the full ciphertext can be used to estimate the length of the keyword. This estimate is given by the following formula, where I is

the index of coincidence for the full ciphertext, and n is the length of the ciphertext.

$$k \approx \frac{0.0270n}{0.0655 - I + n(I - 0.0385)}$$

(a) Use this formula to estimate the length of the keyword for the first ciphertext in Example 7.4. Is your answer what you expected? Why?

(b) Use this formula to estimate the length of the keyword for the second ciphertext in Example 7.4. Is your answer what you expected? Why?

8. For a particular Vigenère ciphertext, suppose that n is the number of cosets, and let I_j be the index of coincidence for the jth coset. Also, for each n, let $\overline{I_{1:n}}$ be the average of the indices I_j for the cosets. For the ciphertext, the following table contains the values of I_j for $j = 1, 2, \ldots, n$ and the value of $\overline{I_{1:n}}$ for $n = 1, 2, \ldots, 9$.

n	I_1	I_2	I_3	I_4	I_5	I_6	I_7	I_8	I_9	$\overline{I_{1:n}}$
1	0.043									0.043
2	0.042	0.042								0.042
3	0.043	0.044	0.037							0.042
4	0.039	0.036	0.036	0.049						0.040
5	0.062	0.068	0.064	0.048	0.072					0.063
6	0.038	0.038	0.040	0.038	0.049	0.036				0.040
7	0.034	0.037	0.057	0.049	0.041	0.036	0.057			0.044
8	0.058	0.028	0.048	0.051	0.032	0.039	0.025	0.039		0.040
9	0.034	0.037	0.031	0.040	0.051	0.040	0.048	0.054	0.046	0.042

(a) What is the likely keyword length for the cipher? Why?

(b) If this table were continued for larger n, what should the next value of n be after your answer to part (a) for which $\overline{I_{1:n}}$ will be noticeably larger than for the surrounding values of n? Why? Can you predict some other values of n for which $\overline{I_{1:n}}$ should be noticeably larger than for the surrounding values of n?

9. Prove the Cauchy-Schwarz Inequality (Theorem 7.1).

10. Let $\mathbf{x}_0$ be a vector containing the relative frequencies shown in Table 6.1 for letters in ordinary English, and let $\mathbf{x}_i$ for $i = 1, 2, \ldots, 25$ be the vectors that result from shifting the entries in $\mathbf{x}_0$ to the right i positions, with the entries at the end wrapping around to the start. Also, let $\mathbf{y}$ be a vector containing the relative frequencies for the letters in a particular coset of a ciphertext that was formed using a Vigenère

cipher with a keyword of length five letters. The following table contains the values of $\mathbf{x}_i \cdot \mathbf{y}$ for $i = 0, 1, \ldots, 25$ and for $\mathbf{y}$ constructed from each of the five cosets in the ciphertext.

Shift	Key	$\mathbf{x}_i \cdot \mathbf{y}$				
i	Letter	Coset 1	Coset 2	Coset 3	Coset 4	Coset 5
0	A	0.030	0.026	0.066	0.032	0.033
1	B	0.035	0.035	0.046	0.042	0.036
2	C	0.038	0.049	0.034	0.034	0.032
3	D	0.035	0.038	0.030	0.034	0.029
4	E	0.027	0.042	0.048	0.047	0.069
5	F	0.046	0.051	0.041	0.039	0.037
6	G	0.027	0.045	0.032	0.037	0.045
7	H	0.037	0.036	0.028	0.039	0.034
8	I	0.033	0.034	0.049	0.030	0.031
9	J	0.034	0.038	0.026	0.033	0.047
10	K	0.037	0.037	0.036	0.037	0.032
11	L	0.036	0.029	0.037	0.042	0.034
12	M	0.038	0.024	0.040	0.036	0.030
13	N	0.039	0.038	0.034	0.050	0.035
14	O	0.048	0.039	0.036	0.039	0.039
15	P	0.040	0.036	0.044	0.041	0.028
16	Q	0.045	0.042	0.040	0.036	0.041
17	R	0.034	0.032	0.045	0.043	0.033
18	S	0.059	0.040	0.028	0.030	0.025
19	T	0.030	0.064	0.042	0.059	0.038
20	U	0.037	0.034	0.037	0.041	0.055
21	V	0.035	0.054	0.039	0.030	0.047
22	W	0.033	0.030	0.030	0.039	0.034
23	X	0.046	0.032	0.040	0.029	0.050
24	Y	0.037	0.041	0.037	0.038	0.042
25	Z	0.040	0.032	0.031	0.038	0.034

What is the likely keyword for the cipher? Why?

11. Let $\mathbf{x} = (x_0, x_1, \ldots, x_{m-1})$ be a vector in R^m representing a plaintext, and let $\mathbf{k} = (k_0, k_1, \ldots, k_{p-1})$ and $\mathbf{l} = (l_0, l_1, \ldots, l_{q-1})$ be vectors in R^p and R^q, respectively, representing keywords for Vigenère ciphers. Choose an encryption function $f : R \to R$ as follows for $i = 0, 1, \ldots, m - 1$.

$$f(x_i) = \left(x_i + k_{(i \bmod p)} + l_{(i \bmod q)} \right) \bmod |R|$$

The vector $f(\mathbf{x}) = (f(x_0), f(x_1), \ldots, f(x_{m-1}))$ forms the ciphertext.

(a) Use an encryption function of this form with the keywords WEST and POINT to encrypt the message US MILITARY ACADEMY.

(b) Decrypt the ciphertext CJLMVHJVWAS, which has been formed using an encryption function of this form with the keywords ELAINE and SUSIE.

(c) Does an encryption function of this form give more security than a normal Vigenère encryption function? If so, how much more? Completely explain your answer, and be as specific as possible.

Computer Exercises

12. Use a Vigenère cipher to encrypt the message THOMAS JEFFERSON HAD MANY MATHEMATICAL INTERESTS INCLUDING CRYPTOGRAPHY.

13. Decrypt the ciphertext RFBANLXGSDZODWBRFMDQMGYVVOPZSEKQRDGOE-HIVPKCFYXPEW, which has been formed using a Vigenère cipher with the keyword ORVILLE.

14. Decrypt the ciphertext EVKEF XYELZ XTZUK MGKLE WUIPR EMNWN VIGKL CWHKY KPEPW LJSGE EMWXK LHDES TYRPX SYRAY UKHCH AFXAI PVPSO IIXAR LUYYJ YFFJM GHVCG VERTY UJHIH VALXM GZEBW UCIXE GRQJK IWKLE DITEM ZSNGZ KLXKV ESMLV XRRDA NJFXE IFSWK SKJMN LBIIX USCMG VRMJS NDSFR XFJTZ YJIWF GUEYE XLYES WPVVU VINVY TVRGX EVNYI XEGRQ JKIWU SCMGV RMFYT DCEMG XXHWJ IIVZW EDITE MZSNG ZKLXK VESML VXYES FYMIK SIEFX VGKPT TWXVZ XEXHG OXLFR RYHYF TEVMN UFLHB EKSGG VJKFQ TZYEW TYEVW NIMXU, which has been formed using a Vigenère cipher.

15. Decrypt the ciphertext FWPSI DTSAG SDODY MGYUI PXJRS GLPXH HQSEC THRJV ACNMC YSLTH MCIRK TSAAY DBPWN JSZSO JUGLU ELYBV WEUXG YEXNO OTYFX YAEXI HRURT LBGXY IDRYN RJTZY BREME CFART VRDTH VJPOF LIGSR FFSYE ECYZZ LRFVT EJLBJ WIYYB RGRSV JNGLR NOTCA KKHPJ OYSXY, which has been formed using a Vigenère cipher.

16. Write a single Maple or MATLAB procedure that will encrypt or decrypt a given message using a Vigenère cipher. This procedure should have three input parameters. The first input parameter should be a string containing a plaintext or ciphertext, the second input parameter should be string containing a keyword for the system, and the third input parameter should be an indicator for whether the message is to be encrypted or decrypted.

17. Write a single Maple or MATLAB procedure that will perform cryptanalysis on a given Vigenère ciphertext. This procedure should have only one input parameter, a string containing the ciphertext to be decrypted, and should return the keyword as well as the plaintext.

Research Exercises

18. Investigate the treatise written by Blaise de Vigenère in the late 1500s in which Vigenère ciphers were introduced, and write a summary of your findings. Include in your summary a description of some of the other topics about which Vigenère wrote in his treatise.

19. Investigate the life and career of Charles Babbage, a 19th century British mathematician, who developed an effective attack against Vigenère ciphers, and write a summary of your findings. Include in your summary a description of Babbage's attack against Vigenère ciphers, and how it differs from and is similar to the attack that we used in this chapter.

20. Investigate the life and career of Friedrich Kasiski, a 19th century Prussian military officer, who developed an effective attack against Vigenère ciphers, and write a summary of your findings. Include in your summary a description of Kasiski's attack against Vigenère ciphers, and how it differs from and is similar to the attack that we used in this chapter.

21. Find a copy of a short article that appeared on page 61 of *Scientific American, Supplement LXXXIII*, published January 27, 1917, in which the security of Vigenère ciphers was discussed, and write a summary of this article.

22. Investigate the life and career of William Friedman, the so-called Founder of Modern Cryptography, who first presented the idea of the index of coincidence for a sample of text, and write a summary of your findings. Include in your summary a brief overview of Friedman's education, his work at Riverbank Laboratories and with the U.S. military during World Wars I and II, his work with the U.S. government's "Black Chamber" and Armed Forces and National Security Agencies, and some information about the cryptographic achievements of his wife, Elizebeth.

23. Investigate the use of Vigenère ciphers by the Confederacy during the American Civil War, and write a summary of your findings.

24. Investigate the cipher wheel that was created by Thomas Jefferson, and write a summary of your findings. Include in your summary when Jefferson created his cipher wheel, how it was constructed and operated, when and where it was actually used as a U.S. Army field cipher, and why it was not used when Jefferson first created it.

Chapter 8

RSA Ciphers

In this chapter, we will present one of the most well known and popular cryptosystems ever developed, the *RSA* cryptosystem. The system is named for Ron Rivest, Adi Shamir, and Len Adleman, a trio of researchers at the Massachusetts Institute of Technology who developed the system and published it publicly in 1978. More than two decades later it was revealed that Clifford Cocks, while working in a classified environment for the British cryptologic agency GCHQ, discovered the algorithm prior to Rivest, Shamir, and Adleman.

A couple of reasons why the RSA system is so well known and popular are that it can be constructed to be extremely secure while still being very easy to use, and the mathematics that underlies the system is truly fascinating. However, the main reason why the RSA system is so well known and popular is that when it was published, it was the world's first *public-key* cryptosystem. Recall that when analyzing the security or functionality of a particular cryptosystem, we generally assume that everything about the system is public knowledge except for the parameters in the encryption function. These parameters must usually be kept secret to prevent intruders from determining the corresponding decryption function. A public-key cryptosystem is a system in which the parameters in the encryption function can be public knowledge without allowing intruders to determine the corresponding decryption function. That is, using the notation that we introduced in Section 6.1, a public-key cryptosystem is a system in which the encryption function f can be public knowledge without revealing f^{-1}.

Before we formally present the details of the RSA cryptosystem, we will look at a very simple example of the mathematics that underlies the system. Consider primes $p = 5$ and $q = 11$, and let $n = pq = 55$ and $m = (p-1)(q-1) = 40$. Next, consider $a = 27$ and $b = 3$, chosen so that

$ab = 1 \bmod m$. Then for $x = 2$, note the following.

$$x^{ab} = 2^{27 \cdot 3} = 2417851639229258349412352 = 2 \bmod 55 = x \bmod n$$

The important thing to note about the preceding computation is the following.

$$x^{ab} = x \bmod n \qquad (8.1)$$

In fact, this equation will hold for any x in the ring $\mathbb{Z}_n = \{0, 1, 2, \ldots, n-1\}$ with the usual operations of addition and multiplication modulo n because of the fact that a and b satisfy the following equation.

$$ab = 1 \bmod m \qquad (8.2)$$

Thus, if we encrypt a numerical plaintext by raising the plaintext to the power a, then we can decrypt the numerical ciphertext by raising the ciphertext to the power b and reducing modulo n. It is certainly not obvious that (8.1) will hold for any $x \in \mathbb{Z}_n$ provided (8.2) is true. Establishing this result will be one of our goals in Section 8.1.

8.1   Preliminary Mathematics

Before establishing the fact that (8.1) will hold for any $x \in \mathbb{Z}_n$ provided (8.2) is true, we will present some additional preliminaries. We will begin by presenting how values of a and b can be determined that satisfy (8.2). Of course, it is not difficult to choose a, as it must only have a multiplicative inverse modulo m. To guarantee this, we must only require that a satisfy $\gcd(a, m) = 1$, or, in other words, that a and m be *relatively prime*. Once a is chosen, we can find b by constructing the Euclidean algorithm table (see Section 1.7) for a and m. We illustrate this in the following example.

Example 8.1 Consider $a = 27$ and $m = 40$. To find a value for b that satisfies $ab = 1 \bmod m$, we first apply the Euclidean algorithm to a and m as follows.

$$40 = 27 \cdot 1 + 13$$
$$27 = 13 \cdot 2 + 1$$

Thus, as required, $\gcd(a, m) = 1$. It can easily be verified that these equations yield the following Euclidean algorithm table.

Row	Q	R	U	V
−1	−	40	1	0
0	−	27	0	1
1	1	13	1	−1
2	2	1	−2	3

Therefore, $40(-2) + 27(3) = 1$. This immediately gives the result, for it states that $27 \cdot 3 = 1 \bmod 40$, and thus $b = 3$ satisfies $ab = 1 \bmod m$. □

Next, we will show the general relationship between the values of $n = pq$ and $m = (p-1)(q-1)$ in the example in the introduction to this chapter. To do this, we must first prove some general results about the ring $\mathbb{Z}$ of integers.

Let $n \in \mathbb{Z}$ with $n > 1$. Then $\mathbb{Z}_n$ inherits many properties from $\mathbb{Z}$ since it is a quotient ring of $\mathbb{Z}$. Consider the set U_n of units in $\mathbb{Z}_n$. That is, consider the set U_n consisting of all $x \in \mathbb{Z}_n$ that have a multiplicative inverse in $\mathbb{Z}_n$. Note that U_n can be expressed as $U_n = \{x \in \mathbb{Z}_n \mid \gcd(x, n) = 1\}$, and that U_n forms a group with the operation of multiplication. The order of U_n is denoted as a function by $\phi(n)$, called the *Euler-phi* function.

Lemma 8.1 *For prime p, $U_p = \mathbb{Z}_p^*$ (the nonzero elements in $\mathbb{Z}_p$), and $\phi(p) = p - 1$.*

Proof. Exercise. □

As it turns out, it is also not hard to find $\phi(p^k)$ for a positive integer k. If x is in $\mathbb{Z}_{p^k}$ for prime p and positive integer k, then $\gcd(x, p^k) \neq 1$ if and only if $p|x$, which occurs only when $x = p, 2p, 3p, \ldots, p^{k-1}p$, or exactly p^{k-1} times. Thus $\gcd(x, p^k) = 1$ for exactly $p^k - p^{k-1}$ values of x, giving the following result.

Lemma 8.2 *For prime p and positive integer k, $\phi(p^k) = p^k - p^{k-1}$.*

Now, suppose r and s are relatively prime integers, and consider the mapping $\sigma : \mathbb{Z} \mapsto \mathbb{Z}_r \times \mathbb{Z}_s$ given by $\sigma(a) = (a \bmod r, a \bmod s)$. Clearly σ is a homomorphism whose kernel is the ideal in $\mathbb{Z}$ generated by rs. Thus $\mathbb{Z}_{rs}$ and $\mathbb{Z}_r \times \mathbb{Z}_s$ are isomorphic, since the homomorphism between them induced by σ is one-to-one and they have the same order. A positive integer a less than rs is relatively prime to rs if and only if it is relatively prime to both r and s. Thus a is a unit in $\mathbb{Z}_{rs}$ if and only if $a \bmod r$ is a unit in $\mathbb{Z}_r$ and $a \bmod s$ is a unit in $\mathbb{Z}_s$, giving the following result.

Lemma 8.3 *For relatively prime integers r and s, $\phi(rs) = \phi(r)\phi(s)$.*

The reason Lemmas 8.1 and 8.3 are of interest to us is because of the following theorem, which gives the general relationship between the values of $n = pq$ and $m = (p-1)(q-1)$ in the example in the introduction to this chapter.

Theorem 8.4 *For distinct primes p and q, $\phi(pq) = (p-1)(q-1)$.*

Proof. Exercise. □

We will now show that (8.1) will hold for any $x \in \mathbb{Z}_n$ provided (8.2) is true. The main result that we will use to show this is the following theorem.

Theorem 8.5 (Fermat's Little Theorem) *Let p be a prime, and suppose $x \in \mathbb{Z}$ satisfies $\gcd(x, p) = 1$. Then $x^{p-1} = 1 \bmod p$.*

Proof. Consider first the following set S_1.

$$S_1 = \{x \bmod p, \ 2x \bmod p, \ \dots, \ (p-1)x \bmod p\}$$

We claim that the elements in S_1 are a rearrangement of the elements in the following set S_2.

$$S_2 = \{1, 2, \dots, p-1\}$$

To see this, note that if $jx \bmod p = kx \bmod p$ for some positive integers j and k less than p, then $p|(j-k)x$. However, since p does not divide x, this implies that $p|j-k$. Thus, because j and k are less than p, it follows that $j = k$. Now, since $S_1 = S_2$, the product of the elements in S_1 will equal the product of the elements in S_2. That is, $x^{p-1}(p-1)! \bmod p = (p-1)! \bmod p$. Thus, $p|(p-1)!(x^{p-1}-1)$. However, since p does not divide $(p-1)!$, then $p|x^{p-1} - 1$, or, equivalently, $x^{p-1} = 1 \bmod p$. $\qquad \square$

In the following theorem, we establish the fact that (8.1) will hold for any $x \in \mathbb{Z}_n$ provided (8.2) is true.

Theorem 8.6 *Let p and q be distinct primes, and suppose $n = pq$ and $m = \phi(n) = (p-1)(q-1)$. If a and b are integers with $ab = 1 \bmod m$, then $x^{ab} = x \bmod n$ for all $x \in \mathbb{Z}_n$.*

Proof. If $ab = 1 \bmod m$, then $ab = 1 + km$ for some $k \in \mathbb{Z}$, and for all $x \in \mathbb{Z}_n$ the following will hold.

$$x^{ab} = x^{1+km} = x(x^{km}) = x(x^{p-1})^{k(q-1)}$$

If $\gcd(x, p) = 1$, then by Theorem 8.5 we know that $x^{p-1} = 1 \bmod p$. Thus, $x^{ab} = x(1)^{k(q-1)} \bmod p = x \bmod p$. Also, if $\gcd(x, p) \neq 1$, then $x = 0 \bmod p$, and certainly $x^{ab} = x \bmod p$. Similarly, $x^{ab} = x \bmod q$ for all $x \in \mathbb{Z}$. Thus, $p|x^{ab} - x$ and $q|x^{ab} - x$, and so $pq|x^{ab} - x$. That is, $n|x^{ab} - x$, or, equivalently, $x^{ab} = x \bmod n$. $\qquad \square$

Incidentally, Theorem 8.6 holds even if n is only a single prime p, and $m = \phi(n) = p - 1$. That is, the following theorem is true.

Theorem 8.7 *For prime p, if a and b are integers with $ab = 1 \bmod (p-1)$, then $x^{ab} = x \bmod p$ for all $x \in \mathbb{Z}_p$.*

Proof. Exercise. $\qquad \square$

Even before the RSA cryptosystem was published, Theorem 8.7 was used by cryptographers to create ciphers. In such ciphers, plaintext integers x are encrypted by being raised to the power a and reduced modulo p. Resulting ciphertext integers $x^a \bmod p$ can be decrypted by being raised to the power b and reduced modulo p, since $(x^a)^b = x^{ab} = x \bmod p$. These ciphers are called *exponentiation* ciphers. Note that in exponentiation ciphers, a and p must be kept secret from intruders, for otherwise intruders could easily find b using the Euclidean algorithm. Thus, exponentiation ciphers are not public-key. This makes them, while mathematically similar to RSA from the perspective of implementation, ultimately useful only under completely different circumstances. Changing the modulus to be the product of two distinct primes, as in RSA, was a major advancement, indeed leading to a full-scale revolution in modern cryptology, because it allows the resulting ciphers to be public-key. In the next section we will describe why changing the modulus to be the product of two distinct primes allows the resulting ciphers to be public-key.

8.2 Encryption and Decryption

To encrypt a message using the RSA cryptosystem, we first convert the plaintext into a list of nonnegative integers. In this chapter, we will again assume that all messages are written using only the characters in the alphabet $L = \{A, B, C, \ldots, Z\}$, and associate each character with its corresponding element in the ring $R = \mathbb{Z}_{26}$ under the bijection $\alpha : L \to R$ that we used in Chapter 6 (i.e., $A \mapsto 0$, $B \mapsto 1$, $C \mapsto 2, \ldots, Z \mapsto 25$). We then choose distinct primes p and q, and let $n = pq$ and $m = (p-1)(q-1)$. Next, we choose $a \in \mathbb{Z}_m^*$ with $\gcd(a, m) = 1$, and find $b \in \mathbb{Z}_m^*$ that satisfies $ab = 1 \bmod m$. To encrypt a numerical plaintext, we raise the plaintext integers to the power a and reduce modulo n. According to Theorem 8.6, to decrypt the numerical ciphertext, we must only raise the ciphertext integers to the power b and reduce modulo n.

Example 8.2 In this example, we will use the RSA cryptosystem to encrypt the message COSMO and decrypt the resulting ciphertext. We will begin by converting this message into a list of elements in $\mathbb{Z}_{26}$.

	C	O	S	M	O
$\alpha \Rightarrow$	2	14	18	12	14

We then choose primes $p = 5$ and $q = 11$, and let $n = pq = 55$ and $m = (p-1)(q-1) = 40$. Next, we choose an encryption exponent $a = 27$ that satisfies $\gcd(a, m) = 1$. To encrypt the message, we calculate the following.

$$2^{27} \ = \ 18 \bmod 55$$
$$14^{27} \ = \ 9 \bmod 55$$
$$18^{27} \ = \ 17 \bmod 55$$
$$12^{27} \ = \ 23 \bmod 55$$
$$14^{27} \ = \ 9 \bmod 55$$

Thus, the ciphertext is the list of integers 18, 9, 17, 23, 9.[1] From Example 8.1, we know that the decryption exponent that corresponds to the encryption exponent $a = 27$ in this example is $b = 3$. As a result, to decrypt the ciphertext, we must only calculate the following.

$$18^3 \ = \ 2 \bmod 55$$
$$9^3 \ = \ 14 \bmod 55$$
$$17^3 \ = \ 18 \bmod 55$$
$$23^3 \ = \ 12 \bmod 55$$
$$9^3 \ = \ 14 \bmod 55$$

Note that the resulting integers are the original plaintext integers.   □

We still have several topics to address regarding the RSA cryptosystem. First, note that no matter how large we choose the encryption exponent and modulus for the RSA system, the system as illustrated in Example 8.2 will not be secure because it will just yield a substitution cipher. However, we can use the RSA encryption procedure exactly as we presented it to obtain a non-substitution cipher by simply grouping consecutive integers in the plaintext before encrypting. Because our exponentiation operations are done modulo n, we will still be able to convert between plaintext and ciphertext uniquely provided the plaintext integers are grouped into integer blocks that are less than n. We illustrate this in the following example.

Example 8.3 In this example, we will again use the RSA cryptosystem to encrypt the message COSMO and decrypt the resulting ciphertext, but this time we will group consecutive plaintext integers in pairs. We begin by choosing primes $p = 79$ and $q = 151$, and forming $n = pq = 11929$ and $m = (p-1)(q-1) = 11700$. Next, we choose an encryption exponent $a = 473$ that satisfies $\gcd(a, m) = 1$. Recall that our plaintext converts to the list of integers 2, 14, 18, 12, 14, which we will express in pairs as the four-digit integer blocks 0214, 1812, 1400, each of which is less than our value of $n = 11929$. (Note that in the first four-digit integer block we use 02 for 2, which would be necessary for decrypting the resulting ciphertext if the single digit 2 were not at the beginning of the block. Note also that

---

[1]Although we coincidentally could in this example, converting an RSA ciphertext back into a list of letters is not usually possible. To see this, note that the encryption results in this example are reduced modulo 55, and thus could be as large as 54.

we append two extra arbitrary single digits to the end of the last two-digit plaintext integer to form the last four-digit block so that each block will represent a pair of characters.) We can then use the RSA encryption procedure exactly as we presented it. That is, to encrypt the message, we calculate the following.

$$0214^{473} = 2789 \bmod 11929$$
$$1812^{473} = 11174 \bmod 11929$$
$$1400^{473} = 6727 \bmod 11929$$

Thus, the ciphertext is the list of integers 2789, 11174, 6727. To decrypt this ciphertext, we would first use the Euclidean algorithm to find that $b = 8237$ satisfies $ab = 1 \bmod m$. We would then calculate the following.

$$2789^{8237} = 0214 \bmod 11929$$
$$11174^{8237} = 1812 \bmod 11929$$
$$6727^{8237} = 1400 \bmod 11929$$

Note that the resulting integers are the four-digit plaintext integer blocks, which we could then split into the original two-digit plaintext integers. □

Another topic that we must address regarding the RSA cryptosystem is how the system actually progresses between two people wishing to exchange a secret message. We stated in the introduction to this chapter that the RSA cryptosystem is a public-key system. This forces the system to progress in a particular way.

Recall that when analyzing the security or functionality of a particular cryptosystem, we generally assume that almost everything about the system is public knowledge, including the form of the encryption function. This means we would assume an intruder who intercepts an RSA ciphertext would know that each ciphertext integer was formed as $x^a \bmod n$ for some plaintext integer x and positive integers a and n. The fact that the RSA cryptosystem is a public-key system means we would also assume the intruder knows the actual values of a and n used in the calculation. For example, we would assume an intruder who intercepts the ciphertext in Example 8.3 would know that each ciphertext integer was formed as $x^{473} \bmod 11929$ for some plaintext integer x. Although this obviously affects the security of the system, we make this assumption because in practice the RSA system is used with a and n being public knowledge. The benefit to this is that two people wishing to use the RSA system to exchange a secret message across an insecure line of communication do not have to figure out a way to securely exchange an encryption exponent and modulus.

The comments we made in the preceding paragraph imply that the RSA system in Example 8.3 is not mathematically secure. This is because an intruder could mathematically break the system as follows. After factoring to find the values of p and q from $n = pq = 11929$, the intruder could form $m = (p-1)(q-1)$, use the Euclidean algorithm to find that $b = 8237$ satisfies $ab = 1 \bmod m$, and decrypt the ciphertext by raising the ciphertext integers to the power b and reducing modulo n. Thus, none of the calculations necessary to break this system would take an intruder more than just a few minutes. Furthermore, even with significantly larger numbers, the Euclidean algorithm and modular exponentiation could easily be efficiently programmed on a computer. However, the first step in this procedure requires the intruder to factor $n = pq$. It is the apparent difficulty of this problem, provided p and q are both very large, that gives the RSA system its extremely high level of security. For example, if p and q were both hundreds of digits long, then the fastest known factoring algorithms would in general take millions of years to factor $n = pq$, even when programmed on a computer that could perform millions of operations per second.[2] Thus, even if the encryption exponent a was public knowledge, an intruder should not be able to determine the decryption exponent b. This is precisely why the RSA cryptosystem is a *public-key* system. The parameters in the encryption function $f(x) = x^a \bmod n$ can be public knowledge without revealing the parameter b in the decryption function $f^{-1}(x) = x^b \bmod n$.

We now describe how the RSA system actually progresses between two people wishing to exchange a secret message across an insecure line of communication. Since only the intended recipient of the message must be able to decrypt the ciphertext, the intended recipient initiates the process by choosing primes p and q, and forming $n = pq$ and $m = (p-1)(q-1)$. The intended recipient then chooses an encryption exponent $a \in \mathbb{Z}_m^*$ with $\gcd(a, m) = 1$, and, using the Euclidean algorithm if necessary, finds $b \in \mathbb{Z}_m^*$ that satisfies $ab = 1 \bmod m$. The intended recipient then sends the values of a and n to the originator of the message across the insecure line of communication, forcing the assumption that a and n are public knowledge. The originator of the message encrypts the message by applying the function $f(x) = x^a \bmod n$ to the plaintext integers, and then sends the resulting ciphertext integers to the intended recipient across the insecure line of communication. Since only the intended recipient knows b, only the intended recipient can decrypt the ciphertext by applying the function $f^{-1}(x) = x^b \bmod n$ to the ciphertext integers.

Example 8.4 Suppose that we wish to use the RSA cryptosystem to send the secret message COSMO to a colleague across an insecure line of communication. Our colleague initiates the process by choosing primes p and q, and calculating $n = pq = 363794227$ and $m = (p-1)(q-1)$. Next, our colleague

---

[2] We will make some comments on choosing very large primes in Section 8.5.1, and on factoring numbers with very large prime factors in Section 8.5.2.

chooses an encryption exponent $a = 13783$ that satisfies $\gcd(a, m) = 1$. Our colleague then sends the values of a and n to us across the insecure line of communication. Recall that our plaintext converts to the list of integers 02, 14, 18, 12, 14. With the value for n that our colleague has sent us, we can group all five of these two-digit plaintext integers into a single block that will be less than n. That is, we can express our plaintext as 0214181214, and encrypt our message by applying the function $f(x) = x^a \bmod n$ to this single plaintext integer. To encrypt the message, we calculate the following.

$$0214181214^{13783} \quad = \quad 137626763 \bmod 363794227$$

We would then transmit the ciphertext integer 137626763 to our colleague across the insecure line of communication. In order for an intruder who intercepts this ciphertext and the previously transmitted values of a and n to decrypt the ciphertext, the intruder would need to find the decryption exponent b. However, to find b, the intruder would first need to find m. Also, to find m, the intruder would need to find the prime factors of n, a problem that, as we have stated, is essentially impossible provided our colleague has chosen sufficiently large p and q. This would not pose a problem for our colleague, however, because our colleague initiated the process by choosing p and q. Thus, our colleague would know that the prime factors of $n = pq = 363794227$ are $p = 14753$ and $q = 24659$, and would have no difficulty in forming $m = (p-1)(q-1) = 363754816$, and using the Euclidean algorithm to find that $b = 20981287$ satisfies $ab = 1 \bmod m$. To decrypt the ciphertext, our colleague would then only need to calculate the following.[3]

$$137626763^{20981287} \quad = \quad 0214181214 \bmod 363794227$$

$\square$

8.3 RSA Ciphers with Maple

We will begin this section by demonstrating several Maple functions that are useful for finding large primes. The first function that we will demonstrate is **nextprime**, which returns the smallest prime larger than an integer input. For example, the following command returns the smallest prime larger than 400043344212007458000000000000000000.

```
>   nextprime(400043344212007458000000000000000000);
            400043344212007458000000000000000161
```

---

[3]We will make some comments on efficiently raising large numbers to large powers in Section 8.5.3.

A similar function is **prevprime**, which returns the largest prime smaller than an integer input. For example, the following command returns the largest prime smaller than 4000433442120074580000000000000000000.

```
> prevprime(4000433442120074580000000000000000000);
         4000433442120074579999999999999999941
```

A final primality function that we will demonstrate is **isprime**, which returns *true* if a given integer input is prime, and *false* if the integer input is not prime. For example, from the following commands we can conclude that 4000433442120074579999999999999999941 is prime, but that 4000433442120074580000000000000000000 is not.

```
> isprime(4000433442120074579999999999999999941);
```
$$true$$

```
> isprime(4000433442120074580000000000000000000);
```
$$false$$

We should also note that the **nextprime**, **prevprime**, and **isprime** functions are probabilistic routines that employ primality tests (see Section 8.5.1). This means that the output returned by Maple is guaranteed to be correct with an extremely high probability, but not absolutely.

We will now show how Maple can be used to encipher and decipher messages using the RSA cryptosystem. We begin by finding and assigning values for the primes p and q for the system.

```
> p := nextprime(4000433442120074580000000000000000000);
       p := 4000433442120074580000000000000000161
```

```
> q := nextprime(5000300663662690012000000000000000000);
       q := 5000300663662690012000000000000000009
```

Next, we assign the values of $n = pq$ and $m = (p-1)(q-1)$ for the system.

```
> n := p*q;
n :=
    200033699955714283337472000688611033705230782877376315200\
    000000000001449
```

```
> m := (p-1)*(q-1);
m :=
    200033699955714283337472000688611032805157372299099856000\
    000000000001280
```

Also, we will use the following value as the encryption exponent a for the system.

```
>  a := 1009876890098767900009100033:
```

To verify that the preceding value of a is a valid RSA encryption exponent for our value of m, we will use the Maple **igcd** function, which is designed to calculate the greatest common divisor of a pair of integer inputs. The following command returns the greatest common divisor of the integers a and m. Note that, as required, $\gcd(a, m) = 1$.

```
>  igcd(a, m);
```

$$1$$

We will now use the RSA encryption procedure to encipher the message RETURN TO HEADQUARTERS. Because one of the functions that we will use is in the Maple **StringTools** package, we will include this package.

```
>  with(StringTools):
```

Next, we will store our message as the following string.

```
>  message := Select(IsAlpha, "RETURN TO HEADQUARTERS");
```
$$message := \text{``RETURNTOHEADQUARTERS''}$$

To convert this plaintext message into numerical form, we will use the user-written function **tonumber**, which we have written separately from this Maple session and saved as the text file tonumber.mpl. To use this function, we must first read it into our Maple session as follows.

```
>  read "tonumber.mpl";
```

The following command then illustrates how the function **tonumber** can be used. Specifically, by entering the following command, we convert *message* into a list of two-digit integers, and group all of these two-digit integers into a single block.

```
>  plaintext := tonumber(message);
```
$$plaintext := 17041920171319140704000316200017190141718$$

To check whether this numerical representation of the plaintext is less than n, we can use the Maple **evalb** function, which returns *true* if a given input statement is true, *false* if the input statement is false, and *FAIL* if the input statement cannot be determined to be true or false. For example, the following command shows that the numerical representation of the plaintext is less than n.

```
> evalb(plaintext < n);
```
$$true$$

Because the single plaintext block is less than n, we can encipher our message as a single block. That is, we can encipher our message by raising *plaintext* to the power a and reducing modulo n. To do this, we will enter the following command, in which, because this modular exponentiation involves such a large exponent, we use the Maple $\&\hat{}$ function instead of just $\hat{}$ for the exponentiation. By using $\&\hat{}$, we cause Maple to do the exponentiation in a very efficient way, like the technique that we will present in Section 8.5.3.

```
> ciphertext := plaintext &^ a mod n;
```

$ciphertext :=$
      12018676826619146902912210512051163727002429551804772430$6\backslash$
      279524922980340

In order to decrypt this ciphertext, we must first find a decryption exponent b that satisfies $ab = 1 \bmod m$. We can do this by using the Maple **igcdex** function, which, like **igcd**, is designed to calculate the greatest common divisor of a pair of integer inputs. However, in addition to taking inputs a and m for which it is to calculate the greatest common divisor, **igcdex** also takes two additional user-defined variable inputs, which it assigns as integers b and v that satisfy $ab + mv = \gcd(a, m)$. Since $\gcd(a, m) = 1$, these will be values of b and v that satisfy $ab + mv = 1$, or, equivalently, $ab = 1 \bmod m$. Thus, we can find a decryption exponent b for this system by entering the following command.

```
> igcdex(a, m, 'b', 'v');
```
$$1$$

To express the decryption exponent b assigned by the previous command as a positive integer less than m, we enter the following command.

```
> b := b mod m;
```

$b :=$
      14223052108886384125567788273486158075885329391129377808$4\backslash$
      122869663461377

An alternative method for finding a decryption exponent b that satisfies $ab = 1 \bmod m$ is to simply raise a to the power -1 and reduce the result modulo m. This is illustrated in the following command.

```
>  b := a^(-1) mod m;
```
$b :=$
  $14223052108886384125567788273486158075885329391129377 8084\backslash$
  122869663461377

By entering the following command, we verify that this value of b satisfies $ab = 1 \bmod m$.

```
>  a*b mod m;
```
$$1$$

To recover the plaintext integer, we must only raise *ciphertext* to the power b and reduce modulo n.

```
>  plaintext := ciphertext &^ b mod n;
```
  $plaintext := 170419201713191407040003162000171904 1718$

In order to see the original plaintext letters, we must split this single plain-text integer into a list of two-digit integers, and then convert this list of two-digit integers back into a list of letters. To do this, we will use the user-written function **toletter**, which we have written separately from this Maple session and saved as the text file toletter.mpl. To use this function, we must first read it into our Maple session as follows.

```
>  read "toletter.mpl";
```

The following command then illustrates how the function **toletter** can be used. Specifically, by entering the following command, we convert *plaintext* back into a list of letters.

```
>  toletter(plaintext);
```
                "RETURNTOHEADQUARTERS"

For a plaintext whose numerical representation is larger than n, we can split the plaintext into shorter blocks of characters whose numerical representations are all less than n, and then encipher each block separately. For example, consider the plaintext RETURN TO HEADQUARTERS NO LATER THAN TWELVE O'CLOCK NOON THE TWENTY-FIFTH OF THE MONTH OF DECEMBER. We first store this message as the following string.

```
>  message := Select(IsAlpha, "RETURN TO HEADQUARTERS NO

   LATER THAN TWELVE O'CLOCK NOON THE TWENTY-FIFTH OF THE

   MONTH OF DECEMBER");
```

message :=

"RETURNTOHEADQUARTERSNOLATERTHANTWELVEOC\
LOCKNOONTHETWENTYFIFTHOFTHEMONTHOFDECEM\
BER"

Next, we use **tonumber** to convert *message* into a list of two-digit integers, and group all of these two-digit integers into a single block.

```
> plaintext := tonumber(message);
```

plaintext :=

1704192017131914070400031620001719041718131411100191041719\
0700131922041121041402111402101314141319070419220413 92405\
0805190714051907041214131907140503040204120104 17

The next command shows that this numerical representation of the plain-text is larger than n.

```
> evalb(plaintext < n);
```
$$false$$

Because *plaintext* is larger than n, it cannot be encrypted as a single block, since the corresponding decryption operation done modulo n would have to result in a value less than n. To encrypt this message, we must first split it into shorter blocks of characters whose numerical representations are all less than n. To do this, we first identify the number of digits in n as follows, by using the Maple **convert** function to convert n from an integer into a string, and then the Maple **length** function to find the number of characters in the resulting string.

```
> length(convert(n, string));
```
$$72$$

Since each character in our message is represented numerically by a two-digit integer, any block with fewer than 36 characters would have a numerical representation with fewer than 72 digits, which would thus be less than n. The Maple **LengthSplit** function, which is in the **StringTools** package (which we have previously included in this Maple session), allows us to split our message into appropriate blocks as follows. The output from this command is a sequence of blocks containing the characters from the first parameter *message* in order, with the length of each block until the last given by the second parameter 35, and the last block ending when the characters in *message* have all been used.

```
> messblocks := [LengthSplit(message, 35)];
```

messblocks :=
    ["RETURNTOHEADQUARTERSNOLATERTHANTWEL",
    "VEOCLOCKNOONTHETWENTYFIFTHOFTHEMONT",
    "HOFDECEMBER"]

Next, we convert each of these character blocks into numerical form.

> `plaintext := tonumber(messblocks);`

plaintext :=
    [
    17041920171319140704000316200017190417181314110019041719\0
    7001319220411,
    21041402111402101314141319070419220413192405080519071405\1
    9070412141319, 71405030402041201041 7]

We next use the **map** and **evalb** functions to show that each numerical
plaintext block is less than n.

> `map(x -> evalb(x < n), plaintext);`

$$[true, true, true]$$

We now use the **map** function to encrypt each numerical plaintext block.

> `ciphertext := map(x -> x &^ a mod n, plaintext);`

ciphertext :=
    [
    19703396137288274108909612870367135168392942487700116319\1
    632702996886750,
    13953095406847145447445534072630962611332175507695398143\0
    24799503409023,
    16177171603718971903105692681896751036431484608034629182\2
    505488116883006]

We can then decrypt this message as follows.

> `plaintext := map(x -> x &^ b mod n, ciphertext);`

plaintext :=
    [
    17041920171319140704000316200017190417181314110019041719\0
    7001319220411,
    21041402111402101314141319070419220413192405080519071405\1
    9070412141319, 71405030402041201041 7]

```
>  toletter(plaintext);
```

"RETURNTOHEADQUARTERSNOLATERTHANTWELVEOCLO\
    CKNOONTHETWENTYFIFTHOFTHEMONTHOFDECEMBE\
    R"

A final Maple function that we will demonstrate in this section is **ifac-tor**, which returns the prime factorization of an integer input. For example, the following command very quickly returns the prime factorization of the 73-digit integer 53390460378134204240999353215010288928336816186750543955-75633065999923549.

```
>  ifactor(533904603781342042409993532150102889283368161861867505
   43955756336065999923549);
```
$$(23)^3(107)(45001200019828331)^4$$

Recall that the security of the RSA cryptosystem is based on the apparent difficulty of factoring the value of $n = pq$ used in the system. Thus, in order for the last RSA system that we used in this section to be secure, it should be very difficult for an intruder to factor the 72-digit value of n from the system. Although this value of n is one digit shorter than the integer used in the preceding command, because both of the prime factors of the 72-digit value of n are very large, it will take **ifactor** much more time to return these prime factors. For example, the reader may wish to enter the preceding and following commands to see the difference.

```
>  ifactor(200033699955714283337472000688611033705230782877
   37637631520000000000000001449);
```

Also, recall as we mentioned in Section 8.2, if p and q were both hundreds of digits long, then the fastest known factoring algorithms, including the ones employed by the **ifactor** function, would in general take millions of years to factor $n = pq$, even when programmed on a computer that could perform millions of operations per second.

8.4   RSA Ciphers with MATLAB

We will begin this section by demonstrating several functions that are useful for finding large primes in MATLAB. The first is the MuPAD **nextprime** function, which returns the smallest prime larger than or equal to an integer input. For example, the following command returns the smallest prime larger than 400043344212007458000000000000000000. Note that in this command, the input parameter is enclosed in single quotes and declared in symbolic format. This same syntax will also be required in the primality commands that follow.

```
>> feval(symengine, 'nextprime', ...
   sym('400043344212007458000000000000000000000'))

ans =

400043344212007458000000000000000000000161
```

A similar function is the MuPAD **prevprime** function, which is designed to return the largest prime smaller than or equal to a given integer input. For example, the following command returns the largest prime smaller than 400043344212007458000000000000000000000.

```
>> feval(symengine, 'prevprime', ...
   sym('400043344212007458000000000000000000000'))

ans =

400043344212007457999999999999999999941
```

A final primality function that we will demonstrate is the MuPAD **isprime** function, which returns *TRUE* if a given integer input is prime, and *FALSE* if the integer input is not prime. For example, from the following commands we can conclude that 400043344212007457999999999999999999941 is prime, but 400043344212007458000000000000000000000 is not.

```
>> feval(symengine, 'isprime', ...
   sym('400043344212007457999999999999999999941'))

ans =

TRUE

>> feval(symengine, 'isprime', ...
   sym('400043344212007458000000000000000000000'))

ans =

FALSE
```

We should also note that the **nextprime**, **prevprime**, and **isprime** functions are probabilistic routines that employ primality tests (see Section

8.5.1). This means that the output returned by MATLAB is guaranteed to
be correct with an extremely high probability, but not absolutely.

We will now show how MATLAB can be used to encipher and decipher
messages using the RSA cryptosystem. We begin by finding and assigning
values for the primes p and q for the system.

```
>> p = feval(symengine, 'nextprime', ...
   sym('40004334421200745800000000000000000000'))


p =

40004334421200745800000000000000000161

>> q = feval(symengine, 'nextprime', ...
   sym('50003006636626900120000000000000000000'))


q =

50003006636626900120000000000000000009
```

Next, we assign the values of $n = pq$ and $m = (p-1)(q-1)$ for the system.

```
>> n = p*q

n =

20003369995571428333747200068861103370523078287737631520000
00000000001449

>> m = (p-1)*(q-1)

m =

20003369995571428333747200068861103280515737229909985600000
00000000001280
```

Also, we will use the following value as the encryption exponent a for the
system.

```
>> a = sym('1009876890098767900091000033')

a =

1009876890098767900091000033
```

To verify that the preceding value of a is a valid RSA encryption exponent given our value of m, we will use the MATLAB **gcd** function, which is designed to calculate the greatest common divisor of a pair of integer inputs. The following command returns the greatest common divisor of the integers a and m. Note that, as required, $\gcd(a, m) = 1$.

```
>> gcd(a, m)

ans =

1
```

We will now use the RSA encryption procedure to encipher the message RETURN TO HEADQUARTERS. We begin by storing our message as the following string.

```
>> message = 'RETURN TO HEADQUARTERS';
>> message(findstr(message, ' ')) = []

message =

RETURNTOHEADQUARTERS
```

To convert this plaintext message into numerical form, we will use the user-written function **tonumber**, which we have written separately from this MATLAB session and saved as the M-file **tonumber.m**. The following command illustrates how the function **tonumber** can be used. Specifically, by entering the following command, we convert *message* into a list of two-digit integers, and group all of these two-digit integers into a single block.

```
>> plaintext = tonumber(message)

plaintext =

1704192017131914070400031620001719041718
```

To check whether this numerical representation of the plaintext is less than n, we can use the MuPAD **bool** function within MATLAB, which returns *TRUE* if a given input statement is true, *FALSE* if the input statement is false, and *UNKNOWN* if the input statement cannot be determined to be true or false. For example, the following command shows that the numerical representation of the plaintext is less than n.

```
>> feval(symengine, 'bool', strcat(char(plaintext), ...
   '<', char(n)))
```

```
ans =
```

TRUE

Because this single plaintext block is less than n, we can encipher our message as a single block. That is, we can encipher our message by raising *plaintext* to the power a and reducing modulo n. To do this, we will enter the following command, in which, because this modular exponentiation involves such a large exponent, we use the MuPAD **powermod** function within MATLAB. By using **powermod**, we cause MATLAB to do the exponentiation in a very efficient way, like the technique that we will present in Section 8.5.3.

```
>> ciphertext = feval(symengine, 'powermod', plaintext, ...
   a, n)
```

```
ciphertext =
```

1201867682661914690291221051205116372700242955180477243062
79524922980340

In order to decrypt this ciphertext, we must first find a decryption exponent b that satisfies $ab = 1 \bmod m$. We can do this by using the MuPAD **igcdex** function within MATLAB, which, like **gcd**, is designed to calculate the greatest common divisor of a pair of integer inputs. However, **igcdex** also finds integers b and v that satisfy $ab + mv = \gcd(a, m)$. Since $\gcd(a, m) = 1$, these will be values of b and v that satisfy $ab + mv = 1$, or, equivalently, $ab = 1 \bmod m$. Thus, we can find a decryption exponent b for this system by entering the following command.

```
>> g = feval(symengine, 'igcdex', a, m)
```

```
g =
```

[1, −578031788668504420817941179537494520463040783878060?
7915877130336539903, 291821300734832493626674310]

The output of the previous command is a vector with three components. The first component is $\gcd(a, m)$, and the second and third are values of b and v, respectively, that satisfy $ab + mv = 1$. To express the decryption exponent b assigned by the previous command as a positive integer less than m, we enter the following command.

```
>> b = mod(g(2), m)

b =

14223052108886384125567788273486158075885329391129377780841
22869663461377
```

An alternative method for finding a decryption exponent b that satisfies $ab = 1 \bmod m$ is to simply raise a to the power -1 and reduce the result modulo m. This is illustrated in the following command.

```
>> b = mod(a^(-1), m)

b =

14223052108886384125567788273486158075885329391129377780841
22869663461377
```

By entering the following command, we verify that this value of b satisfies $ab = 1 \bmod m$.

```
>> mod(a*b, m)

ans =

1
```

To recover the plaintext integer, we must only raise *ciphertext* to the power b and reduce modulo n.

```
>> plaintext = feval(symengine, 'powermod', ciphertext, ...
   b, n)

plaintext =

170419201713191407040003162000171904171718
```

In order to see the original plaintext letters, we must split this single plain-text integer into a list of two-digit integers, and then convert this list of two-digit integers back into a list of letters. To do this, we will use the user-written function **toletter**, which we have written separately from this MATLAB session and saved as the M-file toletter.m. The following command illustrates how the function **toletter** can be used. Specifically, by entering the following command, we convert *plaintext* back into a list of letters.

```
>> toletter(plaintext)

ans =

RETURNTOHEADQUARTERS
```

For a plaintext whose numerical representation is larger than n, we can split the plaintext into shorter blocks of characters whose numerical representations are all less than n, and then encipher each block separately. For example, consider the plaintext RETURN TO HEADQUARTERS NO LATER THAN TWELVE OCLOCK NOON THE TWENTY FIFTH OF THE MONTH OF DECEMBER. We first store this message as the following string.

```
>> message = ['RETURN TO HEADQUARTERS NO LATER THAN ', ...
   'TWELVE OCLOCK NOON THE TWENTY FIFTH OF THE MONTH ', ...
   'OF DECEMBER'];
>> message(findstr(message, ' ')) = []

message =

RETURNTOHEADQUARTERSNOLATERTHANTWELVEOCLOCKNOONTHETWENTYFI
FTHOFTHEMONTHOFDECEMBER
```

Next, we use **tonumber** to convert *message* into a list of two-digit integers, and group all of these two-digit integers into a single block.

```
>> plaintext = tonumber(message)

plaintext =

170419201713191407040000316200017190417181314110019041719 07
001319220411210414021114021013141413190704192204131924050 8
051907140519070412141319071405030402041201041 7
```

The next command shows that this numerical representation of the plaintext is larger than n.

```
>> feval(symengine, 'bool', [char(plaintext), '<', ...
   char(n)])

ans =

FALSE
```

Because *plaintext* is larger than n, it cannot be encrypted as a single block, since the corresponding decryption operation done modulo n would have to result in a value less than n. To encrypt this message, we must first split it into shorter blocks of characters whose numerical representations are all less than n. To do this, we first identify the number of digits in n as follows, by using the MATLAB **char** function to convert n from an integer into a string, and then the MATLAB **length** function to find the number of characters in the resulting string.

```
>> length(char(n))

ans =

     72
```

Since each character in our message is represented numerically by a two-digit integer, any block with fewer than 36 characters would have a numerical representation with fewer than 72 digits, which would thus be less than n. To split our message into appropriate blocks, we will use the user-written function **LengthSplitString**, which we have written separately from this MATLAB session and saved as the M-file LengthSplitString.m. The output from this command is a cell array containing the characters from the first parameter *message* in order, with the length of each block until the last given by the second parameter 35, and the last block ending when the characters in *message* have all been used.

```
>> messblocks = LengthSplitString(message, 35)

messblocks =

    'RETURNTOHEADQUARTERSNOLATERTHANTWEL'    'VEOCLOCKNOONT
    HETWENTYFIFTHOFTHEMONT'    'HOFDECEMBER'
```

Next, we convert each of these character blocks into numerical form.

```
>> plaintext = tonumber(messblocks)

plaintext =

[ 17041920171319140704000316200017190417181314110019041719
07001319220411, 21041402111402101314141319070419220413192405080519071405190704
12141319, 714050304020412010417]
```

We next use a **for** loop with the **bool** function to show that each numerical plaintext block is less than n.

```
>> for i = 1:numel(plaintext)
       bpt(i) = feval(symengine, 'bool', ...
       [char(plaintext(i)), '<', char(n)]);
    end
>> bpt

bpt =

[ TRUE, TRUE, TRUE]
```

We now use a **for** loop with the **powermod** function to encrypt each numerical plaintext block.

```
>> for i = 1:numel(plaintext)
       ciphertext(i) = feval(symengine, 'powermod', ...
       plaintext(i), a, n);
    end
>> ciphertext

ciphertext =

[ 197033961372882741089096128703671351683929424877001163 19
1632702996886750, 1395309540684714544744553407263096261133
2175507695398143024799503409023, 1617717160371897190310569
2681896751036431484608034629182250548811688 3006]
```

We can then decrypt this message as follows.

```
>> for i = 1:numel(ciphertext)
       plaintext(i) = feval(symengine, 'powermod', ...
       ciphertext(i), b, n);
    end
>> plaintext

plaintext =

[ 170419201713191407040003162000171904171813141100190417 19
07001319220411, 210414021114021013141413190704192204131924
050805190714051907041214 1319, 714050304020412010417]

>> toletter(plaintext)
```

```
ans =
```

```
RETURNTOHEADQUARTERSNOLATERTHANTWELVEOCLOCKNOONTHETWENTYFI
FTHOFTHEMONTHOFDECEMBER
```

A final MuPAD function that we will demonstrate in this section is **factor**, which returns the prime factorization of an integer input. For example, the following command very quickly returns the prime factorization of the 73-digit integer 5339046037813420424099935321501028892833668161867505 4395575633065999923549.

```
>> factor(sym(['5339046037813420424099935321501028892', ...
   '8336816186750543955756330659999235 49']))
```

```
ans =
```

```
23^3*107*45001200019828331^4
```

Recall that the security of the RSA cryptosystem is based on the apparent difficulty of factoring the value of $n = pq$ used in the system. Thus, in order for the last RSA system that we used in this section to be secure, it should be very difficult for an intruder to factor the 72-digit value of n from the system. Although this value of n is one digit shorter than the integer used in the preceding command, because both of the prime factors of the 72-digit value of n are very large, it will take **factor** much more time to return these prime factors. For example, the reader may wish to enter the preceding and following commands to see the difference.

```
>> factor(sym(['200033699955714283337472000688611033 70' ...
   '5230782877376315200000000000000001449']))
```

Also, recall as we mentioned in Section 8.2, if p and q were both hundreds of digits long, then the fastest known factoring algorithms, including the ones employed by the **factor** function, would in general take millions of years to factor $n = pq$, even when programmed on a computer that could perform millions of operations per second.

8.5 Efficiency and Security Issues

8.5.1 Primality Testing

Recall that in order for the RSA cryptosystem to be secure, the primes p and q chosen for the system must both be very large. For example, as we mentioned in Section 8.2, if p and q were both hundreds of digits long, then it would in general take an intruder millions of years to break the system, even when using a computer that could perform millions of operations per second. However, implementing the RSA system with such large primes is not particularly easy, for it is not particularly easy to find such large primes. In fact, motivated in part by the development of public-key cryptosystems like RSA, much research has been done over the past few decades in the area of primality testing.

Contrary to what the name *primality test* suggests, a primality test is normally a criterion that can be used to determine if a given integer is *not* prime. The conclusions that can usually be drawn from applying a primality test to an integer n are either that n "fails" the test and is definitely not prime, or that n "passes" the test and is probably prime, with probability depending on the power of the test.

The most direct and conclusive way to determine if a large odd integer n is prime is to try to find nontrivial factors of n by trial and error. We could do this systematically by checking if $m|n$ as m takes on odd integer values starting with $m = 3$ and ending when m reaches $\sqrt{n}$. While this would reveal with certainty whether n was prime, it would also require many more divisions than could reasonably be done if n was of any significant size. In the remainder of this section, we will present a very well known and simple primality test based on Fermat's Little Theorem (Theorem 8.5).

If n is a prime integer, then as a consequence of Fermat's Little Theorem the following will hold for all $a \in \mathbb{Z}_n^*$.

$$a^{n-1} \; = \; 1 \bmod n \tag{8.3}$$

In addition, if $a^{n-1} \neq 1 \bmod n$ for any $a \in \mathbb{Z}_n^*$, then we can conclude that n is definitely not prime. Thus, we can test the primality of an integer n by checking if (8.3) holds for some values of a in $\mathbb{Z}_n^*$, with the power of the test increasing as we check more values of a. While this test is very easy to perform, there are some values of a for which (8.3) holds even when $\gcd(a, n) = 1$ and n is not prime. In such cases, n is called a *pseudoprime* to the base a. For example, $2^{340} = 1 \bmod 341$ even though 341 is not prime, and so 341 is a pseudoprime to the base 2. However, since $3^{340} \neq 1 \bmod 341$, then 341 is not a pseudoprime to the base 3.

Pseudoprimes are scarce relative to the primes. For example, there are only 245 pseudoprimes to the base 2 less than one million, while there are 78498 primes less than one million. Also, most pseudoprimes to the base 2 are not pseudoprimes to many other bases. However, there do exist non-prime integers n that are pseudoprime to every positive base $a < n$ with $\gcd(a, n) = 1$. Such numbers are called *Carmichael* numbers. There are 2163 Carmichael numbers less than 25 billion. The smallest Carmichael number is 561.

There are many primality tests that are more definitive in their conclusion than the test that we just described. For example, a further primality test based on Fermat's Little Theorem that is also very easy to perform fails only for an extremely small number of non-primes called *strong* pseudoprimes. There is only one strong pseudoprime to the bases 2, 3, 5, and 7 less than 25 billion, and there is no strong pseudoprime analogue to Carmichael numbers.

8.5.2 Integer Factorization

Recall that the security of the RSA cryptosystem is based on the apparent difficulty of factoring a number that is the product of two very large distinct primes. As in the area of primality testing, the development of public-key cryptosystems like RSA has motivated much research over the past few decades in the area of integer factorization. In this section, we will present a very simple technique for integer factorization called *Fermat factorization*. Despite the fact that this factorization technique has been known for a very long time, it is still a useful technique for factoring integers that are the product of two very large distinct primes that are close together.

Let $n = pq$ be the product of two distinct primes, and suppose that we would like to determine the values of p and q from n. The most direct way to determine p and q from n would be by trial and error. This could certainly not be done in a reasonable amount of time though if p and q were both significantly large. However, if p and q were relatively close together, then even if they were both very large, we could determine them quickly through the following procedure. Let $x = \frac{p+q}{2}$ and $y = \frac{p-q}{2}$. Then $n = pq = x^2 - y^2 = (x+y)(x-y)$. Thus, since n has prime factors p and q, it follows that p and q would have to be $x + y$ and $x - y$. So to determine p and q, we would only have to find the values of x and y. To do this, we could begin by assuming that x is the smallest integer larger than $\sqrt{n}$. Since $n = x^2 - y^2$, if we have assumed the correct value of x, then it will follow that $x^2 - n$ will be the perfect square y^2. If $x^2 - n$ is not a perfect square, then we would know that we had assumed an incorrect value for x, and we could simply increase x by one and repeat. We could continue to repeat this process as many times as necessary, each time increasing x by one, until $x^2 - n$ is a perfect square. Note that if p and q were relatively close together, then the number of times that this process would have to be repeated would be relatively small.

Example 8.5 Suppose that we would like to find the two prime factors of $n = pq = 64349$. Since the smallest integer larger than $\sqrt{64349}$ is 254, we would begin by letting $x = 254$. However, since $254^2 - n = 167$ is not a perfect square, then 254 is not the correct value for x. Next, we would try

$x = 255$. Since $255^2 - n = 676 = 26^2$ is a perfect square, then the correct values of x and y are $x = 255$ and $y = 26$, and the prime factors of n are $x + y = 281$ and $x - y = 229$.   □

In comparison of the problems of primality testing and integer factorization, we should note that factoring a known non-prime integer is in general significantly more time-consuming than finding a prime of approximately the same size. We have stated several times that the security of the RSA cryptosystem is based on the apparent difficulty of factoring a number that is the product of two very large distinct primes. To be more precise, the security of the RSA cryptosystem is based on the fact that it would apparently be much more time-consuming for an intruder to factor the publicly known value of $n = pq$ than for the intended recipient of the message to choose p and q.[4]

8.5.3   Modular Exponentiation

Securely encrypting and decrypting messages using the RSA cryptosystem generally requires modular exponentiation with extremely large bases and exponents. For example, to decrypt the ciphertext in Sections 8.3 and 8.4, we raised the number 397056677510513368122841363348174734852890 to the power 542993009508418269900718536789979854000035 and reduced the result modulo 20003369995571428334517252158400846898963 9. Even using the world's fastest computer, performing this calculation by actually multiplying 3970566775105133681228413633481747348528 9 by itself repeatedly with 542993009508418269900718536789979854000035 total factors would essentially take forever. In this section, we will present a technique that can be used to perform this modular exponentiation in a very efficient way.

For convenience, we will demonstrate this technique for efficient modular exponentiation in the following calculation that was necessary for decrypting the ciphertext in Example 8.4.

$$137626763^{20981287} \ = \ 0214181214 \bmod 363794227 \qquad (8.4)$$

This calculation can be done in a much more efficient way than actually multiplying 137626763 by itself repeatedly with 20981287 total factors. To perform this calculation more efficiently, we can begin by finding the values of $137626763^{2^i} \bmod 363794227$ for $i = 1, 2, \ldots, 24$. That is, for $P = 137626763$ and $M = 363794227$, we can begin by computing $P^2, P^4, P^8, \ldots, P^{2^{24}}$, and reducing each result modulo M. Note that each

---

[4]We use the word "apparently" in this sentence because it has never been *proven* that the factorization would be significantly more time-consuming. Evidence, however, very strongly suggests this.

$P^{2^i} \mod M$ can be found by squaring $P^{2^{i-1}} \mod M$, and so finding these values would require a total of only 24 multiplications. The modular exponentiation in (8.4) could then be completed by calculating the following.

$$
\begin{aligned}
P^{20981287} \mod M &= P^{16777216+4194304+8192+1024+512+32+4+2+1} \mod M \\
&= P^{2^{24}+2^{22}+2^{13}+2^{10}+2^9+2^5+2^2+2^1+2^0} \mod M \\
&= P^{2^{24}} P^{2^{22}} P^{2^{13}} P^{2^{10}} P^{2^9} P^{2^5} P^{2^2} P^{2^1} P^{2^0} \mod M
\end{aligned}
$$

This would require only 8 additional multiplications. Thus, this technique could be used to perform the modular exponentiation in (8.4) with a total of only $24 + 8 = 32$ multiplications. This is, of course, much fewer than the number of multiplications necessary to multiply P by itself repeatedly with 20981287 total factors.

It is not difficult to see that this technique for efficiently calculating $P^a \mod M$ will in general require at most $2 \cdot \log_2(a)$ multiplications. Thus, to perform even the massive modular exponentiation that we mentioned at the beginning of this section, this technique would require at most only the following number of multiplications.

$$
2 \cdot \log_2(54299300950841826990071853678997985400035) \approx 270
$$

8.5.4 Digital Signatures

When the idea of public-key cryptography was published, one way in which it was envisioned that it could be used was as follows. Suppose that a group of people would all like to be able to communicate spontaneously across a series of insecure lines of communication. For illustration, suppose that they would like to use the RSA cryptosystem to encrypt their messages. To use the RSA system most effectively, each person in the group could choose his or her own secret primes p and q, form his or her own personal value for $n = pq$, and choose his or her own personal encryption exponent a. Each person in the group could then make his or her personal values of n and a public knowledge. Then whenever a person in the group wanted to send another person in the group a secret message, he or she could use the intended recipient's public values of n and a to encrypt the message. That way, only the intended recipient would be able to decrypt the ciphertext. However, this leads to the problem for the intended recipient that he or she would have no way of verifying that the received ciphertext had actually been sent by the person claiming to have sent it. This problem can be avoided in the following way.

Suppose we would like to use the RSA system to send the secret message P to a colleague across an insecure line of communication. Assume we have made public our personal RSA modulus n_1 and encryption exponent

a_1 while keeping our decryption exponent b_1 secret, and our colleague has made public his or her personal RSA modulus n_2 and encryption exponent a_2 while keeping his or her decryption exponent b_2 secret. Assume also $n_1 < n_2$. Suppose that in the encipherment of the plaintext, before applying our colleague's encryption exponent and modulus, we first apply our own decryption exponent and modulus. That is, suppose that instead of sending our colleague the ciphertext $P^{a_2} \bmod n_2$, we compute $P_1 = P^{b_1} \bmod n_1$, and then send our colleague the ciphertext $C_1 = P_1^{a_2} \bmod n_2$. Our colleague could easily decrypt this ciphertext by first applying his or her decryption exponent and modulus to obtain $P_1 = C_1^{b_2} \bmod n_2$, and then our publicly known encryption exponent and modulus to obtain $P = P_1^{a_1} \bmod n_1$. Since the decryption exponent b_1 that we used in the encipherment of the message is known only to us, our colleague would then know that only we could have encrypted the message. Because it has the effect of authenticating the message, applying our own decryption exponent and modulus in the encipherment of a message is sometimes called *signing* the message. We will leave the case when $n_1 > n_2$ as an exercise.

Authentication of messages has been a very important and highly studied branch of cryptography for many years. In fact, it is interesting to note that in the title of the article *A Method for Obtaining Digital Signatures and Public-Key Cryptosystems* in which Rivest, Shamir, and Adleman published the RSA system, the notion of a digital signature was given precedence over that of a public-key cryptosystem.

8.6   The Diffie-Hellman Key Exchange with RSA[5]

Recall that the reason we call the RSA cryptosystem a *public-key* system is because the system can be secure even if the encryption exponent and modulus for the system are public knowledge. However, for two people who would like to use the RSA system to exchange a secret message across an insecure line of communication, it would obviously be desirable to be able to keep either the encryption exponent or modulus for the system secret. In this section, we will present a technique that could be used by two people to securely determine an RSA encryption exponent while communicating only across an insecure line of communication.

There are actually several techniques through which two people can agree upon a cryptographic key securely without having a secure way to communicate. One such technique is the *Diffie-Hellman key exchange*, a

---

[5]Copyright 1999 by COMAP, Inc. This material appeared in the Spring 1999 issue of *The UMAP Journal* (see [19]).

process presented by Whitfield Diffie and Martin Hellman in their seminal paper *New Directions in Cryptography* in which they publicly introduced the idea of public-key cryptography. In order to describe a way to incorporate this key exchange system with the RSA system, suppose that we would like to use the RSA system to receive a secret message from a colleague across an insecure line of communication. Furthermore, suppose that we and our colleague would like to agree upon our encryption exponent securely while communicating only across the insecure line of communication. We could accomplish this by using the following steps in the Diffie-Hellman key exchange with RSA.

1. We choose primes p and q and form $n = pq$, and then choose a positive integer $k < n$ with $\gcd(k, n) = 1$. We then send the values of k and n to our colleague across the insecure line of communication (which forces the assumption that k and n are public knowledge).

2. We choose a positive integer $r < n$ and compute $k^r \bmod n$, and send the result to our colleague while keeping r secret (which forces the assumption that $k^r \bmod n$ is public knowledge). Meanwhile, our colleague chooses a positive integer $s < n$ and computes $k^s \bmod n$, and sends the result to us while keeping s secret (which forces the assumption that $k^s \bmod n$ is public knowledge).

3. Both we and our colleague form the candidate encryption exponent $a = k^{rs} \bmod n$, which we compute as $(k^s)^r \bmod n$, and our colleague computes as $(k^r)^s \bmod n$. Since we know the values of p and q, we can form $m = (p-1)(q-1)$, and then check if a is a valid RSA encryption exponent by determining if $\gcd(a, m) = 1$. If a is not a valid RSA encryption exponent, we must repeat the process. We could continue to repeat this process as many times as necessary until we obtain a valid RSA encryption exponent.

After we obtain a valid RSA encryption exponent a, our colleague could then encrypt his or her message using the usual RSA encryption procedure with encryption exponent a and modulus n.

Example 8.6 Suppose we choose primes $p = 83$ and $q = 101$ and form $n = pq = 8383$ and $m = (p - 1)(q - 1) = 8200$. Suppose we also choose $k = 256$, and send the values of k and n to our colleague. Suppose we then choose $r = 91$ and compute $256^{91} \bmod 8383 = 2908$, and send the result to our colleague while keeping r secret. Meanwhile, suppose our colleague chooses $s = 4882$ and computes $256^{4882} \bmod 8383 = 1754$, and sends the result to us while keeping s secret. We and our colleague form the candidate encryption exponent $a = 6584$, which we find as $1754^{91} \bmod 8383 = 6584$, and our colleague computes as $2908^{4882} \bmod 8383 = 6584$. However, since this value of a is clearly not relatively prime with m (they are both even), we would inform our colleague that we must repeat the process. For the second attempt, suppose we choose the same values for p, q, and k. Suppose we then choose $r = 17$ and compute $256^{17} \bmod 8383 = 5835$, and send the result to our colleague while keeping r secret. Meanwhile, our colleague chooses $s = 109$ and computes $256^{109} \bmod 8383 = 1438$, and sends the result to us while keeping s secret. We and our colleague form the candidate encryption exponent $a = 3439$, which we find as $1438^{17} \bmod 8383 = 3439$, and our colleague computes as $5835^{109} \bmod 8383 = 3439$. Since this value of a is relatively prime with m (as we could easily verify), we would confirm to our colleague that he or she could proceed with the usual RSA encryption procedure. $\square$

Since our description of the Diffie-Hellman key exchange with RSA indicates that the process may need to be repeated an unspecified number of times, it is natural to wonder the number of times one should expect to repeat the process before achieving success. To estimate this, we simulated the process 777,000 times, split as 111,000 for seven different maximum sizes of $n = pq$. These seven different maximum sizes of n were 5, 10, 20, 40, 80, 160, and 320 digits, and for each the success rate, with success being that the Diffie-Hellman key exchange with random k, r, s, and $n = pq$ resulted in a valid encryption exponent a for an RSA cipher with modulus n, ranged from a low of 30.016% to a high of 31.394%. As such, the probability of success on a single trial of the Diffie-Hellman key exchange with RSA seems to be around 30% and independent of the size of n.

Note that in the Diffie-Hellman key exchange, we are forced to assume that k, n, $k^r \bmod n$, and $k^s \bmod n$ are known to intruders since they were all transmitted across an insecure line of communication. Thus, in order for this key exchange system to be secure, it should be an essentially impossible problem for an intruder to determine the candidate encryption exponent $k^{rs} \bmod n$ from the knowledge of k, n, $k^r \bmod n$, and $k^s \bmod n$. This problem is called the *Diffie-Hellman problem*. It has been conjectured that the only way to solve the Diffie-Hellman problem in general is to solve the discrete logarithm problem.

The discrete logarithm problem is important to consider when studying the Diffie-Hellman problem because the solution to a particular discrete logarithm problem leads directly to the solution to a corresponding Diffie-Hellman problem. To see this, suppose that we intercept transmissions between our enemy as they perform a Diffie-Hellman key exchange. That is, using the variables that we have defined, suppose that we intercept values of k, n, $k^r \bmod n$, and $k^s \bmod n$. Consider now the problem of determining r from the knowledge of k, n, and $k^r \bmod n$. In this scenario, r is called a *discrete logarithm* of $k^r \bmod n$ to the base k, and the problem of determining r from the knowledge of k, n, and k^r mod n is called the *discrete logarithm problem*. Note that if we could solve this problem, then a corresponding Diffie-Hellman problem would also be solved, for we could determine r from k^r mod n, and then compute $a = (k^s)^r$ mod n. However, solving the discrete logarithm problem is not an easy method for solving the Diffie-Hellman problem, since it has been argued that the fastest way to solve the discrete logarithm problem with a non-prime modulus n would require factoring n. Thus, the factorization problem that provides security to the RSA cryptosystem also provides security to the Diffie-Hellman key exchange (as we have presented it in this section).

Many algorithms for finding discrete logarithms have been presented in literature. For small values of n, and some special large values of n (for example, powers of a small base), many mathematics software packages have predefined functions for finding discrete logarithms. For example, see Section 8.7 or 8.8.

8.7 Discrete Logarithms with Maple

The Maple function for finding discrete logarithms is **mlog**, which is part of the **numtheory** package. We will begin by including this package.

```
>  with(numtheory):
```

The **mlog** function takes three integer inputs, say y, k, and n, in order, and returns either an integer r with the property that $y = k^r$ mod n, or *false* if no such integer exists. For example, the following command indicates that 256^{109} mod $8383 = 1438$, a fact that we used in Example 8.6.

```
>  mlog(1438, 256, 8383);
```
$$109$$

Using **mlog** and provided n is small, an intruder who intercepts Diffie-Hellman key exchange transmissions could easily determine the resulting cryptographic key. For example, suppose that an intruder intercepts the second set of transmissions $k = 256$, $n = 8383$, k^r mod $n = 5835$, and k^s mod $n = 1438$ from Example 8.6. The intruder could determine the resulting RSA encryption exponent a by using **mlog** to find that $s = 109$ satisfies k^s mod $n = 1438$, and then finding $a = 5835^{109}$ mod $8383 = 3439$.

In addition to the fact that **mlog** would in general run essentially forever for very large values of n (as would all known algorithms for finding discrete logarithms), there is another small problem with using **mlog** to "undo" the operation of modular exponentiation. While it is true that using **mlog** with integer inputs y, k, and n in order will cause Maple to return an integer r with the property that $y = k^r$ mod n (if such an integer exists), this integer might not be the one that was actually used in the modular exponentiation being "undone." For example, in the first part of Example 8.6, we used the fact that $256^{4882} \bmod 8383 = 1754$. However, the following command indicates that also $256^{782} \bmod 8383 = 1754$.

```
>  mlog(1754, 256, 8383);
```
$$782$$

Thus, the integer returned by Maple is not the exponent that we actually used in the example. However, this would not pose a problem for an intruder who intercepts the first set of transmissions in Example 8.6, for the intruder could still find the resulting candidate encryption exponent a by computing $2908^{782} \bmod 8383 = 6584$. So despite the fact that the integer returned by Maple is not the exponent that we actually used in the original modular exponentiation, it can still be used in the intruder's general procedure for determining the candidate encryption exponent. To see that this will be true in general, suppose that an intruder uses **mlog** to try to find the exponent in a Diffie-Hellman key exchange transmission $k^s \bmod n$. Even if the integer returned by Maple was $s' \neq s$, since it would be the case that $k^s \bmod n = k^{s'} \bmod n$, the intruder could still find the candidate encryption exponent a in the way that we have demonstrated since it would also be the case that $a = (k^s)^r \bmod n = (k^{s'})^r \bmod n$.

8.8 Discrete Logarithms with MATLAB

To find discrete logarithms in MATLAB, we will use the user-written function **mlog**, which we have written separately from this MATLAB session and saved as the M-file mlog.m. The **mlog** function takes three integer inputs, say y, k, and n, in order, and returns either an integer r with the property that $y = k^r \bmod n$, or *FAIL* if no such integer exists.[6] For example, the following command indicates that $256^{109} \bmod 8383 = 1438$, a fact that we used in Example 8.6. Note that the input parameters in this function must be expressed in symbolic form.

```
>> mlog(sym('1438'), sym('256'), sym('8383'))

ans =

109
```

Using **mlog** and provided n is small, an intruder who intercepts Diffie-Hellman key exchange transmissions could easily determine the resulting

---

[6]The **mlog** function employs the *Baby-Step, Giant-Step* algorithm.

cryptographic key. For example, suppose that an intruder intercepts the second set of transmissions $k = 256$, $n = 8383$, $k^r \bmod n = 5835$, and $k^s \bmod n = 1438$ from Example 8.6. The intruder could determine the resulting RSA encryption exponent a by using **mlog** to find that $s = 109$ satisfies $k^s \bmod n = 1438$, and then finding $a = 5835^{109} \bmod 8383 = 3439$.

In addition to the fact that **mlog** would in general run essentially forever for very large values of n (as would all known algorithms for finding discrete logarithms), there is another small problem with using **mlog** to "undo" the operation of modular exponentiation. While it is true that using **mlog** with integer inputs y, k, and n in order will cause MATLAB to return an integer r with the property that $y = k^r \bmod n$ (if such an integer exists), this integer might not be the one that was actually used in the modular exponentiation being "undone." For example, in the first part of Example 8.6, we used the fact that $256^{4882} \bmod 8383 = 1754$. However, the following command indicates that also $256^{782} \bmod 8383 = 1754$.

```
>> mlog(sym('1754'), sym('256'), sym('8383'))

ans =

782
```

Thus, the integer returned by MATLAB is not the exponent that we actually used in the example. However, this would not pose a problem for an intruder who intercepts the first set of transmissions in Example 8.6, for the intruder could still find the resulting candidate encryption exponent a by computing $2908^{782} \bmod 8383 = 6584$. So despite the fact that the integer returned by MATLAB is not the exponent that we actually used in the original modular exponentiation, it can still be used in the intruder's general procedure for determining the candidate encryption exponent. To see that this will be true in general, suppose that an intruder uses **mlog** to try to find the exponent in a Diffie-Hellman key exchange transmission $k^s \bmod n$. Even if the integer returned by MATLAB was $s' \neq s$, since it would be the case that $k^s \bmod n = k^{s'} \bmod n$, the intruder could still find the candidate encryption exponent a in the way that we have demonstrated since it would also be the case that $a = (k^s)^r \bmod n = (k^{s'})^r \bmod n$.

Exercises

1. Consider the message ATTACK LEFT FLANK.

   (a) Use the RSA cryptosystem with primes $p = 11$ and $q = 23$ and encryption exponent $a = 7$ to encrypt this message, with each plaintext character encrypted separately as in Example 8.2.

(b) Use the Euclidean algorithm to find the decryption exponent that corresponds to the encryption exponent in part (a).

(c) Use the RSA cryptosystem with primes $p = 83$ and $q = 131$ and encryption exponent $a = 3$ to encrypt this message, with the plaintext integers grouped into blocks with four digits as in Example 8.3 before being encrypted.

2. Suppose that you wish to use the RSA cryptosystem to receive a secret message from a colleague across an insecure line of communication, and that you initiate the process by choosing primes $p = 17$ and $q = 29$ and encryption exponent $a = 153$. Verify that this value of a is a valid RSA encryption exponent for the given values of p and q, and then use the Euclidean algorithm to find the corresponding decryption exponent.

3. Suppose that your enemy is using the RSA cryptosystem to exchange a secret message with a colleague, and that you intercept their modulus $n = 33$ and encryption exponent $a = 7$, and the ciphertext 27, 8, 20, 29, 16, 16, 9, 13, 20, 13, 0, 8, 30, 16, 13, in which each plaintext character was encrypted separately as in Example 8.2. Decrypt the ciphertext.

4. Suppose that you wish to use the RSA cryptosystem to receive a secret message from a colleague across an insecure line of communication, and that you initiate the process by choosing primes $p = 47$ and $q = 59$ and encryption exponent $a = 1779$. Suppose that you also determine the corresponding decryption exponent $b = 3$, and receive from your colleague the ciphertext 0792, 2016, 0709, 0464, 1497, 1086, 2366, 0524, in which the plaintext integers were grouped into blocks with four digits as in Example 8.3 before being encrypted. Decrypt the ciphertext.

5. Show that 15 is a pseudoprime to the base 4, but not to the base 2.

6. Show that 91 is a pseudoprime to the base 4, but not to the base 2.

7. Consider the following calculation that was necessary for encrypting the message in Example 8.4.

$$0214181214^{13783} \quad = \quad 137626763 \bmod 363794227$$

Find the exact number of multiplications the technique for efficient modular exponentiation that we demonstrated in Section 8.5.3 would require to perform this calculation.

8. Use Fermat factorization to find the two prime factors of the integer $n = pq = 321179$.

9. Using primes $p = 5$ and $q = 7$, act as both people in the Diffie-Hellman key exchange and agree upon a valid RSA encryption exponent a. List the results from all trials in the process, including trials that do not result in a valid encryption exponent.

10. Prove that the set U_n of units in $\mathbb{Z}_n$ forms a group with the operation of multiplication.

11. Prove Lemma 8.1 without using Lemma 8.2.

12. For distinct primes $p_1, p_2, \ldots, p_t$ and positive integers $k_1, k_2, \ldots, k_t$, prove that $\phi(p_1^{k_1} p_2^{k_2} \cdots p_t^{k_t})$ is given by the following formula.

$$\phi(p_1^{k_1} p_2^{k_2} \cdots p_t^{k_t}) = (p_1^{k_1} - p_1^{k_1-1})(p_2^{k_2} - p_2^{k_2-1}) \cdots (p_t^{k_t} - p_t^{k_t-1})$$

13. Prove Theorem 8.4.

14. Prove Theorem 8.7.

15. Recall from Exercise 7 in Chapter 6 the concept of *superencryption*. Suppose that you wish to use the RSA cryptosystem to receive a secret message from a colleague across an insecure line of communication, and that you initiate the process by choosing primes p and q and forming $n = pq$. However, suppose also that in an attempt to increase the overall security of the system, you choose two valid encryption exponents a_1 and a_2, and inform your colleague that he or she should first encrypt the message by raising the plaintext to the power a_1 and reducing modulo n, and then superencrypt the message by raising the resulting ciphertext to the power a_2 and reducing modulo n. Would this superencryption increase the overall mathematical security of the system? Completely explain your answer, and be as specific as possible.

16. Consider an analogue to the RSA cryptosystem that uses a modulus n that is the product of three distinct primes instead of just two. As compared to how the parameters n, m, a, and b relate to each other in the normal two-prime RSA system, briefly discuss the differences in how the analogous parameters in the three-prime system would relate to each other. Do you think that the mathematical security of the three-prime system would be more than, less than, or the same as the mathematical security of the two-prime system? Completely explain your answer, and be as specific as possible.

17. Show that the technique for efficiently calculating $P^a \bmod M$ that we demonstrated in Section 8.5.3 would in general require at most $2 \cdot \log_2(a)$ multiplications.

18. Suppose that you would like to use the RSA system to send the secret message P to a colleague across an insecure line of communication. Assume that you have made public your personal RSA modulus n_1 and encryption exponent a_1 while keeping your decryption exponent b_1 secret, and that your colleague has made public his or her personal RSA modulus n_2 and encryption exponent a_2 while keeping his or her decryption exponent b_2 secret. Assume also that $n_1 > n_2$.

   (a) Explain how the method that we described in Section 8.5.4 for digitally signing your message could fail.

   (b) Devise a method similar to the one that we described in Section 8.5.4 for digitally signing your message that could not fail.

19. One method through which the RSA cryptosystem can be broken without a cryptanalyst having to factor the modulus n is the result of what is called a *common modulus protocol failure*. This cryptanalysis method relies on human error on the part of users of the RSA system, specifically if the originator of a message uses RSA ciphers to encrypt the same plaintext for two intended recipients who happen to share a common modulus n. Suppose the originator of a message uses RSA ciphers to encrypt the same plaintext x for two intended recipients, one having modulus n (with $n > x$) and encryption exponent a_1, and the other having the same modulus n but a different encryption exponent a_2 that satisfies $\gcd(a_1, a_2) = 1$. If the originator forms ciphertexts $y_1 = x^{a_1} \bmod n$ and $y_2 = x^{a_2} \bmod n$, and sends each to its respective intended recipient, then an intruder who intercepts both ciphertexts can find the plaintext x by finding integers s and t for which $sa_1 + ta_2 = 1$, and then forming $x = y_1^s y_2^t \bmod n$.

   (a) Consider a pair of intended recipients of a ciphertext formed using an RSA cipher, one having modulus $n = 74663$ and encryption exponent $a_1 = 41$, and the other having the same modulus n but encryption exponent $a_2 = 71$. Suppose the originator of a message uses RSA ciphers to encrypt the same plaintext for these two intended recipients, with resulting ciphertexts 33879 and 62113, respectively. Without factoring n, find the plaintext.

   (b) Show that this cryptanalysis method works. That is, show that if $y_1 = x^{a_1} \bmod n$ and $y_2 = x^{a_2} \bmod n$, and $sa_1 + ta_2 = 1$, then $x = y_1^s y_2^t \bmod n$.

Computer Exercises

20. Consider the message GO TECH.

   (a) Use the RSA cryptosystem with three-digit primes p and q and a valid two-digit encryption exponent a of your choice to encrypt this message, with the plaintext integers grouped into blocks with four digits as in Example 8.3 before being encrypted.

   (b) Use the RSA cryptosystem with four-digit primes p and q and a valid three-digit encryption exponent a of your choice to encrypt this message, with the plaintext integers grouped into blocks with six digits before being encrypted.

   (c) Use the RSA cryptosystem with seven-digit primes p and q and a valid four-digit encryption exponent a of your choice to encrypt this message, with the plaintext integers grouped into a single block as in Example 8.4 before being encrypted.

21. Suppose that your enemy is using the RSA cryptosystem to exchange a secret message with a colleague, and that you intercept their modulus $n = 86722637$ and encryption exponent $a = 679$, and the ciphertext 35747828, 20827476, 55134021, 85009695, in which the plaintext integers were grouped into blocks with six digits before being encrypted. Decrypt the ciphertext.

22. Set up a parameterization of the RSA cryptosystem with primes p and q of at least 30 digits each. Choose a valid encryption exponent a, and determine a corresponding decryption exponent b. Then use this parameterization of the RSA system to encrypt and decrypt the message GEORGIA INSTITUTE OF TECHNOLOGY, with the plaintext integers grouped into a single block as in Example 8.4 before being encrypted.

23. Using primes $p = 503$ and $q = 751$, act as both people in the Diffie-Hellman key exchange and agree upon a valid RSA encryption exponent a. List the results from all trials in the process, including trials that do not result in a valid encryption exponent. Also, show how an intruder could find the value of a.

24. Write a single Maple or MATLAB procedure that will encrypt a given message using the RSA cryptosystem. This procedure should have four input parameters. The first input parameter should be a string of characters representing a plaintext, the second input parameter should be an integer representing the modulus n for the system, the

third input parameter should be an integer representing the encryption exponent a for the system, and the fourth input parameter should be an integer indicating the number of digits in the blocks into which the plaintext integers should be grouped before being encrypted.

25. Write a single Maple or MATLAB procedure that will decrypt a given ciphertext using the RSA cryptosystem. This procedure should have four input parameters. The first input parameter should be a vector representing the ciphertext, the second and third input parameters should be integers representing the primes p and q for the system, the fourth input parameter should be an integer representing the encryption exponent a for the system, and the procedure should return the plaintext as a string of characters.

26. Write a single Maple or MATLAB loop to find the smallest base to which the number 3215031751 is not a pseudoprime.

27. Write a single Maple or MATLAB loop to verify that the number 561 is a Carmichael number.

28. Write a single Maple or MATLAB procedure that will perform the process of Fermat factorization on a given integer. This procedure should have as its only input parameter the integer that is to be factored.

29. Write a single Maple or MATLAB procedure that will use trial and error to find the smallest positive integer that is a discrete logarithm of a given integer to a given base with a given modulus. This procedure should have three input parameters. The first input parameter should be an integer representing k^r mod n for some unknown integer r, the second input parameter should be an integer representing the base k, and the third input parameter should be an integer representing the modulus n. Then compare the time it takes your procedure to run with the time it takes the **mlog** function for finding discrete logarithms in Maple or MATLAB to run.

30. Recall from Exercise 19 the concept of a *common modulus protocol failure*. Consider a pair of intended recipients of a ciphertext formed using an RSA cipher, one having modulus $n = 74297336244080669$ and encryption exponent $a_1 = 41567809$, and the other having the same modulus n but encryption exponent $a_2 = 71243809$. Suppose the originator of a message uses RSA ciphers to encrypt the same plaintext for these two intended recipients, with resulting ciphertexts 18856576535386462 and 57539771675158649, respectively. Without factoring n, find the plaintext.

Research Exercises

31. Investigate the lives and careers of the people for whom the RSA cryptosystem is named, and write a summary of your findings. Include in your summary some highlights of their academic careers, some of their other contributions to cryptography besides the RSA system, some of their professional accomplishments outside the area of cryptography, and some of their honors and awards.

32. Find a copy of the article *A Method for Obtaining Digital Signatures and Public-Key Cryptosystems* in which Rivest, Shamir, and Adleman published the RSA cryptosystem, and write a summary of the article.

33. (a) When Rivest, Shamir, and Adleman published the RSA cryptosystem, they exhibited their faith in its security by presenting a challenge. This challenge was to break a given ciphertext that they had formed using the RSA system, and they offered a small cash prize to anyone who could do so by an imposed deadline. Investigate this ciphertext and the corresponding plaintext, as well as when, how, and by whom the ciphertext was first broken, and write a summary of your findings.

    (b) After Rivest, Shamir, and Adleman's original challenge remained unsolved for more than a decade, RSA Laboratories, a corporation formed by Rivest, Shamir, and Adleman to market RSA ciphers, presented a broader challenge. Investigate this broader challenge, and write a summary of your findings.

    (c) Investigate the RSA Laboratories corporation, including its history, the types of customers it typically serves, and the products and services that it provides, and write a summary of your findings.

34. Investigate several actual real-life uses of the RSA cryptosystem, and write a summary of your findings.

35. Investigate the current standards for the size of the primes believed to be necessary for the RSA cryptosystem to be secure, and write a summary of your findings.

36. Investigate some other primality tests besides the one that we presented in Section 8.5.1, and write a summary of your findings.

37. Investigate some other integer factorization techniques besides the one that we presented in Section 8.5.2, and write a summary of your findings.

38. Investigate some current attacks against the RSA cryptosystem, and write a summary of your findings.

39. Investigate the lives and careers of the people who publicly introduced the idea of public-key cryptography (Whitfield Diffie and Martin Hellman), and write a summary of your findings.

40. Find a copy of the article *New Directions in Cryptography* in which Whitfield Diffie and Martin Hellman publicly introduced the idea of public-key cryptography, and write a summary of the article.

41. Investigate the life and career in cryptology of Ralph Merkle, and write a summary of your findings.

42. Investigate the lives and careers in cryptology of James Ellis, Clifford Cocks, and Malcolm Williamson, and write a summary of your findings.

43. Find a copy of the article in which Clifford Cocks first described RSA ciphers, and write a summary of the article. Include in your summary a comparison of how RSA ciphers were described by Cocks with how they were described by Rivest, Shamir, and Adleman in their article in which RSA ciphers were first publicly described.

44. The isomorphism between $\mathbb{Z}_{rs}$ and $\mathbb{Z}_r \times \mathbb{Z}_s$ induced by the mapping σ described in the paragraph between Lemmas 8.2 and 8.3 suggests the following problem: Given $(a, b) \in \mathbb{Z}_r \times \mathbb{Z}_s$, find $c \in \mathbb{Z}_{rs}$ such that $\sigma(c) = (a, b)$. The solution to this problem is the *Chinese Remainder Theorem*. Investigate the Chinese Remainder Theorem, and write a summary of your findings. Include in your summary some applications of the Chinese Remainder Theorem to RSA and to *secret sharing*.

45. Investigate the *RSA signature scheme*, and write a summary of your findings.

46. Investigate the *man-in-the-middle* attack on public-key ciphers, and write a summary of your findings.

47. Investigate the McEliece cryptosystem, which uses error-correcting codes to form public-key ciphers, and write a summary of your findings.

Chapter 9

Elliptic Curve Cryptography

Recall that the security of the Diffie-Hellman key exchange is based on the difficulty an intruder would encounter when solving the discrete logarithm problem. In this chapter, we will present a public-key cryptosystem whose security is based on the difficulty an intruder would encounter when solving the discrete logarithm problem. This system, called the *ElGamal* cryptosystem in honor of Taher Elgamal[1] who first published the system in 1985, has formed an important area of recent cryptographic research due to how elliptic curves can naturally be incorporated into the system.

9.1   ElGamal Ciphers

Before describing elliptic curves and how they can naturally be incorporated into the ElGamal system, we will first present the system in general and give a couple of simple examples of the system. In order to describe the ElGamal system, suppose that two people would like to exchange a secret message across an insecure line of communication. They could accomplish this by using the following steps in the ElGamal cryptosystem.

1. The intended recipient of the message initiates the process by choosing a finite abelian group G and an element $a \in G$. The intended recipient then chooses a positive integer $n < |G|$, computes $b = a^n$ in G, and sends a and b and information sufficient to determine G to the originator of the message across the insecure line of communication.

---

[1] Elgamal prefers to spell his surname with a lowercase g, to discourage its mispronunciation in English. However, the type of cipher he published, as well as a system he developed for obtaining digital signatures, are usually denoted using a capital G.

2. Using some method of conversion, the originator of the message converts his or her message into an equivalent element or list of elements in G. Suppose that the message converts to the element $w \in G$. The originator of the message then chooses a positive integer k, computes $y = a^k$ and $z = wb^k$ in G, and sends y and z to the intended recipient of the message across the insecure line of communication.

3. Because the intended recipient of the message knows n, he or she can recover w by computing zy^{-n} in G due to the following, where e represents the identity element in G.

$$zy^{-n} = wb^k(a^k)^{-n} = w(ba^{-n})^k = w(e)^k = w$$

Also, note that as a consequence of Lagrange's Theorem (Theorem 1.4), y^{-n} can be determined as $y^{|G|-n}$, which will have a positive exponent.

Although the preceding steps are specific to the ElGamal cryptosystem, the system can appear in many different forms due to the various types of groups that can be used for G. This is precisely how we will incorporate elliptic curves into the system. We will show in Section 9.4 that an elliptic curve over a finite field forms an abelian group with a specially defined operation. However, it is not necessary to use a group of this type in the system. The ElGamal system is especially easy to implement if G is chosen to be a group such as $\mathbb{Z}_p^*$ for prime p with the operation of multiplication.

Example 9.1 Suppose that we wish to use the ElGamal cryptosystem to send the secret message USMC to a colleague across an insecure line of communication.

1. Our colleague initiates the process by choosing $G = \mathbb{Z}_p^*$ for prime $p = 100000007$ and $a = 180989$. Next, our colleague chooses a value for n, computes $b = a^n \bmod p = 10524524$, and sends the values of p, a, and b to us across the insecure line of communication.

2. Suppose we use the correspondence α we introduced in Chapter 6 (i.e., A $\mapsto$ 00, B $\mapsto$ 01, C $\mapsto$ 02, ..., Z $\mapsto$ 25) to convert our message USMC into an equivalent single numerical block. That is, suppose we convert our message into the numerical equivalent $w = 20181202$. We then choose $k = 3638997$, compute $y = a^k \bmod p = 73133845$ and $z = wb^k \bmod p = 88730910$, and send the values of y and z to our colleague across the insecure line of communication.

3. Since our colleague would know the value of $n = 5124541$, our colleague would be able to recover the original numerical plaintext w by computing $zy^{-n} \bmod p = zy^{(p-1)-n} \bmod p = 20181202$.

$\square$

Note that in Example 9.1, we would have to assume that the values of p, a, b, y, and z were all public knowledge since they were all transmitted across an insecure line of communication. Also, in order for the system to be secure, an intruder must not be able to determine $zy^{-n} \bmod p$. Thus, an intruder must not be able to determine n from intercepted values of p, a, and $b = a^n \bmod p$. However, this is precisely the statement of the discrete logarithm problem we presented in Section 8.6 but with a prime modulus. That is, the security of the ElGamal system in Example 9.1 is based on the difficulty an intruder would encounter when solving the discrete logarithm problem that we presented in Section 8.6 but with a prime modulus. We stated in Section 8.6 that the discrete logarithm problem with a large non-prime modulus having only large prime factors is in general very difficult to solve. This is true with a large prime modulus as well.

The discrete logarithm problem can actually be defined much more generally than how we presented it in Section 8.6. More generally, for an element x in a finite group G and an element $y \in G$ that is a power of x, any integer r that satisfies $x^r = y$ is called a discrete logarithm of y to the base x, and the problem of determining an integer r that satisfies $x^r = y$ is called the discrete logarithm problem. As we stated in Section 8.6, many algorithms for finding discrete logarithms have been presented in literature. However, in groups of extremely large order, even the fastest known algorithms for finding discrete logarithms would be very time-consuming. For example, the fastest known algorithms for finding discrete logarithms would in general take millions of years to find discrete logarithms in cyclic groups in which the order is hundreds of digits long, even when programmed on a computer that can perform millions of operations per second.

As we stated, the ElGamal system can appear in many different forms due to the various types of groups that can be used for G. Recall that the group $\mathbb{Z}_p^*$ that we used in Example 9.1 is the group of nonzero elements in the finite field $\mathbb{Z}_p$ with the operation of multiplication. We will close this section with the following example of the ElGamal system in which G is chosen to be the group of nonzero elements in a more general finite field with the operation of multiplication.

Example 9.2 Suppose that we wish to use the ElGamal system to send a secret message to a colleague across an insecure line of communication.

1. Our colleague initiates the process by choosing the primitive polynomial $p(x) = x^5 + 4x + 2 \in \mathbb{Z}_5[x]$, and letting G be the group of nonzero elements in the finite field $F = \mathbb{Z}_5[x]/(p(x))$ with the operation of multiplication. Note that since $|F| = 5^5 = 3125$, then $|G| = |F^*| = 3124$. Next, our colleague chooses $a = x$ and also a value for n, computes $b = a^n = 2x^4 + 4x^3 + x^2 + 4x + 2$ in G, and sends $p(x)$, a, and b to us across the insecure line of communication.

2. Using some method of conversion, we convert our message into the equivalent field element $w = x^4 + x^3 + 3 \in G$. We then choose $k = 537$, compute $y = a^k = 2x^4 + x^3 + 4x + 4$ and $z = wb^k = x^4 + 3x^3 + 2x^2 + 3$ in G, and send y and z to our colleague across the insecure line of communication.

3. Since our colleague would know the value of $n = 1005$, our colleague would be able to recover the original plaintext w by computing $zy^{-n} = zy^{3124-n} = x^4 + x^3 + 3$ in G.

$\square$

Note that an intruder could break the ElGamal cryptosystem in Example 9.2 by solving the discrete logarithm problem in G. Specifically, an intruder could break the system by finding a discrete logarithm of $b = 2x^4 + 4x^3 + x^2 + 4x + 2$ to the base $a = x$ in G.

9.2   ElGamal Ciphers with Maple

In this section, we will show how Maple can be used to perform the computations in Examples 9.1 and 9.2.

In Example 9.1, our colleague initiated the process by choosing $G = \mathbb{Z}_p^*$ for prime $p = 100000007$. Our colleague also chose the value $a = 180989$.

```
>   p := nextprime(100000000);
                    p := 100000007
```

```
>   a := 180989:
```

Next, our colleague chose the value $n = 5124541$.

```
>   n := 5124541:
```

Our colleague then formed $b = a^n \bmod p$. Recall that we can cause Maple to do this modular exponentiation in a very efficient way by using the &^ function as follows.

```
>   b := a &^ n mod p;
                    b := 10524524
```

Our colleague then sent the values of p, a, and b to us. We converted our message into the equivalent numerical block $w = 20181202$, and then chose the value $k = 3638997$.

```
>   w := 20181202:
```

```
>   k := 3638997:
```

Next, we computed $y = a^k \bmod p$ and $z = wb^k \bmod p$.

```
>  y := a &^ k mod p;
```
$$y := 73133845$$

```
>  z := w*(b &^ k) mod p;
```
$$z := 88730910$$

We then sent the values of y and z to our colleague. Our colleague recovered w by computing $zy^{(p-1)-n} \bmod p$.

```
>  z*(y &^ (p-1-n)) mod p;
```
$$20181202$$

In Example 9.2, our colleague initiated the process by choosing the primitive polynomial $p(x) = x^5 + 4x + 2 \in \mathbb{Z}_5[x]$, and letting G be the group of nonzero elements in the finite field $F = \mathbb{Z}_5[x]/(p(x))$ with the operation of multiplication.

```
>  p := x -> x^5+4*x+2:
```

```
>  Primitive(p(x)) mod 5;
```
$$true$$

Next, our colleague chose $a = x$ and the value $n = 1005$.

```
>  a := x:
```

```
>  n := 1005:
```

Our colleague then formed $b = a^n$ in G. Recall that we can perform this computation by using the Maple **Powmod** function as follows.

```
>  b := Powmod(a, n, p(x), x) mod 5;
```
$$b := 2\,x^4 + 4\,x^3 + x^2 + 4\,x + 2$$

Our colleague then sent $p(x)$, a, and b to us. We converted our message into the equivalent field element $w = x^4 + x^3 + 3$, and then chose the value $k = 537$.

```
>  w := x^4+x^3+3:
```

```
>  k := 537:
```

Next, we computed $y = a^k$ in G.

```
>  y := Powmod(a, k, p(x), x) mod 5;
```
$$y := 2\,x^4 + x^3 + 4\,x + 4$$

We then computed $z = wb^k$ in G. To perform this computation, we can use the Maple **Powmod** and **Rem** functions as follows.

```
>  bk := Powmod(b, k, p(x), x) mod 5;
```
$$bk := 3\,x^4 + 2\,x^2 + 3\,x + 1$$

```
>  z := Rem(w*bk, p(x), x) mod 5;
```
$$z := x^4 + 3\,x^3 + 2\,x^2 + 3$$

We then sent y and z to our colleague. Recall that our colleague could recover w by computing zy^{3124-n} in G. To perform this computation, we can use the **Powmod** and **Rem** functions as follows.

```
>  yn := Powmod(y, 3124-n, p(x), x) mod 5;
```
$$yn := x^4 + 4\,x^3 + 4\,x^2 + 4\,x + 2$$

```
>  Rem(z*yn, p(x), x) mod 5;
```
$$x^4 + x^3 + 3$$

9.3   ElGamal Ciphers with MATLAB

In this section, we will show how MATLAB can be used to perform the computations in Examples 9.1 and 9.2.

In Example 9.1, our colleague initiated the process by choosing $G = \mathbb{Z}_p^*$ for prime $p = 100000007$. Our colleague also chose the value $a = 180989$.

```
>> p = feval(symengine, 'nextprime', sym('100000000'))

p =

100000007

>> a = sym('180989');
```

Next, our colleague chose the value $n = 5124541$. Note that in the following and preceding commands, the input parameters are enclosed in single quotes and declared in symbolic form. This same syntax will also be required in some of the commands that follow.

```
>> n = sym('5124541');
```

Our colleague then formed $b = a^n \bmod p$. Recall that we can cause MATLAB to do this modular exponentiation in a very efficient way by using the MuPAD **powermod** function within MATLAB as follows.

```
>> b = feval(symengine, 'powermod', a, n, p)

b =

10524524
```

Our colleague then sent the values of p, a, and b to us. We converted our message into the equivalent numerical block $w = 20181202$, and then chose the value $k = 3638997$.

```
>> w = sym('20181202');
>> k = sym('3638997');
```

Next, we computed $y = a^k \bmod p$ and $z = wb^k \bmod p$.

```
>> y = feval(symengine, 'powermod', a, k, p)

y =

73133845

>> z = mod(w*(feval(symengine, 'powermod', b, k, p)), p)

z =

88730910
```

We then sent the values of y and z to our colleague. Our colleague recovered w by computing $zy^{(p-1)-n} \bmod p$.

```
>> mod(z*(feval(symengine, 'powermod', y, p-1-n, p)), p)

ans =

20181202
```

In Example 9.2, our colleague initiated the process by choosing the primitive polynomial $p(x) = x^5 + 4x + 2 \in \mathbb{Z}_5[x]$, and letting G be the group of nonzero elements in the finite field $F = \mathbb{Z}_5[x]/(p(x))$ with the operation of multiplication. As we first did in Section 1.6, we will establish that $p(x)$ is primitive using the user-written function **Primitive**, which we have written separately from this MATLAB session and saved as the M-file Primitive.m.

```
>> syms x
>> p = @(x) x^5+4*x+2;
>> Primitive(p(x), 5)

ans =

TRUE
```

Next, our colleague chose $a = x$ and the value $n = 1005$.

```
>> a = x;
>> n = sym('1005');
```

Our colleague then formed $b = a^n$ in G. Recall that we can perform this computation using the user-written function **Powmod**, which we first used in Section 1.6, and have written separately from this MATLAB session and saved as the M-file Powmod.m. The following command shows how the **Powmod** function can be used to compute $b = a^n$ in G.

```
>> b = Powmod(a, n, p(x), x, 5)

b =

2*x^4 + 4*x^3 + x^2 + 4*x + 2
```

Our colleague then sent $p(x)$, a, and b to us. We converted our message into the equivalent field element $w = x^4 + x^3 + 3$, and then chose the value $k = 537$.

```
>> w = x^4+x^3+3;
>> k = sym('537');
```

Next, we computed $y = a^k$ in G.

```
>> y = Powmod(a, k, p(x), x, 5)

y =

2*x^4 + x^3 + 4*x + 4
```

We then computed $z = wb^k$ in G. To perform this computation, in addition to the user-written **Powmod** function, we use the user-written **Rem** function, which we also first used in Section 4.4, and have written separately from this MATLAB session and saved as the M-file Rem.m. The following commands show how the **Powmod** and **Rem** functions can be used to compute $z = wb^k$ in G.

```
>> bk = Powmod(b, k, p(x), x, 5)

bk =

3*x^4 + 2*x^2 + 3*x + 1

>> z = Rem(w*bk, p(x), x, 5)

z =

x^4 + 3*x^3 + 2*x^2 + 3
```

We then sent y and z to our colleague. Recall that our colleague could recover w by computing zy^{3124-n} in G. To perform this computation, we can use the **Powmod** and **Rem** functions as follows.

```
>> yn = Powmod(y, 3124-n, p(x), x, 5)

yn =

x^4 + 4*x^3 + 4*x^2 + 4*x + 2

>> Rem(z*yn, p(x), x, 5)

ans =

x^4 + x^3 + 3
```

9.4 Elliptic Curves

Elliptic curves have figured prominently in several important mathematical problems. For example, the recent proof of Fermat's Last Theorem by Andrew Wiles employed elliptic curves. Elliptic curves have also played an important role in integer factorization, primality testing, and, more recently, public-key cryptography. The idea of using elliptic curves in cryptography was first proposed by Neal Koblitz and Victor Miller in 1985.

Let F be a field of characteristic greater than 3, and suppose c and d are elements in F for which $x^3 + cx + d$ has no multiple roots, or, equivalently, for which $4c^3 + 27d^2 \neq 0$. Consider the following equation.

$$y^2 = x^3 + cx + d \tag{9.1}$$

An *elliptic curve* is the set of ordered pairs $(x, y) \in F \times F$ of solutions to (9.1) together with a special element denoted by $\mathcal{O}$ and called the *point at infinity*. As we have stated, an elliptic curve forms an abelian group with a specially defined operation. This operation is initially best viewed geometrically as applied to an elliptic curve over $\mathbb{R}$. For example, consider the following graph of the ordered pairs $(x, y) \in \mathbb{R} \times \mathbb{R}$ of solutions to the equation $y^2 = x^3 - 6x$ over $\mathbb{R}$.

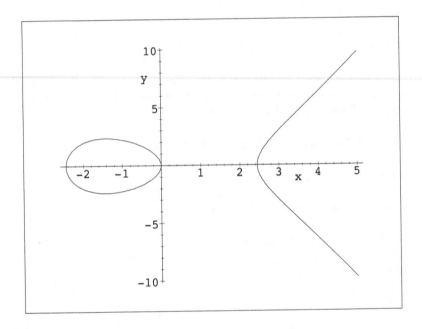

Note first that, as we would expect from the form of (9.1), this graph is symmetric about the x-axis. We will now describe the operation that gives the elliptic curve E consisting of the points on this graph and point at infinity $\mathcal{O}$ the structure of an abelian group. This operation is an addition operation, and can be summarized as follows.

1. The point at infinity serves as the identity element in the group. Thus, by definition, $P + \mathcal{O} = \mathcal{O} + P = P$ for all $P \in E$.

2. Suppose that $P = (x, y)$ is a point on the graph of $y^2 = x^3 - 6x$. We then define the negative of P as $-P = (x, -y)$. This is illustrated in the following graph.

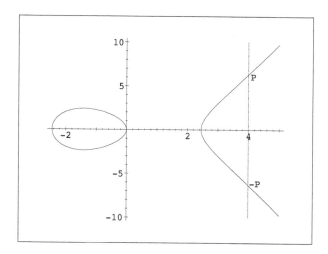

3. Suppose that P and Q are points on the graph of $y^2 = x^3 - 6x$ with $P \neq \pm Q$, and that the line connecting P and Q is not tangent to the graph at P or Q. Then the line connecting P and Q must intersect the graph at a unique third point R. We then define $P + Q = -R$. This is illustrated in the following graph.

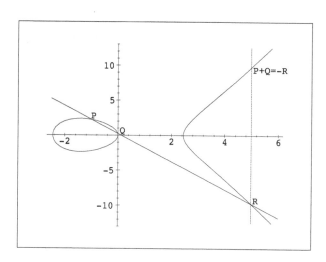

4. Suppose that P and Q are points on the graph of $y^2 = x^3 - 6x$ with $P \neq \pm Q$, and that the line connecting P and Q is tangent to the graph at P. We then define $P + Q = -P$. This is illustrated in the following graph.

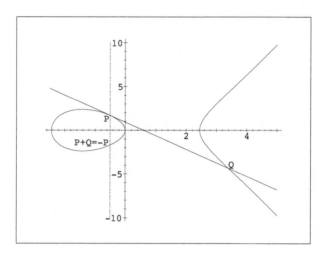

5. Suppose that $P = (x, y)$ is a point on the graph of $y^2 = x^3 - 6x$ with $x \neq 0$, and that the point $P = (x, y)$ is not a point of inflection for the graph. Then it is not too difficult to show that the line tangent to the graph at P must intersect the graph at a unique second point R. We then define $P + P = -R$. This is illustrated in the following graph.

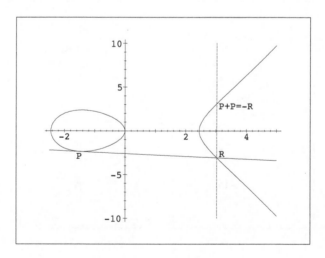

6. Suppose that P is a point on the graph of $y^2 = x^3 - 6x$, and that P is a point of inflection for the graph. We then define $P + P = -P$. This is illustrated in the following graph.

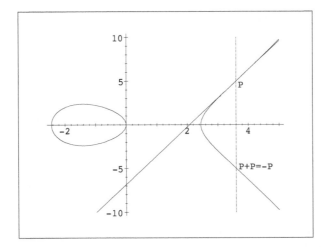

This addition operation on E is clearly commutative. The fact that it is associative is less obvious, and will be assumed.

Recall for the ElGamal cryptosystem, we need a finite abelian group. Thus, an elliptic curve over $\mathbb{R}$ like the one demonstrated in this section is not a group we could use in the ElGamal system. However, for an elliptic curve over a finite field, the elliptic curve will also be finite. For example, consider an elliptic curve over the field $\mathbb{Z}_p$ for some prime $p > 3$. Although the addition operation for elliptic curves we have described in this section geometrically is applied specifically to an elliptic curve over $\mathbb{R}$, this general operation gives any elliptic curve the structure of an abelian group. For an elliptic curve over $\mathbb{Z}_p$, this operation cannot be described in the same way geometrically. However, the operation can be expressed algebraically.

Let p be a prime with $p > 3$, and suppose c and d are elements in $\mathbb{Z}_p$ for which $x^3 + cx + d$ has no multiple roots, or, equivalently, for which $4c^3 + 27d^2 \neq 0 \bmod p$. Let E be the elliptic curve of ordered pairs $(x, y) \in \mathbb{Z}_p \times \mathbb{Z}_p$ of solutions to (9.1) modulo p together with point at infinity $\mathcal{O}$. It can be shown that the addition operation that we have described geometrically which gives E the structure of an abelian group can be expressed algebraically as follows. Recall first that $\mathcal{O}$ serves as the identity in the group. Now, let $P = (x_1, y_1)$ and $Q = (x_2, y_2)$ be ordered pairs in E. If $P = -Q$, then $P + Q = \mathcal{O}$. Otherwise, if $P = Q$, then $P + Q = (x_3, y_3)$, where x_3 and y_3 are defined as follows.

$$x_3 = \left(\left(\frac{3x_1^2 + c}{2y_1} \right)^2 - x_1 - x_2 \right) \bmod p \tag{9.2}$$

$$y_3 = \left(\left(\frac{3x_1^2 + c}{2y_1} \right)(x_1 - x_3) - y_1 \right) \bmod p \tag{9.3}$$

Also, if $P \neq \pm Q$, then $P + Q = (x_3, y_3)$, where x_3 and y_3 are defined as follows.

$$x_3 = \left(\left(\frac{y_2 - y_1}{x_2 - x_1} \right)^2 - x_1 - x_2 \right) \bmod p \tag{9.4}$$

$$y_3 = \left(\left(\frac{y_2 - y_1}{x_2 - x_1} \right)(x_1 - x_3) - y_1 \right) \bmod p \tag{9.5}$$

For small primes p, we can construct elliptic curves over $\mathbb{Z}_p$ by trial and error. Let p be a prime with $p > 3$, and suppose c and d are elements in $\mathbb{Z}_p$ for which $4c^3 + 27d^2 \neq 0 \bmod p$. We can then use the following steps to construct the solutions to (9.1) modulo p.

1. Determine the values of $x \in \mathbb{Z}_p$ for which $z = (x^3 + cx + d) \bmod p$ is a perfect square in $\mathbb{Z}_p$.

2. Determine the values of $y \in \mathbb{Z}_p$ for which $y^2 = z \bmod p$.

The elements in $\mathbb{Z}_p^*$ that are perfect squares in $\mathbb{Z}_p^*$ are called *quadratic residues*. Thus, the values of z determined in the preceding two steps are 0 and the quadratic residues in $\mathbb{Z}_p^*$.

For the first preceding step, we consider the homomorphism $s(y) = y^2$ on $\mathbb{Z}_p^*$. Note that the kernel of $s(y)$ is the set $K = \{x \mid x^2 = 1\} = \{1, -1\}$. Thus, $|K| = 2$, and the set $Q = \{z \in \mathbb{Z}_p^* \mid z = s(y) \text{ for some } y \in \mathbb{Z}_p^*\}$ of quadratic residues in $\mathbb{Z}_p^*$ has order $t = \frac{p-1}{2}$. Next, we consider the function $g(x) = x^t - 1$. If $z \in Q$, then $z = y^2 \bmod p$ for some $y \in \mathbb{Z}_p^*$, and so $g(z) = z^t - 1 = y^{2t} - 1 = y^{p-1} - 1 = 0 \bmod p$ by Lagrange's Theorem (Theorem 1.4). Thus, the t roots of $g(x)$ are precisely the t elements in Q. We summarize this in the following lemma.

Lemma 9.1 *An element z in $\mathbb{Z}_p^*$ is a quadratic residue in $\mathbb{Z}_p^*$ if and only if $z^{\frac{p-1}{2}} = 1 \bmod p$. Thus, z is a perfect square in $\mathbb{Z}_p$ if and only if $z = 0$ or $z^{\frac{p-1}{2}} = 1 \bmod p$.*

For the second preceding step, note that if $z = y^2 \bmod p$, then it will follow that $\left(z^{\frac{p+1}{4}} \right)^2 = y^{p+1} = y^2 = z \bmod p$. Thus, if $p = 3 \bmod 4$, then we can find a square root of z by computing $z^{\frac{p+1}{4}} \bmod p$. We summarize this in the following lemma.

Lemma 9.2 *Suppose $p = 3 \bmod 4$. If z is a quadratic residue in $\mathbb{Z}_p^*$, then $y = z^{\frac{p+1}{4}} \bmod p$ will be a square root of z in $\mathbb{Z}_p^*$. The only other square root of z in $\mathbb{Z}_p^*$ is $-y$.*

In summary, let p be a prime with $p > 3$ and $p = 3 \bmod 4$, and suppose c and d are elements in $\mathbb{Z}_p$ for which $4c^3 + 27d^2 \neq 0 \bmod p$. Let E be the elliptic curve of ordered pairs $(x, y) \in \mathbb{Z}_p \times \mathbb{Z}_p$ of solutions to (9.1) modulo p together with point at infinity $\mathcal{O}$. Then for the set Q of quadratic residues in $\mathbb{Z}_p^*$, the following will be true.

$$
\begin{aligned}
E \;=\; & \{(x, \pm y) \mid z = x^3 + cx + d \in Q \text{ and } y = z^{\frac{p+1}{4}} \bmod p\} \\
& \cup \; \{(x, 0) \mid x^3 + cx + d = 0\} \\
& \cup \; \{\mathcal{O}\}
\end{aligned}
$$

Example 9.3 Suppose $p = 19$, and let E be the elliptic curve of ordered pairs $(x, y) \in \mathbb{Z}_p \times \mathbb{Z}_p$ of solutions to $y^2 = (x^3 + x + 6) \bmod p$ with point at infinity $\mathcal{O}$. We can construct the ordered pairs in E as follows. First, by trial and error, we find the values of $x \in \mathbb{Z}_p$ for which $z = (x^3 + x + 6) \bmod p$ is a quadratic residue in $\mathbb{Z}_p^*$. For example, for $x = 0$, $z = (0^3 + 0 + 6) \bmod p = 6$. Then since $z^{\frac{p-1}{2}} = 6^9 = 1 \bmod p$, by Lemma 9.1 we know $z = 6$ is a quadratic residue in $\mathbb{Z}_p^*$. Also, for $x = 1$, $z = (1^3 + 1 + 6) \bmod p = 8$. Then since $z^{\frac{p-1}{2}} = 8^9 \neq 1 \bmod p$, by Lemma 9.1 we know $z = 8$ is not a quadratic residue in $\mathbb{Z}_p^*$. By continuing this process, we can find that the values of $x \in \mathbb{Z}_p$ for which $z = (x^3 + x + 6) \bmod p$ is a quadratic residue in $\mathbb{Z}_p^*$ are $x = 0, 2, 3, 4, 10, 12, 14,$ and 18. Next, for each quadratic residue $z \in \mathbb{Z}_p^*$, we must find the values of $y \in \mathbb{Z}_p^*$ for which $y^2 = z \bmod p$. Since $p = 3 \bmod 4$, we can use Lemma 9.2. For example, for the quadratic residue $z = 6$ that results from $x = 0$, by Lemma 9.2 we know the square roots of z are $z^{\frac{p+1}{4}} \bmod p = 6^5 \bmod p = 5$ and -5. Then since $z = 6$ results from $x = 0$, it follows that the ordered pairs $(0, \pm 5)$ are in E. By repeating this process for each of the quadratic residues in $\mathbb{Z}_p^*$, we can find that the ordered pairs $(0, \pm 5)$, $(2, \pm 4)$, $(3, \pm 6)$, $(4, \pm 6)$, $(10, \pm 16)$, $(12, \pm 6)$, $(14, \pm 16)$, and $(18, \pm 17)$ are all in E. Also, by trial and error, we can find that the only value of $x \in \mathbb{Z}_p$ for which $z = x^3 + x + 6 = 0 \bmod p$ is $x = 6$. Thus, the only additional ordered pair in E is $(6, 0)$. Now, suppose we wish to find the sum of the elements $P = (x_1, y_1) = (2, 4)$ and $Q = (x_2, y_2) = (10, 16)$ in E. Denote this sum by $P + Q = (x_3, y_3)$. Since $P \neq \pm Q$, we can use (9.4) and (9.5) to find x_3 and y_3. We first compute $\dfrac{y_2 - y_1}{x_2 - x_1} \bmod p$ as follows. (Note that $8^{-1} = 12 \bmod p$ because $8 \cdot 12 = 96 = 1 \bmod p$.)

$$
\frac{y_2 - y_1}{x_2 - x_1} = \frac{16 - 4}{10 - 2} = 12 \cdot (8)^{-1} = 12 \cdot 12 = 11 \bmod p
$$

Then using (9.4) and (9.5), we can compute x_3 and y_3 as follows.

$$
\begin{aligned}
x_3 &= 11^2 - 2 - 10 &= 14 \bmod p \\
y_3 &= 11 \cdot (2 - 14) - 4 &= 16 \bmod p
\end{aligned}
$$

Thus, in this elliptic curve, $(2,4) + (10,16) = (14,16)$. Next, for the element $P = (x_1, y_1) = (0,5)$ in E, suppose that we wish to compute the sum $P + P = (x_3, y_3)$. Since $P \neq -P$, then we can use (9.2) and (9.3) to find x_3 and y_3. We first compute $\dfrac{3x_1^2 + c}{2y_1}$ mod p as follows. (Note that $10^{-1} = 2$ mod p because $10 \cdot 2 = 20 = 1$ mod p.)

$$\frac{3x_1^2 + c}{2y_1} = \frac{3 \cdot (0)^2 + 1}{2 \cdot 5} = 1 \cdot (10)^{-1} = 1 \cdot 2 = 2 \text{ mod } p$$

Then using (9.2) and (9.3), we can compute x_3 and y_3 as follows.

$$
\begin{aligned}
x_3 &= 2^2 - 0 - 0 &= 4 \text{ mod } p \\
y_3 &= 2 \cdot (0 - 4) - 5 &= 6 \text{ mod } p
\end{aligned}
$$

Thus, in this elliptic curve, $(0,5) + (0,5) = (4,6)$. $\qquad\square$

Although elliptic curves over $\mathbb{Z}_p$ for prime p are not particularly easy to construct, their general structure is remarkably simple and specific. We summarize this general structure in the following theorem, which we will include without proof.

Theorem 9.3 *Let p be a prime with $p > 3$, and suppose E is an elliptic curve over $\mathbb{Z}_p$. Then E is isomorphic to the direct product $\mathbb{Z}_m \times \mathbb{Z}_n$ of the groups $\mathbb{Z}_m$ and $\mathbb{Z}_n$ with the operation of addition for some integers m and n with $n | m$ and $n | (p-1)$.*

As an example of Theorem 9.3, consider the elliptic curve E in Example 9.3. For this elliptic curve, $|E| = 18$, and so the only possibilities for the values of m and n in Theorem 9.3 are $m = 18$ and $n = 1$, and $m = 6$ and $n = 3$. Since it can be verified that $(0,5) \in E$ generates all of the elements in E (as do several of the other elements in E), then E is cyclic. Thus, the correct values of m and n are $m = 18$ and $n = 1$, and E is isomorphic to the cyclic group $\mathbb{Z}_{18}$ with the operation of addition.

Consider now an elliptic curve E defined over $\mathbb{Z}_p$ for a very large prime p. While $|E|$ could be so large that constructing all of the elements in E would not be possible, it might still be possible to determine the exact value of $|E|$ using a well known algorithm from René Schoof. Although Schoof's algorithm is beyond the scope of this book, we will include the following theorem (without proof), a well known result from Helmut Hasse that yields upper and lower bounds on $|E|$.

Theorem 9.4 (Hasse's Theorem) *Let E be an elliptic curve defined over $\mathbb{Z}_p$ for prime p. Then $p + 1 - 2\sqrt{p} \leq |E| \leq p + 1 + 2\sqrt{p}$.*

We will close this section with one additional fact regarding elliptic curves. Recall that we began this presentation of elliptic curves by assuming that the underlying field F was of characteristic greater than 3, and that the cubic polynomial on the right in (9.1) had no multiple roots. Elliptic curves can also be defined over fields of characteristic 2 or 3; they are just defined to include the set of solutions to an equation with a form slightly different from (9.1). Specifically, if F is a field of characteristic 2, then an elliptic curve defined over F is the set of ordered pairs $(x, y) \in F \times F$ of solutions to an equation of the following form together with point at infinity $\mathcal{O}$, where c and d are elements in F and the cubic polynomial on the right is allowed to have multiple roots.

$$y^2 + y = x^3 + cx + d$$

Also, if F is a field of characteristic 3, then an elliptic curve defined over F is the set of ordered pairs $(x, y) \in F \times F$ of solutions to an equation of the following form together with point at infinity $\mathcal{O}$, where b, c, and d are elements in F and the cubic polynomial on the right is not allowed to have multiple roots.

$$y^2 = x^3 + bx^2 + cx + d$$

Results analogous to those mentioned in this section hold for elliptic curves defined over fields of characteristic 2 or 3 as well.

9.5   Elliptic Curves with Maple

In this section, we will show how Maple can be used to construct the elliptic curve E in Example 9.3 and perform the elliptic curve addition operation.

We will begin by assigning the prime $p = 19$ and the values of $c = 1$ and $d = 6$ for the elliptic curve equation (9.1), and verifying that these values of c and d satisfy $4c^3 + 27d^2 \neq 0 \bmod p$.

```
>  p := 19:

>  c := 1:

>  d := 6:

>  4*c^3+27*d^2 mod p;
```

Next, we will store the right side of (9.1) as the variable *eqn*.

```
>   eqn := x^3+c*x+d:
```

We are now ready to generate the elements in E that are ordered pairs $(x, y) \in \mathbb{Z}_p \times \mathbb{Z}_p$ of solutions to (9.1) modulo p. To do this, we will use the user-written function **epoints**, which we have written separately from this Maple session and saved as the text file epoints.mpl. To use this function, we must first read it into our Maple session as follows.

```
>   read "epoints.mpl";
```

The following command then illustrates how the function **epoints** can be used. Specifically, by entering the following command, we generate the ordered pairs $(x, y) \in \mathbb{Z}_p \times \mathbb{Z}_p$ of solutions to (9.1) modulo p.

```
>   ecurve := epoints(eqn, x, infinity, p);
```

ecurve := [0, 5], [0, 14], [2, 4], [2, 15], [3, 6], [3, 13], [4, 6], [4, 13], [6, 0], [10, 16], [10, 3], [12, 6], [12, 13], [14, 16], [14, 3], [18, 17], [18, 2]

The first parameter in the preceding command is the right side of (9.1), and the second parameter is the variable used in the first parameter. The third parameter is a numerical value that indicates the number of ordered pairs of solutions to (9.1) that are to be generated. If this parameter exceeds the total number of ordered pairs of solutions to (9.1), then the function will generate all of the ordered pairs of solutions to (9.1). By using *infinity* for this parameter, we guarantee that the function will generate all of the ordered pairs of solutions to (9.1). The final parameter is the prime p in the underlying field $\mathbb{Z}_p$.

Recall that the ordered pairs of solutions to (9.1) form all of the elements in E except the point at infinity $\mathcal{O}$. By entering the following command, we attach the representation 0 for the point at infinity to the list *ecurve* of elements in E.

```
>   ecurve := ecurve, 0;
```

ecurve := [0, 5], [0, 14], [2, 4], [2, 15], [3, 6], [3, 13], [4, 6], [4, 13], [6, 0], [10, 16], [10, 3], [12, 6], [12, 13], [14, 16], [14, 3], [18, 17], [18, 2], 0

The following command returns the number of elements in E.

```
>   nops([ecurve]);
```
18

To perform the addition operation that we presented in Section 9.4, we will use the user-written function **addec**, which we have written separately from this Maple session and saved as the text file addec.mpl. To use this function, we must first read it into our Maple session as follows.

```
>   read "addec.mpl";
```

The following command then illustrates how the function **addec** can be used. Specifically, by entering the following command, we find the sum of the elements $(2, 4)$ and $(10, 16)$ in E.

```
>   addec([2, 4], [10, 16], c, p);
```
$$[14, 16]$$

The first two parameters in the preceding command are the elements in E that are to be added. The final two parameters are the value of c in (9.1) and the prime p in the underlying field $\mathbb{Z}_p$.

As another example of the **addec** function, in the following command we add the element $(0, 5)$ in E to itself.

```
>   addec([0, 5], [0, 5], c, p);
```
$$[4, 6]$$

Next, we add the element $(0, 5)$ in E and the point at infinity.

```
>   addec([0, 5], 0, c, p);
```
$$[0, 5]$$

Finally, we add the elements $(0, 5)$ and $(0, 14)$ in E.

```
>   addec([0, 5], [0, 14], c, p);
```
$$0$$

The preceding output shows that, as expected, $(0, 5)$ and $(0, 14)$ are inverses of each other in E.

We can verify that $(0, 5)$ is a generator for E as follows. We begin by assigning the element $(0, 5)$ as the variable *gen*.

```
>   gen := [0, 5]:
```

We will now construct the cyclic subgroup of E generated by *gen*. To do this, we first assign $(0, 5)$ also as the variable *temp*, and store this element as the first entry in a table *csubgroup*.

```
>   temp := [0, 5]:
>   pct := 1:
>   csubgroup[pct] := temp:
```

Then by entering the following **while** loop, we construct the cyclic subgroup of E generated by *gen*, and place the elements in this cyclic subgroup as the entries in *csubgroup*. More specifically, by entering the following **while** loop, we compute multiples of *gen* using **addec**, and place these multiples as the entries in *csubgroup*. The loop terminates when the identity element 0 is obtained.

```
>   while temp <> 0 do
>           temp := addec(temp, gen, c, p):
>           pct := pct+1:
>           csubgroup[pct] := temp:
>   od:
>   seq(csubgroup[i], i=1..pct);
```

$[0, 5]$, $[4, 6]$, $[2, 4]$, $[3, 6]$, $[14, 3]$, $[12, 13]$, $[18, 2]$, $[10, 3]$, $[6, 0]$, $[10, 16]$, $[18, 17]$, $[12, 6]$, $[14, 16]$, $[3, 13]$, $[2, 15]$, $[4, 13]$, $[0, 14]$, 0

Since the preceding output is all of the elements in E, then we know that $(0, 5)$ is indeed a generator for E.

9.6   Elliptic Curves with MATLAB

In this section, we will show how MATLAB can be used to construct the elliptic curve E in Example 9.3 and perform the elliptic curve addition operation.

We will begin by assigning the prime $p = 19$ and the values of $c = 1$ and $d = 6$ for the elliptic curve equation (9.1), and verifying that these values of c and d satisfy $4c^3 + 27d^2 \neq 0 \bmod p$. Note that in the first three of the following commands, the input parameters are enclosed in single quotes and declared in symbolic form. This same syntax will also be used in some of the commands that follow.

```
>> p = sym('19');
>> c = sym('1');
>> d = sym('6');
>> mod(4*c^3+27*d^2, p)

ans =

7
```

Next, we will store the right side of (9.1) as the variable *eqn*.

```
>> syms x
>> eqn = x^3+c*x+d;
```

We are now ready to generate the elements in E that are ordered pairs $(x, y) \in \mathbb{Z}_p \times \mathbb{Z}_p$ of solutions to (9.1) modulo p. To do this, we will use the user-written function **epoints**, which we have written separately from this MATLAB session and saved as the M-file epoints.m. The following command illustrates how the function **epoints** can be used. Specifically, by entering the following command, we generate the ordered pairs $(x, y) \in \mathbb{Z}_p \times \mathbb{Z}_p$ of solutions to (9.1) modulo p.

```
>> ecurve = epoints(eqn, x, inf, p);
```

The first parameter in the preceding command is the right side of (9.1), and the second parameter is the variable used in the first parameter. The third parameter is a numerical value that indicates the number of ordered pairs of solutions to (9.1) that are to be generated. If this parameter exceeds the total number of ordered pairs of solutions to (9.1), then the function will generate all of the ordered pairs of solutions to (9.1). By using *inf* for this parameter, we guarantee that the function will generate all of the ordered pairs of solutions to (9.1). The final parameter is the prime p in the underlying field $\mathbb{Z}_p$.

The ordered pairs generated by the preceding command are stored in a cell array, which is a MATLAB data structure designed to store dissimilar types of data. Individual elements in a cell array can be accessed by using the element number enclosed in curly braces. For example, by entering the following command we access the second ordered pair stored in the cell array *ecurve*.

```
>> ecurve{2}

ans =

[  0, 14]
```

When the number of items that are stored in a cell array is large, there is no predefined MATLAB function for displaying the results in a concise fashion. To display such output concisely, we will use the user-written functions **estring** and **stringprint**, which we have written separately from this MATLAB session and saved as the M-files estring.m and stringprint.m. This **stringprint** function is the same as the **stringprint** function that we used previously in Section 7.4. The following command shows how the functions **estring** and **stringprint** can be used. Specifically, by entering the following command, we concatenate the ordered pairs stored in the cell array *ecurve* as a string, and display the resulting string in its entirety in rows with fifty characters each.

```
>> stringprint(estring(ecurve), 50)

[0,5],[0,14],[2,4],[2,15],[3,6],[3,13],[4,6],[4,13
],[6,0],[10,16],[10,3],[12,6],[12,13],[14,16],[14,
3],[18,17],[18,2]
```

Recall that the ordered pairs of solutions to (9.1) form all of the elements in E except the point at infinity $\mathcal{O}$. By entering the following command, we attach the representation 0 for the point at infinity to the cell array *ecurve* of elements in E.

```
>> ecurve{numel(ecurve)+1} = 0;
>> stringprint(estring(ecurve), 50)

[0,5],[0,14],[2,4],[2,15],[3,6],[3,13],[4,6],[4,13
],[6,0],[10,16],[10,3],[12,6],[12,13],[14,16],[14,
3],[18,17],[18,2],0
```

The following command returns the number of elements in E.

```
>> numel(ecurve)

ans =

   18
```

To perform the addition operation that we presented in Section 9.4, we will use the user-written function **addec**, which we have written separately from this MATLAB session and saved as the M-file addec.m. The following command illustrates how the function **addec** can be used. Specifically, by entering the following command, we find the sum of the elements $(2, 4)$ and $(10, 16)$ in E.

```
>> addec([sym('2'), sym('4')], [sym('10'), sym('16')], ...
   c, p)

ans =

[ 14, 16]
```

The first two parameters in the preceding command are the elements in E that are to be added. The final two parameters are the value of c in (9.1) and the prime p in the underlying field $\mathbb{Z}_p$. It is worth noting that the components of the elliptic curve elements in the preceding command are not required to be declared in symbolic form if these components are small integers. For example, the same output is returned by the following command.

```
>> addec([2, 4], [10, 16], c, p)

ans =

[ 14, 16]
```

However, due to the limits on storing large integers in MATLAB, the preceding syntax should not be used if the elements being added are in an elliptic curve in which any of the elements have large integer components.

As another example of the **addec** function, in the following command we add the element $(0, 5)$ in E to itself.

```
>> addec([sym('0'), sym('5')], [sym('0'), sym('5')], c, ...
   p)

ans =

[ 4, 6]
```

Next, we add the element $(0, 5)$ in E and the point at infinity.

```
>> addec([sym('0'), sym('5')], sym('0'), c, p)

ans =

[ 0, 5]
```

Finally, we add the elements $(0, 5)$ and $(0, 14)$ in E.

```
>> addec([sym('0'), sym('5')], [sym('0'), sym('14')], ...
   c, p)

ans =

    0
```

The preceding output shows that, as expected, $(0, 5)$ and $(0, 14)$ are inverses of each other in E.

We can verify that $(0, 5)$ is a generator for E as follows. We begin by assigning the element $(0, 5)$ as the variable *gen*.

```
>> gen = [sym('0'), sym('5')];
```

We will now construct the cyclic subgroup of E generated by *gen*. To do this, we first assign $(0, 5)$ also as the variable *temp*, and store this element as the first entry in a cell array *csubgroup*.

```
>> temp = [sym('0'), sym('5')];
>> pct = 1;
>> csubgroup{pct} = temp;
```

Then by entering the following **while** loop, we construct the cyclic subgroup of E generated by *gen*, and place the elements in this cyclic subgroup as the entries in *csubgroup*. More specifically, by entering the following **while** loop, we compute multiples of *gen* using **addec**, and place these multiples as the entries in *csubgroup*. The loop terminates when the identity element 0 is obtained.

```
>> while ~isequal(temp, 0)
          temp = addec(temp, gen, c, p);
          pct = pct+1;
          csubgroup{pct} = temp;
    end
>> stringprint(estring(csubgroup), 50)

[0,5],[4,6],[2,4],[3,6],[14,3],[12,13],[18,2],[10,
3],[6,0],[10,16],[18,17],[12,6],[14,16],[3,13],[2,
15],[4,13],[0,14],0
```

Since the preceding output is all of the elements in E, then we know that $(0, 5)$ is indeed a generator for E.

9.7   Elliptic Curve Cryptography

If an elliptic curve defined over $\mathbb{Z}_p$ for prime p is to be used as the group G in the ElGamal cryptosystem, the value of p would have to be extremely large in order for the system to be secure. More specifically, it is commonly accepted that G should contain a cyclic subgroup of which the order is at least two hundred digits long in order for the ElGamal system to be secure. Of course, constructing all of the elements in such an elliptic curve might not be possible. However, to use an elliptic curve E defined over $\mathbb{Z}_p$ as the group G in the ElGamal cryptosystem, it is not necessary to construct all of the elements in E. It is only necessary to find an element in E that has an extremely large order. Suppose that we wish to use the ElGamal cryptosystem with an elliptic curve defined over $\mathbb{Z}_p$ for some prime p as the group G to send a secret message to a colleague across an insecure line of communication. Then the system could proceed as follows.

1. Our colleague initiates the process by choosing an extremely large prime p and values of c and d in $\mathbb{Z}_p$ with $4c^3 + 27d^2 \neq 0 \bmod p$. Let E be the elliptic curve of ordered pairs $(x, y) \in \mathbb{Z}_p \times \mathbb{Z}_p$ of solutions to (9.1) modulo p together with point at infinity $\mathcal{O}$. Next, our colleague chooses an elliptic curve element $a \in E$ that has an extremely large order. Our colleague then chooses a positive integer $n < |a|$, computes $b = na = a + a + \cdots + a$ (n total terms) in E using the elliptic curve addition operation (note that we use the notation na for b instead of a^n since the elliptic curve operation is an addition), and sends the values of p, c, and d and the elliptic curve elements a and b to us across the insecure line of communication.

2. Using some method of conversion, we convert our message into an equivalent elliptic curve element $w \in E$. We then choose a positive integer k, compute $y = ka$ and $z = w + kb$ in E using the elliptic curve addition operation, and send the elliptic curve elements y and z to our colleague across the insecure line of communication.

3. Because our colleague knows the value of n, he or she can recover the elliptic curve element w by computing $z - ny$ in E using the elliptic curve addition operation, due to the following.

$$z - ny = w + kb - nka = w + kb - kb = w$$

Example 9.4 Suppose that we wish to use the ElGamal cryptosystem with an elliptic curve defined over $\mathbb{Z}_p$ for prime p as the group G to send a secret message to a colleague across an insecure line of communication.

1. Our colleague initiates the process by choosing $p = 19$ (for illustration, we will use a very small value for p in this example), $c = 1$, and $d = 6$. Then the elliptic curve of ordered pairs $(x, y) \in \mathbb{Z}_p \times \mathbb{Z}_p$ of solutions to (9.1) modulo p together with point at infinity $\mathcal{O}$ is the elliptic curve E in Example 9.3. Next, our colleague chooses $a = (0, 5) \in E$, which, recall from Example 9.3, generates all of E. Our colleague then

chooses $n = 4$, computes $b = na = (0, 5) + (0, 5) + (0, 5) + (0, 5) = (3, 6)$ in E using the elliptic curve addition operation, and sends the values of p, c, and d and the elliptic curve elements a and b to us across the insecure line of communication.

2. Using some method of conversion, we convert our plaintext message into the elliptic curve element $w = (18, 17)$. We then choose the integer $k = 3$, compute $y = ka = (0, 5) + (0, 5) + (0, 5) = (2, 4)$ and $z = w + kb = (18, 17) + (3, 6) + (3, 6) + (3, 6) = (14, 3)$ in E using the elliptic curve addition operation, and send the elliptic curve elements y and z to our colleague across the insecure line of communication.

3. Our colleague can recover w as follows by computing $z - ny$ in E using the elliptic curve addition operation.

$$
\begin{aligned}
z - ny &= (14, 3) - ((2, 4) + (2, 4) + (2, 4) + (2, 4)) \\
&= (14, 3) - (12, 6) \\
&= (14, 3) + (12, 13) \\
&= (18, 17)
\end{aligned}
$$

$\square$

Note that to break the ElGamal cryptosystem in Example 9.4, an intruder would need to determine the value of n from the knowledge of the elliptic curve elements a and $b = na$ in E. That is, to break the ElGamal cryptosystem in Example 9.4, an intruder would need to solve the discrete logarithm problem in E (with b expressed using additive notation na instead of a^n). Of course, because the elliptic curve in this example contains so few elements, an intruder could break the system very easily by trial and error. However, if the elliptic curve element chosen for a had a sufficiently large order, then it would take an intruder essentially forever to break the system, even when using the fastest known algorithm for finding discrete logarithms programmed on a computer that could perform millions of operations per second.

There is a practical difficulty with using an elliptic curve E defined over $\mathbb{Z}_p$ for prime p as the group G in the ElGamal cryptosystem. Recall that if the system is to be implemented as we have described, then the plaintext must somehow be converted into one of the elements in E before being encrypted. This obviously limits flexibility in formatting plaintexts, and could possibly require the generation of a very large number of elements in E. This difficulty can be avoided by using a variation of the ElGamal system due to Alfred Menezes and Scott Vanstone. Suppose as before that we wish to use the ElGamal cryptosystem with an elliptic curve defined over $\mathbb{Z}_p$ as the group G to send a secret message to a colleague across an insecure line of communication. The steps in Menezes and Vanstone's variation of the ElGamal system can be described as follows.

1. The first step is identical to the usual ElGamal system. That is, our colleague chooses an extremely large prime p and values of c and d for which $4c^3 + 27d^2 \neq 0 \bmod p$. Then for the elliptic curve E of ordered pairs $(x, y) \in \mathbb{Z}_p \times \mathbb{Z}_p$ of solutions to (9.1) modulo p together with point at infinity $\mathcal{O}$, our colleague chooses an elliptic curve element $a \in E$ that has an extremely large order and a positive integer $n < |a|$, computes $b = na = a + a + \cdots + a$ (n total terms) in E using the elliptic curve addition operation, and sends the values of p, c, and d and the elliptic curve elements a and b to us across the insecure line of communication.

2. Using some method of conversion, we convert our message into an equivalent ordered pair of integers $w = (w_1, w_2) \in \mathbb{Z}_p^* \times \mathbb{Z}_p^*$, which is *not* required to be an element in E. We then choose a positive integer k, compute $y = ka$ and $kb = (c_1, c_2)$ in E using the elliptic curve addition operation, and, provided c_1 and c_2 are both nonzero, encrypt our message as the ordered pair $z = (z_1, z_2) \in \mathbb{Z}_p^* \times \mathbb{Z}_p^*$ by computing the following.

$$z = (z_1, z_2) = (c_1 w_1 \bmod p, c_2 w_2 \bmod p)$$

We then send the ordered pairs y and z to our colleague across the insecure line of communication.

3. Because our colleague knows the value of n, he or she can recover the ordered pair $kb = (c_1, c_2)$ by computing ny in E using the elliptic curve addition operation, since $ny = nka = kna = kb$. Our colleague can then recover the plaintext message $w = (w_1, w_2)$ by computing $(c_1^{-1} z_1 \bmod p, c_2^{-1} z_2 \bmod p)$, due to the following.

$$(c_1^{-1} z_1 \bmod p, c_2^{-1} z_2 \bmod p) = (c_1^{-1} c_1 w_1 \bmod p, c_2^{-1} c_2 w_2 \bmod p)$$
$$= (w_1, w_2)$$

Example 9.5 Suppose that we wish to use Menezes and Vanstone's variation of the ElGamal cryptosystem with an elliptic curve defined over $\mathbb{Z}_p$ for prime p as the group G to send a secret message to a colleague across an insecure line of communication.

1. Our colleague initiates the process by choosing $p = 19$ (we will again use a very small value for p in this example), $c = 1$, and $d = 6$. Then the elliptic curve of ordered pairs $(x, y) \in \mathbb{Z}_p \times \mathbb{Z}_p$ of solutions to (9.1) modulo p together with point at infinity $\mathcal{O}$ is the elliptic curve E in

Example 9.3. Next, our colleague chooses $a = (0,5) \in E$, which, recall from Example 9.3, generates all of E. Our colleague then chooses $n = 4$, computes $b = na = (0,5) + (0,5) + (0,5) + (0,5) = (3,6) \in E$ using the elliptic curve addition operation, and sends the values of p, c, and d and the elliptic curve elements a and b to us across the insecure line of communication.

2. Using some method of conversion, we convert our plaintext message into the ordered pair $w = (5,13) \in \mathbb{Z}_p^* \times \mathbb{Z}_p^*$, which, as we can see from Example 9.3, is not an element in E. We then choose the integer $k = 3$, compute $y = ka = (0,5) + (0,5) + (0,5) = (2,4)$ and also $kb = (3,6) + (3,6) + (3,6) = (12,6)$ in E using the elliptic curve addition operation, and encrypt our plaintext message as the ordered pair $z = (z_1, z_2) \in \mathbb{Z}_p^* \times \mathbb{Z}_p^*$ by computing the following.

$$z = (12 \cdot 5 \bmod p, \ 6 \cdot 13 \bmod p) = (3,2)$$

We then send the ordered pairs y and z to our colleague across the insecure line of communication.

3. Our colleague can first recover the ordered pair kb by computing $ny = (2,4) + (2,4) + (2,4) + (2,4) = (12,6)$ in E using the elliptic curve addition operation. Our colleague can then recover w by computing the following.

$$(12^{-1} \cdot 3 \bmod p, \ 6^{-1} \cdot 2 \bmod p) = (8 \cdot 3 \bmod p, \ 16 \cdot 2 \bmod p)$$
$$= (5, 13)$$

(Note that $12^{-1} = 8 \bmod p$ because $12 \cdot 8 = 96 = 1 \bmod p$, and $6^{-1} = 16 \bmod p$ because $6 \cdot 16 = 96 = 1 \bmod p$.)

$\square$

Note that with the elliptic curve E in Example 9.3 as the group G, the usual ElGamal cryptosystem would allow only $|E| = 18$ possible plaintexts, while Menezes and Vanstone's variation of the ElGamal system would allow $|\mathbb{Z}_p^*|^2 = 324$ possible plaintexts. This is another advantage of using Menezes and Vanstone's variation of the system.

9.8   Elliptic Curve Cryptography with Maple

In this section, we will show how Maple can be used to do the computations in Menezes and Vanstone's variation of the ElGamal cryptosystem with an elliptic curve defined over $\mathbb{Z}_p$ for prime p as the group G.

Recall that to construct an elliptic curve using Lemma 9.2, which is employed by the user-written function **epoints** that we used in Section 9.5 and will use again in this section, the prime p must satisfy $p = 3 \bmod 4$. In order to find a prime that satisfies this requirement, we will use the user-written function **p3mod4**, which we have written separately from this Maple session and saved as the text file p3mod4.mpl. To use this function, we must first read it into our Maple session as follows.

```
>  read "p3mod4.mpl";
```

The following command then illustrates how the function **p3mod4** can be used. Specifically, this function takes as input an integer s, and returns as output the smallest prime p that is larger than s and which satisfies $p = 3 \bmod 4$. For example, the following command assigns as p the smallest prime larger than 22053249629377880580000 which satisfies $p = 3 \bmod 4$. We will use this value as the prime p in our examples in this section.

```
>  p := p3mod4(22053249629377880580000);
```
$$p := 22053249629377880580119$$

For this prime p, let E be the elliptic curve of ordered pairs $(x, y) \in \mathbb{Z}_p \times \mathbb{Z}_p$ of solutions to (9.1) modulo p for $c = 1$ and $d = 6$ together with point at infinity $\mathcal{O}$. In the following commands, we assign these values of c and d, and verify that they satisfy $4c^3 + 27d^2 \neq 0 \bmod p$.

```
>  c := 1:
```

```
>  d := 6:
```

```
>  4*c^3+27*d^2 mod p;
```
$$976$$

Next, we will store the right side of (9.1) as the variable *eqn*.

```
>  eqn := x^3+c*x+d:
```

For the ordered pair a in the system, we will use the first solution to (9.1) generated by the user-written function **epoints** that we used in Section 9.5. Recall that we have written this function separately from this Maple session and saved it as the text file epoints.mpl. In the following commands, we read this function into our Maple session, and use it to generate one of the ordered pairs $(x, y) \in \mathbb{Z}_p \times \mathbb{Z}_p$ of solutions to (9.1) modulo p, which we will assign as the variable a. We will use this ordered pair as a in our examples in this section.

```
>  read "epoints.mpl";
```

```
>   a := epoints(eqn, x, 1, p);
```
$$a := [0, 40675210938100805459]$$

Next, we will assign the following value for n, which we will use in constructing the ordered pair $b = na$ in E.

```
>   n := 91530873521338:
```

To expedite the process of adding a to itself repeatedly with n total terms using the elliptic curve addition operation, we will use the user-written function **elgamal**, which we have written separately from this Maple session and saved as the text file elgamal.mpl. This function uses the user-written function **addec** that we used in Section 9.5, which we have written separately from this Maple session and saved as the text file addec.mpl. To use these functions, we must first read them into our Maple session as follows.

```
>   read "addec.mpl";
```

```
>   read "elgamal.mpl";
```

The following command then illustrates how the function **elgamal** can be used. Specifically, by entering the following command, we construct the ordered pair $b = na$ in E by adding a to itself repeatedly with n total terms using the elliptic curve addition operation.

```
>   b := elgamal(a, n, c, p);
```
$$b := [1575028246746093464921, 8523832800260348129369]$$

The parameters in the preceding command are the ordered pair a, the number n of terms in the sum, the value of c in (9.1), and the prime p.

Next, we will assign the following value for k, which we will use in constructing the ordered pairs $y = ka$ and kb in E.

```
>   k := 431235145514:
```

We can then construct the ordered pairs $y = ka$ and kb in E using the elliptic curve addition operation by entering the following commands.

```
>   y := elgamal(a, k, c, p);
```
$$y := [1869466810570308236390, 10420072039502224215692]$$

```
>   kb := elgamal(b, k, c, p);
```
$$kb := [15261622170740304569981, 16218536420977465416666]$$

We will now use the ordered pair kb to encrypt the message RENDEZVOUS AT NOON. Because several of the functions that we will use are in the Maple **StringTools** package, we will include this package.

```
>  with(StringTools):
```

Next, we will store our message as the following string.

```
>  message := Select(IsAlpha, "RENDEZVOUS AT NOON");
```
$$message := \text{``RENDEZVOUSATNOON''}$$

We can identify the number of digits in the modulus p to be used in the encryption as follows, by using the Maple **convert** function to convert p from an integer into a string, and then the Maple **length** function to find the number of characters in the resulting string.

```
>  length(convert(p, string));
```
$$23$$

Since the correspondence α that we introduced in Chapter 6 (i.e., A $\mapsto$ 00, B $\mapsto$ 01, C $\mapsto$ 02,..., Z $\mapsto$ 25) converts characters into two-digit integers, our 23-digit value of p is guaranteed to work for encrypting messages containing at most 11 characters. The Maple **LengthSplit** function, which is contained in the **StringTools** package, is designed to split strings of characters into shorter blocks containing a designated number of characters. The following commands demonstrate how **LengthSplit** can be used. Specifically, the following commands cause our message to be split into blocks containing 11 characters each, with the last block ending when all of the characters in the message have been used. Since the message must be encrypted as an ordered pair, these commands check to make sure the number of blocks is even, with a extra character X appended as an additional block if necessary to make the number of blocks be even.

```
>  blen := 11:
>  if nops([LengthSplit(message, blen)]) mod 2 = 0 then
>      messblocks := [LengthSplit(message, blen)];
>  else
>      messblocks := [LengthSplit(message, blen), "X"];
>  fi:
>  messblocks;
```
$$[\text{``RENDEZVOUSA''}, \text{``TNOON''}]$$

To convert this plaintext message into numerical form, we will use the user-written function **tonumber**, which we first used in Section 8.3, and which we have written separately from this Maple session and saved as the text file tonumber.mpl. To use this function, we must first read it into our Maple session as follows.

```
>  read "tonumber.mpl";
```

Then in the following command, we use **tonumber** to convert each block of *messblocks* into a list of two-digit integers, and group all of the two-digit integers within each block into a single integer.

```
>  w := tonumber(messblocks);
```
$$w := [1704130304252114201800, 1913141413]$$

Recall that as we did in Section 8.3, we can use the Maple **map** and **evalb** functions as follows to show that each numerical plaintext block in w is less than p.

```
>  map(x -> evalb(x < p), w);
```
$$[true, true]$$

We can then encrypt this message by entering the following command. This command uses the Maple *~ function, which causes the two lists kb and w to be multiplied componentwise, with the results stored as a list.

```
>  z := map(x -> x mod p, kb *~ w);
```
$$z := [5865493398298422326683, 11514808043310828036997]$$

To decrypt this ciphertext, we must first recover the ordered pair kb. We can do this as follows by computing ny in E using the elliptic curve addition operation.

```
>  ny := elgamal(y, n, c, p);
```
$$ny := [1526162217074030456981, 1621853642097746541666]$$

We can then decrypt the ciphertext by entering the following command.

```
>  w := map(x -> x mod p, ny ^~ (-1) *~ z);
```
$$w := [1704130304252114201800, 1913141413]$$

To convert the plaintext integers in w back into a list of letters, we will use the user-written function **toletter**, which we first used in Section 8.3, and which we have written separately from this Maple session and saved as the text file toletter.mpl. To use this function, we must first read it into our Maple session as follows.

```
>  read "toletter.mpl";
```

Then in the following command, we apply **toletter** to the components of w to convert these plaintext integers back into a list of letters.

```
>  toletter(w);
```

"RENDEZVOUSATNOON"

We will now demonstrate how a longer message can be encrypted and decrypted. Consider the plaintext message RENDEZVOUS AT NOON AT THE TRADING POST LOCATED AT THE RAILROAD STATION, which we begin by storing as the following string.

```
>  message := Select(IsAlpha, "RENDEZVOUS AT NOON AT THE

   TRADING POST LOCATED AT THE RAILROAD STATION");
```

message :=
   "RENDEZVOUSATNOONATTHETRADINGPOSTLOCATED\
   ATTHERAILROADSTATION"

We next use the following commands to split this message into an even number of blocks containing 11 characters each.

```
>  if nops([LengthSplit(message, blen)]) mod 2 = 0 then

>     messblocks := [LengthSplit(message, blen)];

>  else

>     messblocks := [LengthSplit(message, blen), "X"];

>  fi:

>  messblocks;
```

["RENDEZVOUSA", "TNOONATTHET", "RADINGPOSTL",
   "OCATEDATTHE", "RAILROADSTA", "TION"]

To convert this plaintext message into numerical form most efficiently, we will use some functions contained in the Maple **ListTools** package. Thus we include this package.

```
>  with(ListTools):
```

Next, we will use the **LengthSplit**[2] function contained in the Maple **ListTools** package to convert each block of *messblocks* into a list of two-digit integers, group all of the two-digit integers within each block into a single integer, and then group these integers in pairs in lists of length two (as specified by the final parameter in the command).

```
>  w := [LengthSplit(tonumber(messblocks), 2)];
```

$$w := [[1704130304252114201800, 1913141413001919070419],$$
$$[1700030813061514181911, 1402001904030019190704],$$
$$[1700081117140003181900, 19081413]]$$

The next command shows that each numerical plaintext block in w is less than p.

```
>  [seq(map(x -> evalb(x < p), w[i]), i = 1..nops(w))];
```

$$[[true, true], [true, true], [true, true]]$$

We can encrypt this message by entering the following command.

```
>  z := map(x -> kb *~ x mod p, w);
```

$$z := [[5865493398298422326683, 1852776028085116876739 4],$$
$$[2022645709381378514620 4, 1667553149736117941778 9],$$
$$[1836558755965069112081 7, 1252641964464228505628]]$$

We can then decrypt the ciphertext by entering the following command.

```
>  w := map(x -> ny ^~ (-1) *~ x mod p, z);
```

$$w := [[1704130304252114201800, 1913141413001919070419],$$
$$[1700030813061514181911, 1402001904030019190704],$$
$$[1700081117140003181900, 19081413]]$$

---

[2]There are **LengthSplit** functions in both the Maple **StringTools** and **ListTools** packages. When this function is called, Maple will use the one from the package most recently included.

Finally, we can convert these plaintext integers back into letters as follows, by first using the Maple **Flatten** function contained in the **ListTools** package to convert the integers into a single list, and then the user-written **toletter** function to convert this list back into a string of letters.

```
>  toletter(Flatten(w));
```

"RENDEZVOUSATNOONATTHETRADINGPOSTLOCATEDATT\
   HERAILROADSTATION"

9.9   Elliptic Curve Cryptography with MATLAB

In this section, we will show how MATLAB can be used to do the computations in Menezes and Vanstone's variation of the ElGamal cryptosystem with an elliptic curve defined over $\mathbb{Z}_p$ for prime p as the group G.

Recall that to construct an elliptic curve using Lemma 9.2, which is employed by the user-written function **epoints** that we used in Section 9.6 and will use again in this section, the prime p must satisfy $p = 3 \bmod 4$. In order to find a prime that satisfies this requirement, we will use the user-written function **p3mod4**, which we have written separately from this MATLAB session and saved as the M-file p3mod4.m. The following command illustrates how the function **p3mod4** can be used. Specifically, this function takes as input an integer s, and returns as output the smallest prime p that is larger than s and which satisfies $p = 3 \bmod 4$. For example, the following command assigns as p the smallest prime larger than 2205324962937788058000 which satisfies $p = 3 \bmod 4$. We will use this value as the prime p in our examples in this section.

```
>> p = p3mod4(sym('2205324962937788058000'))

p =

2205324962937788058119
```

Note that in the preceding command, the input parameter is enclosed in single quotes and declared in symbolic form. This same syntax will also be required in some of the commands that follow.

For the prime p assigned by the preceding command, let E be the elliptic curve of ordered pairs $(x, y) \in \mathbb{Z}_p \times \mathbb{Z}_p$ of solutions to (9.1) modulo p for $c = 1$ and $d = 6$ together with point at infinity $\mathcal{O}$. In the following commands, we assign these values of c and d, and verify that they satisfy $4c^3 + 27d^2 \neq 0 \bmod p$.

```
>> c = sym('1');
>> d = sym('6');
>> mod(4*c^3+27*d^2, p)

ans =

976
```

Next, we will store the right side of (9.1) as the variable *eqn*.

```
>> syms x
>> eqn = x^3+c*x+d;
```

For the ordered pair a in the system, we will use the first solution to (9.1) generated by the user-written function **epoints** that we used in Section 9.6. Recall that we have written this function separately from this MATLAB session and saved it as the M-file epoints.m. In the following command, we use **epoints** to generate one of the ordered pairs $(x, y) \in \mathbb{Z}_p \times \mathbb{Z}_p$ of solutions to (9.1) modulo p, which we will assign as the variable a. We will use this ordered pair as a in our examples in this section.

```
>> a = epoints(eqn, x, 1, p);
```

Recall that results returned by **epoints** are cell arrays, in which individual elements are accessed using curly braces. For example, we can see the result of the preceding command by entering the following command.

```
>> a{1}

ans =

[ 0,  40675210938100805459]
```

For notational convenience, we will make the following assignment.

```
>> a = a{1}

a =

[ 0,  40675210938100805459]
```

Next, we will assign the following value for n, which we will use in constructing the ordered pair $b = na$ in E.

```
>> n = sym('91530873521338');
```

To expedite the process of adding a to itself repeatedly with n total terms using the elliptic curve addition operation, we will use the user-written function **elgamal**, which we have written separately from this MATLAB session and saved as the M-file elgamal.m. This function uses the user-written function **addec** that we used in Section 9.6, which we have written separately from this MATLAB session and saved as the M-file addec.m. The following command illustrates how the function **elgamal** can be used. Specifically, by entering the following command, we construct the ordered pair $b = na$ in E by adding a to itself repeatedly with n total terms using the elliptic curve addition operation.

```
>> b = elgamal(a, n, c, p)

b =

[ 1575028246746093464921, 8523832800260348129369]
```

The parameters in the preceding command are the ordered pair a, the number n of terms in the sum, the value of c in (9.1), and the prime p.

Next, we will assign the following value for k, which we will use in constructing the ordered pairs $y = ka$ and kb in E.

```
>> k = sym('431235145514');
```

We can then construct the ordered pairs $y = ka$ and kb in E using the elliptic curve addition operation by entering the following commands.

```
>> y = elgamal(a, k, c, p)

y =

[ 18694668105703808236390, 10420072039502224215692]

>> kb = elgamal(b, k, c, p)

kb =

[ 15261622170740304569981, 16218536420977465416666]
```

We will now use the ordered pair kb to encrypt the message RENDEZVOUS AT NOON. We begin by storing our message as the following string.

```
>> message = 'RENDEZVOUS AT NOON';
>> message(findstr(message, ' ')) = []

message =

RENDEZVOUSATNOON
```

We can identify the number of digits in the modulus p to be used in the encryption as follows, by using the MATLAB **char** function to convert p from an integer into a string, and then the MATLAB **length** function to find the number of characters in the resulting string.

```
>> length(char(p))

ans =

23
```

Since the correspondence α that we introduced in Chapter 6 (i.e., A $\mapsto$ 00, B $\mapsto$ 01, C $\mapsto$ 02, ..., Z $\mapsto$ 25) converts messages into two-digit integers, our 23-digit value of p is guaranteed to work for encrypting messages containing at most 11 characters. The user-written function **LengthSplitString**, which we first used in Section 8.4, is designed to split strings of characters into shorter blocks containing a designated number of characters. Recall that we have written **LengthSplitString** separately from this MATLAB session and saved it as the M-file LengthSplitString.m. The following commands demonstrate how **LengthSplitString** can be used. Specifically, the following commands cause our message to be split into blocks containing 11 characters each, with the last block ending when all of the characters in the message have been used. Since the message must be encrypted as an ordered pair, these commands check to make sure the number of blocks is even, with a extra character X appended as an additional block if necessary to make the number of blocks be even.

```
>> blen = 11;
>> if mod(numel(LengthSplitString(message, blen)), 2) == 0
       messblocks = LengthSplitString(message, blen);
   else
       messblocks = [LengthSplitString(message, blen), 'X'];
   end
```

```
>> messblocks

messblocks =

   'RENDEZVOUSA'    'TNOON'
```

To convert this plaintext message into numerical form, we will use the user-written function **tonumber**, which we first used in Section 8.4, and which we have written separately from this MATLAB session and saved as the M-file tonumber.m. In the following commands, we store the result of **tonumber** as the first component in a cell array w, and then use the user-written functions **estring** and **stringprint** that we described previously in Section 9.6 to display the result.

```
>> w{1} = tonumber(messblocks);
>> stringprint(estring(w), 50)

ans =

[17041303042521142018800,1913141413]
```

The next command shows how to display the second part of the ordered pair stored in the first component of w.

```
>> w{1}(2)

ans =

1913141413
```

In the next pair of commands we use the MuPAD **bool** function within MATLAB to show that each numerical plaintext block in the first component of w is less than p.

```
>> t{1} = [feval(symengine, 'bool', [char(w{1}(1)), ...
   '<', char(p)]), feval(symengine, 'bool', ...
   [char(w{1}(2)), '<', char(p)])];
>> stringprint(estring(t), 50)

ans =

[TRUE,TRUE]
```

We can then encrypt this message by entering the following command. This command uses the MATLAB .* function, which causes the two lists kb and the first component of w to be multiplied componentwise, with the results stored as the first component in a cell array z

```
>> z{1} = mod(kb .* w{1}, p);
>> stringprint(estring(z), 50)

ans =

[5865493398298422326683,11514808043310828036997]
```

To decrypt this ciphertext, we must first recover the ordered pair kb. We can do this as follows by computing ny in E using the elliptic curve addition operation.

```
>> ny = elgamal(y, n, c, p)

ny =

[ 15261622170740304569981, 16218536420977465416666]]
```

We can then decrypt the ciphertext by entering the following command. In this command, the notation $1./ny$ causes MATLAB to form and use a vector whose components are the multiplicative inverses of the components of ny.

```
>> w{1} = mod(1./ny .* z{1}, p);
>> stringprint(estring(w), 50)

ans =

[1704130304252114201800,1913141413]]
```

To convert the plaintext integers in w back into a list of letters, we will use the user-written function **toletter**, which we first used in Section 8.4, and which we have written separately from this MATLAB session and saved as the M-file toletter.m. In the following command, we apply **toletter** to the components of w to convert these plaintext integers back into a list of letters.

```
>> toletter([w{:}])

ans =

RENDEZVOUSATNOON
```

We will now demonstrate how a longer message can be encrypted and decrypted. We begin by clearing the contents of the variables w, t, and z that we used previously.

```
>> clearvars w t z
```

Consider the plaintext message RENDEZVOUS AT NOON AT THE TRADING POST LOCATED AT THE RAILROAD STATION, which we will store as the following string.

```
>> message = ['RENDEZVOUS AT NOON AT THE TRADING ', ...
   'POST LOCATED AT THE RAILROAD STATION'];
>> message(findstr(message, ' ')) = []

message =

RENDEZVOUSATNOONATTHETRADINGPOSTLOCATEDATTHERAILROADSTATION
```

We next use the following commands to split this message into an even number of blocks containing 11 characters each.

```
>> if mod(numel(LengthSplitString(message, blen)), 2) == 0
      messblocks = LengthSplitString(message, blen);
   else
      messblocks = [LengthSplitString(message, blen), 'X'];
   end
>> messblocks

messblocks =

   'RENDEZVOUSA'    'TNOONATTHET'    'RADINGPOSTL'
   'OCATEDATTHE'    'RAILROADSTA'    'TION'
```

To convert this plaintext message into numerical form, we will use the user-written function **LengthSplitList**, which we have written separately from this MATLAB session and saved as the M-file LengthSplitList.m. The next command demonstrates how **LengthSplitList** can be used. Specifically, the following command converts each block of *messblocks* into a list of two-digit integers, groups all of the two-digit integers within each block into a single integer, and then group these integers in pairs in lists of length two (as specified by the final parameter in the command) in a cell array w.

```
>> w = LengthSplitList(tonumber(messblocks), 2);
```

```
>> stringprint(estring(w), 50)
```

```
[1704130304252114201800,1913141413001919070419],[1
700030813061514181911,1402001904030019190704],[170
0081117140003181900,19081413]
```

The next commands show that each numerical plaintext block in w is less than p.

```
>> for i = 1:numel(w)
       t{i} = [feval(symengine, 'bool', [char(w{i}(1)), ...
       '<', char(p)]), feval(symengine, 'bool', ...
       [char(w{i}(2)), '<', char(p)])];
   end
>> stringprint(estring(t), 50)
```

```
ans =
```

```
[TRUE,TRUE],[TRUE,TRUE],[TRUE,TRUE]
```

We can encrypt this message by entering the following commands.

```
>> for i = 1:numel(w)
       z{i} = mod(kb .* w{i}, p);
   end
>> stringprint(estring(z), 50)
```

```
[58654933982984223266883,18527760280851168767394],[
20226457093813785146204,16675531497361179417789],[
18365587559650691120817,12526419644464228505628]
```

We can then decrypt the ciphertext by entering the following commands.

```
>> for i = 1:numel(z)
       w{i} = mod(1./ny .* z{i}, p);
   end
>> stringprint(estring(w), 50)
```

```
[1704130304252114201800,1913141413001919070419],[1
700030813061514181911,1402001904030019190704],[170
0081117140003181900,19081413]
```

Finally, we can convert these plaintext integers back into letters as follows.

```
>> toletter([w{:}])

ans =

RENDEZVOUSATNOONATTHETRADINGPOSTLOCATEDATTHERAILROADSTATION
```

Exercises

1. Suppose you wish to use the ElGamal cryptosystem with $\mathbb{Z}_p^*$ under multiplication modulo a prime p as the group G to send a secret message to a colleague across an insecure line of communication. Your colleague sends you $p = 31$, $a = 13$, and $b = 9$, and you convert your message into the numerical equivalent $w = 20$. Using $k = 6$, construct the values y and z you would then send to your colleague.

2. Suppose you wish to use the ElGamal cryptosystem with $\mathbb{Z}_p^*$ under multiplication modulo a prime p as the group G to receive a secret message from a colleague across an insecure line of communication. You send your colleague $p = 13$, $a = 2$, and $b = 2^3 = 8 \mod p$, and your colleague converts his or her message into a numerical equivalent w and sends you $y = 5$ and $z = 2$. Recover w.

3. Suppose you wish to use the ElGamal cryptosystem with the group of nonzero elements in a finite field under multiplication as the group G to send a secret message to a colleague across an insecure line of communication. Your colleague sends you the primitive polynomial $p(x) = x^2 + x + 2 \in \mathbb{Z}_5[x]$ and $a = x$ and $b = 4x$ in G, and you convert your message into the equivalent field element $w = 2x + 4 \in G$. Using $k = 6$, construct the polynomials y and z you would then send to your colleague.

4. Suppose you wish to use the ElGamal cryptosystem with the group of nonzero elements in a finite field under multiplication as the group G to receive a secret message from a colleague across an insecure line of communication. You send your colleague the primitive polynomial $p(x) = x^2 + x + 2 \in \mathbb{Z}_5[x]$ and $a = x$ and $b = x^8 = 3x + 1$ in G, and your colleague converts his or her message into an equivalent field element $w \in G$ and sends you $y = 2x$ and $z = 4x + 4$. Recover w.

5. Let E be the elliptic curve of ordered pairs $(x, y) \in \mathbb{Z}_{11} \times \mathbb{Z}_{11}$ of solutions to $y^2 = (x^3 + x + 1) \mod 11$ with point at infinity $\mathcal{O}$.

   (a) Construct the elements in E.

   (b) Find $(3, 8) + (4, 6)$ in E using elliptic curve addition.

(c) Find $(1,6) + (1,6)$ in E using elliptic curve addition.

(d) Find $(1,6) + (1,5)$ in E using elliptic curve addition.

6. Recall for the ordered pairs $(x, y) \in \mathbb{Z}_p \times \mathbb{Z}_p$ of solutions to (9.1) modulo a prime $p > 3$ together with point at infinity $\mathcal{O}$ to be an elliptic curve, c and d must satisfy $4c^3 + 27d^2 \neq 0 \bmod p$. To demonstrate the importance of this condition, use the elliptic curve addition operation to add the ordered pairs $(0, 1)$ and $(14, 0)$ of solutions to the equation $y^2 = (x^3 + x + 1) \bmod 31$. Explain why your answer shows the importance of the condition $4c^3 + 27d^2 \neq 0 \bmod p$ in the definition of an elliptic curve over $\mathbb{Z}_p$ for prime $p > 3$.

7. Let E be the elliptic curve of ordered pairs $(x, y) \in \mathbb{Z}_{11} \times \mathbb{Z}_{11}$ of solutions to $y^2 = (x^3 + 2x) \bmod 11$ with point at infinity $\mathcal{O}$.

   (a) Construct the elements in E.

   (b) Is E cyclic? Find the structure of E given by Theorem 9.3.

8. Let E be the elliptic curve of ordered pairs $(x, y) \in \mathbb{Z}_{23} \times \mathbb{Z}_{23}$ of solutions to $y^2 = (x^3 + x + 7) \bmod 23$ with point at infinity $\mathcal{O}$.

   (a) Use Hasse's Theorem (Theorem 9.4) to find upper and lower bounds on $|E|$.

   (b) The actual value of $|E|$ is 18. Use this fact and Theorem 9.3 to show that E is cyclic.

9. Suppose you wish to use the ElGamal cryptosystem with an elliptic curve over $\mathbb{Z}_p$ for prime p as the group G to send a secret message to a colleague across an insecure line of communication. Your colleague sends you $a = (8, 9)$ and $b = (1, 6)$ in the elliptic curve E in Exercise 5, and you convert your message into the equivalent elliptic curve element $w = (4, 6) \in E$. Using $k = 2$, construct the elliptic curve elements y and z in E you would then send to your colleague.

10. Suppose you wish to use the ElGamal cryptosystem with an elliptic curve over $\mathbb{Z}_p$ for prime p as the group G to receive a secret message from a colleague across an insecure line of communication. You send your colleague $a = (8, 9)$ and $b = 2a = (0, 1)$ in the elliptic curve E in Exercise 5, and your colleague converts his or her message into an equivalent elliptic curve element w in E and sends you $y = (3, 3)$ and $z = (2, 0)$. Recover w.

11. Suppose you wish to use Menezes and Vanstone's variation of the ElGamal cryptosystem with an elliptic curve over $\mathbb{Z}_p$ for prime p as the group G to send a secret message to a colleague across an insecure line of communication. Your colleague sends you $a = (8, 9)$ and $b = (1, 6)$ in the elliptic curve E in Exercise 5, and you convert your message into the equivalent ordered pair $w = (5, 7)$. Using $k = 2$, construct the ordered pairs y and z you would then send to your colleague.

12. Suppose you wish to use Menezes and Vanstone's variation of the ElGamal cryptosystem with an elliptic curve over $\mathbb{Z}_p$ for prime p as the group G to receive a secret message from a colleague across an insecure line of communication. You send your colleague $a = (4, 6)$ and $b = 2a = (6, 6)$ in the elliptic curve E in Exercise 5, and your colleague converts his or her message into an equivalent ordered pair w and sends you $y = (1, 6)$ and $z = (10, 10)$. Recover w.

13. Let E be the elliptic curve of ordered pairs $(x, y) \in \mathbb{Z}_p \times \mathbb{Z}_p$ of solutions to (9.1) modulo a prime p together with point at infinity $\mathcal{O}$. Also, suppose $P = (x_1, y_1)$ and $Q = (x_2, y_2)$ are ordered pairs in E with $P \neq -Q$, and denote the sum of P and Q by $P + Q = (x_3, y_3)$.

(a) Assuming $P = Q$, verify algebraically that the formulas for x_3 and y_3 given in (9.2) and (9.3) yield values that satisfy (9.1).

(b) Assuming $P \neq Q$, verify algebraically that the formulas for x_3 and y_3 given in (9.4) and (9.5) yield values that satisfy (9.1).

Computer Exercises

14. Suppose you wish to use the ElGamal cryptosystem with $\mathbb{Z}_p^*$ under multiplication modulo a prime p as the group G to send a secret message to a colleague across an insecure line of communication. Your colleague sends you $p = 10000000019$, $a = 132$, and $b = 240246247$, and then you convert your message into the numerical equivalent $w = 704111114$. Using $k = 398824116$, construct the values y and z you would then send to your colleague.

15. Suppose you wish to use the ElGamal cryptosystem with $\mathbb{Z}_p^*$ under multiplication modulo a prime p as the group G to receive a secret message from a colleague across an insecure line of communication. You send your colleague $p = 10000000019$, $a = 132$, and $b = a^{121314333} = 5803048419 \bmod p$, and your colleague converts his or her message into a numerical equivalent w and sends you $y = 9054696956$ and $z = 1323542761$. Recover w.

16. Suppose you wish to use the ElGamal cryptosystem with the group of nonzero elements in a finite field under multiplication as the group G to send a secret message to a colleague across an insecure line of communication. Your colleague sends you the primitive polynomial $p(x) = 3x^7 + 4x + 1 \in \mathbb{Z}_5[x]$ and $a = x$ and $b = 3x^5 + x^4 + 2x^3 + 4x$ in G, and you convert your message into the equivalent field element $w = 2x^6 + 4x^5 + x^2 + x + 1 \in G$. Using $k = 1851$, construct the polynomials y and z you would then send to your colleague.

17. Suppose you wish to use the ElGamal cryptosystem with the group of nonzero elements in a finite field under multiplication as the group G to receive a secret message from a colleague across an insecure line of communication. You send your colleague the primitive polynomial $p(x) = 3x^7 + 4x + 1 \in \mathbb{Z}_5[x]$ and $a = x$ and $b = x^{51801} = 2x^6 + 3x^5 + 1$ in G, and your colleague converts his or her message into an equivalent field element $w \in G$ and sends you $y = x^6 + 4x^5 + 3x^4 + x^3 + x + 2$ and $z = 2x^6 + 2x^5 + 3x^4 + 3x^3 + 2x^2 + 3x + 2$. Recover w.

18. Let E be the elliptic curve of ordered pairs $(x, y) \in \mathbb{Z}_{59} \times \mathbb{Z}_{59}$ of solutions to $y^2 = (x^3 + 31x + 21) \bmod 59$ with point at infinity $\mathcal{O}$.

    (a) Construct the elements in E.

    (b) Find $(42, 3) + (54, 6)$ in E using elliptic curve addition.

    (c) Find $(42, 3) + (42, 3)$ in E using elliptic curve addition.

    (d) Find $(42, 3) + (42, 56)$ in E using elliptic curve addition.

    (e) Pick any ordered pair in E, and construct the cyclic subgroup of E that it generates.

    (f) Use Theorem 9.3 to show that E is cyclic. Then explain why every element in E except $\mathcal{O}$ must be a generator for E.

19. Suppose you wish to use Menezes and Vanstone's variation of the ElGamal cryptosystem with the elliptic curve $y^2 = x^3 + cx + d$ with $c = 2342421$ and $d = 92374976$ over $\mathbb{Z}_p$ for prime $p = 98374975897$ 359837598213355599 as the group G to send a secret message to a colleague across an insecure line of communication. Your colleague sends you $a = (2, 94393919677540627614364024443)$ and $b = (29614$ $185402086121556719356641, 32189993543457709982023417371)$. Using $k = 10123454323$, encrypt the following messages using the largest possible plaintext block sizes.

    (a) THE MESSAGE WAS CLEARLY STATED

    (b) THE MESSAGE WAS CLEARLY STATED AND I WILL FOLLOW THE ORDERS IMMEDIATELY

20. Suppose you wish to use Menezes and Vanstone's variation of the ElGamal cryptosystem with the elliptic curve $y^2 = x^3 + cx + d$ with $c = 2342421$ and $d = 92374976$ over $\mathbb{Z}_p$ for prime $p = 98374975897$ 359837598213355599 as the group G to receive a secret message from a colleague across an insecure line of communication. You send your colleague $a = (2, 943939196775406276143640244443)$ and $b = 971871$ $841a = (48820353267635843200474647175, 80171271653192257615561993668)$. Decrypt the following ordered pairs received from your colleague.

    (a) $y = (92436137518719269139671865147, 8785263336972049961597 3016617)$ and $z = (68628237843079846972088689971, 36655234 464795342752545473306)$

    (b) $y = (92436137518719269139671865147, 8785263336972049961597 3016617)$ and $z = \{(27278930476997761410745631010, 8073722 4183179971800887798669), (83943589878874339313717168141, 6 89434512414251353043344554841)\}$

21. Set up a parameterization of Menezes and Vanstone's variation of the ElGamal cryptosystem with an elliptic curve defined over $\mathbb{Z}_p$ for some prime p of at least 25 digits as the group G. Then use this parameterization of the ElGamal system to encrypt and decrypt the message TARGET HIT SEND NEW ORDERS.

22. Show how an intruder knowing only the values of p, a, b, y, and z in Example 9.1 could use the **mlog** function in Section 8.7 or 8.8 to determine the value of w.

23. Show how an intruder knowing only the values of p, a, b, y, and z in Exercise 15 could use the **mlog** function in Section 8.7 or 8.8 to determine the value of w.

24. Write a single Maple or MATLAB procedure that will encrypt a given message using the ElGamal system with an elliptic curve over $\mathbb{Z}_p$ for prime p as the group G. This procedure should have six input parameters. The first input parameter should be a string of characters representing a plaintext, the second input parameter should be an integer of approximately the same size as the prime p for the system, the third and fourth input parameters should be integers representing the values of c and d in (9.1), and the fifth and sixth input parameters should be integers representing the values of n and k for the system.

25. Write a single Maple or MATLAB procedure that will decrypt a given ciphertext using the ElGamal system with an elliptic curve over $\mathbb{Z}_p$ for prime p as the group G. This procedure should have six input

parameters. The first input parameter should be an ordered pair of integers representing the ciphertext, the second input parameter should be an integer representing the prime p for the system, the third input parameter should be an integer representing the value of c in (9.1), the fourth input parameter should be an integer representing the value of n for the system, and the fifth and sixth input parameters should be ordered pairs of integers representing y and kb.

Research Exercises

26. Investigate the life and career of the person for whom ElGamal ciphers are named, and write a summary of your findings.

27. Investigate some specific details about the *ElGamal signature scheme* and its variant, the *digital signature algorithm*, and write a summary of your findings.

28. Investigate some specific details about the *PGP* computer program, and write a summary of your findings. Include in your summary a description of how the PGP program uses the RSA and ElGamal cryptosystems.

29. Investigate some specific details about how Andrew Wiles used elliptic curves in his proof of Fermat's Last Theorem, and write a summary of your findings.

30. Investigate some specific details about how elliptic curves are used in integer factorization, and write a summary of your findings.

31. Investigate an actual real-life use of elliptic curve cryptography, and write a summary of your findings.

32. Investigate the current standards for the size of the prime believed to be necessary for an elliptic curve cryptosystem to be secure, and write a summary of your findings.

33. Investigate the general opinion among professional mathematicians and cryptologists regarding the comparison between elliptic curve cryptosystems and the RSA cryptosystem from both security and implementation perspectives, and write a summary of your findings.

Chapter 10

The Advanced Encryption Standard

From 1977 through 2001, the Data Encryption Standard (DES) was the standard algorithm for data encryption by the U.S. federal government. Throughout this time, DES was used extensively in the banking industry and in other types of electronic commerce both in the U.S. and worldwide. However, as predicted, DES eventually became outdated, as over time many successful attacks against the system were developed. As a result of this, in 1997 the U.S. National Institute of Standards and Technology made an open request for candidates to replace DES. Five finalists were selected and analyzed, and from these finalists the algorithm Rijndael, named for its creators Joan Daemen and Vincent Rijmen, was chosen. In November 2001, Rijndael was adopted as the Advanced Encryption Standard (AES), and became the new standard algorithm for data encryption by the U.S. federal government.

In practice, both DES and AES are implemented using binary arithmetic. However, unlike DES, AES has a precise mathematical description that can be described using the ideas that we have presented in previous chapters. Our goal for this chapter is to describe AES in its mathematical form using the ideas that we have presented in previous chapters.

10.1 Text Setup

When using AES, information is transmitted in binary form. Eight binary digits, or *bits*, strung together form a *byte*, which for our purposes will be used to represent a single character. In this chapter, we will again assume that all messages are written using only the characters in the alphabet

$L = \{\text{A, B, C}, \ldots, \text{Z}\}$, and associate each character with its corresponding element in the ring $R = \mathbb{Z}_{26}$ under the bijection $\alpha : L \to R$ that we used in Chapter 6 (i.e., $\text{A} \mapsto 0$, $\text{B} \mapsto 1$, $\text{C} \mapsto 2, \ldots, \text{Z} \mapsto 25$). However, in order to use bytes to represent these characters, we will extend this bijection α as described in the following example.

Example 10.1 Consider G, which under α maps to 6. Since 6 can be written as $6 = 0 \cdot 2^7 + 0 \cdot 2^6 + 0 \cdot 2^5 + 0 \cdot 2^4 + 0 \cdot 2^3 + 1 \cdot 2^2 + 1 \cdot 2^1 + 0 \cdot 2^0$, then as a byte G can be expressed as 00000110. Furthermore, since this byte can be represented as $0 \cdot x^7 + 0 \cdot x^6 + 0 \cdot x^5 + 0 \cdot x^4 + 0 \cdot x^3 + 1 \cdot x^2 + 1 \cdot x^1 + 0 \cdot x^0 = x^2 + x$, then as a polynomial G can be represented by $x^2 + x$. Similarly, under α the letter X maps to 23, which as a byte can be expressed as 000010111, and as a polynomial can be represented by $x^4 + x^2 + x + 1$. Thus, X can be expressed as the byte 00010111, and represented by the polynomial $x^4 + x^2 + x + 1$. The following table shows the polynomial representations of each of the letters in L.

Letter	Polynomial	Letter	Polynomial
A	0	N	$x^3 + x^2 + 1$
B	1	O	$x^3 + x^2 + x$
C	x	P	$x^3 + x^2 + x + 1$
D	$x + 1$	Q	x^4
E	x^2	R	$x^4 + 1$
F	$x^2 + 1$	S	$x^4 + x$
G	$x^2 + x$	T	$x^4 + x + 1$
H	$x^2 + x + 1$	U	$x^4 + x^2$
I	x^3	V	$x^4 + x^2 + 1$
J	$x^3 + 1$	W	$x^4 + x^2 + x$
K	$x^3 + x$	X	$x^4 + x^2 + x + 1$
L	$x^3 + x + 1$	Y	$x^4 + x^3$
M	$x^3 + x^2$	Z	$x^4 + x^3 + 1$

□

The total number of bytes that are possible is $2^8 = 256$. Operations in the AES algorithm are performed on bytes. Specific bytes are represented by elements in the finite field $F = \mathbb{Z}_2[x]/(p(x))$ of order 2^8, where $p(x)$ is an irreducible polynomial of degree 8 over $\mathbb{Z}_2$. The choice for $p(x)$ that we will use throughout this chapter is $p(x) = x^8 + x^4 + x^3 + x + 1$, which was the polynomial that was actually used in the original Rijndael algorithm. It is worth noting that this polynomial is irreducible in $\mathbb{Z}_2[x]$ but not primitive. However, once the alphabet assignment is established, there is no real need to generate the entire field F, which is the reason why choosing a primitive polynomial would be useful in the first place. It is also worth noting that other choices for $p(x)$ would yield equally good encryption algorithms, but we will only use $p(x) = x^8 + x^4 + x^3 + x + 1$ since this was the polynomial that was actually used in the original Rijndael algorithm.

The AES algorithm is designed to encrypt messages that contain 16 characters, or, equivalently, 128 bits. Consider a plaintext that contains 16 characters. In order to describe how this message would be stored for the AES algorithm, consider the characters in the message labeled in order as follows.

$$a_{0,0} \ a_{1,0} \ a_{2,0} \ a_{3,0} \ a_{0,1} \ a_{1,1} \ \cdots \ a_{2,3} \ a_{3,3}$$

We would then convert each character $a_{i,j}$ into its polynomial equivalent $p_{i,j}$ using the table in Example 10.1, and arrange the resulting polynomials in the following 4×4 plaintext matrix A.

$$A = \begin{bmatrix} p_{0,0} & p_{0,1} & p_{0,2} & p_{0,3} \\ p_{1,0} & p_{1,1} & p_{1,2} & p_{1,3} \\ p_{2,0} & p_{2,1} & p_{2,2} & p_{2,3} \\ p_{3,0} & p_{3,1} & p_{3,2} & p_{3,3} \end{bmatrix}$$

Example 10.2 Suppose that we wish to use the AES algorithm to encrypt the message RENDEZVOUS AT SIX. We would begin by adding an extra arbitrary letter, say A, to this message so that the resulting plaintext RENDEZVOUS AT SIX A contains exactly 16 letters. We would then convert each of these letters into its polynomial equivalent using the table in Example 10.1, and arrange the resulting polynomials in the following 4×4 plaintext matrix A.

$$A = \begin{bmatrix} x^4 + 1 & x^2 & x^4 + x^2 & x^4 + x \\ x^2 & x^4 + x^3 + 1 & x^4 + x & x^3 \\ x^3 + x^2 + 1 & x^4 + x^2 + 1 & 0 & x^4 + x^2 + x + 1 \\ x + 1 & x^3 + x^2 + x & x^4 + x + 1 & 0 \end{bmatrix}$$

We are now ready to begin the AES encryption process using this matrix as the plaintext matrix for the system. We will describe this encryption process and continue this example later in this chapter. □

Recall, as described in Example 10.1, the polynomial entries in the matrix A in Example 10.2 can be viewed as representations for bytes. Converting each of the polynomial entries in this matrix into the bytes of which they can be viewed as representations yields the following alternative representation for A.

$$A = \begin{bmatrix} 00010001 & 00000100 & 00010100 & 00010010 \\ 00000100 & 00011001 & 00010010 & 00001000 \\ 00001101 & 00010101 & 00000000 & 00010111 \\ 00000011 & 00001110 & 00010011 & 00000000 \end{bmatrix}$$

Throughout the remainder of this chapter, we will frequently convert the polynomial entries in matrices defined over $F = \mathbb{Z}_2[x]/(p(x))$ into the bytes of which the polynomial entries can be viewed as representations.

10.2   The S-Box

The S-box is used in the AES algorithm to transform a given element in $F = \mathbb{Z}_2[x]/(p(x))$ into another unique element in F. This is done by using Table 10.1, which itself is called the AES *S-box*.

	0	1	2	3	4	5	6	7	8	9	10	11	12	13	14	15
0	99	124	119	123	242	107	111	197	48	1	103	43	254	215	171	118
1	202	130	201	125	250	89	71	240	173	212	162	175	156	164	114	192
2	183	253	147	38	54	63	247	204	52	165	229	241	113	216	49	21
3	4	199	35	195	24	150	5	154	7	18	128	226	235	39	178	117
4	9	131	44	26	27	110	90	160	82	59	214	179	41	227	47	132
5	83	209	0	237	32	252	177	91	106	203	190	57	74	76	88	207
6	208	239	170	251	67	77	51	133	69	249	2	127	80	60	159	168
7	81	163	64	143	146	157	56	245	188	182	218	33	16	255	243	210
8	205	12	19	236	95	151	68	23	196	167	126	61	100	93	25	115
9	96	129	79	220	34	42	144	136	70	238	184	20	222	94	11	219
10	224	50	58	10	73	6	36	92	194	211	172	98	145	149	228	121
11	231	200	55	109	141	213	78	169	108	86	244	234	101	122	174	8
12	186	120	37	46	28	166	180	198	232	221	116	31	75	189	139	138
13	112	62	181	102	72	3	246	14	97	53	87	185	134	193	29	158
14	225	248	152	17	105	217	142	148	155	30	135	233	206	85	40	223
15	140	161	137	13	191	230	66	104	65	153	45	15	176	84	187	22

Table 10.1   The AES S-box.

The AES S-box contains 16 rows and 16 columns, which are each numbered from 0 through 15. We describe how Table 10.1 can be used to apply the S-box transformation in the following example.

Example 10.3 Consider the polynomial $x^4 + x^2 + x$. This polynomial can be viewed as a representation for the byte 00010110. The first four and last four bits in this byte can each be viewed as the binary expression of an integer between 0 and 15. These integers identify for us a specific row and column, respectively, in Table 10.1. Specifically, since the first four bits 0001 in this byte can be viewed as the binary expression of the integer $0 \cdot 2^3 + 0 \cdot 2^2 + 0 \cdot 2^1 + 1 \cdot 2^0 = 1$, and the last four bits 0110 can be viewed as the binary expression of the integer $0 \cdot 2^3 + 1 \cdot 2^2 + 1 \cdot 2^1 + 0 \cdot 2^0 = 6$, this tells us that to apply the S-box transformation to the polynomial $x^4 + x^2 + x$, we would use the entry in the row labeled as 1 and the column labeled as 6 in Table 10.1. The entry in this position in Table 10.1 is 71, which as a byte can be expressed as 010000111, and as a polynomial can be represented by $x^6 + x^2 + x + 1$. Thus, for the input $x^4 + x^2 + x$, the output of the S-box transformation is the polynomial $x^6 + x^2 + x + 1$. $\square$

Although the S-box transformation can be implemented using Table 10.1, it also has a very simple mathematical structure that can be described as follows. Let $f(x)$ be an element in $F = \mathbb{Z}_2[x]/(p(x))$. Assuming $f(x) \neq 0$, then $f(x)$ has a multiplicative inverse $f^{-1}(x)$ in F. Suppose that $f^{-1}(x)$ can be viewed as the polynomial representation of the byte $y_7 y_6 y_5 y_4 y_3 y_2 y_1 y_0$, and let S be defined as the following matrix.

$$S = \begin{bmatrix} 1 & 0 & 0 & 0 & 1 & 1 & 1 & 1 \\ 1 & 1 & 0 & 0 & 0 & 1 & 1 & 1 \\ 1 & 1 & 1 & 0 & 0 & 0 & 1 & 1 \\ 1 & 1 & 1 & 1 & 0 & 0 & 0 & 1 \\ 1 & 1 & 1 & 1 & 1 & 0 & 0 & 0 \\ 0 & 1 & 1 & 1 & 1 & 1 & 0 & 0 \\ 0 & 0 & 1 & 1 & 1 & 1 & 1 & 0 \\ 0 & 0 & 0 & 1 & 1 & 1 & 1 & 1 \end{bmatrix}$$

For the vectors $\mathbf{y} = [\, y_0\ y_1\ y_2\ y_3\ y_4\ y_5\ y_6\ y_7\,]^T$ and $\mathbf{k} = [\, 1\ 1\ 0\ 0\ 0\ 1\ 1\ 0\,]^T$, let $\mathbf{z}$ be the vector $\mathbf{z} = [\, z_0\ z_1\ z_2\ z_3\ z_4\ z_5\ z_6\ z_7\,]^T$ that results from the following formula.

$$\mathbf{z} = S\mathbf{y} + \mathbf{k} \qquad (10.1)$$

If we let $g(x)$ be the polynomial representation of the byte $z_7 z_6 z_5 z_4 z_3 z_2 z_1 z_0$, then it will follow that for the input $f(x)$ into the S-box transformation, the output will be $g(x)$. Also, if $f(x) = 0$, then since $f^{-1}(x)$ would not exist, we would use $\mathbf{y} = [\, 0\ 0\ 0\ 0\ 0\ 0\ 0\ 0\,]^T$ in (10.1) to determine the output of the S-box transformation that corresponds to the input $f(x)$.

Example 10.4 Let $f(x) = x^4 + x^2 + x$, the same polynomial to which we applied the S-box transformation using Table 10.1 in Example 10.3. In this example, we will show how the output $x^6 + x^2 + x + 1$ of the S-box transformation that corresponds to the input $f(x)$ can be determined mathematically. We begin by using the Euclidean algorithm to find the multiplicative inverse of $f(x)$ in $F = \mathbb{Z}_2[x]/(p(x))$. To do this, we first form the following equations.

$$\begin{aligned} p(x) &= f(x)(x^4 + x^2 + x) + (x^3 + x^2 + x + 1) \\ f(x) &= (x^3 + x^2 + x + 1)(x + 1) + (x^2 + x + 1) \\ x^3 + x^2 + x + 1 &= (x^2 + x + 1)(x) + 1 \\ x^2 + x + 1 &= 1(x^2 + x + 1) + 0 \end{aligned}$$

Thus, $\gcd(p(x), f(x)) = 1$, which will be true for all nonzero $f(x)$ in F, of course, since $p(x)$ is irreducible. However, forming these equations is still necessary, for the quotients and remainders that it gives are necessary for constructing the following Euclidean algorithm table.

Row	Q	R	U	V
-1	$-$	$p(x)$	1	0
0	$-$	$f(x)$	0	1
1	$x^4 + x^2 + x$	$x^3 + x^2 + x + 1$	1	$x^4 + x^2 + x$
2	$x + 1$	$x^2 + x + 1$	$x + 1$	$x^5 + x^4 + x^3 + x + 1$
3	x	1	$x^2 + x + 1$	$x^6 + x^5$

Therefore, we can conclude that $p(x)(x^2 + x + 1) + f(x)(x^6 + x^5) = 1$. This shows that $f^{-1}(x) = x^6 + x^5$, for it states that $f(x)(x^6 + x^5) = 1$ in F. Next, note that $f^{-1}(x)$ can be viewed as the polynomial representation of the byte 011000000. Then for $\mathbf{y} = [\,0\,0\,0\,0\,0\,1\,1\,0\,]^T$, we can use (10.1) to compute $\mathbf{z} = [\,1\,1\,1\,0\,0\,0\,1\,0\,]^T$. It follows that the output of the S-box transformation that corresponds to the input $f(x)$ is the polynomial representation $x^6 + x^2 + x + 1$ of the byte 01000111.   □

The S-box transformation can be inverted by using another table similar to Table 10.1, or by reversing the steps necessary for using Table 10.1 to apply the S-box transformation. We illustrate this process in the following example.

Example 10.5 Consider the output polynomial $x^6 + x^5 + x^4 + x^2 + x$ that was produced from some input polynomial $f(x)$ using the S-box transformation. To find $f(x)$, note first that $x^6 + x^5 + x^4 + x^2 + x$ can be viewed as the polynomial representation for the byte 01110110, which is the binary expression of the integer 118. The integer 118 occurs in the row labeled as 0 and the column labeled as 15 in Table 10.1. The binary expression of this row label is 0000, and the binary expression of this column label is 1111. Adjoining these binary expressions yields 00001111, which as a polynomial can be represented by $x^3 + x^2 + x + 1$. Thus, $x^3 + x^2 + x + 1$ is the input $f(x)$ into the S-box transformation that has corresponding output $x^6 + x^5 + x^4 + x^2 + x$.   □

To invert the S-box transformation mathematically, we must solve for $\mathbf{y}$ in (10.1) over $\mathbb{Z}_2$. This yields the following equation.

$$\mathbf{y} = S^{-1}(\mathbf{z} + \mathbf{k}) \tag{10.2}$$

It can be shown that the inverse of the matrix S in (10.1) over $\mathbb{Z}_2$ is the following matrix S^{-1}.

$$S^{-1} = \begin{bmatrix} 0 & 0 & 1 & 0 & 0 & 1 & 0 & 1 \\ 1 & 0 & 0 & 1 & 0 & 0 & 1 & 0 \\ 0 & 1 & 0 & 0 & 1 & 0 & 0 & 1 \\ 1 & 0 & 1 & 0 & 0 & 1 & 0 & 0 \\ 0 & 1 & 0 & 1 & 0 & 0 & 1 & 0 \\ 0 & 0 & 1 & 0 & 1 & 0 & 0 & 1 \\ 1 & 0 & 0 & 1 & 0 & 1 & 0 & 0 \\ 0 & 1 & 0 & 0 & 1 & 0 & 1 & 0 \end{bmatrix}$$

We illustrate the process of inverting the S-box transformation mathematically in the following example.

Example 10.6 Consider again the output polynomial $x^6 + x^5 + x^4 + x^2 + x$ that was produced from some input polynomial $f(x)$ using the S-box transformation. In Example 10.5, we reversed the steps necessary for using Table 10.1 to apply the S-box transformation to determine that the polynomial $f(x) = x^3 + x^2 + x + 1$ is the input into the S-box transformation that has corresponding output $x^6 + x^5 + x^4 + x^2 + x$. In this example, we will show how this input polynomial $f(x)$ can be determined from $x^6 + x^5 + x^4 + x^2 + x$ mathematically. Note that $x^6 + x^5 + x^4 + x^2 + x$ can be viewed as the polynomial representation of the byte 01110110. Then for $\mathbf{z} = [\, 0\, 1\, 1\, 0\, 1\, 1\, 1\, 0\,]^T$, we can use (10.2) to find $\mathbf{y} = [\, 1\, 1\, 1\, 0\, 0\, 0\, 1\, 1\,]^T$, which is the vector that would be constructed from the byte 11000111. As a polynomial the byte 11000111 can be represented by $x^7 + x^6 + x^2 + x + 1$, which has inverse $x^3 + x^2 + x + 1$ in $F = \mathbb{Z}_2[x]/(p(x))$, as we could determine using the Euclidean algorithm. It follows that the input $f(x)$ into the S-box transformation that has corresponding output $x^6 + x^5 + x^4 + x^2 + x$ is $f(x) = x^3 + x^2 + x + 1$. □

10.3 Key Generation

The AES encryption process executes a certain number of repeated steps, or *rounds*, which is dependent upon the size of an initial key. A recursive key generation process, beginning with an initial key, is implemented to generate keys for the rounds in the process. In this section, we will present the form of the initial key and the recursive key generation process that is used to generate keys for the rounds in the process.

10.3.1 The Initial Key

The AES encryption process is designed to use initial keys of lengths 128, 192, or 256 bits, stored one byte at a time as the columns in an initial key matrix that contains four rows. Thus, a 128-bit initial key yields a 4×4

initial key matrix, a 192-bit initial key yields a 4×6 initial key matrix, and a 256-bit initial key yields a 4×8 initial key matrix. For simplicity, we will limit our examples to 128-bit initial keys. Opportunities for working with 192-bit and 256-bit initial keys will be left as exercises.

AES initial keys are usually understood to be formed from actual characters in a keyword. Thus, a 128-bit initial key would require a 16-character keyword. To use a 16-character keyword to form an initial key matrix for the AES encryption process, we would consider the characters in the keyword labeled in order as follows.

$$k_{0,0} \ k_{1,0} \ k_{2,0} \ k_{3,0} \ k_{0,1} \ k_{1,1} \ \ldots \ k_{2,3} \ k_{3,3}$$

We would then convert each character $k_{i,j}$ into its polynomial equivalent $q_{i,j}$ using the table in Example 10.1, and arrange the resulting polynomials in the following 4×4 initial key matrix K.

$$K = \begin{bmatrix} q_{0,0} & q_{0,1} & q_{0,2} & q_{0,3} \\ q_{1,0} & q_{1,1} & q_{1,2} & q_{1,3} \\ q_{2,0} & q_{2,1} & q_{2,2} & q_{2,3} \\ q_{3,0} & q_{3,1} & q_{3,2} & q_{3,3} \end{bmatrix}$$

We illustrate this process in the following example.

Example 10.7 Suppose that we wish to construct a 4×4 initial key matrix for the AES encryption process using the keyword RIJNDAEL AES KEY. We would begin by adding two extra arbitrary letters, say AA, to this keyword so that the resulting keyword RIJNDAEL AES KEY AA contains exactly 16 letters. We would then convert each of these letters into its polynomial equivalent using the table in Example 10.1, and arrange the resulting polynomials in the following 4×4 initial key matrix K.

$$K = \begin{bmatrix} x^4 + 1 & x + 1 & 0 & x^2 \\ x^3 & 0 & x^2 & x^4 + x^3 \\ x^3 + 1 & x^2 & x^4 + x & 0 \\ x^3 + x^2 + 1 & x^3 + x + 1 & x^3 + x & 0 \end{bmatrix}$$

We are now ready to begin the AES encryption process using this matrix as the initial key matrix for the system. We will describe this encryption process and continue this example later in this chapter. □

10.3.2   The Key Schedule

The key schedule is designed to generate key matrices for the rounds in the AES encryption process. Let r represent the number of rounds after the initial round in the AES encryption process. The value of r varies depending on the length of the initial key for the system. For 128-bit initial keys, $r = 10$, for 192-bit initial keys, $r = 12$, and for 256-bit initial keys, $r = 14$. Also, let c represent the number of columns in the initial key matrix K for the system. Recall that for 128-bit initial keys, $c = 4$, for 192-bit initial keys, $c = 6$, and for 256-bit initial keys, $c = 8$. Label the columns in K as $\mathbf{w}_0, \mathbf{w}_1, \ldots, \mathbf{w}_{c-1}$. In order to form key matrices for the rounds in the AES encryption process, we would first construct new columns $\mathbf{w}_c, \mathbf{w}_{c+1}, \ldots, \mathbf{w}_{4r+3}$ from the existing columns recursively in the following way. Suppose that we have already constructed columns $\mathbf{w}_0, \mathbf{w}_1, \ldots, \mathbf{w}_{j-1}$, where $c - 1 \le j - 1 \le 4r + 2$. Then the column $\mathbf{w}_j$ is determined by the following formula, with the calculations done in $F = \mathbb{Z}_2[x]/(p(x))$, using a transformation T that we will describe next.

$$\mathbf{w}_j = \mathbf{w}_{j-c} + T(\mathbf{w}_{j-1}) \tag{10.3}$$

In order to describe the transformation T in (10.3), denote column $\mathbf{w}_{j-1}$ by $\mathbf{w}_{j-1} = \begin{bmatrix} w_{0,j-1} & w_{1,j-1} & w_{2,j-1} & w_{3,j-1} \end{bmatrix}^T$. The result of $T(\mathbf{w}_{j-1})$ depends on the value of c. Consider first the case $c \le 6$. If $j \bmod c \ne 0$, then $T(\mathbf{w}_{j-1}) = \mathbf{w}_{j-1}$. If $j \bmod c = 0$, then $T(\mathbf{w}_{j-1})$ is found by the following three steps.

1. The entries in $\mathbf{w}_{j-1}$ are shifted up by one position, with the entry at the top wrapping around to the bottom, yielding the new column $\begin{bmatrix} w_{1,j-1} & w_{2,j-1} & w_{3,j-1} & w_{0,j-1} \end{bmatrix}^T$.

2. The S-box transformation is applied to each of the entries in the column that results from the first step, yielding the new column $\begin{bmatrix} v_{1,j-1} & v_{2,j-1} & v_{3,j-1} & v_{0,j-1} \end{bmatrix}^T$.

3. Finally, for the *round constant* $x^{\frac{j-c}{c}}$ in F, $T(\mathbf{w}_{j-1})$ is defined as follows.

$$T(\mathbf{w}_{j-1}) = \begin{bmatrix} \left(v_{1,j-1} + x^{\frac{j-c}{c}} \right) & v_{2,j-1} & v_{3,j-1} & v_{0,j-1} \end{bmatrix}^T$$

If $c > 6$, then $T(\mathbf{w}_{j-1})$ is defined in the same way as for $c \le 6$ with one exception. If $c > 6$ and $j \bmod c = 4$, then $T(\mathbf{w}_{j-1})$ is defined to be the result of just applying the S-box transformation directly to each of the entries in $\mathbf{w}_{j-1}$. That is, with the notation that we used in the second step for $c \le 6$, if $c > 6$ and $j \bmod c = 4$, then $T(\mathbf{w}_{j-1})$ is defined as follows.

$$T(\mathbf{w}_{j-1}) = \begin{bmatrix} v_{0,j-1} & v_{1,j-1} & v_{2,j-1} & v_{3,j-1} \end{bmatrix}^T$$

For the AES encryption process, we will denote the key matrix used in the initial round as K_0, and the key matrix used in the ith round after the initial round as K_i for $i = 1, 2, \ldots, r$. The key matrix K_i for $i = 0, 1, \ldots, r$ is defined as the 4×4 matrix containing the columns $\mathbf{w}_{4i}$, $\mathbf{w}_{4i+1}$, $\mathbf{w}_{4i+2}$, and $\mathbf{w}_{4i+3}$ that we just described. That is, the key matrix K_i for $i = 0, 1, \ldots, r$ is defined as follows.

$$K_i = \begin{bmatrix} \mathbf{w}_{4i} & \mathbf{w}_{4i+1} & \mathbf{w}_{4i+2} & \mathbf{w}_{4i+3} \end{bmatrix}$$

We illustrate this process in the following example.

Example 10.8 Continuing Example 10.7, note first that the key matrix used in the initial round in the AES encryption process will be $K_0 = K$. We will now construct the key matrix K_1 that would be used in the first round after the initial round in the AES encryption process. We begin by labeling the columns in the matrix K in Example 10.7 as $\mathbf{w}_0, \mathbf{w}_1, \mathbf{w}_2$, and $\mathbf{w}_3$, and we must now construct the new columns $\mathbf{w}_4, \mathbf{w}_5, \mathbf{w}_6$, and $\mathbf{w}_7$. Using (10.3), we see that $\mathbf{w}_4 = \mathbf{w}_0 + T(\mathbf{w}_3)$. Since $c = j = 4$, then $j \bmod c = 0$, and we must determine $T(\mathbf{w}_3)$ using the three steps that we described in the transformation process for T. Assigning $\mathbf{w}_3 = \begin{bmatrix} x^2 & x^4 + x^3 & 0 & 0 \end{bmatrix}^T$, for the first step in the transformation process we would shift the entries in $\mathbf{w}_3$ to obtain the new column $\begin{bmatrix} x^4 + x^3 & 0 & 0 & x^2 \end{bmatrix}^T$. Next, we would apply the S-box transformation to each of the entries in this new column to obtain the following new column.

$$\begin{bmatrix} x^7 + x^5 + x^3 + x^2 + 1 \\ x^6 + x^5 + x + 1 \\ x^6 + x^5 + x + 1 \\ x^7 + x^6 + x^5 + x^4 + x \end{bmatrix}$$

Finally, with the round constant $x^{(4-4)/4} = x^0 = 1$, we have the following.

$$T(\mathbf{w}_3) = \begin{bmatrix} x^7 + x^5 + x^3 + x^2 + 1 + 1 \\ x^6 + x^5 + x + 1 \\ x^6 + x^5 + x + 1 \\ x^7 + x^6 + x^5 + x^4 + x \end{bmatrix}$$

$$= \begin{bmatrix} x^7 + x^5 + x^3 + x^2 \\ x^6 + x^5 + x + 1 \\ x^6 + x^5 + x + 1 \\ x^7 + x^6 + x^5 + x^4 + x \end{bmatrix}$$

Thus, since $\mathbf{w}_4 = \mathbf{w}_0 + T(\mathbf{w}_3)$, we have the following.

$$\mathbf{w}_4 = \begin{bmatrix} x^4 + 1 \\ x^3 \\ x^3 + 1 \\ x^3 + x^2 + 1 \end{bmatrix} + \begin{bmatrix} x^7 + x^5 + x^3 + x^2 \\ x^6 + x^5 + x + 1 \\ x^6 + x^5 + x + 1 \\ x^7 + x^6 + x^5 + x^4 + x \end{bmatrix}$$

$$= \begin{bmatrix} x^7 + x^5 + x^4 + x^3 + x^2 + 1 \\ x^6 + x^5 + x^3 + x + 1 \\ x^6 + x^5 + x^3 + x \\ x^7 + x^6 + x^5 + x^4 + x^3 + x^2 + x + 1 \end{bmatrix}$$

Next, using (10.3), we see that $\mathbf{w}_5 = \mathbf{w}_1 + T(\mathbf{w}_4)$. Since $c = 4$ and $j = 5$, then $j \bmod c \neq 0$, and so $T(\mathbf{w}_4) = \mathbf{w}_4$. Thus, we have the following.

$$\begin{aligned} \mathbf{w}_5 &= \mathbf{w}_1 + T(\mathbf{w}_4) \\ &= \mathbf{w}_1 + \mathbf{w}_4 \end{aligned}$$

$$= \begin{bmatrix} x + 1 \\ 0 \\ x^2 \\ x^3 + x + 1 \end{bmatrix} + \begin{bmatrix} x^7 + x^5 + x^4 + x^3 + x^2 + 1 \\ x^6 + x^5 + x^3 + x + 1 \\ x^6 + x^5 + x^3 + x \\ x^7 + x^6 + x^5 + x^4 + x^3 + x^2 + x + 1 \end{bmatrix}$$

$$= \begin{bmatrix} x^7 + x^5 + x^4 + x^3 + x^2 + x \\ x^6 + x^5 + x^3 + x + 1 \\ x^6 + x^5 + x^3 + x^2 + x \\ x^7 + x^6 + x^5 + x^4 + x^2 \end{bmatrix}$$

The columns $\mathbf{w}_6$ and $\mathbf{w}_7$ can be constructed similarly, and follow.

$$\mathbf{w}_6 = \begin{bmatrix} x^7 + x^5 + x^4 + x^3 + x^2 + x \\ x^6 + x^5 + x^3 + x^2 + x + 1 \\ x^6 + x^5 + x^4 + x^3 + x^2 \\ x^7 + x^6 + x^5 + x^4 + x^3 + x^2 + x \end{bmatrix}$$

$$\mathbf{w}_7 = \begin{bmatrix} x^7 + x^5 + x^4 + x^3 + x \\ x^6 + x^5 + x^4 + x^2 + x + 1 \\ x^6 + x^5 + x^4 + x^3 + x^2 \\ x^7 + x^6 + x^5 + x^4 + x^3 + x^2 + x \end{bmatrix}$$

Thus, if we express the columns $\mathbf{w}_4, \mathbf{w}_5, \mathbf{w}_6$, and $\mathbf{w}_7$ using the bytes of which the polynomial entries in the columns can be viewed as representations, we obtain the following key matrix K_1 that would be used in the first round after the initial round in the AES encryption process.

$$K_1 = \begin{bmatrix} 10111101 & 10111110 & 10111110 & 10111010 \\ 01101011 & 01101011 & 01101111 & 01110111 \\ 01101010 & 01101110 & 01111100 & 01111100 \\ 11111111 & 11110100 & 11111110 & 11111110 \end{bmatrix}$$

We will describe the rounds in the AES encryption process and continue this example later in this chapter. □

10.4   Encryption

Recall that the AES algorithm is designed to encrypt messages that contain 128 bits, which are arranged one byte at a time in a 4×4 plaintext matrix, whose entries can be represented by polynomials in $F = \mathbb{Z}_2[x]/(p(x))$. The AES encryption process includes the following four basic types of operations, known as the AES *layers*.

1. **ByteSub (BS)**: Each entry in an input matrix is transformed into another entry using the S-box transformation.

2. **ShiftRow (SR)**: The entries in each row of an input matrix are shifted to the left by zero, one, two, or three positions, with the entries at the start wrapping around to the end.

3. **MixColumn (MC)**: An input matrix is multiplied on the left by a fixed matrix of size 4×4, with the calculations done in F.

4. **AddRoundKey (ARK)**: The key matrix for each round is added to an input matrix, with the calculations done in F.

With these AES layers, which we will describe in more detail in Section 10.5, the following are the steps in the AES encryption process.

AES Encryption: *Let r represent the number of rounds after the initial round in the AES encryption process. Given a plaintext that contains 128 bits and an initial key that contains 128, 192, or 256 bits, we encrypt the plaintext using the following steps.*

1. *For the initial round, apply ARK directly to the plaintext matrix using the key matrix for the initial round.*

2. *For the next $r - 1$ rounds, apply BS, SR, MC, and ARK in order, using for ARK the key matrix given by the key schedule for the ith round after the initial round for $i = 1, 2, \ldots, r - 1$.*

3. *For the final round, apply BS, SR, and ARK in order, using for ARK the key matrix given by the key schedule for the rth round after the initial round.*

The output is a 128-bit ciphertext, in the form of a 4×4 ciphertext matrix.

Note that in the AES encryption process, the MixColumn layer is not included in the final round. We will explain the reason for this in Section 10.6.

10.5 The AES Layers

In this section, we will describe the AES layers in more detail. We will also demonstrate these layers in the first round after the initial round in the AES encryption process as a continuation of Examples 10.2 and 10.8.

Suppose we wish to use the AES encryption process on a plaintext of 128 bits. We would begin by using the procedure presented in Section 10.1 to arrange the plaintext one byte at a time in a 4×4 plaintext matrix A, whose entries can be represented by polynomials in $F = \mathbb{Z}_2[x]/(p(x))$. Then for an initial key matrix K and corresponding key matrix K_0 for the initial round in the AES encryption process, this initial round would consist of forming the following matrix A_0, with the calculations done in F.

$$A_0 = K_0 + A \qquad (10.4)$$

We illustrate this process in the following example.

Example 10.9 Suppose that we wish to encrypt the message RENDEZVOUS AT SIX A using the AES encryption process with the keyword RIJNDAEL AES KEY AA. To complete the initial round in the AES encryption process, we would add the matrices K_0 in Example 10.8 and A in Example 10.2, with the calculations done in $F = \mathbb{Z}_2[x]/(p(x))$. This would yield the following matrix A_0 given by (10.4), in which we have expressed the entries using the bytes of which the polynomial entries can be viewed as representations.

$$A_0 = \begin{bmatrix} 00000000 & 00000111 & 00010100 & 00010110 \\ 00001100 & 00011001 & 00010110 & 00010000 \\ 00000100 & 00010001 & 00010010 & 00010111 \\ 00001110 & 00000101 & 00011001 & 00000000 \end{bmatrix}$$

We are now ready to begin the first round after the initial round in the AES encryption process with A_0 as the input matrix into the first layer. □

10.5.1 ByteSub

For the ith round after the initial round in the AES encryption process for $i = 1, 2, \ldots, r$, we would begin by applying the ByteSub transformation to the matrix A_{i-1} that was output at the end of the previous round. ByteSub converts each entry in the matrix A_{i-1} into another entry using the S-box transformation, yielding the following matrix B_i.

$$B_i = \begin{bmatrix} b_{i:0,0} & b_{i:0,1} & b_{i:0,2} & b_{i:0,3} \\ b_{i:1,0} & b_{i:1,1} & b_{i:1,2} & b_{i:1,3} \\ b_{i:2,0} & b_{i:2,1} & b_{i:2,2} & b_{i:2,3} \\ b_{i:3,0} & b_{i:3,1} & b_{i:3,2} & b_{i:3,3} \end{bmatrix} \qquad (10.5)$$

We illustrate this process in the following example.

Example 10.10 As a continuation of Example 10.9, for the first round after the initial round in the AES encryption process, we would begin by applying the ByteSub transformation to the matrix A_0 in Example 10.9 that was output at the end of the initial round. This would yield the following matrix B_1 given by (10.5), in which we have expressed the entries using the bytes of which the polynomial entries can be viewed as representations.

$$B_1 = \begin{bmatrix} 01100011 & 11000101 & 11111010 & 01000111 \\ 11111110 & 11010100 & 01000111 & 11001010 \\ 11110010 & 10000010 & 11001001 & 11110000 \\ 10101011 & 01101011 & 11010100 & 01100011 \end{bmatrix}$$

We are now ready to continue the first round after the initial round in the AES encryption process with B_1 as the input matrix into the next layer. □

10.5.2   ShiftRow

For the ith round after the initial round in the AES encryption process for $i = 1, 2, \ldots, r$, after applying ByteSub, we would apply the ShiftRow transformation to the matrix B_i that was output by ByteSub. ShiftRow leaves the entries in the first row of B_i unchanged, shifts the entries in the second row of B_i to the left by one position, shifts the entries in the third row of B_i to the left by two positions, and shifts the entries in the fourth row of B_i to the left by three positions, with the entries at the start of each row wrapping around to the end. This yields, using the same notation as in (10.5), the following matrix C_i.

$$C_i = \begin{bmatrix} b_{i:0,0} & b_{i:0,1} & b_{i:0,2} & b_{i:0,3} \\ b_{i:1,1} & b_{i:1,2} & b_{i:1,3} & b_{i:1,0} \\ b_{i:2,2} & b_{i:2,3} & b_{i:2,0} & b_{i:2,1} \\ b_{i:3,3} & b_{i:3,0} & b_{i:3,1} & b_{i:3,2} \end{bmatrix} \tag{10.6}$$

We illustrate this process in the following example.

Example 10.11 As a continuation of Example 10.10, for the first round after the initial round in the AES encryption process, after applying Byte-Sub, we would apply the ShiftRow transformation to the matrix B_1 in Example 10.10 that was output by ByteSub. This would yield the following matrix C_1 given by (10.6), in which we have expressed the entries using the bytes of which the polynomial entries can be viewed as representations.

$$C_1 = \begin{bmatrix} 01100011 & 11000101 & 11111010 & 01000111 \\ 11010100 & 01000111 & 11001010 & 11111110 \\ 11001001 & 11110000 & 11110010 & 10000010 \\ 01100011 & 10101011 & 01101011 & 11010100 \end{bmatrix}$$

We are now ready to continue the first round after the initial round in the AES encryption process with C_1 as the input matrix into the next layer. □

10.5.3 MixColumn

For the ith round after the initial round in the AES encryption process for $i = 1, 2, \ldots, r-1$, after applying ShiftRow, we would apply the MixColumn transformation to the matrix C_i that was output by ShiftRow. MixColumn multiplies the matrix C_i on the left by the following matrix M, with the calculations done in $F = \mathbb{Z}_2[x]/(p(x))$.

$$
M = \begin{bmatrix}
x & x+1 & 1 & 1 \\
1 & x & x+1 & 1 \\
1 & 1 & x & x+1 \\
x+1 & 1 & 1 & x
\end{bmatrix}
$$

This yields the following matrix D_i for $i = 1, 2, \ldots, r-1$.

$$
D_i = MC_i \tag{10.7}
$$

We illustrate this process in the following example.

Example 10.12 As a continuation of Example 10.11, for the first round after the initial round in the AES encryption process, after applying ShiftRow, we would apply the MixColumn transformation to the matrix C_1 in Example 10.11 that was output by ShiftRow. This would yield the following matrix D_1 given by (10.7), in which we have expressed the entries using the bytes of which the polynomial entries can be viewed as representations.

$$
D_1 = \begin{bmatrix}
00001011 & 00000011 & 00110011 & 11000001 \\
11110011 & 11101011 & 00010011 & 11101001 \\
10011011 & 10011111 & 01110010 & 11000001 \\
01111110 & 10101110 & 11111011 & 00000110
\end{bmatrix}
$$

As a demonstration of the matrix product that yields D_1, consider the entry in the upper left corner of D_1. To determine this entry, we would form the dot product of the first row of M with the first column of C_1. This dot product is $x(x^6 + x^5 + x + 1) + (x + 1)(x^7 + x^6 + x^4 + x^2) + 1(x^7 + x^6 + x^3 + 1) + 1(x^6 + x^5 + x + 1) = x^8 + x^4$. To determine the element in $F = \mathbb{Z}_2[x]/(p(x))$ that is equivalent to the result of this dot product, we would need to find the remainder when $x^8 + x^4$ is divided by $p(x) = x^8 + x^4 + x^3 + x + 1$. This division yields a quotient of 1 and a remainder of $x^3 + x + 1$. It is this remainder that is the entry in the upper left corner of D_1. We are now ready to continue the first round after the initial round in the AES encryption process with D_1 as the input matrix into the next layer. □

10.5.4   AddRoundKey

For the ith round after the initial round in the AES encryption process
for $i = 1, 2, \ldots, r - 1$, after applying MixColumn, we would add the key
matrix K_i given by the key schedule to the matrix D_i that was output by
MixColumn, with the calculations done in $F = \mathbb{Z}_2[x]/(p(x))$. This yields
the following matrix A_i for $i = 1, 2, \ldots, r - 1$.

$$A_i = K_i + D_i \tag{10.8}$$

We illustrate this process in the following example.

Example 10.13 As a continuation of Example 10.12, for the first round
after the initial round in the AES encryption process, after applying Mix-
Column, we would add the key matrix K_1 in Example 10.8 to the matrix
D_1 in Example 10.12 that was output by MixColumn. This would yield
the following matrix A_1 given by (10.8), in which we have expressed the
entries using the bytes of which the polynomial entries can be viewed as
representations.

$$A_1 = \begin{bmatrix} 10110110 & 10111101 & 10001101 & 01111011 \\ 10011000 & 10000000 & 01111100 & 10011110 \\ 11110001 & 11110001 & 00001110 & 10111101 \\ 10000001 & 01011010 & 00000101 & 11111000 \end{bmatrix}$$

This completes the first round after the initial round in the AES encryption
process. We are now ready begin the second round after the initial round
in the AES encryption process with A_1 as the input matrix into the first
layer.   □

Recall that in the AES encryption process, the MixColumn layer is not
included in the final round. Thus, for the final round r in the AES encryption process, we would add the key matrix K_r given by the key schedule to
the matrix C_r that was output by ShiftRow, with the calculations done in
$F = \mathbb{Z}_2[x]/(p(x))$. This yields the following ciphertext matrix A_r.

$$A_r = K_r + C_r$$

The bytes of which the polynomial entries in A_r can be viewed as representations, when strung together one byte at a time reading down the columns
of A_r, form the 128-bit ciphertext.

10.6   Decryption

To decrypt a ciphertext that has been formed using the AES encryption process, we must use the inverses of the AES encryption layers. The inverses of

the encryption layers ByteSub, ShiftRow, MixColumn, and AddRoundKey can be described as follows.

1. **InvByteSub (IBS)**: To invert ByteSub, each entry in an input matrix is transformed into another entry using the inverse of the S-box transformation.

2. **InvShiftRow (ISR)**: To invert ShiftRow, the entries in the first row of an input matrix are left unchanged, the entries in the second row of an input matrix are shifted to the right by one position, the entries in the third row of an input matrix are shifted to the right by two positions, and the entries in the fourth row of an input matrix are shifted to the right by three positions, with the entries at the end of each row wrapping around to the start.

3. **InvMixColumn (IMC)**: To invert MixColumn, an input matrix is multiplied on the left by a fixed matrix M^{-1} of size 4×4, with the calculations done in $F = \mathbb{Z}_2[x]/(p(x))$. This matrix M^{-1} is the inverse of the matrix M in (10.7). It can be shown that the inverse of M over F is the following matrix M^{-1}.

$$
M^{-1} = \begin{bmatrix}
x^3 + x^2 + x & x^3 + x + 1 & x^3 + x^2 + 1 & x^3 + 1 \\
x^3 + 1 & x^3 + x^2 + x & x^3 + x + 1 & x^3 + x^2 + 1 \\
x^3 + x^2 + 1 & x^3 + 1 & x^3 + x^2 + x & x^3 + x + 1 \\
x^3 + x + 1 & x^3 + x^2 + 1 & x^3 + 1 & x^3 + x^2 + x
\end{bmatrix}
$$

4. AddRoundKey is its own inverse.

Recall that the AES encryption process consists of the following steps, with each of the following lines representing one round in the process, ordered starting with the initial round at the top.

ARK

BS, SR, MC, ARK

$\vdots \quad \vdots \quad \vdots \quad \vdots$

BS, SR, MC, ARK

BS, SR, ARK

In order to decrypt a ciphertext that has been formed using the AES encryption process, we would apply the inverses of these steps in the reverse order. Thus, the AES decryption process would consist of the following steps, with each of the following lines representing one round in the decryption process, ordered starting with the initial decryption round at the top.

ARK, ISR, IBS
ARK, IMC, ISR, IBS

$$\vdots \quad \vdots \quad \vdots \quad \vdots$$

ARK, IMC, ISR, IBS
ARK

However, it is possible to write the steps in the AES decryption process so that they more closely resemble the steps in the encryption process. To see this, note first that since the ByteSub and ShiftRow encryption layers operate on specific bytes in an input matrix, their order can be reversed in the encryption process without changing the process. Similarly, the order of InvShiftRow and InvByteSub can be reversed in the decryption process without changing the process.

It it not possible to directly reverse the order of the AddRoundKey and InvMixColumn layers in the AES decryption process without changing the process. However, it is possible through a little matrix algebra to identify an alternative way to reverse the order of these layers. For the ith round after the initial round in the AES encryption process for any round that includes the MixColumn layer, let C_i represent the matrix output by ShiftRow, D_i the matrix output by MixColumn, K_i the key matrix, and A_i the matrix output by AddRoundKey. Then (10.7) and (10.8) can be combined, yielding the following equation.

$$A_i = K_i + MC_i$$

Solving for C_i in this equation yields the following, where $K_i' = M^{-1}K_i$.

$$C_i = M^{-1}(K_i + A_i) = K_i' + M^{-1}A_i$$

The term $M^{-1}A_i$ represents InvMixColumn applied to A_i. Thus, if we define **InvAddRoundKey** (**IARK**) to be addition of the decryption key matrix $K_i' = M^{-1}K_i$ to the input matrix $M^{-1}A_i$, then we see that we can invert the encryption layers *MixColumn followed by AddRoundKey* with the decryption layers *InvMixColumn followed by InvAddRoundKey*. This yields the following equivalent steps in the AES decryption process, with each of the following lines representing one round in the decryption process, ordered starting with the initial decryption round at the top.

ARK, IBS, ISR
IMC, IARK, IBS, ISR

$$\vdots \quad \vdots \quad \vdots \quad \vdots$$

IMC, IARK, IBS, ISR
ARK

If we regroup these rounds by moving the layers *InvByteSub followed by InvShiftRow* from the end of each of the rounds except the final round to the beginning of the next round, we obtain the following final version of the steps in the AES decryption process.

$$
\begin{array}{cccc}
\text{ARK} & & & \\
\text{IBS, ISR, IMC, IARK} & & & \\
\vdots & \vdots & \vdots & \vdots \\
\text{IBS, ISR, IMC, IARK} & & & \\
\text{IBS, ISR, ARK} & & &
\end{array}
$$

AES Decryption: *Given a ciphertext that contains* 128 *bits and which was formed using the AES encryption process with r rounds after the initial round, we decrypt the ciphertext using the following steps.*

1. *For the initial decryption round, apply ARK directly to the ciphertext matrix using the key matrix given by the key schedule for the rth encryption round after the initial round.*

2. *For the next $r-1$ decryption rounds, apply IBS, ISR, IMC, and IARK in order, using for IARK the product of the matrix M^{-1} and the key matrix given by the key schedule for the $(r-i)$th encryption round after the initial round for $i = 1, 2, \ldots, r - 1$.*

3. *For the final decryption round, apply IBS, ISR, and ARK in order, using for ARK the key matrix for the initial encryption round.*

The output is the 128-*bit plaintext, in the form of the 4×4 plaintext matrix.*

Note that the rounds in the AES decryption process have the same structure as the rounds in the encryption process, but with ByteSub, ShiftRow, and MixColumn replaced by their inverses, and AddRoundKey replaced by InvAddRoundKey in all but the initial and final rounds. Also, in the decryption process the key matrices given by the key schedule are used in the reverse of the order in which they were used in the encryption process. Through a preliminary agreement, the originator and intended recipient of a message would have to have agreed upon an initial key for the system, and thus could form the initial key matrix K and use the key schedule to determine the key matrices necessary for the encryption and decryption processes.

Our discussion in this section indicates why the MixColumn layer is not included in the final round of the AES encryption process. If MixColumn was included in this round, then the final encryption round would be BS,

SR, MC, ARK. This would make the initial decryption round, after reversing orders, IMC, IARK, IBS, ISR, resulting in an extra InvMixColumn at the beginning of the decryption process, and making the algorithm less efficient.

We illustrate the layers in the AES decryption process in the following example. In this example, for demonstration purposes we will suppose that a plaintext has been encrypted by applying only the initial round and first subsequent round in the AES encryption process. That is, we will suppose that a plaintext has been encrypted by applying only the layers ARK, BS, SR, MC, and ARK, in order. To decrypt the ciphertext, we will need to apply only the last two layers in the next-to-final decryption round and the three layers in the final decryption round. That is, to decrypt the ciphertext, we will need to apply only the layers IMC, IARK, IBS, ISR, and ARK, in order.

Example 10.14 Suppose that a 128-bit plaintext was encrypted by applying the initial round and first subsequent round in the AES encryption process using the keyword **RIJNDAEL AES KEY AA**, with the output being the following matrix A_1, in which we have expressed the entries using the bytes of which the polynomial entries can be viewed as representations.

$$A_1 = \begin{bmatrix} 00111000 & 01110101 & 01111101 & 00010010 \\ 11000011 & 01010001 & 00010111 & 01101000 \\ 11111101 & 10110001 & 01010100 & 00001110 \\ 10110001 & 10000011 & 11111001 & 11010001 \end{bmatrix}$$

To recover the plaintext, we will invert the initial round and first subsequent round in the AES encryption process by applying IMC, IARK, IBS, ISR, ARK, in order. First, applying InvMixColumn to the preceding matrix A_1, we compute the following matrix $M^{-1}A_1$, in which we have expressed the entries using the bytes of which the polynomial entries can be viewed as representations.

$$M^{-1}A_1 = \begin{bmatrix} 10010100 & 10100110 & 10101000 & 01101100 \\ 00010100 & 01001001 & 10100001 & 10001100 \\ 00100100 & 01001110 & 00010100 & 00100001 \\ 00010011 & 10110111 & 11011010 & 01100100 \end{bmatrix}$$

Next, in order to apply InvAddRoundKey to the preceding matrix $M^{-1}A_1$, we must first find the decryption key matrix $K_1' = M^{-1}K_1$. Recall that for the given keyword **RIJNDAEL AES KEY AA**, we used the key schedule in Example 10.8 to determine K_1. Thus, with K_1 from Example 10.8, we can compute the following matrix $K_1' = M^{-1}K_1$, in which we have expressed the entries using the bytes of which the polynomial entries can be viewed as representations.

$$K_1' = \begin{bmatrix} 00010110 & 01100011 & 11011111 & 00001111 \\ 01100010 & 00101010 & 11000110 & 01110010 \\ 01000111 & 00101101 & 10111011 & 01010111 \\ 01110000 & 00101011 & 11110001 & 01100101 \end{bmatrix}$$

Applying InvAddRoundKey to the matrix $M^{-1}A_1$, we compute the following matrix $C_1 = K_1' + M^{-1}A_1$, in which we have expressed the entries using the bytes of which the polynomial entries can be viewed as representations.

$$C_1 = \begin{bmatrix} 10000010 & 11000101 & 01110111 & 00001111 \\ 01110110 & 01100011 & 01100111 & 11111110 \\ 01100011 & 01100011 & 10101111 & 01110110 \\ 01100011 & 10011100 & 00101011 & 00000001 \end{bmatrix}$$

Next, applying InvByteSub to the preceding matrix C_1, we obtain the following matrix B_1, in which we have expressed the entries using the bytes of which the polynomial entries can be viewed as representations.

$$B_1 = \begin{bmatrix} 00010001 & 00000111 & 00000010 & 00000000 \\ 00001111 & 00000000 & 00001010 & 00001100 \\ 00000000 & 00000000 & 00011011 & 00001111 \\ 00000000 & 00011100 & 00001011 & 00001001 \end{bmatrix}$$

Applying InvShiftRow to the preceding matrix B_1, we obtain the following matrix A_0, in which we have expressed the entries using the bytes of which the polynomial entries can be viewed as representations.

$$A_0 = \begin{bmatrix} 00010001 & 00000111 & 00000010 & 00000000 \\ 00001100 & 00001111 & 00000000 & 00001010 \\ 00011011 & 00001111 & 00000000 & 00000000 \\ 00011100 & 00001011 & 00001001 & 00000000 \end{bmatrix}$$

Finally, in order to apply AddRoundKey to the preceding matrix A_0, recall that for the given keyword RIJNDAEL AES KEY AA, we noted K_0 in Example 10.8. Thus, with K_0 from Example 10.8, we can compute the following matrix $A = K_0 + A_0$, in which we have expressed the entries using the bytes of which the polynomial entries can be viewed as representations.

$$A = \begin{bmatrix} 0 & x^2 & x & x^2 \\ x^2 & x^3 + x^2 + x + 1 & x^2 & x^4 + x \\ x^4 + x & x^3 + x + 1 & x^4 + x & 0 \\ x^4 + 1 & 0 & x + 1 & 0 \end{bmatrix}$$

Using the table in Example 10.1 to convert the entries in the preceding matrix A into the letters of which the polynomial entries can be viewed as representations, we obtain the plaintext AESREPLACESDESAA. $\square$

We will close this discussion of AES with some very brief comments on security. When the algorithm Rijndael was adopted as AES, the number of encryption rounds was chosen to be at least 10 because beyond six rounds, no attacks against the system were known that were faster than testing all possible initial key matrices by trial and error. It was believed at the time that four extra encryption rounds would prevent all possible attacks against the system for many years. Also, it was shown at the time that the machines that were successfully able to break the DES algorithm would in general take trillions of years to break Rijndael, even when the system was implemented using an initial key of the smallest possible size. It would also be very easy, of course, to increase the security of the system by simply increasing the number of encryption rounds. Barring any future attacks against AES that are faster than testing all possible initial key matrices by trial and error, and even with expected advances in technology, AES has the potential to remain secure well beyond the 20 years that spanned the life of DES.

We should also emphasize the fact that AES is not a public-key system. As we noted previously, when using AES, the originator and intended recipient of a message would have to have agreed upon an initial key for the system through a preliminary agreement. However, there are methods for overcoming this problem, which we will leave to the reader for investigation.

10.7   AES with Maple

In this section, we will show how Maple can be used to encrypt and decrypt messages using the AES algorithm.

10.7.1   Construction of Initial Parameters

Because some of the functions that we will use are in the Maple **String-Tools** and **LinearAlgebra** packages, we will begin by including these packages.

```
> with(StringTools):
```

```
> with(LinearAlgebra):
```

Next, we will define the irreducible polynomial $p(x) = x^8 + x^4 + x^3 + x + 1$ for the underlying finite field $F = \mathbb{Z}_2[x]/(p(x))$.

```
>   p := x -> x^8+x^4+x^3+x+1:
```

```
>   Irreduc(p(x)) mod 2;
```
$$true$$

We will use $p(x)$ to construct a polynomial form of the S-box given in Table 10.1. This form of the S-box will be a 16×16 matrix containing as its entries the polynomials that can be viewed as representations of the binary expressions of the entries in Table 10.1. To construct this form of the S-box, we will use the user-written function **sboxtable**, which we have written separately from this Maple session and saved as the text file sboxtable.mpl. To use this function, we must first read it into our Maple session as follows.

```
>  read "sboxtable.mpl";
```

The following command then illustrates how the function **sboxtable** can be used. Specifically, by entering the following command, we construct the polynomial form of the AES S-box, a 16×16 matrix stored as the first entry in the list *Stab* that results from the command, and the polynomial form of the inverse of this matrix over $F = \mathbb{Z}_2[x]/(p(x))$, a 16×16 matrix stored as the second entry in *Stab*. The second parameter in this command is the variable x used in the first parameter, the polynomial $p(x)$. Also, due to the size of these matrices, we will suppress the output of this command.

```
>  Stab := sboxtable(p(x), x):
```

For later use, in the following two commands we assign the polynomial form of the S-box that is stored as the first entry in *Stab* as the variable *Sbx*, and the inverse of this matrix over F that is stored as the second entry in *Stab* as the variable *inSbx*.

```
>  Sbx := Stab[1]:
```

```
>  inSbx := Stab[2]:
```

For a given polynomial $f(x)$ in $F = \mathbb{Z}_2[x]/(p(x))$, we will need to be able to determine the output of the S-box transformation that corresponds to the input $f(x)$. Also, for a given polynomial $g(x)$ in F, we will need to be able to determine the input into the S-box transformation that has corresponding output $g(x)$. To do these operations, we will use the user-written function **sbox**, which we have written separately from this Maple session and saved as the text file sbox.mpl. To use this function, we must first read it into our Maple session as follows.

```
>  read "sbox.mpl";
```

The following command then illustrates how the function **sbox** can be used. Specifically, by entering the following command, we determine the output of the S-box transformation that corresponds to the input $x^7 + x^3 + 1$. By using *Sbx* as the second parameter in this command, we are specifying that we wish to determine the output of the S-box transformation that corresponds to the input given by the first parameter.

```
>  sbox(x^7+x^3+1, Sbx, p(x), x);
```
$$x^7 + x^5 + x^2 + x + 1$$

Also, by entering the next command, we determine the input into the S-box transformation that has corresponding output $x^7 + x^5 + x^2 + x + 1$. By using *inSbx* as the second parameter in this command, we are specifying that we wish to determine the input into the S-box transformation that has corresponding output given by the first parameter.

```
>  sbox(x^7+x^5+x^2+x+1, inSbx, p(x), x);
```
$$x^7 + x^3 + 1$$

Now, suppose that we wish to use the AES algorithm to encrypt the message RENDEZVOUS AT SIX. First, we will store this message as the following string.

```
>  message := Select(IsAlpha, "RENDEZVOUS AT SIX");
```
$$message := \text{``RENDEZVOUSATSIX''}$$

Recall that the AES encryption algorithm is designed to encrypt plaintext messages that contain 16 characters. To expand our plaintext message RENDEZVOUS AT SIX so that it contains exactly 16 letters, we will use the user-written function **mblocks**, which we have written separately from this Maple session and saved as the text file mblocks.mpl. To use this function, we must first read it into our Maple session as follows.

```
>  read "mblocks.mpl";
```

The following command then illustrates how the function **mblocks** can be used. Specifically, by entering the following command, we expand *message* by adding an extra letter A so that the result contains exactly 16 letters.

```
>  ptext := mblocks(message, 16);
```
$$ptext := \text{``RENDEZVOUSATSIXA''}$$

The second parameter in the preceding command specifies the total number of letters that should be used to express the message string given by the first parameter. If the number of letters in this message string is less than the number of letters specified by the second parameter, then the message string will be expanded by adding extra letters A repeatedly to the end of the message until the result contains exactly the number of letters specified by the second parameter. If the number of letters in this message string is greater than the number of letters specified by the second parameter, then the message string will be divided into blocks that contain exactly the number of letters specified by the second parameter, with the last block expanded by adding extra letters A repeatedly to the end of the block until the result contains exactly the number of letters specified by the second parameter.

Next, we will convert each letter in *ptext* into its polynomial equivalent using the table that is shown in Example 10.1. To do this, we will use the user-written function **topoly**, which we have written separately from this Maple session and saved as the text file topoly.mpl. To use this function, we must first read it into our Maple session as follows.

> read "topoly.mpl";

The following command then illustrates how the function **topoly** can be used. Specifically, by entering the following command, we convert each letter in *ptext* into its polynomial equivalent using the table in Example 10.1. The second parameter in this command is the variable that is to be used in the resulting polynomials.

> ptext := topoly(ptext, x);

$$ptext := [x^4 + 1, x^2, x^3 + x^2 + 1, x + 1, x^2, x^4 + x^3 + 1, x^4 + x^2 + 1,$$
$$x^3 + x^2 + x, x^4 + x^2, x^4 + x, 0, x^4 + x + 1, x^4 + x, x^3, x^4 + x^2$$
$$+ x + 1, 0]$$

Using the Maple **Matrix** function, we can arrange these polynomials in order as the entries in the columns of a 4×4 matrix A as follows.

> A := Matrix(4, 4, (i, j) -> ptext[(j-1)*4+i]);

$$A := \begin{bmatrix} x^4 + 1 & x^2 & x^4 + x^2 & x^4 + x \\ x^2 & x^4 + x^3 + 1 & x^4 + x & x^3 \\ x^3 + x^2 + 1 & x^4 + x^2 + 1 & 0 & x^4 + x^2 + x + 1 \\ x + 1 & x^3 + x^2 + x & x^4 + x + 1 & 0 \end{bmatrix}$$

Now, suppose that we wish to construct a key matrix for the initial round in the AES encryption process with a 128-bit initial key using the keyword RIJNDAEL AES KEY. In the following sequence of commands, we assign and expand this keyword so that it contains exactly 16 letters, convert each letter in the result into its polynomial equivalent using the table in Example 10.1, and arrange the resulting polynomials in order as the entries in the columns of a 4×4 key matrix *K0* for the initial round in the encryption process.

```
> key := Select(IsAlpha, "RIJNDAEL AES KEY"):
```

```
> key := mblocks(key, 16);
```
$$key := \text{``RIJNDAELAESKEYAA''}$$

```
> K0 := Matrix(4, 4, (i, j) -> topoly(key, x)[(j-1)*4+i]);
```

$$K0 := \begin{bmatrix} x^4 + 1 & x + 1 & 0 & x^2 \\ x^3 & 0 & x^2 & x^4 + x^3 \\ x^3 + 1 & x^2 & x^4 + x & 0 \\ x^3 + x^2 + 1 & x^3 + x + 1 & x^3 + x & 0 \end{bmatrix}$$

Recall that for the AES algorithm, the key schedule is used to generate a series of columns, beginning with the columns in an initial key matrix, which are used to form the key matrices for each of the rounds in the AES encryption and decryption processes. To generate this series of columns, we will use the user-written function **keysched**,[1] which we have written separately from this Maple session and saved as the text file keysched.mpl. To use this function, we must first read it into our Maple session as follows.

```
> read "keysched.mpl";
```

The following command then illustrates how the function **keysched** can be used. Specifically, by entering the following command, we construct a matrix *KS* which contains as its columns the series of columns, beginning with the columns in the initial key matrix *K0*, which would be used to form the key matrices for the rounds in both of the AES encryption and decryption processes. In this command, the first parameter is the initial key matrix *K0*, and the S-box matrix *Sbx* is included as a parameter because, recall, the key schedule uses the S-box. Also, due to the size of the resulting matrix *KS*, we will suppress rather than display the output of this command.

```
> KS := keysched(K0, Sbx, p(x), x):
```

With this matrix *KS*, we can now see the key matrix that would be used in any of the rounds in the AES encryption or decryption processes. For example, recall that the fifth, sixth, seventh, and eighth columns of *KS* are the columns in the key matrix that would be used in the first round after the initial round in the AES encryption process. To see these columns, we can enter the following commands.

---

[1] The function **keysched** uses the user-written function **sbox** that we included previously in this Maple session.

```
>  Matrix([Column(KS, 5..6)]);
```

$$
\begin{bmatrix}
x^7 + x^5 + x^4 + x^3 + x^2 + 1 & x^7 + x^5 + x^4 + x^3 + x^2 + x \\
x^6 + x^5 + x^3 + x + 1 & x^6 + x^5 + x^3 + x + 1 \\
x^6 + x^5 + x^3 + x & x^6 + x^5 + x^3 + x^2 + x \\
x^7 + x^6 + x^5 + x^4 + x^3 + x^2 + x + 1 & x^7 + x^6 + x^5 + x^4 + x^2
\end{bmatrix}
$$

```
>  Matrix([Column(KS, 7..8)]);
```

$$
\begin{bmatrix}
x^7 + x^5 + x^4 + x^3 + x^2 + x & x^7 + x^5 + x^4 + x^3 + x \\
x^6 + x^5 + x^3 + x^2 + x + 1 & x^6 + x^5 + x^4 + x^2 + x + 1 \\
x^6 + x^5 + x^4 + x^3 + x^2 & x^6 + x^5 + x^4 + x^3 + x^2 \\
x^7 + x^6 + x^5 + x^4 + x^3 + x^2 + x & x^7 + x^6 + x^5 + x^4 + x^3 + x^2 + x
\end{bmatrix}
$$

We can assign these columns as the key matrix *K1* that would be used in the first round after the initial round in the AES encryption process by entering the following command. Due to the size of this matrix, we will suppress the output of this command.

```
>  K1 := Matrix([Column(KS, 5..8)]):
```

10.7.2 The Encryption Layers

We will now show how Maple can be used to apply the AES encryption layers. We will specifically demonstrate these layers in the first round after the initial round in the AES encryption process using the matrices A and *K0* that we constructed previously in this Maple session.

To complete the initial round in the AES encryption process, we would form the sum of the matrices *K0* and A, with the calculations done in $F = \mathbb{Z}_2[x]/(p(x))$. To perform this operation, we will use the following function **addrkey**.

```
>  addrkey := (M, K) -> map(m -> sort(m) mod 2, M+K):
```

The following command illustrates how the function **addrkey** can be used. Specifically, by entering the following command, we form the sum of the matrices *K0* and A, with the calculations done in F, and assign the resulting matrix as *A0*.

```
>  A0 := addrkey(K0, A);
```

$$A0 := \begin{bmatrix} 0 & x^2 + x + 1 & x^4 + x^2 & x^4 + x^2 + x \\ x^3 + x^2 & x^4 + x^3 + 1 & x^4 + x^2 + x & x^4 \\ x^2 & x^4 + 1 & x^4 + x & x^4 + x^2 + x + 1 \\ x^3 + x^2 + x & x^2 + 1 & x^4 + x^3 + 1 & 0 \end{bmatrix}$$

We are now ready to begin the first round after the initial round in the AES encryption process with $A0$ as the input matrix into the first layer.

We would begin the first round after the initial round in the AES encryption process by applying the ByteSub transformation to the matrix $A0$ that was output at the end of the initial round. To perform this operation, we will use the following function **bytesub**.

```
> bytesub := (M, SBT, p, x) -> map(sbox, M, SBT, p, x):
```

The following command illustrates how the function **bytesub** can be used. Specifically, by entering the following command, we apply the ByteSub transformation to the matrix $A0$, and assign the resulting matrix as $B1$. In this command, the S-box matrix Sbx is included as a parameter because, recall, the ByteSub transformation uses the S-box. Also, due to the size of $B1$, we will suppress the output of this command.

```
> B1 := bytesub(A0, Sbx, p(x), x):
```

Next, we would apply ShiftRow to this matrix $B1$. To do this operation, we will use the user-written function **shiftrow**, which we have written separately from this Maple session and saved as the text file shiftrow.mpl. To use this function, we must first read it into our Maple session as follows.

```
> read "shiftrow.mpl";
```

The following command then illustrates how the function **shiftrow** can be used. Specifically, by entering the following command, we apply the ShiftRow transformation to the matrix $B1$, and assign the resulting matrix as $C1$. Due to the size of $C1$, we will suppress the output of this command.

```
> C1 := shiftrow(B1):
```

Next, we would apply MixColumn to this matrix $C1$, with the calculations done in $F = \mathbb{Z}_2[x]/(p(x))$. To do this operation, we will use the user-written function **mixcolumn**, which we have written separately from this Maple session and saved as the text file mixcolumn.mpl. To use this function, we must first read it into our Maple session as follows.

```
>  read "mixcolumn.mpl";
```

The following command then illustrates how the function **mixcolumn** can be used. Specifically, by entering the following command, we apply the MixColumn transformation to the matrix *C1*, and assign the resulting matrix as *D1*. Due to the size of *D1*, we will suppress the output of this command.

```
>  D1 := mixcolumn(C1, p(x), x):
```

Next, we would add the key matrix *K1* that we constructed previously in this Maple session to this matrix *D1*, with the calculations done in $F = \mathbb{Z}_2[x]/(p(x))$. To perform this operation, we will use the user-written function **addrkey** that we defined previously in this Maple session. In the following command we add *K1* to *D1*, with the calculations done in F, and assign the resulting matrix as *A1*. Due to the size of *A1*, we will suppress the output of this command.

```
>  A1 := addrkey(K1, D1):
```

This completes the first round after the initial round in the AES encryption process.

10.7.3   The Decryption Layers

We will now show how Maple can be used to apply the AES decryption layers. We will specifically demonstrate these layers by beginning with the matrix *A1* that we just constructed as the output of the first round after the initial round in the AES encryption process, and recovering the plaintext from which it resulted.

We would begin by applying InvMixColumn to the matrix *A1*, with the calculations done in $F = \mathbb{Z}_2[x]/(p(x))$. To do this operation, we will use the user-written function **invmixcolumn**, which we have written separately from this Maple session and saved as the text file invmixcolumn.mpl. To use this function, we must first read it into our Maple session as follows.

```
>  read "invmixcolumn.mpl";
```

The following command then illustrates how the function **invmixcolumn** can be used. Specifically, by entering the following command, we apply the InvMixColumn transformation to the matrix *A1*, and assign the resulting matrix as *IMCA1*. Due to the size of *IMCA1*, we will suppress the output of this command.

```
> IMCA1 := invmixcolumn(A1, p(x), x):
```

Next, to apply InvAddRoundKey to this matrix *IMCA1*, we will use the user-written function **invaddrkey**, which we have written separately from this Maple session and saved as the text file invaddrkey.mpl. To use this function, we must first read it into our Maple session as follows.

```
> read "invaddrkey.mpl";
```

The following command then illustrates how the function **invaddrkey** can be used. Specifically, by entering the following command, we apply InvAddRoundKey to the matrix *IMCA1*, and assign the resulting matrix as *C1*. In this command, the matrix *K1* is included as a parameter. This is because, recall, the matrix that would be added to *IMCA1* under InvAddRoundKey would be the product of the matrix M^{-1} given on page 375 and *K1*. Also, due to the size of *C1*, we will suppress the output of this command.

```
> C1 := invaddrkey(IMCA1, K1, p(x), x):
```

Next, we would apply the InvByteSub transformation to this matrix *C1*. To perform this operation, we will use the user-written function **bytesub** that we defined previously in this Maple session. In the following command, we apply InvByteSub to *C1*, and assign the resulting matrix as *B1*. In this command, the inverse *inSbx* of the S-box matrix *Sbx* is included as a parameter because, recall, the InvByteSub transformation uses the inverse of the S-box. Also, due to the size of *B1*, we will suppress the output of this command.

```
> B1 := bytesub(C1, inSbx, p(x), x):
```

Next, we would apply InvShiftRow to this matrix *B1*. To do this, we will use the user-written function **invshiftrow**, which we have written separately from this Maple session and saved as the text file invshiftrow.mpl. To use this function, we must first read it into our Maple session as follows.

```
> read "invshiftrow.mpl";
```

The following command then illustrates how the function **invshiftrow** can be used. Specifically, by entering the following command, we apply the InvShiftRow transformation to the matrix *B1*, and assign the resulting matrix as *A0*.

```
> A0 := invshiftrow(B1);
```

$$
A0 := \begin{bmatrix}
0 & x^2 + x + 1 & x^4 + x^2 & x^4 + x^2 + x \\
x^3 + x^2 & x^4 + x^3 + 1 & x^4 + x^2 + x & x^4 \\
x^2 & x^4 + 1 & x^4 + x & x^4 + x^2 + x + 1 \\
x^3 + x^2 + x & x^2 + 1 & x^4 + x^3 + 1 & 0
\end{bmatrix}
$$

Next, we would add the key matrix $K0$ that we constructed previously in this Maple session to this matrix $A0$, with the calculations done in $F = \mathbb{Z}_2[x]/(p(x))$. To perform this operation, we will use the user-written function **addrkey** that we defined previously in this Maple session. In the following command we add $K0$ to $A0$, with the calculations done in F, and assign the resulting matrix as A.

```
>  A := addrkey(K0, A0);
```

$$
A := \begin{bmatrix}
x^4 + 1 & x^2 & x^4 + x^2 & x^4 + x \\
x^2 & x^4 + x^3 + 1 & x^4 + x & x^3 \\
x^3 + x^2 + 1 & x^4 + x^2 + 1 & 0 & x^4 + x^2 + x + 1 \\
x + 1 & x^3 + x^2 + x & x^4 + x + 1 & 0
\end{bmatrix}
$$

To recover the plaintext from which this matrix A was constructed, we will first use the Maple **convert** function as follows to arrange the polynomial entries in A reading in order down the columns of A into a single list.

```
>  ptext := convert(convert(A, Vector), list);
```

$ptext := [x^4 + 1, x^2, x^3 + x^2 + 1, x + 1, x^2, x^4 + x^3 + 1, x^4 + x^2 + 1,$
$\quad x^3 + x^2 + x, x^4 + x^2, x^4 + x, 0, x^4 + x + 1, x^4 + x, x^3, x^4 + x^2$
$\quad + x + 1, 0]$

Finally, we will use the table in Example 10.1 to convert these polynomials into the letters of which they can be viewed as representations. For this, we will use the user-written function **frompoly**, which we have written separately from this Maple session and saved as the text file frompoly.mpl. To use this function, we must first read it into our Maple session as follows.

```
>  read "frompoly.mpl";
```

The following command then illustrates how the function **frompoly** can be used. Specifically, by entering the following command, we use the table in Example 10.1 to convert each polynomial entry in *ptext* into the letter of which it can be viewed as a representation.

```
>  ptext := frompoly(ptext, x);
```
$$ptext := \text{``RENDEZVOUSATSIXA''}$$

The second parameter in the preceding command is the variable that is used in the entries in the first parameter.

10.7.4   Encryption and Decryption

We will now show how Maple can be used to perform the entire AES encryption and decryption processes.

We will first demonstrate these processes by encrypting the message TRANSFER THE MONEY with the 128-bit keyword AES KEYS ARE SECURE. We begin with the following sequence of commands, in which we assign this message and keyword as *ptext1* and *key16*, respectively, and use the user-written function **mblocks** that we included previously in this Maple session to verify that they each contain exactly 16 letters.

```
>  ptext1 := Select(IsAlpha, "TRANSFER THE MONEY"):
```

```
>  ptext1 := mblocks(ptext1, 16);
```
$$ptext1 := \text{``TRANSFERTHEMONEY''}$$

```
>  key16 := Select(IsAlpha, "AES KEYS ARE SECURE"):
```

```
>  key16 := mblocks(key16, 16);
```
$$key16 := \text{``AESKEYSARESECURE''}$$

To encrypt the plaintext message *ptext1*, we will use the user-written function **aesencipher**,[2] which we have written separately from this Maple session and saved as the text file aesencipher.mpl. To use this function, we must first read it into our Maple session as follows.

```
>  read "aesencipher.mpl";
```

The following command shows how the function **aesencipher** can be used. Specifically, by entering the following command, we encrypt *ptext1* using the AES encryption process with the keyword *key16*. In this command, the S-box matrix *Sbx* is included as a parameter because, recall, the AES en-

---

[2]The function **aesencipher** uses the user-written functions **topoly**, **keysched**, **add-rkey**, **bytesub**, **shiftrow**, and **mixcolumn** that we included or defined previously in this Maple session.

cryption process uses the S-box. Also, the fifth parameter in this command is the variable x used in the fourth parameter, the polynomial $p(x)$.

>   `ctext1 := aesencipher(ptext1, key16, Sbx, p(x), x);`

    ctext1 := [63, 254, 12, 222, 9, 109, 165, 12, 208, 244, 90, 148, 190, 172,
      17, 4]

The output produced by the preceding command is a list of integers whose binary expressions can be represented by the polynomial entries in the matrix output at the end of the final round in the AES encryption process.

To decrypt the ciphertext message *ctext1*, we will use the user-written function **aesdecipher**,[3] which we have written separately from this Maple session and saved as the text file aesdecipher.mpl. To use this function, we must first read it into our Maple session as follows.

>   `read "aesdecipher.mpl";`

The following command then illustrates how the function **aesdecipher** can be used. Specifically, by entering the following command, we decrypt *ctext1* using the AES decryption process with the keyword *key16*. In this command, the S-box matrix *Sbx* and its inverse *inSbx* are included as parameters because, recall, the AES decryption process uses both the S-box and the inverse of the S-box.

>   `aesdecipher(ctext1, key16, Sbx, inSbx, p(x), x);`
              "TRANSFERTHEMONEY"

Recall that the AES encryption process is designed to use initial keys of lengths 128, 192, or 256 bits. In the following sequence of commands, we show how the user-written functions **mblocks**, **aesencipher**, and **aesdecipher** can be used to encrypt and decrypt *ptext1* using a keyword of length 192 bits, or, equivalently, 24 letters.

>   `key24 := Select(IsAlpha, "AES IS MORE SECURE THAN DES"):`

>   `key24 := mblocks(key24, 24);`
    key24 := "AESISMORESECURETHANDESAA"

>   `ctext1 := aesencipher(ptext1, key24, Sbx, p(x), x);`

    ctext1 := [114, 74, 72, 43, 232, 116, 69, 94, 216, 191, 80, 34, 131, 81,
      220, 126]

---

[3]The function **aesdecipher** uses the user-written functions **topoly**, **keysched**, **addrkey**, **bytesub**, **invshiftrow**, **invmixcolumn**, and **invaddrkey** that we included or defined previously in this Maple session.

```
> aesdecipher(ctext1, key24, Sbx, inSbx, p(x), x);
                    "TRANSFERTHEMONEY"
```

Recall that the AES algorithm is designed to encrypt messages that contain exactly 16 characters. Thus, we have expanded messages that are shorter than 16 letters so that they contain exactly 16 letters before being encrypted. For messages that are longer than 16 letters, the AES algorithm can still be used, of course; we just encrypt only 16 plaintext letters at a time. In the following sequence of commands, we show how the user-written functions **mblocks** and **aesencipher** can be used to encrypt the message TRANSFER ONE MILLION DOLLARS FROM STOCKS TO BONDS using the keyword *key16*.

```
> ptext2 := Select(IsAlpha, "TRANSFER ONE MILLION DOLLARS
  FROM STOCKS TO BONDS"):
```

```
> ptext2 := mblocks(ptext2, 16);
```
ptext2 := ["TRANSFERONEMILLI", "ONDOLLARSFROMSTO", "CKSTOBONDSAAAAAA"]

```
> ctext2 := map(aesencipher, ptext2, key16, Sbx, p(x), x);
```
ctext2 := [[40, 160, 254, 160, 157, 64, 227, 62, 199, 221, 215, 116, 34, 164, 99, 129], [65, 1, 106, 251, 129, 88, 115, 108, 52, 13, 39, 208, 218, 190, 132, 111], [82, 28, 245, 229, 60, 190, 131, 229, 61, 232, 15, 173, 5, 91, 10, 210]]

The result *ctext2* defined by the preceding command contains as its three entries the lists of integers whose binary expressions can be represented by the polynomial entries in the matrix output at the end of the final round of the AES encryption process for the plaintext consisting of the first, second, and third blocks of 16 letters in the original message, respectively. In the following command, we show how the user-written function **aesdecipher** can be used to decrypt *ctext2* using the keyword *key16*.

```
> map(aesdecipher, ctext2, key16, Sbx, inSbx, p(x), x);
```
["TRANSFERONEMILLI", "ONDOLLARSFROMSTO", "CKSTOBONDSAAAAAA"]

```
> cat(seq(%[i], i=1..nops(%)));
```
"TRANSFERONEMILLIONDOLLARSFROMSTOCKSTOBONDSA\
    AAAAA"

10.8 AES with MATLAB

In this section, we will show how MATLAB can be used to encrypt and decrypt messages using the AES algorithm.

10.8.1 Construction of Initial Parameters

We will begin by declaring the variable x as a symbolic variable, and defining the irreducible polynomial $p(x) = x^8 + x^4 + x^3 + x + 1$ for the underlying finite field $F = \mathbb{Z}_2[x]/(p(x))$. As we did in Section 1.6, we will establish that the polynomial is irreducible using the user-written function **Irreduc**, which we have written separately from this MATLAB session and saved as the M-file Irreduc.m.

```
>> syms x
>> p = @(x) x^8+x^4+x^3+x+1;
>> Irreduc(p(x), 2)

ans =

TRUE
```

Next, we will use $p(x)$ to construct a polynomial form of the AES S-box given in Table 10.1. This polynomial form of the AES S-box will be a 16×16 matrix containing as its entries the polynomials that can be viewed as representations of the binary expressions of the entries in Table 10.1. To construct this polynomial form of the AES S-box, we will use the user-written function **sboxtable**, which we have written separately from this MATLAB session and saved as the M-file sboxtable.m. The following command illustrates how the function **sboxtable** can be used. Specifically, by entering the following command, we construct the polynomial form of the AES S-box, a 16×16 matrix stored as the first entry *Sbx* in the vector that results from the command, and the polynomial form of the inverse of *Sbx* over $F = \mathbb{Z}_2[x]/(p(x))$, a 16×16 matrix stored as the second entry *inSbx* in the vector that results from the command. Due to the size of these matrices, we will suppress the output of this command.

```
>> [Sbx inSbx] = sboxtable(p(x), x);
```

The second parameter in the preceding command is the variable x used in the first parameter, the polynomial $p(x)$.

For a given polynomial $f(x)$ in $F = \mathbb{Z}_2[x]/(p(x))$, we will need to be able to determine the output of the S-box transformation that corresponds to the input $f(x)$. Also, for a given polynomial $g(x)$ in F, we will need to be able to determine the input into the S-box transformation that has corresponding output $g(x)$. To do these operations, we will use the user-written function **sbox**, which we have written separately from this MATLAB session and saved as the M-file sbox.m. The following command illustrates how the function **sbox** can be used. Specifically, by entering the following command, we determine the output of the S-box transformation that corresponds to the input $x^7 + x^3 + 1$. By using *Sbx* as the second parameter in this command, we are specifying that we wish to determine the output of the S-box transformation that corresponds to the input given by the first parameter.

```
>> sbox(x^7+x^3+1, Sbx, p(x), x)

ans =

x^7 + x^5 + x^2 + x + 1
```

Also, by entering the next command, we determine the input into the S-box transformation that has corresponding output $x^7 + x^5 + x^2 + x + 1$. By using *inSbx* as the second parameter in this command, we are specifying that we wish to determine the input into the S-box transformation that has corresponding output given by the first parameter.

```
>> sbox(x^7+x^5+x^2+x+1, inSbx, p(x), x)

ans =

x^7 + x^3 + 1
```

Now, suppose that we wish to use the AES algorithm to encrypt the message **RENDEZVOUS AT SIX**. First, we will store this message as the following string.

```
>> message = 'RENDEZVOUS AT SIX';
>> message(findstr(message, ' ')) = []

message =

RENDEZVOUSATSIX
```

Recall that the AES encryption algorithm is designed to encrypt plaintext messages that contain 16 characters. To expand our plaintext message **RENDEZVOUS AT SIX** so that it contains exactly 16 letters, we will use the user-written function **mblocks**, which we have written separately from this MATLAB session and saved as the M-file mblocks.m. The following command illustrates how the function **mblocks** can be used. Specifically, by entering the following command, we expand *message* by adding an extra letter A so that the result contains exactly 16 letters.

```
>> ptext = mblocks(message, 16)

ptext =

RENDEZVOUSATSIXA
```

The second parameter in the preceding command specifies the total number of letters that should be used to express the message string given by the first parameter. If the number of letters in this message string is less than the number of letters specified by the second parameter, the message string will be expanded by adding extra letters A repeatedly to the end of the message until the result contains exactly the number of letters specified by the second parameter. If the number of letters in this message string is greater than the number of letters specified by the second parameter, the message string will be divided into blocks that contain exactly the number of letters specified by the second parameter, with the last block expanded by adding extra letters A repeatedly to the end of the block until the result contains exactly the number of letters specified by the second parameter.

Next, we will convert each letter in *ptext* into its polynomial equivalent using the table that is shown in Example 10.1. To do this, we will use the user-written function **topoly**, which we have written separately from this MATLAB session and saved as the M-file topoly.m. The following command shows how the function **topoly** can be used. Specifically, by entering the following, we convert each letter in *ptext* into its polynomial equivalent using the table in Example 10.1. The second parameter in this command is the variable to be used in the resulting polynomials.

```
>> ptext = topoly(ptext, x)

ptext =

[ x^4 + 1, x^2, x^3 + x^2 + 1, x + 1, x^2, x^4 + x^3 + 1,
  x^4 + x^2 + 1, x^3 + x^2 + x, x^4 + x^2, x^4 + x, 0,
  x^4 + x + 1, x^4 + x, x^3, x^4 + x^2 + x + 1, 0]
```

Using the MATLAB **reshape** function, we can arrange these polynomials in order as the entries in the columns of a 4×4 matrix A as follows.

```
>> A = reshape(ptext, 4, 4)

A =

[       x^4+1,           x^2,      x^4+x^2,          x^4+x]
[         x^2,     x^4+x^3+1,        x^4+x,            x^3]
[   x^3+x^2+1,     x^4+x^2+1,            0,    x^4+x^2+x+1]
[         x+1,     x^3+x^2+x,      x^4+x+1,              0]
```

Now, suppose we wish to construct a key matrix for the initial round in the AES encryption process with a 128-bit initial key using the keyword RIJNDAEL AES KEY. In the following sequence of commands, we assign and expand this keyword so that it contains exactly 16 letters, convert each letter in the result into its polynomial equivalent using the table in Example 10.1, and arrange the resulting polynomials in order as the entries in the columns of a 4×4 key matrix $K0$ for the initial round in the process.

```
>> key = 'RIJNDAEL AES KEY';
>> key(findstr(key, ' ')) = [];
>> key = mblocks(key, 16)

key =

RIJNDAELAESKEYAA

>> K0 = reshape(topoly(key,x), 4, 4)

K0 =

[       x^4+1,         x+1,            0,            x^2]
[         x^3,           0,          x^2,      x^4+x^3]
[       x^3+1,         x^2,        x^4+x,              0]
[   x^3+x^2+1,     x^3+x+1,        x^3+x,              0]
```

Recall that for the AES algorithm, the key schedule is used to generate a series of columns, beginning with the columns in an initial key matrix, which are used to form the key matrices for each of the rounds in the AES encryption and decryption processes. To generate this series of columns, we will use the user-written function **keysched**,[4] which we have written separately from this MATLAB session and saved as the M-file keysched.m. The following command shows how the function **keysched** can be used.

---

[4]The function **keysched** uses the user-written function **sbox** that we mentioned previously in this section.

Specifically, by entering the following command, we construct a matrix *KS* which contains as its columns the series of columns, beginning with the columns in the initial key matrix *K0*, which would be used to form the key matrices for the rounds in the AES encryption and decryption processes. In this command, the first parameter is the initial key matrix *K0*, and the S-box matrix *Sbx* is included as a parameter because, recall, the key schedule uses the S-box. Also, due to the size of the resulting matrix *KS*, we will suppress the output of this command.

```
>> KS = keysched(K0, Sbx, p(x), x);
```

With this matrix *KS*, we can now see the key matrix that would be used in any of the rounds in the AES encryption or decryption processes. For example, recall that the fifth, sixth, seventh, and eighth columns of *KS* are the columns in the key matrix that would be used in the first round after the initial round in the AES encryption process. To see these columns, we can enter the following commands.

```
>> KS(:, 5:6)

ans =

[       x^7+x^5+x^4+x^3+x^2+1,    x^7+x^5+x^4+x^3+x^2+x]
[           x^6+x^5+x^3+x+1,        x^6+x^5+x^3+x+1]
[           x^6+x^5+x^3+x,         x^6+x^5+x^3+x^2+x]
[ x^7+x^6+x^5+x^4+x^3+x^2+x+1,    x^7+x^6+x^5+x^4+x^2]

>> KS(:, 7:8)

ans =

[     x^7+x^5+x^4+x^3+x^2+x,        x^7+x^5+x^4+x^3+x]
[       x^6+x^5+x^3+x^2+x+1,      x^6+x^5+x^4+x^2+x+1]
[       x^6+x^5+x^4+x^3+x^2,       x^6+x^5+x^4+x^3+x^2]
[ x^7+x^6+x^5+x^4+x^3+x^2+x,  x^7+x^6+x^5+x^4+x^3+x^2+x]
```

We can assign these columns as the key matrix *K1* that would be used in the first round after the initial round in the AES encryption process by entering the following command. Due to the size of this matrix, we will suppress the output of this command.

```
>> K1 = KS(:, 5:8);
```

10.8.2   The Encryption Layers

We will now show how MATLAB can be used to apply the AES encryption layers. We will specifically demonstrate these layers in the first round after the initial round in the AES encryption process using the matrices A and $K0$ that we constructed previously in this MATLAB session.

To complete the initial round in the AES encryption process, we would form the sum of the matrices $K0$ and A, with the calculations done in $F = \mathbb{Z}_2[x]/(p(x))$. To perform this operation, we will use the user-written function **addrkey**, which we have written separately from this MATLAB session and saved as the M-file addrkey.m. The following command illustrates how the function **addrkey** can be used. Specifically, by entering the following command, we form the sum of the matrices $K0$ and A, with the calculations done in F, and assign the resulting matrix as $A0$.

```
>> AO = addrkey(KO, A)

AO =

[          0,    x^2+x+1,       x^4+x^2,      x^4+x^2+x]
[    x^3+x^2,  x^4+x^3+1,     x^4+x^2+x,            x^4]
[        x^2,      x^4+1,         x^4+x,  x^4+x^2+x+1]
[  x^3+x^2+x,      x^2+1,    x^4+x^3+1,              0]
```

We are now ready to begin the first round after the initial round in the AES encryption process with $A0$ as the input matrix into the first layer.

We would begin the first round after the initial round in the AES encryption process by applying the ByteSub transformation to the matrix $A0$ that was output at the end of the initial round. To perform this operation, we will use the user-written function **bytesub**,[5] which we have written separately from this MATLAB session and saved as the M-file bytesub.m. The following command shows how the function **bytesub** can be used. Specifically, by entering the following command, we apply the ByteSub transformation to the matrix $A0$, and assign the resulting matrix as $B1$. In this command, the S-box matrix Sbx is included as a parameter because, recall, the ByteSub transformation uses the S-box. Also, due to the size of $B1$, we will suppress the output of this command.

```
>> B1 = bytesub(AO, Sbx, p(x), x);
```

---

[5]The function **bytesub** uses the user-written function **sbox** that we mentioned previously in this section.

Next, we would apply ShiftRow to this matrix *B1*. To do this operation, we will use the user-written function **shiftrow**, which we have written separately from this MATLAB session and saved as the M-file shiftrow.m. The following command illustrates how the function **shiftrow** can be used. Specifically, by entering the following command, we apply the ShiftRow transformation to the matrix *B1*, and assign the resulting matrix as *C1*. Due to the size of *C1*, we will suppress the output of this command.

```
>> C1 = shiftrow(B1);
```

Next, we would apply MixColumn to this matrix *C1*, with the calculations done in $F = \mathbb{Z}_2[x]/(p(x))$. To do this operation, we will use the user-written function **mixcolumn**, which we have written separately from this MAT-LAB session and saved as the M-file mixcolumn.m. The following command illustrates how the function **mixcolumn** can be used. Specifically, by entering the following command, we apply the MixColumn transformation to the matrix *C1*, and assign the resulting matrix as *D1*. Due to the size of *D1*, we will suppress the output of this command.

```
>> D1 = mixcolumn(C1, p(x), x);
```

Next, we would add the key matrix *K1* that we constructed previously in this MATLAB session to this matrix *D1*, with the calculations done in $F = \mathbb{Z}_2[x]/(p(x))$. To perform this operation, we will use the user-written function **addrkey** that we mentioned previously in this MATLAB session. In the following command we add *K1* to *D1*, with the calculations done in *F*, and assign the resulting matrix as *A1*. Due to the size of *A1*, we will suppress the output of this command.

```
>> A1 = addrkey(K1, D1);
```

This completes the first round after the initial round in the AES encryption process.

10.8.3 The Decryption Layers

We will now show how MATLAB can be used to apply the AES decryption layers. We will specifically demonstrate these layers by beginning with the matrix *A1* that we just constructed as the output of the first round after the initial round in the AES encryption process, and recovering the plaintext from which it resulted.

We would begin by applying InvMixColumn to the matrix *A1*, with the calculations done in $F = \mathbb{Z}_2[x]/(p(x))$. To do this operation, we will use the user-written function **invmixcolumn**, which we have written separately from this MATLAB session and saved as the M-file invmixcolumn.m. The following command illustrates how the function **invmixcolumn** can be used. Specifically, by entering the following command, we apply the InvMixColumn transformation to the matrix *A1*, and assign the resulting matrix as *IMCA1*. Due to the size of *IMCA1*, we will suppress the output of this command.

```
>> IMCA1 = invmixcolumn(A1, p(x), x);
```

Next, to apply InvAddRoundKey to this matrix *IMCA1*, we will use the user-written function **invaddrkey**, which we have written separately from this MATLAB session and saved as the M-file invaddrkey.m. The following command illustrates how the function **invaddrkey** can be used. Specifically, by entering the following command, we apply InvAddRound-Key to the matrix *IMCA1*, and assign the resulting matrix as *C1*. In this command, the matrix *K1* is included as a parameter. This is because, recall, the matrix that would be added to *IMCA1* under InvAdd-RoundKey would be the product of the matrix M^{-1} given on page 375 and *K1*. Also, due to the size of *C1*, we will suppress the output of this command.

```
>> C1 = invaddrkey(IMCA1, K1, p(x), x);
```

Next, we would apply the InvByteSub transformation to this matrix *C1*. To perform this operation, we will use the user-written function **bytesub** that we mentioned previously in this MATLAB session. In the following command, we apply InvByteSub to *C1*, and assign the resulting matrix as *B1*. In this command, the inverse *inSbx* of the S-box matrix *Sbx* is included as a parameter because, recall, the InvByteSub transformation uses the inverse of the S-box. Also, due to the size of *B1*, we will suppress the output of this command.

```
>> B1 = bytesub(C1, inSbx, p(x), x);
```

Next, we would apply InvShiftRow to this matrix *B1*. To do this, we will use the user-written function **invshiftrow**, which we have written separately from this MATLAB session and saved as the M-file invshiftrow.m. The following command shows how the function **invshiftrow** can be used. Specifically, by entering the following command, we apply the InvShift-Row transformation to the matrix *B1*, and assign the resulting matrix as *A0*.

```
>> A0 = invshiftrow(B1)
```

```
A0 =

[            0,     x^2+x+1,       x^4+x^2,     x^4+x^2+x]
[      x^3+x^2,   x^4+x^3+1,     x^4+x^2+x,           x^4]
[          x^2,       x^4+1,       x^4+x,   x^4+x^2+x+1]
[    x^3+x^2+x,       x^2+1,   x^4+x^3+1,             0]
```

Next, we would add the key matrix *K0* that we constructed previously in this MATLAB session to this matrix *A0*, with the calculations done in $F = \mathbb{Z}_2[x]/(p(x))$. To perform this operation, we will use the user-written function **addrkey** that we mentioned previously in this MATLAB session. In the following command we add *K0* to *A0*, with the calculations done in F, and assign the resulting matrix as A.

```
>> A = addrkey(K0, A0)

A =

[        x^4+1,            x^2,     x^4+x^2,         x^4+x]
[          x^2,     x^4+x^3+1,       x^4+x,           x^3]
[    x^3+x^2+1,     x^4+x^2+1,           0,   x^4+x^2+x+1]
[          x+1,     x^3+x^2+x,     x^4+x+1,             0]
```

To recover the plaintext from which this matrix A was constructed, we will first use the MATLAB **reshape** function as follows to arrange the polynomial entries in A reading in order down the columns of A into a single list.

```
>> ptext = reshape(A, 1, numel(A))

ptext =

[ x^4 + 1, x^2, x^3 + x^2 + 1, x + 1, x^2, x^4 + x^3 + 1,
  x^4 + x^2 + 1, x^3 + x^2 + x, x^4 + x^2, x^4 + x, 0,
  x^4 + x + 1, x^4 + x, x^3, x^4 + x^2 + x + 1, 0]
```

Finally, we will use the table in Example 10.1 to convert these polynomials into the letters of which they can be viewed as representations. To do this, we will use the user-written function **frompoly**, which we have written separately from this MATLAB session and saved as the M-file frompoly.m. The following command illustrates how the function **frompoly** can be used. Specifically, by entering the following command, we use the table in Example 10.1 to convert each polynomial entry in *ptext* into the letter of which it can be viewed as a representation.

```
>> ptext = frompoly(ptext, x)

ptext =

RENDEZVOUSATSIXA
```

The second parameter in the preceding command is the variable that is used in the entries in the first parameter.

10.8.4   Encryption and Decryption

We will now show how MATLAB can be used to perform the entire AES encryption and decryption processes.

We will first demonstrate these processes by encrypting the message TRANSFER THE MONEY using the AES encryption process with the 128-bit keyword AES KEYS ARE SECURE. We begin with the following sequence of commands, in which we assign this message and keyword as *ptext1* and *key16*, respectively, and use the user-written function **mblocks** that we mentioned previously in this MATLAB session to verify that they each contain exactly 16 letters.

```
>> ptext1 = 'TRANSFER THE MONEY';
>> ptext1(findstr(ptext1, ' ')) = [];
>> ptext1 = mblocks(ptext1, 16)

ptext1 =

TRANSFERTHEMONEY

>> key16 = 'AES KEYS ARE SECURE';
>> key16(findstr(key16, ' ')) = [];
>> key16 = mblocks(key16, 16)

key16 =

AESKEYSARESECURE
```

To encrypt the plaintext message *ptext1*, we will use the user-written function **aesencipher**,[6] which we have written separately from this MATLAB

---

[6]The function **aesencipher** uses the user-written functions **topoly**, **keysched**, **addrkey**, **bytesub**, **shiftrow**, and **mixcolumn** that we mentioned previously in this section.

session and saved as the M-file aesencipher.m. The following command illustrates how the function **aesencipher** can be used. Specifically, by entering the following command, we encrypt *ptext1* using the AES encryption process with the keyword *key16*. In this command, the S-box matrix *Sbx* is included as a parameter because, recall, the AES encryption process uses the S-box. Also, the fifth parameter in this command is the variable x used in the fourth parameter, the polynomial $p(x)$.

```
>> ctext1 = aesencipher(ptext1, key16, Sbx, p(x), x)

ctext1 =

  Columns 1 through 11

    63   254    12   222     9   109   165    12   208   244    90

  Columns 12 through 16

   148   190   172    17     4
```

The output produced by the preceding command is a list of integers whose binary expressions can be represented by the polynomial entries in the matrix output at the end of the final round in the AES encryption process.

To decrypt the ciphertext message *ctext1*, we will use the user-written function **aesdecipher**,[7] which we have written separately from this MATLAB session and saved as the M-file aesdecipher.m. The following command illustrates how the function **aesdecipher** can be used. Specifically, by entering the following command, we decrypt *ctext1* using the AES decryption process with the keyword *key16*. In this command, the S-box matrix *Sbx* and its inverse *inSbx* are included as parameters because, recall, the AES decryption process uses both the S-box and the inverse of the S-box.

```
>> aesdecipher(ctext1, key16, Sbx, inSbx, p(x), x)

ans =

TRANSFERTHEMONEY
```

---

[7]The function **aesdecipher** uses the user-written functions **topoly**, **keysched**, **addrkey**, **bytesub**, **invshiftrow**, **invmixcolumn**, and **invaddrkey** that we mentioned previously in this section.

Recall that the AES encryption process is designed to use initial keys of lengths 128, 192, or 256 bits. In the following sequence of commands, we show how the user-written functions **mblocks**, **aesencipher**, and **aesdecipher** can be used to encrypt and decrypt *ptext1* using a keyword of length 192 bits, or, equivalently, 24 letters.

```
>> key24 = 'AES IS MORE SECURE THAN DES';
>> key24(findstr(key24, ' ')) = [];
>> key24 = mblocks(key24, 24)

key24 =

AESISMORESECURETHANDESAA

>> ctext1 = aesencipher(ptext1, key24, Sbx, p(x), x)

ctext1 =

  Columns 1 through 11

   114    74    72    43   232   116    69    94   216   191    80

  Columns 12 through 16

    34   131    81   220   126

>> aesdecipher(ctext1, key24, Sbx, inSbx, p(x), x)

ans =

TRANSFERTHEMONEY
```

Recall that the AES algorithm is designed to encrypt messages that contain exactly 16 characters. Thus, we have expanded messages that are shorter than 16 letters so that they contain exactly 16 letters before being encrypted. For messages that are longer than 16 letters, the AES algorithm can still be used, of course; we just encrypt only 16 plaintext letters at a time. In the following sequence of commands, we show how the user-written functions **mblocks** and **aesencipher** can be used to encrypt the message TRANSFER ONE MILLION DOLLARS FROM STOCKS TO BONDS using the keyword *key16*.

```
>> ptext2 = ['TRANSFER ONE MILLION DOLLARS FROM STOCKS' ...
   'TO BONDS'];
```

```
>> ptext2(findstr(ptext2, ' ')) = [];
>> ptext2 = mblocks(ptext2, 16)

ptext2 =

  'TRANSFERONEMILLI' 'ONDOLLARSFROMSTO' 'CKSTOBONDSAAAAAA'
>> ctext2 = [];
>> for i = 1:numel(ptext2)
        ctext2 = [ctext2; aesencipher(char(ptext2(i)), ...
        key16, Sbx, p(x), x)];
   end
>> ctext2

ctext2 =

  Columns 1 through 11

    40  160  254  160  157   64  227   62  199  221  215
    65    1  106  251  129   88  115  108   52   13   39
    82   28  245  229   60  190  131  229   61  232   15

  Columns 12 through 16

   116   34  164   99  129
   208  218  190  132  111
   173    5   91   10  210
```

The resulting matrix *ctext2* contains as its rows the lists of integers whose binary expressions can be represented by the polynomial entries in the matrix output at the end of the final round of the AES encryption process for the plaintext consisting of the first, second, and third blocks of 16 letters in the original message, respectively. In the following sequence of commands, we show how the user-written function **aesdecipher** can be used to decrypt *ctext2* using the keyword *key16*.

```
>> ptext2 = [];
>> for i = 1:size(ctext2, 1)
        ptext2 = [ptext2; aesdecipher(ctext2(i, 1:16), ...
        key16, Sbx, inSbx, p(x), x)];
   end
>> ptext2

ptext2 =
```

```
TRANSFERONEMILLI
ONDOLLARSFROMSTO
CKSTOBONDSAAAAAA

>> reshape(ptext2', 1, numel(ptext2))

ans =

TRANSFERONEMILLIONDOLLARSFROMSTOCKSTOBONDSAAAAAA
```

Exercises

1. For the polynomial $f(x) = x^4 + x^3$, find the output of the S-box transformation that corresponds to the input $f(x)$ both by using Table 10.1 and by using (10.1).

2. For the polynomial $g(x) = x^7 + x^5 + x^3 + x + 1$, find the input into the S-box transformation that has corresponding output $g(x)$ both by using Table 10.1 and by using (10.2).

3. As a continuation of Example 10.8, construct the key matrix K_2 that would be used in the second round after the initial round in the AES encryption process.

4. Suppose that you wish to encrypt a message using the AES encryption process with a 192-bit initial key using the keyword LAISSEZ LES BON TEMPS ROULER.

    (a) Construct the initial key matrix K for the process.
    (b) Construct the key matrix K_0 that would be used in the initial round in the process.
    (c) Construct the key matrix K_1 that would be used in the first round after the initial round in the process.
    (d) Construct the key matrix K_2 that would be used in the second round after the initial round in the process.

5. Suppose that you wish to encrypt a message using the AES encryption process with a 256-bit initial key using the keyword THE SEA WAS ANGRY THAT DAY MY FRIENDS.

    (a) Construct the initial key matrix K for the process.
    (b) Construct the key matrix K_0 that would be used in the initial round in the process.

(c) Construct the key matrix K_1 that would be used in the first round after the initial round in the process.

(d) Construct the key matrix K_2 that would be used in the second round after the initial round in the process.

(e) Construct the key matrix K_3 that would be used in the third round after the initial round in the process.

6. Suppose that you wish to encrypt the message ROCHELLE ROCHELLE using the AES encryption process with a 128-bit initial key using the keyword FROM MILAN TO MINSK.

   (a) Construct the plaintext matrix A for the process.

   (b) Construct the key matrix K_0 that would be used in the initial round in the process.

   (c) Construct the key matrix K_1 that would be used in the first round after the initial round in the process.

   (d) Construct the key matrix K_2 that would be used in the second round after the initial round in the process.

   (e) Construct the matrix A_0 that would be output at the end of the initial round in the process.

   (f) Construct the matrix B_1 that would be output by ByteSub in the first round after the initial round in the process.

   (g) Construct the matrix C_1 that would be output by ShiftRow in the first round after the initial round in the process.

   (h) Construct the matrix D_1 that would be output by MixColumn in the first round after the initial round in the process.

   (i) Construct the matrix A_1 that would be output at the end of the first round after the initial round in the process.

7. As a continuation of Exercise 6, construct the matrix A_2 that would be output at the end of the second round after the initial round in the AES encryption process.

8. As a continuation of Example 10.13, construct the matrix A_2 that would be output at the end of the second round after the initial round in the AES encryption process.

9. Suppose that a 128-bit plaintext was encrypted by applying the initial round and first subsequent round in the AES encryption process using the 128-bit keyword RIJNDAEL AES KEY AA, with the output

being the following matrix A_1, in which we have expressed the entries using the bytes of which the polynomial entries can be viewed as representations.

$$A_1 = \begin{bmatrix} 11101001 & 01111011 & 01110110 & 00010111 \\ 11010100 & 00000101 & 11110110 & 00110011 \\ 00000111 & 11001111 & 01011001 & 00101110 \\ 11011111 & 01101111 & 00100001 & 10101010 \end{bmatrix}$$

Recover the plaintext that was encrypted by using the following steps.

(a) Construct the matrix $M^{-1}A_1$ that would result from applying InvMixColumn to A_1.

(b) Construct the matrix C_1 that would result from applying InvAddRoundKey to $M^{-1}A_1$.

(c) Construct the matrix B_1 that would result from applying InvByteSub to C_1.

(d) Construct the matrix A_0 that would result from applying InvShiftRow to B_1.

(e) Construct the plaintext matrix A that would result from applying AddRoundKey to A_0.

(f) Recover the plaintext that was encrypted.

10. Suppose that a 128-bit plaintext was encrypted by applying the initial round and first and second subsequent rounds in the AES encryption process using the 128-bit keyword RIJNDAEL AES KEY AA, with the output being the following matrix A_2, in which we have expressed the entries using the bytes of which the polynomial entries can be viewed as representations.

$$A_2 = \begin{bmatrix} 10101100 & 11011111 & 00001001 & 10100100 \\ 01011001 & 01111000 & 00100110 & 00010000 \\ 11000110 & 00000111 & 01001010 & 00010100 \\ 11001010 & 11001000 & 00000100 & 11010111 \end{bmatrix}$$

Recover the plaintext that was encrypted.

11. Show that the inverse of the matrix S in (10.1) over $\mathbb{Z}_2$ is the matrix S^{-1} given on page 365.

12. Using (6.1), determine the inverse of the matrix M in (10.7) over $F = \mathbb{Z}_2[x]/(p(x))$, and note that the result is the matrix M^{-1} given on page 375.

Computer Exercises

13. Suppose that you wish to encrypt the message NEED MORE COWBELL using the AES encryption process with a 128-bit initial key using the keyword BLUE OYSTER CULT.

    (a) Construct the matrix A_1 that would be output at the end of the first round after the initial round in the process.

    (b) Beginning with the matrix A_1 in part (a), show how the AES decryption layers can be used to recover the plaintext.

    (c) Construct the matrix A_2 that would be output at the end of the second round after the initial round in the process.

    (d) Beginning with the matrix A_2 in part (c), show how the AES decryption layers can be used to recover the plaintext.

14. Encrypt the message MARINE BIOLOGIST using the AES encryption process ...

    (a) ... with a 128-bit initial key.

    (b) ... with a 192-bit initial key.

    (c) ... with a 256-bit initial key.

15. Encrypt the message LIKE AN OLD MAN TRYING TO SEND BACK SOUP IN A DELI using the AES encryption process ...

    (a) ... with a 128-bit initial key.

    (b) ... with a 192-bit initial key.

    (c) ... with a 256-bit initial key.

16. Decrypt the following ciphertexts, each of which is a list of integers whose binary expressions can be represented by the polynomial entries in the matrix output at the end of the final round in the AES encryption process.

    (a) The ciphertext 4, 236, 215, 152, 180, 163, 61, 11, 183, 58, 230, 28, 189, 147, 173, 43, which was formed using the 128-bit keyword YOU ARE A RICH MAN AA.

    (b) The ciphertext 97, 58, 97, 135, 48, 98, 141, 208, 41, 54, 40, 50, 145, 216, 153, 183, which was formed using the 192-bit keyword YOU ARE A RICHER MAN THAN MOST.

    (c) The ciphertext 196, 242, 93, 186, 255, 54, 53, 0, 101, 63, 117, 33, 44, 30, 107, 57, which was formed using the 256-bit keyword YOU ARE RICHER THAN MOST ON EARTH AAAAA.

17. Decrypt the following ciphertexts, each of which is a list of integers whose binary expressions can be represented by the polynomial entries in the matrix output at the end of the final round in the AES encryption process for the first, second, and third blocks of 16 letters in the plaintext, in order.

    (a) The ciphertext 239, 108, 18, 31, 8, 187, 67, 3, 91, 217, 167, 193, 230, 49, 111, 46, 71, 177, 164, 177, 229, 92, 20, 248, 151, 113, 184, 102, 27, 197, 227, 202, 113, 253, 254, 18, 249, 117, 220, 191, 32, 43, 234, 58, 224, 71, 139, 35, which was formed using the 128-bit keyword CRYPTOGRAPHY MAN A.

    (b) The ciphertext 56, 14, 106, 177, 213, 136, 25, 55, 165, 194, 96, 105, 66, 204, 89, 226, 29, 112, 191, 0, 56, 206, 54, 53, 248, 114, 234, 65, 9, 121, 125, 147, 47, 45, 37, 144, 194, 191, 169, 85, 134, 225, 220, 222, 87, 111, 103, 92, which was formed using the 192-bit keyword CODE BREAKING IN WAR TIME AAA.

    (c) The ciphertext 172, 157, 75, 231, 68, 31, 59, 163, 55, 83, 0, 142, 86, 239, 251, 227, 201, 210, 117, 112, 173, 160, 199, 220, 194, 72, 214, 240, 20, 18, 229, 101, 202, 128, 185, 179, 239, 249, 202, 231, 251, 217, 148, 99, 32, 223, 122, 2, which was formed using the 256-bit keyword PEOPLE CAN INSPIRE AND MENTOR OTHERS A.

18. The AES system is an example of a *symmetric* cipher system, meaning that the originator and intended recipient of a message would have to have agreed upon or have some secure way to agree upon an initial key for the system. To overcome this problem, two people wishing to exchange a secret message across an insecure line of communication using the AES system, but having no secure way to agree upon an initial key for the system, could begin by using a public-key cryptosystem to agree upon an initial key.

    (a) Suppose a 16-character keyword is split into blocks with four characters each, and then encrypted using the RSA cryptosystem with primes $p = 7927$ and $q = 129793$ and encryption exponent $a = 365109131$, resulting in the ciphertext 1011730103, 386581625, 94657192, 0. Find this keyword.

    (b) Decrypt the ciphertext 28, 86, 218, 125, 91, 163, 244, 133, 149, 21, 161, 52, 71, 24, 175, 72, 248, 221, 30, 124, 21, 116, 125, 18, 207, 2, 230, 172, 127, 12, 195, 154, which is a list of integers whose binary expressions can be represented by the polynomial entries in the matrix output at the end of the final round in the AES encryption process for the first and second blocks of 16 letters in the plaintext, in order, formed using a 128-bit initial key resulting from the keyword that is the answer to part (a).

(c) Suppose a 16-character keyword is split into blocks containing four characters each, and encrypted using the ElGamal cryptosystem with the group $\mathbb{Z}_p^*$ under multiplication modulo prime $p = 809094427$, giving four ciphertexts, each one formed using $n = 676943009$: $(y_1, z_1) = (417687279, 731640305)$, $(y_2, z_2) = (290850181, 522248088)$, $(y_3, z_3) = (41125975, 786985104)$, and $(y_4, z_4) = (614715416, 0)$. Find this keyword.

(d) Decrypt the ciphertext 192, 92, 218, 224, 6, 164, 197, 124, 183, 14, 235, 217, 248, 159, 153, 186, 125, 8, 136, 85, 139, 222, 224, 127, 155, 98, 15, 191, 214, 26, 86, 172, 182, 127, 207, 201, 207, 153, 114, 40, 163, 99, 40, 174, 128, 49, 71, 251, which is a list of integers whose binary expressions can be represented by the polynomial entries in the matrix output at the end of the final round in the AES encryption process for the first, second, and third blocks of 16 letters in the plaintext, in order, formed using a 128-bit initial key resulting from the keyword that is the answer to part (c).

19. Recall from Exercise 18 that the AES system is an example of a symmetric cipher system. Modify the user-written functions and other commands in Sections 10.7 or 10.8 so that they can be used to implement the AES system with some public-key cryptosystem used to securely agree upon an initial key for the system.

20. Recall that the AES algorithm is designed to encrypt messages that contain 128 bits. The original Rijndael algorithm that was submitted as a candidate to be AES actually allowed plaintexts that contained 128, 192, or 256 bits, each of which could be encrypted using initial keys of lengths 128, 192, or 256 bits. Let c_1 represent the number of columns in a plaintext matrix (always designed to have four rows) for the original Rijndael algorithm, and let c_2 represent the number of columns in an initial key matrix (also always designed to have four rows) for the original Rijndael algorithm. Then the number r of rounds after the initial round in the original Rijndael algorithm is shown in the following table.

	$c_1 = 4$	$c_1 = 6$	$c_1 = 8$
$c_2 = 4$	10	12	14
$c_2 = 6$	12	12	14
$c_2 = 8$	14	14	14

Also, the number of positions the entries in input matrices were to be shifted by the ShiftRow transformation in the original Rijndael algorithm depended on c_1. This number of positions to be shifted in the original Rijndael algorithm is shown in the following table.

	First Row	Second Row	Third Row	Fourth Row
$c_1 = 4$	0	1	2	3
$c_1 = 6$	0	1	2	3
$c_1 = 8$	0	1	3	4

In addition, the size of the matrices to be input into and output from each of the layers in the original Rijndael algorithm was $4 \times c_1$. Using all of these facts, modify the user-written functions and other commands in Sections 10.7 or 10.8 so that they can be used to encrypt and decrypt messages using the original Rijndael algorithm.

Research Exercises

21. Investigate the lives and careers of Joan Daemen and Vincent Rijmen, the creators of the Rijndael algorithm for whom the algorithm is named, and write a summary of your findings.

22. ByteSub, ShiftRow, and MixColumn were each included in Rijndael for specific reasons. The key schedule was constructed in its way for a specific reason as well. Investigate each of these reasons, and write a summary of your findings.

23. Investigate some cryptanalytic attacks against AES, and write a summary of your findings.

24. Investigate some techniques for securely exchanging the key for AES besides the method presented in Exercise 18, and write a summary of your findings.

25. Investigate one or more real-life uses of AES, and write a summary of your findings.

26. Rijndael was selected from five finalists to be the Advanced Encryption Standard. Investigate the other finalists, including who created them and why they were not selected over Rijndael, and write a summary of your findings.

27. Recall that AES was created as a replacement for the outdated DES. Investigate DES, including its development, how it functions, and some of the reasons why a replacement was necessary, and write a summary of your findings. Include in your summary some information about the controversy surrounding the S-boxes used in DES.

Chapter 11

Pólya Theory

In this chapter, we will present some results for counting equivalence classes when a group acts on a set. Because the most celebrated result that we will present in this chapter is the Pólya Enumeration Theorem, we will refer to the theory in this chapter as *Pólya theory*.

11.1 Group Actions

We begin with an example of the type of problem we will consider in this chapter. Suppose we wish to construct a necklace with four colored beads, and each bead can be blue or green. Assuming the beads can be rotated around the necklace, and that the necklace can be flipped over, then how many different necklaces can we construct? To answer this, suppose we stretch the necklace into the shape of a square with one bead at each corner. The following shows the set X of 16 possible arrangements for the beads.

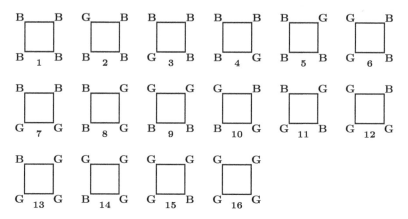

413

Of course, not all of these possible arrangements for the beads represent different necklaces. By rotating the beads around the necklace, we can see that the arrangements in X labeled as 2 through 5 all represent the same necklace. Similarly, by flipping the necklace over, we can see that the arrangements in X labeled as 7 and 9 represent the same necklace as well (which we can also see by rotating the beads around the necklace). These operations of rotating the beads around the necklace and flipping the necklace over are called *rigid motions* of the necklace. We can also view these rigid motions as motions of the following single figure, or, more specifically, motions of the set S of corners, or *vertices*, of this figure.

Note that each rigid motion of the necklace permutes the elements in X and the elements in S. Thus, we can represent these rigid motions by their permutations on X or S. We will use the representations as permutations on S to answer questions like the one posed at the beginning of this section. The advantage to using the representations as permutations on S rather than X is that while there are 16 elements in X, there are only four elements in S. This reduction would be much more dramatic and important if the necklace contained more beads or if there were more colors available for the beads. For example, if the necklace contained four beads but there were five colors available, then there would be $5^4 = 625$ possible arrangements of the beads, but still only four vertices of the preceding general figure.

To formalize this discussion, recall that the set of rigid motions of a square with the composition operation forms a group (see Example 1.3). In the following table, we list the elements in this group G with their representations as permutations on S expressed using cycle notation (rotations are counterclockwise). Note that in this table we include all cycles of length one. The reason for this will be apparent later in this chapter.

Element in G	Permutation on S
$\pi_1 = 0°$ rotation (identity)	$(1)(2)(3)(4)$
$\pi_2 = 90°$ rotation	(1234)
$\pi_3 = 180°$ rotation	$(13)(24)$
$\pi_4 = 270°$ rotation	(1432)
$\pi_5 = $ reflection across horizontal	$(12)(34)$
$\pi_6 = $ reflection across vertical	$(14)(23)$
$\pi_7 = $ reflection across 1–3 diagonal	$(1)(24)(3)$
$\pi_8 = $ reflection across 2–4 diagonal	$(13)(2)(4)$

Representing the elements in G as permutations on X would, in general, require much more extensive notation. For example, the 90° counterclockwise rotation in G would be represented as a permutation on X by $(1)(2,3,4,5)(6,7,8,9)(10,11)(12,13,14,15)(16)$. Since it is true that each element $\pi_i \in G$ corresponds to unique permutations on S and X, in this chapter we will often write π_i when we mean to refer to one of these permutations. For example, by writing π_2, we could mean the $90°$ counterclockwise rotation in G, the permutation (1234) on S, or the permutation $(1)(2,3,4,5)(6,7,8,9)(10,11)(12,13,14,15)(16)$ on X. The context will make it clear which one we intend.

We will now formalize our necklace example using the terminology of group actions. Let S be a collection of objects, and let R be a set of elements called *colors* (not necessarily colors in the usual sense). A *coloring* of S by R is an assignment of a unique color to each element in S. That is, a coloring of S by R can be viewed as a function $f : S \to R$. Note that if $|S| = n$ and $|R| = m$, then there will be m^n distinct colorings of S by R. We will denote this set of all possible colorings of S by R as X. The set X of 16 possible arrangements for the beads in our necklace example is the set of 16 possible colorings of $S = \{$vertices of a square$\}$ by $R = \{$blue, green$\}$.

Now, consider a group G and a set Y. An *action* of G on Y is a mapping from $G \times Y$ to Y that satisfies the following two properties.

1. $(g_1 g_2)(y) = g_1(g_2(y))$ for all $g_1, g_2 \in G$ and $y \in Y$.

2. $e(y) = y$ for all $y \in Y$, where e represents the identity element in G.

In our necklace example, the group $G = \{$rigid motions of a square$\}$ acts on both $S = \{$vertices of a square$\}$ and $X = \{$colorings of S by $R\}$, where $R = \{$blue, green$\}$. As illustrated in this example, when a group G acts on a set Y, each element in G can be represented as a permutation on Y.

Lemma 11.1 *Suppose that a group G acts on a set Y, and for any elements $x, y \in Y$, define $x \sim y$ if there exists $g \in G$ for which $g(x) = y$. Then $\sim$ is an equivalence relation.*

Proof. Exercise. (See Chapter 3 Exercise 24 for the definition of an equivalence relation.) □

As a consequence of Lemma 11.1, when a group acts on a set, the set is decomposed into equivalence classes of elements that can be mapped to each other by elements in the group. These equivalence classes are called *orbits*. For a set of colorings, these orbits are also called *patterns*. The general type of problem of interest to us in this chapter can be viewed as counting the number of patterns when a group acts on a set of colorings.

In summary, suppose S is a set, R is a set of colors, and X is the set of all possible colorings of S by R. When a group G acts on S, then G also acts on

X by $\pi(f(x)) = f(\pi(x))$ for all $\pi \in G$, $f \in X$, and $x \in S$. A pair of colorings $f, g \in X$ are equivalent if there exists $\pi \in G$ such that $\pi(f(a)) = g(a)$ for all $a \in S$. Thus, two of the 16 colorings in our necklace example are equivalent if there is a rigid motion of a square that maps one to the other. To answer the question that began our necklace example, we must only count the number of patterns under this equivalence relation. From the set of 16 colorings shown at the beginning of this section, we can easily see that there are six such patterns: $\{1\}$, $\{2, 3, 4, 5\}$, $\{6, 7, 8, 9\}$, $\{10, 11\}$, $\{12, 13, 14, 15\}$, and $\{16\}$. With a total of only 16 possible arrangements for the beads, it is not necessary to consider the group action of G on S or X to count the patterns. However, it would certainly not be practical to list all of the possible arrangements for the beads if the necklace had 10 beads and there were 12 colors available for each bead. In this chapter, we will consider how the idea of a group action can be used to count patterns without having to actually construct the patterns themselves.

11.2   Burnside's Theorem

Our goal in this chapter is to count the number of patterns when a group acts on a set of colorings. Counting the number of orbits when a group acts on a set is the focus of a fundamental result due to W. Burnside. Before establishing this result, we will first define some additional terms.

Suppose a group G acts on a set Y. Then for each $\pi \in G$, we denote the elements in Y fixed by π as $\mathrm{Fix}(\pi)$. That is, $\mathrm{Fix}(\pi) = \{y \in Y \mid \pi(y) = y\}$.

Example 11.1 Consider $G = \{\text{rigid motions of a square}\}$ acting on X in our necklace example. Using the notation π_i that we defined previously for the elements in G, and the enumeration from the beginning of Section 11.1 for the colorings in X, we list $\mathrm{Fix}(\pi_i)$ for each $\pi_i \in G$ in the following table.

$\pi_i \in G$	$\mathrm{Fix}(\pi_i)$	$|\mathrm{Fix}(\pi_i)|$
π_1	X	16
π_2	1, 16	2
π_3	1, 10, 11, 16	4
π_4	1, 16	2
π_5	1, 6, 8, 16	4
π_6	1, 7, 9, 16	4
π_7	1, 2, 4, 10, 11, 13, 15, 16	8
π_8	1, 3, 5, 10, 11, 12, 14, 16	8

□

Suppose again that a group G acts on a set Y. Then for each element $y \in Y$, we denote the subgroup of elements in G that fix y as $\mathrm{Stab}(y)$. That is, $\mathrm{Stab}(y) = \{\pi \in G \mid \pi(y) = y\}$.

Example 11.2 Consider again $G = \{$rigid motions of a square$\}$ acting on X in our necklace example. Again using the notation π_i for the elements in G, and the enumeration from the beginning of Section 11.1 for the colorings in X, we list $\mathrm{Stab}(x)$ for each $x \in X$ in the following table.

$x \in X$	$\mathrm{Stab}(x)$	$|\mathrm{Stab}(x)|$
1	G	8
2	π_1, π_7	2
3	π_1, π_8	2
4	π_1, π_7	2
5	π_1, π_8	2
6	π_1, π_5	2
7	π_1, π_6	2
8	π_1, π_5	2
9	π_1, π_6	2
10	$\pi_1, \pi_3, \pi_7, \pi_8$	4
11	$\pi_1, \pi_3, \pi_7, \pi_8$	4
12	π_1, π_8	2
13	π_1, π_7	2
14	π_1, π_8	2
15	π_1, π_7	2
16	G	8

$\square$

Note that the sums of the entries in the $|\mathrm{Fix}(\pi_i)|$ column in Example 11.1 and the $|\mathrm{Stab}(x)|$ column in Example 11.2 are both 48. The equality of these sums is guaranteed in general by the following lemma.

Lemma 11.2 *Suppose that a group G acts on a set Y. Then it will follow that* $\sum_{\pi \in G} |\mathrm{Fix}(\pi)| = \sum_{y \in Y} |\mathrm{Stab}(y)|.$

Proof. Exercise. (Hint: Let $S = \{(\pi, y) \mid \pi \in G, \ y \in Y, \ \pi(y) = y\}$, and count $|S|$ in two ways, first by ranging through the possibilities for π, and then by ranging through the possibilities for y.) $\square$

Suppose again that a group G acts on a set Y. Then for each element $y \in Y$, we denote the equivalence class, or orbit, containing y as $\mathrm{Orb}(y)$. That is, $\mathrm{Orb}(y) = \{x \in Y \mid x = \pi(y) \text{ for some } \pi \in G\}$.

Lemma 11.3 *Suppose that a group G acts on a set Y. Then for each $y \in Y$ it will follow that $|G| = |\mathrm{Stab}(y)| \cdot |\mathrm{Orb}(y)|$.*

Proof. Suppose that g_1 and g_2 are in the same left coset of $\mathrm{Stab}(y)$. Then $g_1 = g_2\pi$ for some $\pi \in \mathrm{Stab}(y)$, and $g_1(y) = (g_2\pi)(y) = g_2(\pi(y)) = g_2(y)$.

On the other hand, suppose that $g_1(y) = g_2(y)$ for some $g_1, g_2 \in G$. Then $y = (g_1^{-1}g_2)(y)$, and so $g_1^{-1}g_2 \in \text{Stab}(y)$. Therefore, $g_1^{-1}g_2 = \pi$ for some $\pi \in \text{Stab}(y)$. From this it follows that $g_2 = g_1\pi$, and so g_1 and g_2 are in the same left coset of $\text{Stab}(y)$. In summary, g_1 and g_2 are in the same left coset of $\text{Stab}(y)$ if and only if $g_1(y) = g_2(y)$. Thus, there is a bijection between the left cosets of $\text{Stab}(y)$ and the elements in $\text{Orb}(y)$. Using this and Lagrange's Theorem (Theorem 1.4), we can conclude the following.

$$\begin{aligned} |G| &= |\text{Stab}(y)| \cdot (\text{number of left cosets of } \text{Stab}(y)) \\ &= |\text{Stab}(y)| \cdot |\text{Orb}(y)| \end{aligned}$$

$\square$

We will now establish the following fundamental result due to W. Burnside for counting the number of orbits when a group acts on a set.

Theorem 11.4 (Burnside's Theorem) *Suppose that a group G acts on a set Y. Then the number of orbits in Y is* $\dfrac{1}{|G|} \displaystyle\sum_{\pi \in G} |Fix(\pi)|$.

Proof. By Lemma 11.2, we know $\dfrac{1}{|G|} \displaystyle\sum_{\pi \in G} |\text{Fix}(\pi)| = \dfrac{1}{|G|} \displaystyle\sum_{y \in Y} |\text{Stab}(y)|$.

In addition, by Lemma 11.3, we know $\dfrac{1}{|G|} \displaystyle\sum_{y \in Y} |\text{Stab}(y)| = \displaystyle\sum_{y \in Y} \dfrac{1}{|\text{Orb}(y)|}$.

Say there are s orbits in Y, denoted by $O_1, O_2, \ldots, O_s$. Then for $x, y \in O_i$, it follows that $|\text{Orb}(x)| = |\text{Orb}(y)| = |O_i|$. We then have the following.

$$\sum_{y \in O_i} \frac{1}{|\text{Orb}(y)|} = \frac{1}{|O_i|} + \frac{1}{|O_i|} + \cdots + \frac{1}{|O_i|} = 1$$

Thus, $\displaystyle\sum_{y \in Y} \dfrac{1}{|\text{Orb}(y)|} = s$, and the result follows. $\square$

To see how Burnside's Theorem can be applied, consider again the group $G = \{\text{rigid motions of a square}\}$ acting on X in our necklace example. From Example 11.1, we know that $\displaystyle\sum_{\pi \in G} |\text{Fix}(\pi)| = 48$. Then since $|G| = 8$, it follows by Theorem 11.4 that the number of orbits in X is $48/8 = 6$. That is, as we have already noted, there are 6 distinct necklaces in our necklace example. While this result is certainly correct, in practice the result from applying Burnside's Theorem is not usually determined exactly in this manner. Specifically, in practice the value of $\displaystyle\sum_{\pi \in G} |\text{Fix}(\pi)|$ is not usually determined by actually constructing the sets $\text{Fix}(\pi)$ as we did in Example 11.1. To construct the table in Example 11.1, we made reference to the list of all possible arrangements for the necklace beads shown at the beginning of Section 11.1. However, recall that in general we would like to be able to count orbits without having to actually construct all of these possible arrangements. We will present a method for doing this next.

11.3 The Cycle Index

In our necklace example, consider the motion $\pi_7 = $ reflection across 1–3 diagonal. Note that if $x \in \text{Fix}(\pi_7)$, then x must have the same color bead at vertices 2 and 4, but can have any color bead at vertices 1 and 3. Thus, since there are two colors available for the beads, there will be $2 \cdot 2 \cdot 2 = 8$ colorings fixed by π_7. So we see that we can determine $|\text{Fix}(\pi_7)| = 8$ without having to make reference to the list of all possible arrangements for the necklace beads shown at the beginning of Section 11.1. Also, we could determine $|\text{Fix}(\pi_7)|$ in the exact same way if there were more than two colors available for the beads. For example, if there were five colors available for the beads instead of just two, then there would be $5 \cdot 5 \cdot 5 = 125$ colorings fixed by π_7. Note that $|\text{Fix}(\pi_7)|$ depends only on the number of colors that are available for the beads and the number of collections of vertices that can take arbitrary colors. Specifically, if there are a colors available for the beads, then $|\text{Fix}(\pi_7)|$ will be equal to a^3.

To generalize this discussion, suppose π is a rigid motion for which there are k collections of vertices that can take arbitrary colors. Then if there are a colors available for the beads, it follows that $|\text{Fix}(\pi)|$ will be equal to a^k. Furthermore, it would be easy to see the number of collections of vertices that could take arbitrary colors under π from the representation of π as a permutation on the set S of vertices. For example, recall in our necklace example, the rigid motion π_7 can be represented as the permutation $(1)(24)(3)$ on S. Since there are three disjoint cycles in this representation for π_7, there will be three factors of a in $|\text{Fix}(\pi_7)|$. More generally, if there are k disjoint cycles in the representation of π as a permutation on S (including all cycles of length one) and there are a colors available for the beads, then $|\text{Fix}(\pi)|$ will be equal to a^k. This relates to our discussion in Section 11.2 in that it provides us with a way to use Burnside's Theorem for counting orbits without having to actually construct all of the possible arrangements for the beads like we did at the beginning of Section 11.1. Specifically, it states that the sum in the formula in Burnside's Theorem can be expressed as $\sum_{\pi \in G} |\text{Fix}(\pi)| = \sum_{\pi \in G} a^{k_\pi}$, where k_π is the number of disjoint cycles in the representation of π as a permutation on S.

Example 11.3 Consider $G = \{$rigid motions of a square$\}$ acting on S in our necklace example. From the table in Section 11.1 in which we listed the elements in G along with their representations as permutations on S expressed using cycle notation, we can see that the number of disjoint cycles in the representations of π_i for $i = 1, 2, \ldots, 8$ as permutations on S are 4, 1, 2, 1, 2, 2, 3, and 3, respectively. Thus, since there are two colors available for the beads in this example, we have the following.

$$\sum_{\pi_i \in G} |\text{Fix}(\pi_i)| = 2^4 + 2^1 + 2^2 + 2^1 + 2^2 + 2^2 + 2^3 + 2^3 = 48$$

Also, if there were five colors available for the beads instead of just two, then we would have the following.

$$\sum_{\pi_i \in G} |\text{Fix}(\pi_i)| = 5^4 + 5^1 + 5^2 + 5^1 + 5^2 + 5^2 + 5^3 + 5^3 = 960$$

Thus, if there were five colors available for the beads instead of just two, according to Burnside's Theorem the number of distinct necklaces that we could construct would be $\dfrac{1}{|G|} \displaystyle\sum_{\pi_i \in G} |\text{Fix}(\pi_i)| = \dfrac{1}{8} \cdot 960 = 120.$ □

This process for determining $\displaystyle\sum_{\pi \in G} |\text{Fix}(\pi)|$ can be refined as follows. Suppose $\pi \in G$ is a rigid motion for which the representation of π as a permutation on S is the product of disjoint cycles of lengths $i_1, i_2, \ldots, i_t$ (including all cycles of length one). We then associate with π the monomial $f_\pi = x_{i_1} x_{i_2} \cdots x_{i_t}$. For example, since the rigid motion π_7 in our necklace example can be represented as $(1)(24)(3)$ on S, we would associate with π_7 the monomial $f_{\pi_7} = x_1 x_2 x_1 = (x_1)^2 x_2$. We then define the *cycle index* $f(x_1, x_2, \ldots, x_w)$ of G acting on S as follows, where w is the length of the longest cycle in the representation of any $\pi \in G$ as a permutation on S.

$$f(x_1, x_2, \ldots, x_w) = \frac{1}{|G|} \sum_{\pi \in G} f_\pi$$

The cycle index is of interest to us because of the following theorem, which states how it can be used to count orbits when a group acts on a set.

Theorem 11.5 *Suppose that S is a set, R is a set of colors, and X is the set of all possible colorings of S by R. If a group G acts on S with cycle index $f(x_1, x_2, \ldots, x_w)$, and $|R| = a$, then the number of patterns in X under the corresponding action of G on X is given by $f(a, a, \ldots, a)$.*

Proof. Exercise. □

Example 11.4 Consider $G = \{$rigid motions of a square$\}$ in our necklace example. We list the monomial f_{π_i} for each $\pi_i \in G$ in the following table.

$\pi_i \in G$	f_{π_i}
$(1)(2)(3)(4)$	$(x_1)^4$
(1234)	x_4
$(13)(24)$	$(x_2)^2$
(1432)	x_4
$(12)(34)$	$(x_2)^2$
$(14)(23)$	$(x_2)^2$
$(1)(24)(3)$	$(x_1)^2 x_2$
$(13)(2)(4)$	$(x_1)^2 x_2$

The resulting cycle index is the following.

$$f(x_1, x_2, x_3, x_4) = \frac{1}{8} \left(x_1^4 + 2x_4 + 3x_2^2 + 2x_1^2 x_2 \right)$$

Thus, if there were two colors available for the beads, then we could construct $f(2, 2, 2, 2) = \frac{1}{8}(16 + 4 + 12 + 16) = 6$ distinct necklaces. Also, if there were five colors available for the beads instead of just two, then we could construct $f(5, 5, 5, 5) = \frac{1}{8}(625 + 10 + 75 + 250) = 120$ distinct necklaces. $\square$

Example 11.5 Suppose that we wish to construct a necklace with six colored beads instead of just four, again assuming that the beads can be rotated around the necklace, and that the necklace can be flipped over. In this example, we will use a cycle index to determine the number of distinct necklaces that we could construct in this case with a specified number of colors available for each bead. To do this, suppose that we stretch the necklace into the shape of a regular hexagon with one bead at each corner. Consider the following resulting general shape for the necklace.

Let G be the group of rigid motions of a regular hexagon, and let S be the set of vertices of the preceding general figure. In the following table, we list the elements $\pi \in G$ along with their representations as permutations on S expressed using cycle notation (rotations are counterclockwise) and the associated monomials f_π.

$\pi \in G$	Permutation on S	f_π
0° rotation	$(1)(2)(3)(4)(5)(6)$	$(x_1)^6$
60° rotation	(123456)	x_6
120° rotation	$(135)(246)$	$(x_3)^2$
180° rotation	$(14)(25)(36)$	$(x_2)^3$
240° rotation	$(153)(264)$	$(x_3)^2$
300° rotation	(165432)	x_6
reflection across horizontal	$(13)(2)(46)(5)$	$(x_1)^2(x_2)^2$
reflection across vertical	$(16)(25)(34)$	$(x_2)^3$
reflection across 1–4 diagonal	$(1)(26)(35)(4)$	$(x_1)^2(x_2)^2$
reflection across 3–6 diagonal	$(15)(24)(3)(6)$	$(x_1)^2(x_2)^2$
reflection across diagonal	$(12)(36)(45)$	$(x_2)^3$
reflection across diagonal	$(14)(23)(56)$	$(x_2)^3$

The resulting cycle index is the following.

$$f(x_1, x_2, x_3, x_4, x_5, x_6) = \frac{1}{12}\left(x_1^6 + 2x_6 + 2x_3^2 + 4x_2^3 + 3x_1^2 x_2^2\right)$$

Thus, if there were two colors available for the beads, we could construct $f(2, 2, 2, 2, 2, 2) = \frac{1}{12}(64 + 4 + 8 + 32 + 48) = 13$ distinct necklaces. Also, if five colors were available for the beads instead of two, we could construct $f(5, 5, 5, 5, 5, 5) = \frac{1}{12}(15625 + 10 + 50 + 500 + 1875) = 1505$ necklaces. $\quad\square$

11.4   The Pattern Inventory

From Example 11.5, we see we could construct 1505 distinct six-bead necklaces if five colors were available for the beads. We now consider how related questions can be answered. For example, how many of these 1505 necklaces would contain beads that use only four of the five possible colors? More specifically, if the colors available were blue, green, red, white, and yellow, then how many of these 1505 distinct six-bead necklaces would contain exactly one blue bead, one red bead, three white beads, and one yellow bead? In this section, we present a method for answering such questions.

Suppose that S is a set, R is the set $\{C_1, C_2, \ldots, C_t\}$ of colors, and X is the set of all possible colorings of S by R. If a group acts on S with cycle index $f(x_1, x_2, \ldots, x_w)$, then the following simplified symbolic expression is called the *pattern inventory* of X.

$$f(C_1 + C_2 + \cdots + C_t, \; C_1^2 + C_2^2 + \cdots + C_t^2, \; \ldots, \; C_1^w + C_2^w + \cdots + C_t^w)$$

As stated in the following theorem, which is named for G. Pólya, the pattern inventory of X allows us to answer questions like the ones that we posed at the beginning of this section.

Theorem 11.6 (Pólya Enumeration Theorem) *Suppose the monomial* $k\, C_1^{i_1} C_2^{i_2} \cdots C_t^{i_t}$ *appears in the pattern inventory of X. Then there are k patterns in X in which C_1 appears i_1 times, C_2 appears i_2 times, ... , and C_t appears i_t times.*

We will include a partial verification of this theorem later, but first we will show how it can be applied in our four-bead necklace example.

Example 11.6 Consider our four-bead necklace example with the following cycle index.

$$f(x_1, x_2, x_3, x_4) = \frac{1}{8}\left(x_1^4 + 2x_4 + 3x_2^2 + 2x_1^2 x_2\right)$$

Suppose each bead can be either $\mathcal{B} = $ blue or $\mathcal{G} = $ green. Then the pattern inventory of the set X of all colorings of $S = \{$vertices of a square$\}$ by $R = \{\mathcal{B}, \mathcal{G}\}$ is given by the following.

$$f(\mathcal{B} + \mathcal{G}, \mathcal{B}^2 + \mathcal{G}^2, \mathcal{B}^3 + \mathcal{G}^3, \mathcal{B}^4 + \mathcal{G}^4)$$

$$= \frac{1}{8}\left((\mathcal{B} + \mathcal{G})^4 + 2(\mathcal{B}^4 + \mathcal{G}^4) + 3(\mathcal{B}^2 + \mathcal{G}^2)^2 + 2(\mathcal{B} + \mathcal{G})^2(\mathcal{B}^2 + \mathcal{G}^2)\right)$$

$$= \mathcal{B}^4 + \mathcal{B}^3\mathcal{G} + 2\mathcal{B}^2\mathcal{G}^2 + \mathcal{B}\mathcal{G}^3 + \mathcal{G}^4$$

From this pattern inventory, we can easily see the number of distinct four-bead necklaces we could construct with prescribed numbers of blue and green beads. For example, because $\mathcal{B}\mathcal{G}^3$ appears in this pattern inventory with a coefficient of 1 and exponents of 1 on $\mathcal{B}$ and 3 on $\mathcal{G}$, we could only construct one distinct four-bead necklace with one blue bead and three green beads. Also, because $2\mathcal{B}^2\mathcal{G}^2$ appears in this pattern inventory with a coefficient of 2 and exponents of 2 on both $\mathcal{B}$ and $\mathcal{G}$, we could construct two distinct four-bead necklaces with two blue beads and two green beads. Now, suppose each bead could also be $\mathcal{R} = $ red. Then the pattern inventory of the set of all colorings of S by $R = \{\mathcal{B}, \mathcal{G}, \mathcal{R}\}$ is given by the following.

$$f(\mathcal{B} + \mathcal{G} + \mathcal{R}, \mathcal{B}^2 + \mathcal{G}^2 + \mathcal{R}^2, \mathcal{B}^3 + \mathcal{G}^3 + \mathcal{R}^3, \mathcal{B}^4 + \mathcal{G}^4 + \mathcal{R}^4)$$

$$= \frac{1}{8}\left((\mathcal{B} + \mathcal{G} + \mathcal{R})^4 + 2(\mathcal{B}^4 + \mathcal{G}^4 + \mathcal{R}^4) + 3(\mathcal{B}^2 + \mathcal{G}^2 + \mathcal{R}^2)^2\right.$$
$$\left. + 2(\mathcal{B} + \mathcal{G} + \mathcal{R})^2(\mathcal{B}^2 + \mathcal{G}^2 + \mathcal{R}^2)\right)$$

$$= \mathcal{B}^4 + \mathcal{B}^3\mathcal{G} + 2\mathcal{B}^2\mathcal{G}^2 + \mathcal{B}\mathcal{G}^3 + \mathcal{G}^4 + \mathcal{B}^3\mathcal{R} + 2\mathcal{B}^2\mathcal{R}^2 + \mathcal{B}\mathcal{R}^3 + \mathcal{R}^4$$
$$+ \mathcal{G}^3\mathcal{R} + 2\mathcal{G}^2\mathcal{R}^2 + \mathcal{G}\mathcal{R}^3 + 2\mathcal{B}^2\mathcal{G}\mathcal{R} + 2\mathcal{B}\mathcal{G}^2\mathcal{R} + 2\mathcal{B}\mathcal{G}\mathcal{R}^2$$

From this pattern inventory, we can easily see the number of distinct four-bead necklaces that we could construct with prescribed numbers of blue, green, and red beads. For example, because the term $2\mathcal{BGR}^2$ appears in this pattern inventory, then we could construct two distinct four-bead necklaces with one blue bead, one green bead, and two red beads. Note also that by adding the coefficients of all of the terms in this pattern inventory, we can see that if there were three colors available for the beads, then we could construct a total of 21 distinct four-bead necklaces. In addition, note that by adding the coefficients of just the last ten terms in this pattern inventory, we can see that 15 of these 21 distinct four-bead necklaces would contain at least one red bead. Finally, note that each term in the two-color pattern inventory is in the three-color pattern inventory, as we would expect.   □

A pattern inventory can be used to answer questions like the ones that we posed at the beginning of this section. Specifically, consider our six-bead necklace example with the following cycle index.

$$f(x_1, x_2, x_3, x_4, x_5, x_6) = \frac{1}{12}\left(x_1^6 + 2x_6 + 2x_3^2 + 4x_2^3 + 3x_1^2 x_2^2\right)$$

Suppose that each bead can be colored either $\mathcal{B}$ = blue, $\mathcal{G}$ = green, $\mathcal{R}$ = red, $\mathcal{W}$ = white, or $\mathcal{Y}$ = yellow. We showed in Example 11.5 that with these five colors available for the beads, we could construct 1505 distinct six-bead necklaces. Of these 1505 distinct necklaces, the number that would contain exactly one blue bead, one red bead, three white beads, and one yellow bead would be the coefficient of $\mathcal{BRW}^3\mathcal{Y}$ in the following pattern inventory.

$$f(\mathcal{B} + \mathcal{G} + \mathcal{R} + \mathcal{W} + \mathcal{Y}, \ldots, \mathcal{B}^6 + \mathcal{G}^6 + \mathcal{R}^6 + \mathcal{W}^6 + \mathcal{Y}^6)$$

Of course, it would not be particularly easy to construct this pattern inventory by hand. However, with the help of a computer, this pattern inventory would be very easy to construct.

We will close this section with a discussion of why the Pólya Enumeration Theorem is true.[1] Rather than including a formal proof of the theorem, which would be complicated and not intuitive, we will give an informal discussion of why the theorem is true with two colors. This discussion can be generalized in an obvious way to apply for more than two colors.

Suppose that S is a set of s vertices, R is the set $\{\mathcal{B}$ = blue, $\mathcal{G}$ = green$\}$ of colors, and X is the set of all possible colorings of S by R. Suppose also that a group G of rigid motions acts on S with cycle index $f(x_1, x_2, \ldots, x_w)$. If the representation of $\pi \in G$ as a permutation on S expressed using cycle notation includes only a single cycle of length s, then for an element in X

---

[1]Because this discussion of why the Pólya Enumeration Theorem is true is extensive, the reader may wish to postpone this discussion until completing the remainder of this chapter, or skip this discussion altogether, as understanding why the Pólya Enumeration Theorem is true is not a prerequisite for the remainder of this chapter or the subsequent chapters.

to be fixed by π, each of the s vertices in the single cycle must be assigned the same color. We will keep a record of this by writing $\mathcal{B}^s + \mathcal{G}^s$, which we would interpret as representing the fact that for a coloring to be fixed by π, the s vertices in the single cycle must be all blue or all green. For example, with $\pi_2 = (1234)$ in our four-bead necklace example, we would write $\mathcal{B}^4 + \mathcal{G}^4$, representing the fact that for a coloring to be fixed by π_2, the four vertices in the cycle (1234) must be all blue or all green.

Now, suppose that the representation of $\pi \in G$ as a permutation on S expressed using cycle notation contains two disjoint cycles of lengths s_1 and s_2 with $s_1 + s_2 = s$. Then for an element in X to be fixed by π, each of the s_1 vertices in the cycle of length s_1 must be assigned the same color, and each of the s_2 vertices in the cycle of length s_2 must also be assigned the same color. We will keep a record of this by writing $(\mathcal{B}^{s_1} + \mathcal{G}^{s_1})(\mathcal{B}^{s_2} + \mathcal{G}^{s_2})$, representing the fact that for a coloring to be fixed by π, the s_1 vertices in the cycle of length s_1 must be all blue or all green (thus the first factor), while the s_2 vertices in the cycle of length s_2 must also be all blue or all green (thus the second factor). Note that if we expand this expression symbolically for the variables $\mathcal{B}$ and $\mathcal{G}$, we obtain the following.

$$(\mathcal{B}^{s_1} + \mathcal{G}^{s_1})(\mathcal{B}^{s_2} + \mathcal{G}^{s_2}) \;=\; \mathcal{B}^{s_1+s_2} + \mathcal{B}^{s_1}\mathcal{G}^{s_2} + \mathcal{B}^{s_2}\mathcal{G}^{s_1} + \mathcal{G}^{s_1+s_2}$$

We would interpret the terms on the right side of this equation as representing the fact that for a coloring to be fixed by π, either the s vertices in either cycle must be all blue (thus the first term), the s_1 vertices in the cycle of length s_1 must be all blue and the s_2 vertices in the cycle of length s_2 must be all green (thus the second term), the s_1 vertices in the cycle of length s_1 must be all green and the s_2 vertices in the cycle of length s_2 must be all blue (thus the third term), or the s vertices in either cycle must be all green (thus the fourth term). For example, with $\pi_3 = (13)(24)$ in our four-bead necklace example, we would write the following.

$$(\mathcal{B}^2 + \mathcal{G}^2)(\mathcal{B}^2 + \mathcal{G}^2) \;=\; \mathcal{B}^4 + \mathcal{B}^2\mathcal{G}^2 + \mathcal{B}^2\mathcal{G}^2 + \mathcal{G}^4$$

The first and last terms on the right side of this equation indicate that the colorings in our four-bead necklace example with four blue beads and four green beads are fixed by π_3. The middle two terms indicate that there are also two colorings in our four-bead necklace example with two blue beads and two green beads that are fixed by π_3.

More generally, suppose the representation of $\pi \in G$ as a permutation on S using cycle notation contains j disjoint cycles of lengths $s_1, s_2, \ldots, s_j$ with $s_1 + s_2 + \cdots + s_j = s$. Then for an element in X to be fixed by π, each of the s_i vertices in the cycle of length s_i must be assigned the same color for $i = 1, 2, \ldots, j$. We will keep a record of this by writing the following.

$$(\mathcal{B}^{s_1} + \mathcal{G}^{s_1})(\mathcal{B}^{s_2} + \mathcal{G}^{s_2}) \cdots (\mathcal{B}^{s_j} + \mathcal{G}^{s_j}) \tag{11.1}$$

We would interpret (11.1) as representing the fact that for a coloring to be fixed by π, the s_i vertices in the cycle of length s_i must be all blue or all green for $i = 1, 2, \ldots, j$. Recall for the cycle index of G acting on S, we would associate with π the monomial $f_\pi = x_{s_1} x_{s_2} \cdots x_{s_j}$. Note that (11.1) can be viewed as $f_\pi(\mathcal{B}+\mathcal{G}, \mathcal{B}^2+\mathcal{G}^2, \ldots, \mathcal{B}^t+\mathcal{G}^t)$, where t is the length of the longest cycle in the representation of π as a permutation on S. As we have demonstrated, if we expand (11.1) symbolically for the variables $\mathcal{B}$ and $\mathcal{G}$, each term in the expansion represents a coloring fixed by π, in which the distribution of colors is given by the bases, and the number of vertices by the exponents. Thus, if we simplify this symbolic expansion by combining terms that are similar, the coefficient of each resulting term will show the total number of colorings fixed by π, in which the distribution of colors is given by the bases, and the number of vertices is given by the exponents. Finally, if we sum over all $\pi \in G$ and combine similar terms, the coefficient of each resulting term will show the total number of colorings fixed by any $\pi \in G$, in which the distribution of colors is given by the bases, and the number of vertices is given by the exponents.

We now claim that if the monomial $k\,\mathcal{B}^{i_1}\mathcal{G}^{i_2}$ appears in the pattern inventory of X, there will be k patterns in X in which $\mathcal{B}$ appears i_1 times and $\mathcal{G}$ appears i_2 times. To see this, let Y be the subset of X that contains all of the colorings in which $\mathcal{B}$ appears i_1 times and $\mathcal{G}$ i_2 times. Since G acts on Y, Burnside's Theorem states the following, with G acting on Y.

$$\text{Number of patterns in } Y = \frac{1}{|G|} \sum_{\pi \in G} |\text{Fix}(\pi)| \tag{11.2}$$

It is precisely this number of patterns that we wish to determine. As we have shown, for any $\pi \in G$, the coefficients in the symbolic expansion of $f_\pi(\mathcal{B} + \mathcal{G}, \mathcal{B}^2 + \mathcal{G}^2, \ldots, \mathcal{B}^t + \mathcal{G}^t)$, where t is the length of the longest cycle in the representation of π as a permutation on S, show the total number of colorings fixed by π (i.e., $|\text{Fix}(\pi)|$). Thus, if we combine the $\mathcal{B}^{i_1}\mathcal{G}^{i_2}$ terms in the symbolic expansion of $f_\pi(\mathcal{B} + \mathcal{G}, \mathcal{B}^2 + \mathcal{G}^2, \ldots, \mathcal{B}^t + \mathcal{G}^t)$ for all $\pi \in G$, the coefficient of the resulting term will be the sum in (11.2). If we then divide this coefficient by $|G|$, this will show the number of patterns in Y. More generally, we can find the number of patterns for all distributions of the colors $\mathcal{B}$ and $\mathcal{G}$ by combining the similar terms in the symbolic expansion of $f_\pi(\mathcal{B}+\mathcal{G}, \mathcal{B}^2+\mathcal{G}^2, \ldots, \mathcal{B}^t+\mathcal{G}^t)$ for all $\pi \in G$ and dividing the resulting terms by $|G|$. The coefficients of the resulting expression will show the numbers of patterns in X for all distributions of the colors $\mathcal{B}$ and $\mathcal{G}$, in which the distributions of colors are given by the bases, and the numbers of vertices by the exponents. Finally, note that adding $f_\pi(\mathcal{B}+\mathcal{G}, \mathcal{B}^2+\mathcal{G}^2, \ldots, \mathcal{B}^t+\mathcal{G}^t)$ for all $\pi \in G$ and dividing the resulting terms by $|G|$ will yield exactly the cycle index $f(x_1, x_2, \ldots, x_w)$ of G acting on S evaluated at the inputs

$\mathcal{B} + \mathcal{G}, \mathcal{B}^2 + \mathcal{G}^2, \ldots, \mathcal{B}^w + \mathcal{G}^w$, respectively. Since this is exactly how we defined the pattern inventory of X, the result is shown.

11.5 The Pattern Inventory with Maple

In this section, we will show how Maple can be used to count patterns and construct pattern inventories. We will consider our six-bead necklace example in Example 11.5 with the following cycle index.

$$ f(x_1, x_2, x_3, x_4, x_5, x_6) = \frac{1}{12} \left(x_1^6 + 2x_6 + 2x_3^2 + 4x_2^3 + 3x_1^2 x_2^2 \right) $$

To find this cycle index and the cycle index in similar examples, we will use the user-written function **cidn**, which we have written separately from this Maple session and saved as the text file cidn.mpl. To use this function, we must first read it into our Maple session as follows.

```
>   read "cidn.mpl";
```

The following command then illustrates how the function **cidn** can be used. Specifically, by entering the following command, we find the cycle index for our six-bead necklace example in terms of the variable x.

```
>   f := cidn(6, x);
```

$$ f := \frac{1}{12} x_1^6 + \frac{1}{3} x_2^3 + \frac{1}{6} x_3^2 + \frac{1}{6} x_6 + \frac{1}{4} x_1^2 x_2^2 $$

We can then use the Maple **unapply** function as follows to convert this expression for f into a function.

```
>   f := unapply(f, seq(x[i], i=1..6));
```

$$ f := (x\_1, x\_2, x\_3, x\_4, x\_5, x\_6) \rightarrow \frac{1}{12} x\_1^6 + \frac{1}{3} x\_2^3 + \frac{1}{6} x\_3^2 $$
$$ + \frac{1}{6} x\_6 + \frac{1}{4} x\_1^2 x\_2^2 $$

Although the preceding command causes the variables that are displayed as output for f to be changed from x_i to x_i, this has no effect on how f can be used. For example, to find the number of distinct six-bead necklaces that we could construct if there were two colors available for the beads, we can enter the following command.

```
>   f(2, 2, 2, 2, 2, 2);
```
$$ 13 $$

Thus, as we noted in Example 11.5, we could construct 13 distinct six-bead necklaces if there were two colors available for the beads. Suppose that the colors available for the beads were $\mathcal{B}$ = blue and $\mathcal{G}$ = green. To see how many of these 13 distinct six-bead necklaces would contain prescribed numbers of blue and green beads, we form the following pattern inventory.

```
> simplify(f(B+G, B^2+G^2, B^3+G^3, B^4+G^4, B^5+G^5,
    B^6+G^6));
```

$$B^6 + B^5G + 3\,B^4G^2 + 3\,B^3G^3 + 3\,B^2G^4 + BG^5 + G^6$$

Thus, for example, because the term $3\,B^4G^2$ appears in this pattern inventory, then we could construct three distinct six-bead necklaces with four blue beads and two green beads.

Now, suppose that each bead could also be $\mathcal{R}$ = red. To find the number of distinct six-bead necklaces that we could construct if there were three colors available for the beads, we can enter the following command.

```
> f(seq(3, i=1..6));
```
$$92$$

Thus, we could construct 92 distinct six-bead necklaces if there were three colors available for the beads. To see how many of these 92 distinct six-bead necklaces would contain prescribed numbers of blue, green, and red beads, we form the following pattern inventory.

```
> pinv := simplify(f(seq(B^i+G^i+R^i, i=1..6)));
```

$$\begin{aligned}
pinv := \; & B^6 + B^5G + B^5R + 3\,B^4G^2 + 3\,B^4GR + 3\,B^4R^2 + 3\,B^3G^3 \\
& + 6\,B^3G^2R + 6\,B^3GR^2 + 3\,B^3R^3 + 3\,B^2G^4 + 6\,B^2G^3R \\
& + 11\,B^2G^2R^2 + 6\,B^2GR^3 + 3\,B^2R^4 + BG^5 + 3\,BG^4R + 6\,BG^3R^2 \\
& + 6\,BG^2R^3 + 3\,BGR^4 + BR^5 + G^6 + G^5R + 3\,G^4R^2 + 3\,G^3R^3 \\
& + 3\,G^2R^4 + GR^5 + R^6
\end{aligned}$$

Note that each term in the two-color pattern inventory is in the three-color pattern inventory. To help with finding information from pattern inventories containing a larger number of terms, we will use the user-written function **cpinv**, which we have written separately from this Maple session and saved as the text file cpinv.mpl. To use this function, we must first read it into our Maple session as follows.

```
> read "cpinv.mpl";
```

The following command illustrates how we can use the **cpinv** function to find that we could construct six distinct six-bead necklaces with two blue beads, one green bead, and three red beads.

```
> cpinv(pinv, B^2*G*R^3);
```
$$6$$

Finally, suppose that each bead could also be $\mathcal{W} =$ white. To find the number of distinct six-bead necklaces that we could construct if there were four colors available for the beads, we can enter the following command.

```
>  f(seq(4, i=1..6));
```
$$430$$

Thus, we could construct 430 distinct six-bead necklaces if there were four colors available for the beads. To see how many of these 430 distinct six-bead necklaces would contain prescribed numbers of blue, green, red, and white beads, we form the following pattern inventory.

```
>  pinv := simplify(f(seq(B^i+G^i+R^i+W^i, i=1..6)));
```

$$
\begin{aligned}
pinv := {}& B^6 + B^5G + B^5R + B^5W + 3\,B^4G^2 + 3\,B^4GR + 3\,B^4GW \\
& + 3\,B^4R^2 + 3\,B^4RW + 3\,B^4W^2 + 3\,B^3G^3 + 6\,B^3G^2R + 6\,B^3G^2W \\
& + 6\,B^3GR^2 + 10\,B^3GRW + 6\,B^3GW^2 + 3\,B^3R^3 + 6\,B^3R^2W \\
& + 6\,B^3RW^2 + 3\,B^3W^3 + 3\,B^2G^4 + 6\,B^2G^3R + 6\,B^2G^3W \\
& + 11\,B^2G^2R^2 + 16\,B^2G^2RW + 11\,B^2G^2W^2 + 6\,B^2GR^3 \\
& + 16\,B^2GR^2W + 16\,B^2GRW^2 + 6\,B^2GW^3 + 3\,B^2R^4 + 6\,B^2R^3W \\
& + 11\,B^2R^2W^2 + 6\,B^2RW^3 + 3\,B^2W^4 + BG^5 + 3\,BG^4R \\
& + 3\,BG^4W + 6\,BG^3R^2 + 10\,BG^3RW + 6\,BG^3W^2 + 6\,BG^2R^3 \\
& + 16\,BG^2R^2W + 16\,BG^2RW^2 + 6\,BG^2W^3 + 3\,BGR^4 \\
& + 10\,BGR^3W + 16\,BGR^2W^2 + 10\,BGRW^3 + 3\,BGW^4 + BR^5 \\
& + 3\,BR^4W + 6\,BR^3W^2 + 6\,BR^2W^3 + 3\,BRW^4 + BW^5 + G^6 \\
& + G^5R + G^5W + 3\,G^4R^2 + 3\,G^4RW + 3\,G^4W^2 + 3\,G^3R^3 \\
& + 6\,G^3R^2W + 6\,G^3RW^2 + 3\,G^3W^3 + 3\,G^2R^4 + 6\,G^2R^3W \\
& + 11\,G^2R^2W^2 + 6\,G^2RW^3 + 3\,G^2W^4 + GR^5 + 3\,GR^4W \\
& + 6\,GR^3W^2 + 6\,GR^2W^3 + 3\,GRW^4 + GW^5 + R^6 + R^5W \\
& + 3\,R^4W^2 + 3\,R^3W^3 + 3\,R^2W^4 + RW^5 + W^6
\end{aligned}
$$

In the following command we use **cpinv** to find that we could construct 16 distinct six-bead necklaces with two blue beads, one green bead, one red bead, and two white beads.

```
>  cpinv(pinv, B^2*G*R*W^2);
```
$$16$$

Recall that it can also sometimes be useful to find the cycle index for a necklace whose colored beads can be rotated around the necklace, but which cannot be flipped over. To do this, we will use the user-written function **cicn**, which we have written separately from this Maple session and saved as the text file cicn.mpl. To use this function, we must first read it into our Maple session as follows.

```
>  read "cicn.mpl";
```

The following command illustrates how the **cicn** function can be used to find the cycle index for a six-bead necklace, in terms of the variable x, whose colored beads can be rotated around the necklace, but which cannot be flipped over.

```
>  f := cicn(6, x);
```

$$f := \frac{1}{6} x_1^6 + \frac{1}{6} x_2^3 + \frac{1}{3} x_3^2 + \frac{1}{3} x_6$$

11.6   The Pattern Inventory with MATLAB

In this section, we will show how MATLAB can be used to count patterns and construct pattern inventories. We will consider our six-bead necklace example in Example 11.5 with the following cycle index.

$$f(x_1, x_2, x_3, x_4, x_5, x_6) \;=\; \frac{1}{12} \left(x_1^6 + 2x_6 + 2x_3^2 + 4x_2^3 + 3x_1^2 x_2^2 \right)$$

To find this cycle index and the cycle index in similar examples, we will use the user-written function **cidn**, which we have written separately from this MATLAB session and saved as the text file cidn.m. The following commands illustrate how the function **cidn** can be used. Specifically, by entering the following commands, we find the cycle index for our six-bead necklace example in terms of the variable x.

```
>> syms x
>> cires = cidn(6, x);
```

The **cidn** function returns two entries in a cell array that we have assigned as *cires*. The first entry in *cires* is the resulting cycle index. In the following command we assign this cycle index as the variable f.

```
>> f = cires{1};
```

We can then use the MATLAB **pretty** function as follows to display this cycle index in a form that resembles typeset mathematics.

```
>> pretty(f)

     6      2   2      3      2
    x1     x1  x2     x2     x3     x6
    --- + ------- + --- + --- + --
    12       4        3      6      6
```

Next, we will convert this expression for f into a function. To do this, we will use the second entry in *cires*, which is a list of the variables used in the cycle index.

```
>> fvar = cires{2};
>> [fvar{:}]

ans =

[ x1, x2, x3, x4, x5, x6]
```

We can then convert the cycle index stored as f from an expression into a function by entering the following command.

```
>> f = @(x) sym(subs(f, {fvar{:}}, x));
```

The following command illustrates how this function can be used. Specifically, to find the number of distinct six-bead necklaces that we could construct if there were two colors available for the beads, we enter the following command.

```
>> f({2, 2, 2, 2, 2, 2})

ans =

13
```

Thus, as we noted in Example 11.5, we could construct 13 distinct six-bead necklaces if there were two colors available for the beads. Suppose that the colors available for the beads were $\mathcal{B}$ = blue and $\mathcal{G}$ = green. To see how many of these 13 distinct six-bead necklaces would contain prescribed numbers of blue and green beads, we form the following pattern inventory.

```
>> syms B G
>> pretty(expand(f({B+G, B^2+G^2, B^3+G^3, B^4+G^4, ...
   B^5+G^5, B^6+G^6})))

  6     5         4  2      3  3      2  4      5      6
  B  + B  G + 3 B  G  + 3 B  G  + 3 B  G  + B G  + G
```

Thus, for example, because the term $3\,\mathcal{B}^4\,\mathcal{G}^2$ appears in this pattern inventory, then we could construct three distinct six-bead necklaces with four blue beads and two green beads.

Now, suppose that each bead could also be $\mathcal{R}$ = red. To find the number of distinct six-bead necklaces that we could construct if there were three colors available for the beads, we can enter the following command.

```
>> f({sym('3')*ones(1, 6)})

ans =

92
```

In the previous command we used the MATLAB **ones** function, which generates a row vector of the specified size made up of ones, with each entry in this case multiplied by three, to efficiently generate the input for the cycle index. The result shows that we could construct 92 distinct six-bead necklaces if there were three colors available for the beads. To see how many of these 92 distinct six-bead necklaces would contain prescribed numbers of blue, green, and red beads, we form the following pattern inventory.

```
>> syms B G R
>> pinv = expand(f({B.^(1:6) + G.^(1:6) + R.^(1:6)}));
>> pretty(pinv)
```

$$B^6 + B^5 G + B^5 R + 3 B^4 G^2 + 3 B^4 G R + 3 B^4 R^2$$

$$+ 3 B^3 G^3 + 6 B^3 G^2 R + 6 B^3 G R^2 + 3 B^3 R^3 + 3 B^2 G^4$$

$$+ 6 B^2 G^3 R + 11 B^2 G^2 R^2 + 6 B^2 G R^3 + 3 B^2 R^4$$

$$+ B G^5 + 3 B G^4 R + 6 B G^3 R^2 + 6 B G^2 R^3 + 3 B G R^4$$

$$+ B R^5 + G^6 + G^5 R + 3 G^4 R^2 + 3 G^3 R^3 + 3 G^2 R^4$$

$$+ G R^5 + R^6$$

In this command, we used the MATLAB period function to exponentiate and create a vector for each variable raised to the sequence of powers, which allows the input for the cycle index to be generated more efficiently. Note that each term in the two-color pattern inventory is in the three-color pattern inventory. To help with finding information for pattern inventories containing a larger number of terms, we will use the user-written function **cpinv**, which we have written separately from this MATLAB session and saved as the text file cpinv.m. The following command illustrates how we can use the **cpinv** function to find that we could construct six distinct six-bead necklaces with two blue beads, one green bead, and three red beads.

```
>> cpinv(pinv, B^2*G*R^3)

ans =

6
```

Finally, suppose that each bead could also be $\mathcal{W}$ = white. To find the number of distinct six-bead necklaces that we could construct if there were four colors available for the beads, we can enter the following command.

```
>> f( {sym('4')*ones(1,6)} )

ans =

430
```

Thus, we could construct 430 distinct six-bead necklaces if there were four colors available for the beads. To see how many of these 430 distinct six-bead necklaces would contain prescribed numbers of blue, green, red, and white beads, we form the following pattern inventory.

```
>> syms B G R W
>> pinv = expand(f({B.^(1:6) + G.^(1:6) + R.^(1:6) ...
   + W.^(1:6)}));
>> pretty(pinv)

    6     5       5       5         4  2      4           4
   B   + B  G + B  R + B  W + 3 B  G  + 3 B  G R + 3 B  G W

          4  2      4           4  2      3  3       3  2
     + 3 B  R  + 3 B  R W + 3 B  W  + 3 B  G  + 6 B  G  R

        3  2        3    2        3           3    2
     + 6 B  G  W + 6 B  G R + 10 B  G R W + 6 B  G W

        3  3      3  2          3    2      3  3      2  4
     + 3 B  R  + 6 B  R  W + 6 B  R W  + 3 B  W  + 3 B  G

        2  3        2  3          2  2  2        2  2
     + 6 B  G  R + 6 B  G  W + 11 B  G  R  + 16 B  G  R W
```

$$+ 11\,B^2 G^2 W^2 + 6\,B^2 G R^3 + 16\,B^2 G R^2 W$$

$$+ 16\,B^2 G R W + 6\,B^2 G W^3 + 3\,B^2 R^4 + 6\,B^2 R^3 W$$

$$+ 11\,B^2 R^2 W^2 + 6\,B^2 R W^3 + 3\,B^2 W^4 + B G^5 + 3\,B G^4 R$$

$$+ 3\,B G^4 W + 6\,B G^3 R^2 + 10\,B G^3 R W + 6\,B G^3 W^2$$

$$+ 6\,B G^2 R^3 + 16\,B G^2 R^2 W + 16\,B G^2 R W^2 + 6\,B G^2 W^3$$

$$+ 3\,B G R^4 + 10\,B G R^3 W + 16\,B G R^2 W^2 + 10\,B G R W^3$$

$$+ 3\,B G W^4 + B R^5 + 3\,B R^4 W + 6\,B R^3 W^2 + 6\,B R^2 W^3$$

$$+ 3\,B R W^4 + B W^5 + G^6 + G^5 R + G^5 W + 3\,G^4 R^2$$

$$+ 3\,G^4 R W + 3\,G^4 W^2 + 3\,G^3 R^3 + 6\,G^3 R^2 W$$

$$+ 6\,G^3 R W^2 + 3\,G^3 W^3 + 3\,G^2 R^4 + 6\,G^2 R^3 W$$

$$+ 11\,G^2 R^2 W^2 + 6\,G^2 R W^3 + 3\,G^2 W^4 + G R^5 + 3\,G R^4 W$$

$$+ 6\,G R^3 W^2 + 6\,G R^2 W^3 + 3\,G R W^4 + G W^5 + R^6 + R^5 W$$

$$+ 3\,R^4 W^2 + 3\,R^3 W^3 + 3\,R^2 W^4 + R W^5 + W^6$$

In the next command we use **cpinv** to find that we could construct 16 distinct six-bead necklaces with two blue beads, one green bead, one red bead, and two white beads.

```
>> cpinv(pinv, B^2*G*R*W^2)

ans =

16
```

Recall that it can also sometimes be useful to find the cycle index for a necklace whose colored beads can be rotated around the necklace, but which cannot be flipped over. To do this, we will use the user-written function **cicn**, which we have written separately from this MATLAB session and saved as the text file cicn.m. The following commands illustrate how the **cicn** function can be used to find the cycle index for a six-bead necklace, in terms of the variable x, whose colored beads can be rotated around the necklace, but which cannot be flipped over.

```
>> cires = cicn(6, x);
>> f = cires{1};
>> pretty(f)

    6     3     2
   x1    x2    x3    x6
   --- + --- + --- + --
    6     6     3     3
```

11.7 Switching Functions

In this section, we will show how Pólya theory can be applied to the classification of switching functions. A *switching function* is a function $f : \mathbb{Z}_2^n \to \mathbb{Z}_2$ for a positive integer n.[2] Switching functions were first studied in the early part of the twentieth century as a way to handle the increasingly large number of telephone calls being placed through local switchboards. The way that switching functions were subsequently used by telephone companies led to their more recent use in the design of digital computers. It is in this area that functions that map from $\mathbb{Z}_2^n$ to $\mathbb{Z}_2$ are most useful because of how computers store and transmit information as bit strings.

Although we will not describe any specific applications of switching functions, we will state that in general it is desirable to keep a record of all possible switching functions for a given value of n. However, this poses a problem, since even for very small values of n, switching functions

---

[2]More generally, a switching function is a process that can take any number of inputs, but has only two possible outputs. However, the definition that a switching function is a function $f : \mathbb{Z}_2^n \to \mathbb{Z}_2$ for a positive integer n is sufficient for our purposes.

are numerous. Specifically, for each positive integer n, since $|\mathbb{Z}_2^n| = 2^n$, then there are 2^{2^n} switching functions. Thus, even for a value of n as small as five, there are more than four billion switching functions. Because switching functions are so numerous, it would not be practical to literally keep a record of all of them. What is done instead is that an equivalence relation is defined on the set of all possible switching functions. This divides the set of all possible switching functions into equivalence classes, and then a record can be kept of just one function from each equivalence class. In order to define the equivalence relation that is used, suppose n is a fixed positive integer, and let X be the set of all possible colorings of $\mathbb{Z}_2^n$ by $\mathbb{Z}_2$. Then X is precisely the set of all possible switching functions for the fixed value of n. Note that the symmetric group S_n acts on X as follows for all $\pi \in S_n$ and $f \in X$.

$$\pi f(x_1, x_2, \ldots, x_n) = f(x_{\pi(1)}, x_{\pi(2)}, \ldots, x_{\pi(n)}) \tag{11.3}$$

Then for any $f, g \in X$, we define the equivalence relation $f \sim g$ if there exists $\pi \in S_n$ with $\pi f = g$.

Example 11.7 Let $n = 2$, so that $\mathbb{Z}_2^n = \mathbb{Z}_2^2 = \{00, 01, 10, 11\}$. Suppose that we define $f : \mathbb{Z}_2^2 \to \mathbb{Z}_2$ as follows.

$$
\begin{aligned}
f(0,0) &= 1 \\
f(0,1) &= 0 \\
f(1,0) &= 1 \\
f(1,1) &= 0
\end{aligned}
$$

Let π be the permutation $(12) \in S_2$. Then using the group action that we just stated, we see that $\pi f(x_1, x_2) = f(x_{\pi(1)}, x_{\pi(2)}) = f(x_2, x_1)$. Thus, we have the following.

$$
\begin{aligned}
\pi f(0,0) &= 1 \\
\pi f(0,1) &= 1 \\
\pi f(1,0) &= 0 \\
\pi f(1,1) &= 0
\end{aligned}
$$

Thus, suppose that we define $g : \mathbb{Z}_2^2 \to \mathbb{Z}_2$ as follows.

$$
\begin{aligned}
g(0,0) &= 1 \\
g(0,1) &= 1 \\
g(1,0) &= 0 \\
g(1,1) &= 0
\end{aligned}
$$

Then $f \sim g$.   □

Since we keep a record of all possible switching functions by recording one function from each equivalence class, it is obviously important to know the number of equivalence classes for each value of n. This is where Pólya theory can be applied. To count the number of equivalence classes, we can use a cycle index. Also, to see how many of these equivalence classes contain functions that produce prescribed numbers of zeros and ones, we can use a pattern inventory. We illustrate these ideas in the following example.

Example 11.8 In this example, we will consider the switching functions $f : \mathbb{Z}_2^n \to \mathbb{Z}_2$ for $n = 3$. We begin by noting that the elements in the symmetric group S_3 can be expressed using cycle notation as follows.

$\pi_i \in S_3$	Cycle Representation
π_1	$(1)(2)(3)$
π_2	$(12)(3)$
π_3	$(13)(2)$
π_4	$(1)(23)$
π_5	(123)
π_6	(132)

Next, we will apply each of the permutations in S_3 to each of the elements in $\mathbb{Z}_2^3 = \{000, 001, 010, 011, 100, 101, 110, 111\}$. For example, applying π_2 to 011 yields 101, since π_2 swaps the first and second bits and leaves the third bit fixed. In the following table, we list the results from applying each of the permutations in S_3 to each of the elements in $\mathbb{Z}_2^3$. Also, in the first column in the following table, we attach numeric labels to the elements in $\mathbb{Z}_2^3$. We will use these labels to express the actions of the permutations in S_3 on the elements in $\mathbb{Z}_2^3$ using cycle notation.

Label	$\mathbb{Z}_2^3$	π_1	π_2	π_3	π_4	π_5	π_6
1	000	000	000	000	000	000	000
2	001	001	001	100	010	100	010
3	010	010	100	010	001	001	100
4	011	011	101	110	011	101	110
5	100	100	010	001	100	010	001
6	101	101	011	101	110	110	011
7	110	110	110	011	101	011	101
8	111	111	111	111	111	111	111

Now, in the following table, we list the actions of each of the permutations in S_3 on each of the labels of the elements in $\mathbb{Z}_2^3$ in the preceding table. For example, the action of π_2 on the labels of the elements in $\mathbb{Z}_2^3$ in the preceding table is $(1)(2)(35)(46)(7)(8)$, since π_2 leaves elements 1, 2, 7, and

8 fixed, sends elements 3 and 5 to each other, and sends elements 4 and 6 to each other. Also, in the third column in the following table, we list the monomials for the resulting cycle index.

$\pi \in S_3$	Action on $\mathbb{Z}_2^3$	Monomial
π_1	$(1)(2)(3)(4)(5)(6)(7)(8)$	$(x_1)^8$
π_2	$(1)(2)(35)(46)(7)(8)$	$(x_1)^4(x_2)^2$
π_3	$(1)(25)(3)(47)(8)$	$(x_1)^4(x_2)^2$
π_4	$(1)(23)(4)(5)(67)(8)$	$(x_1)^4(x_2)^2$
π_5	$(1)(253)(467)(8)$	$(x_1)^2(x_3)^2$
π_6	$(1)(235)(476)(8)$	$(x_1)^2(x_3)^2$

The resulting cycle index is the following.

$$f(x_1, x_2, x_3) = \frac{1}{6}\left(x_1^8 + 3x_1^4x_2^2 + 2x_1^2x_3^2\right)$$

Since there are two colors in this example (the numbers zero and one), then the total number of equivalence classes of switching functions for $n = 3$ is given by $f(2,2,2) = \frac{1}{6}(256 + 192 + 32) = 80$. To see how many of these 80 equivalence classes contain functions that produce prescribed numbers of zeros and ones, we form the following pattern inventory, with the colors denoted by by $\mathcal{Z} = $ zero and $\mathcal{O} = $ one.

$$f(\mathcal{Z} + \mathcal{O}, \mathcal{Z}^2 + \mathcal{O}^2, \mathcal{Z}^3 + \mathcal{O}^3)$$

$$= \frac{1}{6}\left((\mathcal{Z} + \mathcal{O})^8 + 3(\mathcal{Z} + \mathcal{O})^4(\mathcal{Z}^2 + \mathcal{O}^2)^2 + 2(\mathcal{Z} + \mathcal{O})^2(\mathcal{Z}^3 + \mathcal{O}^3)^2\right)$$

$$= \mathcal{Z}^8 + 4\mathcal{Z}^7\mathcal{O} + 9\mathcal{Z}^6\mathcal{O}^2 + 16\mathcal{Z}^5\mathcal{O}^3 + 20\mathcal{Z}^4\mathcal{O}^4 + 16\mathcal{Z}^3\mathcal{O}^5 + 9\mathcal{Z}^2\mathcal{O}^6$$
$$+ 4\mathcal{Z}\mathcal{O}^7 + \mathcal{O}^8 \ .$$

Thus, for example, because the term $16\mathcal{Z}^5\mathcal{O}^3$ appears in this pattern inventory, then there are 16 equivalence classes that contain functions that produce five zeros and three ones.   □

Exercises

1. Suppose that you wish to construct a necklace with three colored beads, and that each bead can be either red or white. Assume that the beads can be rotated around the necklace, and that the necklace can be flipped over. Consider the following general shape for the

necklace, stretched into the shape of an equilateral triangle with one bead at each corner.

Let G be the group of rigid motions of an equilateral triangle, let S be the set of vertices of the preceding general figure, and let X_1 be the set of all possible colorings of S by the colors $\mathcal{R}$ = red and $\mathcal{W}$ = white.

(a) For each rigid motion $\pi \in G$, find $\text{Fix}(\pi)$.

(b) For each coloring $x \in X_1$, find $\text{Stab}(x)$.

(c) Find the cycle index of G acting on S. Then use this cycle index to determine the number of distinct three-bead necklaces you could construct if there were two colors available for the beads.

(d) Find the pattern inventory of X_1.

(e) Suppose each bead could also be blue, and let X_2 be the set of all possible colorings of S by the colors $\mathcal{R}$ = red, $\mathcal{W}$ = white, and $\mathcal{B}$ = blue. Determine the number of distinct three-bead necklaces that you could construct with these three colors available for the beads. Also, find the pattern inventory of X_2. According to this pattern inventory of X_2, of the total number of distinct three-bead necklaces that you could construct with these three colors available for the beads, how many would contain at least one blue bead?

2. How many distinct three-bead necklaces could you construct if there were 10 colors available for the beads? Assume the beads can be rotated around the necklace, and the necklace can be flipped over.

3. Suppose you wish to construct a necklace with five colored beads, and each bead can be either red or white. Assume the beads can be rotated around the necklace, but the necklace cannot be flipped over.

(a) How many distinct necklaces could you construct?

(b) Among the total number of distinct necklaces that you could construct as found in part (a), how many of these necklaces would contain two red beads and three white beads?

(c) Among the total number of distinct necklaces that you could construct as found in part (a), how many of these necklaces would contain at least three white beads?

4. Repeat Exercise 3 assuming that the necklace can also be flipped over.

5. Suppose that you wish to construct a building in the shape of a regular pentagon, and that you will paint each side of the building one of 10 different colors. Assume that two buildings are equivalent if one could be rotated to look like the other, but not flipped over to look like the other. How many nonequivalent buildings could you construct?

6. Repeat Exercise 5 assuming that two buildings are also equivalent if one could be flipped over to look like the other.

7. Suppose you wish to construct a perfectly symmetrical six-pointed star, and that you will paint the tip of each point of the star either blue, green, or red. Assume that two six-pointed stars are equivalent if one could be rotated to look like the other, but not flipped over to look like the other.

   (a) How many nonequivalent six-pointed stars could you construct?

   (b) Among the total number of nonequivalent six-pointed stars that you could construct as found in part (a), how many of these six-pointed stars would have each of the three colors painted on the tip of exactly two of the points?

   (c) Among the total number of nonequivalent six-pointed stars that you could construct as found in part (a), how many of these six-pointed stars would have each of the three colors painted on the tip of at least one of the points?

8. Repeat Exercise 7 assuming that two six-pointed stars are also equivalent if one could be flipped over to look like the other.

9. Prove Lemma 11.1.

10. Suppose that a group G acts on a set Y. For an element $y \in Y$, prove that the set $\text{Stab}(y)$ is a subgroup of G.

11. Prove Lemma 11.2.

12. Prove Theorem 11.5.

13. Let X be the set of all possible switching functions $f : \mathbb{Z}_2^n \to \mathbb{Z}_2$ for a fixed positive integer n, and consider the following operation for π in the symmetric group S_n and $f \in X$.

$$\pi f(x_1, x_2, \ldots, x_n) = f(x_{\pi(1)}, x_{\pi(2)}, \ldots, x_{\pi(n)})$$

Prove that this operation defines a group action of S_n on X.

14. Let X be the set of all possible switching functions $f : \mathbb{Z}_2^n \to \mathbb{Z}_2$ for a fixed positive integer n. For $f, g \in X$, define $f \sim g$ if there exists $\pi \in S_n$ with $\pi f = g$ using the action of S_n on X defined in Exercise 13. Prove that $\sim$ is an equivalence relation.

15. Determine the number of equivalence classes of switching functions $f : \mathbb{Z}_2^n \to \mathbb{Z}_2$ for $n = 4$ under the equivalence relation defined in Exercise 14.

Computer Exercises

16. Let G be the group of rigid motions of a regular pentagon, let S be the set of vertices of a regular pentagon, and let X be the set of all possible colorings of S by the colors $\mathcal{R} = $ red, $\mathcal{W} = $ white, $\mathcal{B} = $ blue, and $\mathcal{Y} = $ yellow. Find the pattern inventory of X. Then use this pattern inventory to determine the number of distinct five-bead necklaces that you could construct with exactly two red beads, one white bead, one blue bead, and one yellow bead, assuming that the beads can be rotated around the necklace, and that the necklace can be flipped over.

17. Let G be the group of rigid motions of a regular octagon, let S be the set of vertices of a regular octagon, and let X be the set of all possible colorings of S by the colors $\mathcal{R} = $ red, $\mathcal{W} = $ white, $\mathcal{B} = $ blue, and $\mathcal{Y} = $ yellow. Find the pattern inventory of X. Then use this pattern inventory to determine the number of distinct eight-bead necklaces that you could construct with exactly three red beads, two white beads, two blue beads, and one yellow bead, assuming that the beads can be rotated around the necklace, and that the necklace can be flipped over.

18. Repeat Exercise 17 assuming that the beads can be rotated around the necklace, but that the necklace cannot be flipped over.

19. Suppose that you wish to construct an analog watch with a ring around its hour indicators containing 12 birthstones, and that each birthstone can be either emerald, garnet, ruby, sapphire, or turquoise. Assume the ring containing the birthstones can be rotated around the watch, but that the ring cannot be flipped over.

    (a) How many distinct watches could you construct?

    (b) Among the total number of distinct watches that you could construct as found in part (a), how many of these watches would contain four emeralds, two garnets, two rubies, three sapphires, and one turquoise?

(c) Among the total number of distinct watches that you could construct as found in part (a), how many of these watches would contain only rubies and/or sapphires?

20. Repeat Exercise 19 assuming that the ring containing the birthstones can also be flipped over.

21. Write a single Maple or MATLAB procedure that will take as input the number of vertices of a regular polygon, and will return as output the cycle index of the group of rigid motions of the polygon acting on the set of vertices of the polygon.

22. Among the total number of equivalence classes of switching functions $f : \mathbb{Z}_2^n \to \mathbb{Z}_2$ for $n = 4$ under the equivalence relation defined in Exercise 14 (see Exercise 15), how many contain functions that produce nine zeros and seven ones?

Research Exercises

23. Investigate the life and career of the person for whom Burnside's Theorem is named, and write a summary of your findings. Include in your summary the various earlier mathematicians who successfully proved Burnside's Theorem.

24. Investigate the life and career of the person for whom the Pólya Enumeration Theorem is named, and write a summary of your findings. Include in your summary a description of what the *George Pólya Award* is given for.

25. Investigate applications of Pólya theory in the enumeration of chemical compounds, and write a summary of your findings.

26. Investigate applications of Pólya theory in the analysis of Rubik's cube, and write a summary of your findings.

27. Investigate some specific details about switching functions and their use in digital computers for information storage and transmission, and write a summary of your findings.

Chapter 12

Graph Theory

In this chapter, we will show how the Pólya theory we presented in Chapter 11 can be applied to the problem of counting undirected graphs. Specifically, we will consider the problem of counting the number of nonequivalent simple undirected graphs with n vertices for an arbitrary positive integer n. As a first step, we will consider the cycle index of the symmetric group S_n acting on the set $\{1, 2, \ldots, n\}$.

12.1   The Cycle Index of S_n

We will begin with the notion of a partition of a positive integer. A *partition* of a positive integer n is a set of positive integers whose sum is n. For example, one of the partitions of the integer 7 is the set $\{1, 3, 3\}$. In a partition P of a positive integer n, let p_i denote the number of times that the integer i appears for $i = 1, 2, \ldots, n$. Then with this notation, we will alternatively express P as $P = [p_1, p_2, \ldots, p_n]$. For example, we will alternatively express the partition $\{1, 3, 3\}$ of the integer 7 as $[1, 0, 2, 0, 0, 0, 0]$. Note that for any partition $P = [p_1, p_2, \ldots, p_n]$ of a positive integer n, it will be true that $\sum_{i=1}^{n} i \cdot p_i = n$.

Recall that the symmetric group S_n is the group of all permutations on the set $\{1, 2, \ldots, n\}$ with the composition operation (see Section 1.1). Using the ideas that we presented in Section 11.3, we will now proceed to determine the cycle index of S_n acting on the set $\{1, 2, \ldots, n\}$ for an arbitrary positive integer n. Note first that the expression of a given partition of a positive integer n in set form can be viewed as containing indicators of the lengths of the cycles in the representation of an element in S_n as a permutation on $\{1, 2, \ldots, n\}$ (including all cycles of length one). For example, the partition $\{1, 3, 3\}$ of the integer 7 can be viewed as indicating, among many others, the element $(137)(4)(256) \in S_7$. Also, note that if we were forming the cycle index of S_7 acting on the set $\{1, 2, \ldots, 7\}$, the monomial that we would associate with $(137)(4)(256)$ would be $x_1(x_3)^2$, whose subscripts are given by the positions of the nonzero entries in the expression $[1, 0, 2, 0, 0, 0, 0]$ of the partition $\{1, 3, 3\}$, and whose exponents are given by the actual nonzero entries in the expression $[1, 0, 2, 0, 0, 0, 0]$. More generally, if $P = [p_1, p_2, \ldots, p_n]$ is a partition of a positive integer n, then the monomial that we would associate with all of the elements in S_n that can be viewed as being indicated by P would be $x_1^{p_1} x_2^{p_2} \cdots x_n^{p_n}$. We demonstrate this in the following example.

Example 12.1 In the following table, we list each partition of the positive integer 5, along with a sample element in S_5 that can be viewed as being indicated by the partition, the monomial that we would associate with all of the elements in S_5 that can be viewed as being indicated by the partition, and total number of elements in S_5 that can be viewed as being indicated by the partition.

Partition	Element in S_5	Monomial	Total Number
$[5, 0, 0, 0, 0]$	$(1)(2)(3)(4)(5)$	$(x_1)^5$	1
$[3, 1, 0, 0, 0]$	$(12)(3)(4)(5)$	$(x_1)^3 x_2$	10
$[1, 2, 0, 0, 0]$	$(12)(34)(5)$	$x_1(x_2)^2$	15
$[2, 0, 1, 0, 0]$	$(123)(4)(5)$	$(x_1)^2 x_3$	20
$[0, 1, 1, 0, 0]$	$(123)(45)$	$x_2 x_3$	20
$[1, 0, 0, 1, 0]$	$(1234)(5)$	$x_1 x_4$	30
$[0, 0, 0, 0, 1]$	(12345)	x_5	24

□

Each entry in the column on the right in the table in Example 12.1 shows the total number of elements in S_5 that can be viewed as being indicated by the partition shown in the column on the left. With a value of n for S_n as small as $n = 5$, these entries can be determined by counting. However, with larger values of n, counting might not be a practical means for determining these entries. In such cases, these entries can be determined by the formula given in the following lemma.

Lemma 12.1 *Suppose $P = [p_1, p_2, \ldots, p_n]$ is a partition of a positive integer n. Then the number of elements in S_n that can be viewed as being indicated by P is given by the following formula.*

$$n! \prod_{j=1}^{n} \frac{1}{j^{p_j} p_j!}$$

Proof. First, the number of ways to choose the cycles of length one is given by $\binom{n}{p_1}$. For later reference, note that this can be expressed as follows.

$$\binom{n}{p_1} = \frac{1}{1^{p_1} p_1!} \cdot \frac{(n)!}{(n - p_1)!}$$

Next, the number of ways to choose the cycles of length two is given by the following formula. (The factor of $p_2!$ appears in this formula because the result does not depend on the order in which the cycles of length two are chosen.)

$$\frac{1}{p_2!} \binom{n - p_1 - 2(0)}{2} \binom{n - p_1 - 2(1)}{2} \cdots \binom{n - p_1 - 2(p_2 - 1)}{2}$$

$$= \frac{1}{2^{p_2} p_2!} \cdot \frac{(n - p_1)!}{(n - p_1 - 2p_2)!}$$

Next, the number of ways to choose the cycles of length three is given by the following formula. (The factor of 2! appears in this formula because for any choice of three integers, there will be 2! distinct cycles of length three that include the integers.)

$$\frac{2!}{p_3!} \binom{n - p_1 - 2p_2}{3} 2! \binom{n - p_1 - 2p_2 - 3}{3} \cdots 2! \binom{n - p_1 - 2p_2 - 3(p_3 - 1)}{3}$$

$$= \frac{1}{3^{p_3} p_3!} \cdot \frac{(n - p_1 - 2p_2)!}{(n - p_1 - 2p_2 - 3p_3)!}$$

The pattern that is evident in each of the three preceding formulas for counting the number of ways to choose the cycles of lengths one, two, and three continues for the number of ways to choose the cycles of length r for all integers r from 4 through n. Specifically, for all integers r from 1 through n, the number of ways to choose the cycles of length r will be given by the following formula.

$$\frac{1}{r^{p_r} p_r!} \cdot \frac{(n - p_1 - 2p_2 - \cdots - (r - 1)p_{r-1})!}{(n - p_1 - 2p_2 - \cdots - rp_r)!}$$

As a result, we see that the number of elements in S_n that can be viewed as being indicated by P is given by the following formula.

$$\prod_{j=1}^{n} \frac{1}{j^{p_j} p_j!} \cdot \frac{(n - p_1 - 2p_2 - \cdots - (j - 1)p_{j-1})!}{(n - p_1 - 2p_2 - \cdots - jp_j)!} = n! \prod_{j=1}^{n} \frac{1}{j^{p_j} p_j!}$$

$\square$

Example 12.2 Consider the partition $P = [2, 0, 1, 0, 0]$ of the integer 5. Using the formula given in Lemma 12.1, we can determine the number of elements in S_5 that can be viewed as being indicated by P as follows.

$$5! \left[\left(\frac{1}{1^2 \cdot 2!} \right) \left(\frac{1}{2^0 \cdot 0!} \right) \left(\frac{1}{3^1 \cdot 1!} \right) \left(\frac{1}{4^0 \cdot 0!} \right) \left(\frac{1}{5^0 \cdot 0!} \right) \right] = 20$$

$\square$

Example 12.3 Consider the partition $P = [0, 3, 1, 0, 0, 0, 0, 0, 0]$ of the integer 9. Using the formula given in Lemma 12.1, we can determine the number of elements in S_9 that can be viewed as being indicated by P as follows.

$$9! \left[(1) \left(\frac{1}{2^3 \cdot 3!} \right) \left(\frac{1}{3^1 \cdot 1!} \right) (1)\,(1)\,(1)\,(1)\,(1)\,(1) \right] = 2520$$

$\square$

Now, using the formula given in Lemma 12.1, we can determine the cycle index of S_n acting on the set $\{1, 2, \ldots, n\}$ for an arbitrary positive integer n. Summing over all partitions $P = [p_1, p_2, \ldots, p_n]$ of the integer n, we see that this cycle index $f(x_1, x_2, \ldots, x_n)$ is given by the following formula.

$$f(x_1, x_2, \ldots, x_n) = \frac{1}{n!} \sum_P \left(\left(\prod_{j=1}^{n} \frac{n!}{j^{p_j} p_j!} \right) x_1^{p_1} x_2^{p_2} \cdots x_n^{p_n} \right) \qquad (12.1)$$

Example 12.4 Using (12.1) and the table in Example 12.1, we find that the cycle index of S_5 acting on the set $\{1, 2, 3, 4, 5\}$ is as follows.

$f(x_1, x_2, x_3, x_4, x_5)$

$$= \frac{1}{120} \left(x_1^5 + 10x_1^3 x_2 + 15x_1 x_2^2 + 20x_1^2 x_3 + 20x_2 x_3 + 30x_1 x_4 + 24x_5 \right)$$

$\square$

We will close this section by considering what the cycle index of S_n acting on the set $\{1, 2, \ldots, n\}$ allows us to count. Suppose that we have r different types of objects, with an unlimited supply of each type, and that each of the integers in the set $S = \{1, 2, \ldots, n\}$ is to be assigned a unique object. Using the terminology that we introduced in Chapter 11, we will call this assignment a *coloring* of S by the r different types of objects. As in Chapter 11, we will consider two colorings f and g to be equivalent if there exists $\pi \in S_n$ such that $\pi(f(a)) = g(a)$ for all $a \in S$. Thus, two colorings f

and g are equivalent if f and g both assign the r different types of objects to the same number of elements in S. If we number the types of objects from $1, 2, \ldots, r$, and use x_j to represent the number of times that an object of type j is assigned to an element in S for $j = 1, 2, \ldots, r$, then counting the number of nonequivalent colorings is identical to counting the number of nonnegative integer solutions to $x_1 + x_2 + \cdots + x_r = n$.[1] It is known that the solution to this problem is $\binom{n+r-1}{n}$. The following Theorem 12.2 states how the cycle index of S_n acting on the set $\{1, 2, \ldots, n\}$ can be used to determine this number as well.

Theorem 12.2 *Suppose that the cycle index of S_n acting on the set $\{1, 2, \ldots, n\}$ is given by $f(x_1, x_2, \ldots, x_n)$. Then the number of nonnegative integer solutions to $x_1 + x_2 + \cdots + x_r = n$ is $f(r, r, \ldots, r)$, which is equal to $\binom{n+r-1}{n}$.*

Proof. Exercise. □

Example 12.5 Suppose we wish to determine the number of nonequivalent colorings of the set $\{1, 2, 3, 4, 5\}$ by three different types of objects, or, equivalently, the number of nonnegative integer solutions to $x_1 + x_2 + x_3 = 5$. Using the cycle index in Example 12.4, we see that this number is the following.

$$f(3, 3, 3, 3, 3) = \frac{1}{120}(243 + 810 + 405 + 540 + 180 + 270 + 72) = 21$$

Note that this result is identical to $\binom{7}{5} = 21$. □

12.2 The Cycle Index of S_n with Maple

In this section, we will show how Maple can be used to generate the partitions of a positive integer n, and construct the cycle index of S_n acting on the set $\{1, 2, \ldots, n\}$.

Because some of the functions that we will use are in the Maple **combinat** package, we will begin by including this package.

```
>   with(combinat):
```

Now, suppose that we wish to generate the partitions of the integer $n = 5$.

---

[1]For a more thorough discussion of this problem, see [4].

```
>  n := 5:
```

In the following command, we use the Maple **numbpart** function to find the number of partitions of n.

```
>  numbpart(n);
```

$$7$$

Next, we will use the Maple **partition** function to generate a list of the partitions of n expressed in set form.

```
>  partition(n);
```

$$[[1, 1, 1, 1, 1], [1, 1, 1, 2], [1, 2, 2], [1, 1, 3], [2, 3], [1, 4], [5]]$$

However, recall that to construct the cycle index of S_n acting on the set $\{1, 2, \ldots, n\}$, we need to have each of these partitions expressed as an ordered list of the number of times that the integer i appears in the set for $i = 1, 2, \ldots, n$. To do this, we will use the following user-written function **partlist**.

```
>  partlist := proc(y, p)

>      p[y] := p[y]+1;

>  end:
```

In order to demonstrate how **partlist** can be used, we will use the Maple **firstpart** function as follows to retrieve the first partition of n that was returned by **partition**.

```
>  part1 := firstpart(n);
```

$$part1 := [1, 1, 1, 1, 1]$$

Next, we will create the following zero vector *P1* of length n positions in which to store the integers in the expression of *part1* as an ordered list of the number of times that the integer i appears in the set for $i = 1, 2, \ldots, n$.

```
>  P1 := Vector[row](n);
```

$$P1 := [0 \quad 0 \quad 0 \quad 0 \quad 0]$$

In the following command, we use the **map** function to apply **partlist** to the partition *part1*. The effect of entering this command is that *P1* is converted into the expression of *part1* as an ordered list of the number of times that the integer i appears in the set for $i = 1, 2, \ldots, n$.

```
>  map(partlist, part1, P1):
```

```
> P1;
```
$$[5 \quad 0 \quad 0 \quad 0 \quad 0]$$

Since the first partition of n that was returned by **partition** is stored as *part1*, we can use the Maple **nextpart** function as follows to retrieve the second partition of n that was returned by **partition**.

```
> part2 := nextpart(part1);
```
$$part2 := [1, 1, 1, 2]$$

Similarly to how we converted *part1* from set form into an expression as an ordered list of the number of times that the integer i appears in the set for $i = 1, 2, \ldots, n$, we can convert *part2* from set form into an expression as an ordered list of the number of times that the integer i appears in the set by entering the following commands.

```
> P2 := Vector[row](n):
> map(partlist, part2, P2):
> P2;
```
$$[3 \quad 1 \quad 0 \quad 0 \quad 0]$$

In the following **for** loop, we convert each partition of n that was returned by **partition** from set form into an expression as an ordered list of the number of times that the integer i appears in the set for $i = 1, 2, \ldots, n$. The resulting ordered lists are stored in the variable *partitions*, which we initialize before entering the loop.

```
> partitions := []:
> for pnum from 1 to numbpart(n) do
>       if pnum = 1 then
>           part := firstpart(n):
>       else
>           part := nextpart(part):
>       fi:
>       P := Vector[row](n):
>       map(partlist, part, P):
>       partitions := [op(partitions), P]:
> od:
> partitions;
```

$$[[5 \quad 0 \quad 0 \quad 0 \quad 0 \,], [3 \quad 1 \quad 0 \quad 0 \quad 0 \,], [1 \quad 2 \quad 0 \quad 0 \quad 0 \,],$$
$$[2 \quad 0 \quad 1 \quad 0 \quad 0 \,], [0 \quad 1 \quad 1 \quad 0 \quad 0 \,], [1 \quad 0 \quad 0 \quad 1 \quad 0 \,],$$
$$[0 \quad 0 \quad 0 \quad 0 \quad 1 \,]]$$

To construct the cycle index of S_n acting on the set $\{1, 2, \ldots, n\}$, we will use the user-written function **cisn**, which we have written separately from this Maple session and saved as the text file cisn.mpl. To use this function, we must first read it into our Maple session as follows.

```
>   read "cisn.mpl";
```

The following command then illustrates how the function **cisn** can be used. Specifically, by entering the following command, we construct the cycle index of S_n acting on the set $\{1, 2, \ldots, n\}$ for our previously assigned value of $n = 5$.

```
>   f := cisn(n, x);
```

$$f := \frac{1}{120}\, x_1^5 + \frac{1}{12}\, x_1^3 x_2 + \frac{1}{8}\, x_1 x_2^2 + \frac{1}{6}\, x_1^2 x_3 + \frac{1}{6}\, x_2 x_3 + \frac{1}{4}\, x_1 x_4 + \frac{1}{5}\, x_5$$

The second parameter in the preceding command specifies the variable to be used in the resulting cycle index. In the following command we use **unapply** to convert this cycle index from an expression into a function.

```
>   f := unapply(f, seq(x[i], i=1..n));
```

$$f := (x\_1, x\_2, x\_3, x\_4, x\_5) \rightarrow \frac{1}{120}\, x\_1^5 + \frac{1}{12}\, x\_1^3\, x\_2$$
$$+ \frac{1}{8}\, x\_1\, x\_2^2 + \frac{1}{6}\, x\_1^2\, x\_3 + \frac{1}{6}\, x\_2\, x\_3 + \frac{1}{4}\, x\_1\, x\_4 + \frac{1}{5}\, x\_5$$

We can now use this cycle index as follows to find the number of nonequivalent colorings of the set $\{1, 2, 3, 4, 5\}$ by three different types of objects.

```
>   f(3, 3, 3, 3, 3);
```

$$21$$

12.3   The Cycle Index of S_n with MATLAB

In this section, we will show how MATLAB can be used to generate the partitions of a positive integer n, and construct the cycle index of S_n acting on the set $\{1, 2, \ldots, n\}$. To begin, suppose that we wish to generate the partitions of the integer $n = 5$.

```
>> n = 5

n =

    5
```

In the following command, we use the MuPAD **partitions** function, which is contained in the MuPAD **combinat** package, within MATLAB to find the number of partitions of n, storing the result as *numbpart*.

```
>> numbpart = feval(symengine, 'combinat::partitions', n)

numbpart =

    7
```

Next, we will use the the user-written function **partition**,[2] which we have written separately from this MATLAB session and saved as the M-file partition.m, to generate a list of the partitions of n expressed in set form. This will return the resulting partitions in a string. To display these partitions effectively, we will use the user-written function **stringprint** that we used previously in Section 7.4, which we have written separately from this MATLAB session and saved as the M-file stringprint.m. The following command illustrates how **partition** and **stringprint** can be used. Specifically, by entering the following command, we generate the partitions of $n = 5$ as a string, displayed in its entirety in rows with fifty characters each.

```
>> stringprint(partition(n), 50)

[[1, 1, 1, 1, 1], [1, 1, 1, 2], [1, 1, 3], [1, 2,
2], [1, 4], [2, 3], [5]]
```

Recall that to construct the cycle index of S_n acting on $\{1, 2, \ldots, n\}$, we need to have each of these partitions expressed as an ordered list of the number of times that the integer i appears in the set for $i = 1, 2, \ldots, n$. To do this, we will use the user-written function **partlist**, which we have written separately from this MATLAB session and saved as the M-file partlist.m. In order to demonstrate how **partlist** can be used, we will first use the user-written function **firstpart**, which we have written separately from this MATLAB session and saved as the M-file firstpart.m, as follows to retrieve

---

[2]The function **partition** uses the user-written functions **firstpart** and **nextpart**, which we will describe later in this section.

the first partition of n that was returned by the **partition** function.

```
>> part1 = firstpart(n)

part1 =

[1, 1, 1, 1, 1]
```

Next, we will create the following zero vector $P1$ of length n positions in which to store the integers in the expression of *part1* as an ordered list of the number of times that the integer i appears in the set for $i = 1, 2, \ldots, n$.

```
>> P1 = zeros(1, n)

P1 =
```

    0   0   0   0   0

In the following command, we apply **partlist** to the partition *part1*. The effect of entering this command is that $P1$ is converted into the expression of *part1* as an ordered list of the number of times that the integer i appears in the set for $i = 1, 2, \ldots, n$.

```
>> P1 = partlist(part1, P1)

P1 =
```

    5   0   0   0   0

With the first partition of n returned by the **partition** function stored as *part1*, we can use the user-written function **nextpart**, which we have written separately from this MATLAB session and saved as the M-file nextpart.m, as follows to retrieve the second partition of n returned by **partition**.

```
>> part2 = nextpart(part1)

part2 =

[1, 1, 1, 2]
```

Similarly to how we converted *part1* from set form into an expression as an ordered list of the number of times that the integer i appears in the set for

$i = 1, 2, \ldots, n$, we can convert *part2* from set form into an expression as an ordered list of the number of times that the integer i appears in the set by entering the following commands.

```
>> P2 = zeros(1, n)

P2 =

     0     0     0     0     0

>> P2 = partlist(part2, P2)

P2 =

     3     1     0     0     0
```

In the following **for** loop, we convert each partition of n that was returned by the **partition** function from set form into an expression as an ordered list of the number of times that the integer i appears in the set for $i = 1, 2, \ldots, n$. In order for the results to be displayed effectively, we will store the result as a string in the variable *partitions*, which we initialize before entering the loop.

```
>> partitions = '';
>> for pnum = 1:numbpart
       P = zeros(1, n);
       if pnum == 1
           part = firstpart(n);
           s = '';
       else
           part = nextpart(part);
           s = ',';
       end
       opart{pnum} = partlist(part, P);
       partitions = strcat(partitions, s, ...
       mat2str(opart{pnum}));
   end
>> stringprint(partitions, 50)

[5 0 0 0 0],[3 1 0 0 0],[2 0 1 0 0],[1 2 0 0 0],[1
   0 0 1 0],[0 1 1 0 0],[0 0 0 0 1]
```

To construct the cycle index of S_n acting on the set $\{1, 2, \ldots, n\}$, we will use the user-written function **cisn**,[3] which we have written separately from this MATLAB session and saved as the M-file cisn.m. The following command illustrates how the function **cisn** can be used. Specifically, by entering the following command, we construct the cycle index of S_n acting on the set $\{1, 2, \ldots, n\}$ for our previously assigned value of $n = 5$. The second parameter in this command specifies the variable to be used in the resulting cycle index, which must be declared as a symbolic variable.

```
>> syms x;
>> cires = cisn(n, x);
```

The **cisn** function returns two entries in a cell array that we have assigned as *cires*. The first entry in *cires* is the resulting cycle index. By entering the following command we store this cycle index as the variable f.

```
>> f = cires{1};
```

We can now use the MATLAB **pretty** function as follows to see the resulting cycle index in a form that resembles typeset mathematics.

```
>> pretty(f)

    5       3             2             2
   x1      x1 x2     x3 x1       x1 x2       x4 x1     x3 x2    x5
   --- + ------ + ------ + ------ + ----- + ----- + --
   120     12        6          8           4         6       5
```

Next, we will convert this cycle index from an expression into a function. To do this, we will use the second entry in *cires*, which is a list of the variables used in the resulting cycle index.

```
>> fvar = cires{2};
>> [fvar{:}]

ans =

[ x1, x2, x3, x4, x5]
```

To convert the cycle index stored as the variable f from an expression into a function, we enter the following command.

---

[3]The function **cisn** uses the user-written function **partlist** that we used previously in this section.

```
>> f = @(x) sym(subs(f, {fvar{:}}, x));
```

We can now use this cycle index as follows to determine the number of nonequivalent colorings of the set $\{1, 2, 3, 4, 5\}$ by three different types of objects.

```
>> f({3, 3, 3, 3, 3})
```

```
ans =
```

```
21
```

12.4 Counting Undirected Graphs

An *undirected graph* with n vertices is a set V of order n and a set T of (distinct) unordered pairs of distinct elements in V. The elements in V are the *vertices* of the graph, and the ordered pairs in T are the *edges*. For example, consider the graph with vertices $V = \{1, 2, 3, 4\}$ and edges $T = \{(1, 2), (1, 3), (3, 4)\}$, which can be represented as follows, in which we have indicated the vertices with a grid of dots, and included a line connecting the dots that represent vertices a and b if and only if $(a, b) \in T$.

For a set V of vertices, let E be the set of all possible unordered pairs of distinct elements in V. For example, for the set $V = \{1, 2, 3, 4\}$, then $E = \{(1, 2), (1, 3), (1, 4), (2, 3), (2, 4), (3, 4)\}$. Let X be the set of all colorings of E by two colors, with one color assigned to the pairs in E in a particular subset T of edges, and the other assigned to the pairs in E that do not appear in T. In the graph represented by the preceding figure, we have indicated the color assigned to the pairs in $T = \{(1, 2), (1, 3), (3, 4)\}$ by connecting with a line the dots that represent the vertices in these pairs, and the color assigned to the pairs that do not appear in T by not connecting with a line the dots that represent the vertices in the pairs.

Our goal in this section is to devise a method for counting the number of nonequivalent undirected graphs with n vertices for a positive integer n, with two graphs equivalent if there is a bijection on their vertices that induces a bijection on their edges. Since a bijection on the vertices of a graph with n vertices can be viewed as an element in S_n acting on the vertices of the graph, the number of nonequivalent undirected graphs with

n vertices will equal the number of distinct equivalence classes in the set of all colorings when S_n acts on the vertices of the graphs. In the following examples we illustrate how Pólya theory can be used to count the number of nonequivalent undirected graphs with a given number of vertices.

Example 12.6 Let $V = \{1, 2, 3\}$, for which the set X of all possible colorings of $E = \{e_1 = (1,2), e_2 = (1,3), e_3 = (2,3)\}$ by two colors can be represented by the following eight figures.

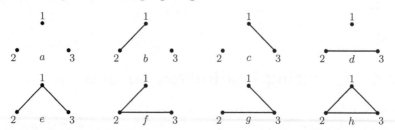

Two colorings in X are equivalent if one can be transformed into the other by the action of an element in S_3 on the vertices of the graph. It is not difficult to see that there are four equivalence classes in X, which indicates that there are four nonequivalent undirected graphs with three vertices. Consider the following table.

Element in S_3	Action on E	Action on X	Monomial for Column 2	Total Number
$(1)(2)(3)$	$(e_1)(e_2)(e_3)$	identity	$(x_1)^3$	1
$(12)(3)$	$(e_1)(e_2e_3)$	$(cd)(ef)$	x_1x_2	3
(123)	$(e_1e_3e_2)$	$(bdc)(efg)$	x_3	2

The values in the column on the right in the preceding table can be determined by using the formula given in Lemma 12.1. The resulting cycle index is the following.

$$f(x_1, x_2, x_3) = \frac{1}{6}(x_1^3 + 3x_1x_2 + 2x_3)$$

Thus, the calculation $f(2,2,2) = \frac{1}{6}(8 + 12 + 4) = 4$ also shows that there are four equivalence classes in X, or four nonequivalent undirected graphs with three vertices. □

Example 12.7 Consider $V = \{1, 2, 3, 4, 5\}$, for which the set X of all possible colorings of $E = \{e_1 = (1,2), e_2 = (1,3), e_3 = (1,4), e_4 = (1,5), e_5 = (2,3), e_6 = (2,4), e_7 = (2,5), e_8 = (3,4), e_9 = (3,5), e_{10} = (4,5)\}$ has order $|X| = 2^{10}$. Consider the following table.

Element in S_5	Action on E	Monomial for Column 2	Total Number
$(1)(2)(3)(4)(5)$	identity	$(x_1)^{10}$	1
$(12)(3)(4)(5)$	$(e_2e_5)(e_3e_6)(e_4e_7)$	$(x_1)^4(x_2)^3$	10
$(12)(34)(5)$	$(e_2e_6)(e_3e_5)(e_4e_7)(e_9e_{10})$	$(x_1)^2(x_2)^4$	15
$(123)(4)(5)$	$(e_1e_5e_2)(e_3e_6e_8)(e_4e_7e_9)$	$x_1(x_3)^3$	20
$(123)(45)$	$(e_1e_5e_2)(e_3e_7e_8e_4e_6e_9)$	$x_1x_3x_6$	20
$(1234)(5)$	$(e_1e_5e_8e_3)(e_2e_6)(e_4e_7e_9e_{10})$	$x_2(x_4)^2$	30
(12345)	$(e_1e_5e_8e_{10}e_4)(e_2e_6e_9e_3e_7)$	$(x_5)^2$	24

As in Example 12.6, the values in the column on the right in the preceding table can be determined by using the formula given in Lemma 12.1. The resulting cycle index is the following.

$$f(x_1, x_2, x_3, x_4, x_5, x_6)$$
$$= \frac{1}{120}\left(x_1^{10} + 10x_1^4x_2^3 + 15x_1^2x_2^4 + 20x_1x_3^3 + 20x_1x_3x_6 + 30x_2x_4^2 + 24x_5^2\right)$$

Thus, the following calculation shows the number of equivalence classes in X, or the number of nonequivalent undirected graphs with five vertices.

$$f(2,2,2,2,2,2) = \frac{1}{120}(1024 + 1280 + 960 + 320 + 160 + 240 + 96) = 34$$

□

We will now determine the cycle index of S_n acting on the set E of all possible edges of an undirected graph with n vertices for an arbitrary positive integer n. Note first that the number of elements in E will be $|E| = \binom{n}{2}$. Also, recall for any $\pi \in S_n$, the action of π on E is naturally induced by the action of π on the set $V = \{1, 2, \ldots, n\}$ of vertices. For example, with the enumeration of E in Example 12.7, the induced action of $(123)(45) \in S_5$ on E is $(e_1e_5e_2)(e_3e_7e_8e_4e_6e_9)$, as noted in the second column and fifth row of the table in Example 12.7. In general, graphs that are equivalent under any $\pi \in S_n$ will arise from the following cases.

Case 1: Let $E_1 = \{(a, b) \mid a, b \text{ are in the same cycle } \tau \text{ of } \pi \text{ acting on } V\}$, where $|\tau| = k$ and k is odd. Then $|E_1| = \binom{k}{2}$, and since each element in E_1 will be in a cycle of length k in the induced action of π on E, the number of cycles of length k in the induced action will be the following.

$$\frac{\binom{k}{2}}{k} = \frac{k-1}{2}$$

For each cycle τ of length k in π acting on V, the monomial for the resulting cycle index will be $(x_k)^{\frac{k-1}{2}}$. Also, if there are p_k cycles of length k in π acting on V, then this monomial will be raised to the power p_k, yielding the entire monomial $(x_k)^{\frac{k-1}{2} \cdot p_k}$ for the resulting cycle index.

Example 12.8 Consider S_5 acting on V with induced action on E for the sets V and E in Example 12.7. For the element $\pi = \tau = (12345) \in S_5$, using the notation in the Case 1 summary, $k = 5$, $E_1 = E$, and in the induced action of π on E there are $\frac{5-1}{2} = 2$ cycles of length 5. The two cycles of length 5 in the induced action of π on E are $(e_1 e_5 e_8 e_{10} e_4)$ and $(e_2 e_6 e_9 e_3 e_7)$, which are shown in the second column and last row of the table in Example 12.7. For the resulting cycle index, this yields the monomial $(x_5)^{\frac{5-1}{2} \cdot 1} = (x_5)^2$, which is shown in the third column and last row of the table in Example 12.7.   □

Case 2: Let $E_2 = \{(a, b) \mid a, b$ are in the same cycle τ of π acting on $V\}$, where $|\tau| = k$ and k is even. Then $|E_2| = \binom{k}{2}$, and in the induced action of π on E, there will be one cycle of length $k/2$, and the following number of cycles of length k.

$$\frac{\binom{k}{2} - \frac{k}{2}}{k} = \frac{k-2}{2}$$

For each cycle τ of length k in π acting on V, the monomial for the resulting cycle index will be $x_{k/2}(x_k)^{\frac{k-1}{2}}$. Also, if there are p_k cycles of length k in π acting on V, then this monomial will be raised to the power p_k, yielding the entire monomial $(x_{k/2})^{p_k}(x_k)^{\frac{k-2}{2} \cdot p_k}$ for the resulting cycle index.

Example 12.9 Consider again S_5 acting on V with induced action on E for the sets V and E in Example 12.7. For the cycle $\tau = (1234)$ in the element $\pi = (1234)(5) \in S_5$, using the notation in the Case 2 summary, $k = 4$, $E_2 = \{e_1 = (1, 2), e_2 = (1, 3), e_3 = (1, 4), e_5 = (2, 3), e_6 = (2, 4), e_8 = (3, 4)\}$, and in the induced action of π on E there is one cycle of length $4/2 = 2$ and $\frac{4-2}{2} = 1$ cycle of length 4. The cycles of lengths 2 and 4 in the induced action of π on E are $(e_2 e_6)$ and $(e_1 e_5 e_8 e_3)$, which are two of the cycles shown in the second column and next-to-last row of the table in Example 12.7. For the resulting cycle index, these cycles yield the monomial $(x_{4/2})^1(x_4)^{\frac{4-2}{2} \cdot 1} = x_2 x_4$, which are two of the factors shown in the third column and next-to-last row of the table in Example 12.7. To complete the induced action of π on E and construct the monomial for the resulting cycle index, we must consider an additional case, which we will do later in Example 12.12.   □

Case 3: Let $E_3 = \{(a, b) \mid a, b$ are in differing cycles τ_1, τ_2, respectively, of π acting on $V\}$, where $|\tau_1| = |\tau_2| = s$. Then $|E_3| = s^2$, and each element in E_3 will be contained in a cycle of length s in the induced action of π on E. Since there will be a total of s such unique cycles in the induced action of π on E, the monomial for the resulting cycle index will be $(x_s)^s$. Also, if there are p_s cycles of length s in π acting on V, then the number of possible pairs of cycles for τ_1 and τ_2 will be $\binom{p_s}{2}$, yielding the entire monomial $(x_s)^{s \cdot \binom{p_s}{2}}$ for the resulting cycle index.

Example 12.10 Consider again S_5 acting on V with induced action on E for the sets V and E in Example 12.7. For the cycles $\tau_1 = (12)$ and $\tau_2 = (34)$ in the element $\pi = (12)(34)(5) \in S_5$, using the notation in the Case 3 summary, $s = 2$, $E_3 = \{e_2 = (1,3), e_3 = (1,4), e_5 = (2,3), e_6 = (2,4)\}$, and in the induced action of π on E there are two cycles of length 2. The two cycles of length 2 in the induced action of π on E are $(e_2 e_6)$ and $(e_3 e_5)$, which are two of the cycles shown in the second column and third row of the table in Example 12.7. For the resulting cycle index, these cycles yield the monomial $(x_2)^2$, which is part of the entry shown in the third column and third row of the table in Example 12.7. To complete the induced action of π on E and construct the monomial for the resulting cycle index, we must consider two additional cases, which we will do later in Example 12.12. $\square$

Case 4: Let $E_4 = \{(a, b) \mid a, b$ are in differing cycles τ_1, τ_2, respectively, of π acting on $V\}$, where $|\tau_1| = s$ and $|\tau_2| = t$ with $s \neq t$. Then $|E_2| = st$, and the action of $\tau_1 \tau_2$ on V induces an action on E that would be represented by a cycle with length given by the least common multiple $\mathrm{lcm}(s, t)$ of s and t. Since there will be a total of $\frac{st}{\mathrm{lcm}(s,t)} = \gcd(s, t)$ such unique cycles in the induced action of π on E, the monomial for the resulting cycle index will be $(x_{\mathrm{lcm}(s,t)})^{\gcd(s,t)}$. Also, if there are p_s cycles of length s and p_t cycles of length t in π acting on V, then the number of possible pairs of cycles for τ_1 and τ_2 will be $p_s \cdot p_t$, yielding the entire monomial $(x_{\mathrm{lcm}(s,t)})^{\gcd(s,t) \cdot p_s \cdot p_t}$ for the resulting cycle index.

Example 12.11 Consider again S_5 acting on V with induced action on E for the sets V and E in Example 12.7. For the cycles $\tau_1 = (123)$ and $\tau_2 = (45)$ in the element $\pi = (123)(45) \in S_5$, using the notation in the Case 4 summary, $s = 3$, $t = 2$, $E_4 = \{e_3 = (1,4), e_4 = (1,5), e_6 = (2,4), e_7 = (2,5), e_8 = (3,4), e_9 = (3,5)\}$, and in the induced action of π on E there is a cycle of length 6. The cycle of length 6 in the induced action of π on E is $(e_3 e_7 e_8 e_4 e_6 e_9)$, which is one of the cycles shown in the second column and fifth row of the table in Example 12.7. For the resulting cycle index, this cycle yields the monomial $(x_{\mathrm{lcm}(3,2)})^{\gcd(3,2) \cdot 1 \cdot 1} = x_6$, which is one of the factors shown in the third column and fifth row of the table in Example 12.7. To complete the induced action of π on E and construct the monomial for the resulting cycle index, we must consider two additional cases, which we will do later in Example 12.12.   □

In summary of these cases, for any $\pi \in S_n$, the action of π on V naturally induces an action on E, which produces the following types of factors in the monomials for the resulting cycle index.

$$(x_k)^{\frac{k-1}{2} \cdot p_k}$$
(when the edge components are in the same cycle of odd length k)

$$(x_{k/2})^{p_k}(x_k)^{\frac{k-2}{2} \cdot p_k}$$
(when the edge components are in the same cycle of even length k)

$$(x_s)^{s \cdot \binom{p_s}{2}}$$
(when the edge components are in different cycles of the same length s)

$$(x_{\mathrm{lcm}(s,t)})^{\gcd(s,t) \cdot p_s \cdot p_t}$$
(when the edge components are in different cycles of differing lengths s and t)

In addition, recall as we presented in Section 12.1, each of the elements in S_n can be viewed as being indicated by a partition of the integer n, and the partition of n that indicates a particular element $\pi \in S_n$ completely determines the expression of π using cycle notation. Thus, with the types of factors for the monomials that we just listed, an element $\pi \in S_n$ whose expression using cycle notation is determined by the partition $P = [p_1, p_2, \ldots, p_n]$ of the integer n contributes the following towards the cycle index of S_n acting on the set E of all possible edges of an undirected graph with n vertices.

$$\prod_{\substack{i \text{ odd}}}^{n} (x_i)^{\frac{i-1}{2} \cdot p_i} \prod_{\substack{i \text{ even}}}^{n} (x_{i/2})^{p_i}(x_i)^{\frac{i-2}{2} \cdot p_i} \prod_{i}^{n} (x_i)^{i \cdot \binom{p_i}{2}} \prod_{s<t}^{n} (x_{\mathrm{lcm}(s,t)})^{\gcd(s,t) \cdot p_s \cdot p_t}$$

Example 12.12 Consider again S_5 acting on V with induced action on E for the sets V and E in Example 12.7. For the partition $[0,0,0,0,1]$ of the integer 5, which indicates the element $(12345) \in S_5$, Case 1 produces the monomial $(x_5)^2$, which is shown in the third column and last row of the table in Example 12.7. For the partition $[1,0,0,1,0]$ of the integer 5, which indicates the element $(1234)(5) \in S_5$, Case 2 produces the monomial $x_2 x_4$, and Case 4 produces the monomial x_4. This results in the entire monomial $x_2(x_4)^2$, which is shown in the third column and next-to-last row of the table in Example 12.7. For the partition $[1,2,0,0,0]$ of the integer 5, which indicates the element $(12)(34)(5) \in S_5$, Case 2 produces the monomial $(x_1)^2$, Case 3 produces the monomial $(x_2)^2$, and Case 4 produces the monomial $(x_2)^2$. This results in the entire monomial $(x_1)^2(x_2)^4$, which is shown in the third column and third row of the table in Example 12.7. Finally, for the partition $[0, 1, 1, 0, 0]$ of the integer 5, which indicates the element $(123)(45) \in S_5$, Case 1 produces the monomial x_3, Case 2 produces the monomial x_1, and Case 4 produces the monomial x_6. This results in the entire monomial $x_1 x_3 x_6$, which is shown in the third column and fifth row of the table in Example 12.7. $\square$

If we use the expression that appears immediately before Example 12.12 and the formula given in Lemma 12.1, and sum over all of the elements in S_n, or, equivalently, all of the partitions P of the positive integer n, we have the following theorem.

Theorem 12.3 *The cycle index of S_n acting on the set E of all possible edges of an undirected graph with n vertices is given by the following formula.*

$$f\left(x_1, x_2, \ldots, x_{max(lcm(s,t))}\right)$$

$$= \sum_P \left(\prod_{j=1}^n \frac{1}{j^{p_j} p_j!} \prod_{i \text{ odd}}^n (x_i)^{\frac{i-1}{2} \cdot p_i} \prod_{i \text{ even}}^n (x_{i/2})^{p_i} (x_i)^{\frac{i-2}{2} \cdot p_i} \prod_i^n (x_i)^{i \cdot \binom{p_i}{2}} \right.$$

$$\left. \prod_{s<t}^n (x_{lcm(s,t)})^{gcd(s,t) \cdot p_s \cdot p_t} \right)$$

Example 12.13 In this example, we will use Theorem 12.3 to construct the cycle index of S_3 acting on the set E of all possible edges of an undirected graph with three vertices. To do this, we will first construct the following table.

Partition of 3	Monomial	Coefficient
$[3, 0, 0]$	$(x_1)^3$	$\frac{1}{6}$
$[1, 1, 0]$	$x_1 x_2$	$\frac{1}{2}$
$[0, 0, 1]$	x_3	$\frac{1}{3}$

Summing over all of the partitions of 3, we obtain the following cycle index.

$$f(x_1, x_2, x_3) = \frac{1}{6}x_1^3 + \frac{1}{2}x_1 x_2 + \frac{1}{3}x_3$$

We can then use this cycle index to determine that the number of nonequivalent undirected graphs with three vertices is $f(2, 2, 2) = 4$. We can also use the cycle index to construct a pattern inventory, which would allow us to determine the number of nonequivalent undirected graphs with three vertices that have prescribed numbers of edges. With only two colors, and since we are ultimately only interested in graphs in which edges occur, we will use the number 1 to represent the color to be assigned to the ordered pairs in E that do not appear in a particular set T of edges, and the symbol $\mathcal{E}$ to represent the color to be assigned to the ordered pairs in E that do appear in T. This yields the following pattern inventory.

$$
\begin{aligned}
f(1+\mathcal{E}, 1+\mathcal{E}^2, 1+\mathcal{E}^3) &= \frac{1}{6}(1+\mathcal{E})^3 + \frac{1}{2}(1+\mathcal{E})(1+\mathcal{E}^2) + \frac{1}{3}(1+\mathcal{E}^3) \\
&= 1 + \mathcal{E} + \mathcal{E}^2 + \mathcal{E}^3
\end{aligned}
$$

The coefficients and exponents in this pattern inventory indicate that of the four nonequivalent undirected graphs with three vertices, there is 1 graph with 0 edges, 1 graph with 1 edge, 1 graph with 2 edges, and 1 graph with 3 edges.   □

Example 12.14 In this example, we will use Theorem 12.3 to construct the cycle index of S_5 acting on the set E of all possible edges of an undirected graph with five vertices. To do this, we will first construct the following table.

Partition of 5	Monomial	Coefficient
$[5,0,0,0,0]$	$(x_1)^{10}$ $\cdot$	$\frac{1}{120}$
$[3,1,0,0,0]$	$(x_1)^4(x_2)^3$	$\frac{1}{12}$
$[1,2,0,0,0]$	$(x_1)^2(x_2)^4$	$\frac{1}{8}$
$[2,0,1,0,0]$	$x_1(x_3)^3$	$\frac{1}{6}$
$[0,1,1,0,0]$	$x_1 x_3 x_6$	$\frac{1}{6}$
$[1,0,0,1,0]$	$x_2(x_4)^2$	$\frac{1}{4}$
$[0,0,0,0,1]$	$(x_5)^2$	$\frac{1}{5}$

Summing over all of the partitions of 5, we obtain the following cycle index.

$f(x_1, x_2, x_3, x_4, x_5, x_6)$

$$
= \frac{1}{120}x_1^{10} + \frac{1}{12}x_1^4 x_2^3 + \frac{1}{8}x_1^2 x_2^4 + \frac{1}{6}x_1 x_3^3 + \frac{1}{6}x_1 x_3 x_6 + \frac{1}{4}x_2 x_4^2 + \frac{1}{5}x_5^2
$$

We can then use this cycle index to determine that the number of nonequivalent undirected graphs with five vertices is $f(2,2,2,2,2,2) = 34$. We can also form the following pattern inventory, which will allow us to determine the number of nonequivalent undirected graphs with five vertices that have prescribed numbers of edges.

$f(1+\mathcal{E}, 1+\mathcal{E}^2, 1+\mathcal{E}^3, 1+\mathcal{E}^4, 1+\mathcal{E}^5, 1+\mathcal{E}^6)$

$$= 1 + \mathcal{E} + 2\mathcal{E}^2 + 4\mathcal{E}^3 + 6\mathcal{E}^4 + 6\mathcal{E}^5 + 6\mathcal{E}^6 + 4\mathcal{E}^7 + 2\mathcal{E}^8 + \mathcal{E}^9 + \mathcal{E}^{10}$$

Thus, for example, from this pattern inventory we can see that of the 34 nonequivalent undirected graphs with 5 vertices, there are 6 graphs with 4 edges, and 2 graphs with 8 edges. $\qquad\square$

12.5 Counting Undirected Graphs with Maple

In this section, we will show how Maple can be used to construct the cycle index of S_n acting on the set E of all possible edges of an undirected graph with n vertices, and the resulting pattern inventory.

In order to construct the cycle index of S_n acting on the set E of all possible edges of an undirected graph with n vertices, we will use the user-written function **igraph**, which we have written separately from this Maple session and saved as the text file igraph.mpl. To use this function, we must first read it into our Maple session as follows.

```
>  read "igraph.mpl";
```

The following command then illustrates how the function **igraph** can be used. Specifically, by entering the following command, we construct the cycle index of S_3 acting on the set E of all possible edges of an undirected graph with three vertices.

```
>  f := igraph(3, x);
```

$$f := \frac{1}{6}{x_1}^3 + \frac{1}{2}x_1 x_2 + \frac{1}{3}x_3$$

The second parameter in the preceding command specifies the variable to be used in the resulting cycle index. In the following command we use **unapply** to convert this cycle index from an expression into a function.

```
>  f := unapply(f, x[1], x[2], x[3]);
```

$$f := (x\_1, \ x\_2, \ x\_3) \rightarrow \frac{1}{6}{x\_1}^3 + \frac{1}{2}x\_1\,x\_2 + \frac{1}{3}x\_3$$

We can now use this cycle index as follows to determine the number of nonequivalent undirected graphs with three vertices.

```
>  f(2, 2, 2);
```

We can also construct the pattern inventory that results from this cycle index by entering the following command.

```
>  sort(simplify(f(1+E, 1+E^2, 1+E^3)));
```
$$E^3 + E^2 + E + 1$$

Next, we will enter the following command to construct the cycle index of S_5 acting on the set E of all possible edges of an undirected graph with five vertices. Note that we include a third parameter *maxsub* in this command. The value returned for this parameter will be the largest subscript on the variables in the resulting cycle index. The single quotes are necessary in this parameter to make sure that the variable name is unassigned before being defined by the function.

```
>  f := igraph(5, x, 'maxsub');
```
$$f := \frac{1}{120} x_1^{10} + \frac{1}{12} x_1^4 x_2^3 + \frac{1}{8} x_1^2 x_2^4 + \frac{1}{6} x_3^3 x_1 + \frac{1}{6} x_3 x_1 x_6 + \frac{1}{4} x_2 x_4^2$$
$$+ \frac{1}{5} x_5^2$$

```
>  maxsub;
```
$$6$$

Having the largest subscript on the variables in the preceding cycle index returned by the **igraph** command is useful because it prevents us from having to search the cycle index to find it. In the following command we use *maxsub* in conjunction with the **seq** and **unapply** functions to convert the cycle index from an expression into a function.

```
>  f := unapply(f, seq(x[i], i=1..maxsub));
```
$$f := (x\_1, x\_2, x\_3, x\_4, x\_5, x\_6) \rightarrow \frac{1}{120} x\_1^{10} + \frac{1}{12} x\_1^4 x\_2^3$$
$$+ \frac{1}{8} x\_1^2 x\_2^4 + \frac{1}{6} x\_3^3 x\_1 + \frac{1}{6} x\_3 x\_1 x\_6 + \frac{1}{4} x\_2 x\_4^2$$
$$+ \frac{1}{5} x\_5^2$$

We can now use this cycle index as follows to determine the number of nonequivalent undirected graphs with five vertices.

```
>  f(seq(2, i=1..maxsub));
```
$$34$$

We can also construct the pattern inventory that results from this cycle index by entering the following command.

```
>   sort(simplify(f(seq(1+E^i, i=1..maxsub))));
```

$$E^{10} + E^9 + 2E^8 + 4E^7 + 6E^6 + 6E^5 + 6E^4 + 4E^3 + 2E^2 + E + 1$$

12.6 Counting Undirected Graphs with MATLAB

In this section, we will show how MATLAB can be used to construct the cycle index of S_n acting on the set E of all possible edges of an undirected graph with n vertices, and the resulting pattern inventory.

We will begin by declaring the variables x and E that we will use in this section as symbolic variables.

```
>> syms x E
```

Next, to construct the cycle index of S_n acting on the set E of all possible edges of an undirected graph with n vertices, we will use the user-written function **igraph**, which we have written separately from this MATLAB session and saved as the M-file igraph.m. This function uses the user-written function **partlist** that we used in Section 12.3, which recall we have written separately from this MATLAB session and saved as the M-file partlist.m. The following command illustrates how the function **igraph** can be used. Specifically, by entering the following command, we construct the cycle index of S_3 acting on the set E of all possible edges of an undirected graph with three vertices.

```
>> cires = igraph(3, x);
```

The second parameter in the preceding command specifies the variable to be used in the resulting cycle index. The **igraph** function returns two entries in a cell array that we have assigned as *cires*. The first entry in *cires* is the resulting cycle index. By entering the following command we store this cycle index as the variable f.

```
>> f = cires{1};
>> pretty(f)
```

```
   3
 x1      x2 x1    x3
 ---  +  -----  + --
  6        2       3
```

Next, we will convert this cycle index from an expression into a function. To do this, we will use the second entry in *cires*, which is a list of the variables used in the resulting cycle index.

```
>> fvar = cires{2};
>> [fvar{:}]

ans =

[ x1, x2, x3]
```

To convert the cycle index stored as the variable f from an expression into a function, we can enter the following command.

```
>> f = @(x) sym(subs(f, {fvar{:}}, x));
```

We can now use this cycle index as follows to determine the number of nonequivalent undirected graphs with three vertices.

```
>> f({2, 2, 2})

ans =

4
```

We can also construct the pattern inventory that results from this cycle index by entering the following command.

```
>> feval(symengine, 'expand', f({1+E, 1+E^2, 1+E^3}))

ans =

E^3 + E^2 + E + 1
```

Next, we will enter the following command to construct the cycle index of S_5 acting on the set E of all possible edges of an undirected graph with five vertices.

```
>> cires = igraph(5, x);
>> f = cires{1};
>> pretty(f)

    10        4   3     2   4       3                      2       2
    x1       x1  x2    x1  x2     x1 x3       x6 x1 x3    x2 x4    x5
   ---- +  ------- +  ------- +  ------ +  --------- +  ------ +  ---
    120        12         8          6          6            4       5
```

Next, we will convert this cycle index from an expression into a function.

```
>> fvar = cires{2};
>> [fvar{:}]

ans =

[ x1, x2, x3, x4, x5, x6]

>> f = @(x) sym(subs(f, {fvar{:}}, x));
```

To find the number of nonequivalent undirected graphs with five vertices, and to construct the pattern inventory that results from the preceding cycle index, we enter the following **for** loop. In this loop, we construct an array g that contains the inputs into the cycle index that result in the number of nonequivalent undirected graphs with five vertices, and an array p that contains the inputs into the cycle index that result in the pattern inventory.

```
>> for i = 1:numel(fvar)
        g{i} = sym('2');
        p{i} = 1+E^i;
    end
>> [g{:}]

ans =

[ 2, 2, 2, 2, 2, 2]

>> [p{:}]

ans =

[ E + 1, E^2 + 1, E^3 + 1, E^4 + 1, E^5 + 1, E^6 + 1]
```

Using g, we can now determine the number of nonequivalent undirected graphs with five vertices by entering the following command.

```
>> f({g{:}})

ans =

34
```

Also, using p, we can construct the resulting pattern inventory by entering the following command.

```
>> feval(symengine, 'expand', f({p{:}}))

ans =

E^10 + E^9 + 2*E^8 + 4*E^7 + 6*E^6 + 6*E^5 + 6*E^4 + 4*E^3
+ 2*E^2 + E + 1
```

Exercises

1. Make a list of all of the partitions of the integer $n = 4$, with each partition expressed both as a set and as an ordered list of the number of times that the integer i appears in the set for $i = 1, 2, \ldots, n$.

2. Make a list of all of the partitions of the integer $n = 6$, with each partition expressed both as a set and as an ordered list of the number of times that the integer i appears in the set for $i = 1, 2, \ldots, n$.

3. Construct the cycle index of S_4 acting on the set $\{1, 2, 3, 4\}$.

4. Construct the cycle index of S_6 acting on the set $\{1, 2, 3, 4, 5, 6\}$.

5. Find the number of nonequivalent colorings of $S = \{1, 2, 3, 4\}$ by three different types of objects, where two colorings are equivalent if they both assign the three different types of objects to the same number of elements in S.

6. Find the number of nonequivalent colorings of $S = \{1, 2, 3, 4, 5, 6\}$ by four different types of objects, where two colorings are equivalent if they both assign the four different types of objects to the same number of elements in S.

7. For the set $V = \{1, 2, 3, 4\}$ of vertices, make a list of all of the elements in the set E of all possible unordered pairs of distinct elements in V. Then draw figures that can be viewed as representations of the elements in the set X of all possible colorings of E by two colors (similarly to how we drew figures in Example 12.6 starting with the set $V = \{1, 2, 3\}$).

8. For the set $V = \{1, 2, 3, 4, 5, 6\}$ of vertices, make a list of all of the elements in the set E of all possible unordered pairs of distinct elements in V. If you drew figures that could be viewed as representations of the elements in the set X of all possible colorings of E by two colors (similarly to how we drew figures in Example 12.6 starting with the set $V = \{1, 2, 3\}$), how many total figures would there be?

9. Determine the cycle index of S_4 acting on the set E of all possible edges of an undirected graph with four vertices, and use this cycle index to find the number of nonequivalent undirected graphs with four vertices. Then divide the figures that you drew in Exercise 7 into this number of subsets, with each subset containing graphs that are all equivalent to each other. Finally, use the cycle index to construct a pattern inventory, and verify that the coefficients and exponents in this pattern inventory are consistent with the subsets into which you divided the figures that you drew in Exercise 7.

10. Determine the cycle index of S_6 acting on the set E of all possible edges of an undirected graph with six vertices, and use this cycle index to find the number of nonequivalent undirected graphs with six vertices.

11. Give one example of each of the 34 different types of nonequivalent undirected graphs with five vertices.

12. Give one example of each of the different types of nonequivalent undirected graphs with six vertices.

13. Prove Theorem 12.2.

Computer Exercises

14. Make a list of all of the partitions of the integer $n = 7$, with each partition expressed as an ordered list of the number of times that the integer i appears in the set form of the partition for $i = 1, 2, \ldots, n$.

15. Make a list of all of the partitions of the integer $n = 10$, with each partition expressed as an ordered list of the number of times that the integer i appears in the set form of the partition for $i = 1, 2, \ldots, n$.

16. Construct the cycle index of S_7 acting on the set $\{1, 2, \ldots, 7\}$, and use this cycle index to determine the number of nonequivalent colorings of the set $S = \{1, 2, \ldots, 7\}$ by four different types of objects, where two colorings are equivalent if they both assign the four different types of objects to the same number of elements in S.

17. Construct the cycle index of S_{10} acting on the set $\{1, 2, \ldots, 10\}$, and use this cycle index to determine the number of nonequivalent colorings of the set $S = \{1, 2, \ldots, 10\}$ by six different types of objects, where two colorings are equivalent if they both assign the six different types of objects to the same number of elements in S.

18. Determine the cycle index of S_7 acting on the set E of all possible edges of an undirected graph with seven vertices, and use this cycle index to find the number of nonequivalent undirected graphs with seven vertices.

19. Determine the cycle index of S_{10} acting on the set E of all possible edges of an undirected graph with ten vertices, and use this cycle index to find the number of nonequivalent undirected graphs with ten vertices.

20. Of the number of nonequivalent undirected graphs with seven vertices that you determined in Exercise 18, how many contain 12 edges?

21. Of the number of nonequivalent undirected graphs with ten vertices that you determined in Exercise 19, how many contain 18 edges?

Research Exercises

22. An *undirected multigraph* with n vertices is a set V of order n and a set T of (possibly repeated) unordered pairs of distinct elements in V. Investigate how the theory that we presented in this chapter can be modified, if possible, for counting the number of nonequivalent undirected multigraphs with n vertices, and write a summary of your findings.

23. An *undirected pseudograph* with n vertices is a set V of order n and a set T of (distinct) unordered pairs of (possibly repeated) elements in V. Investigate how the theory that we presented in this chapter can be modified, if possible, for counting the number of nonequivalent undirected pseudographs with n vertices, and write a summary of your findings.

24. A *directed graph* with n vertices is a set V of order n and a set T of (distinct) ordered pairs of distinct elements in V. Investigate how the theory that we presented in this chapter can be modified, if possible, for counting the number of nonequivalent directed graphs with n vertices, and write a summary of your findings.

25. Investigate some specific details about at least two applications of undirected graphs, and write a summary of your findings.

Chapter 13

Symmetry in Western Music

In this chapter[1] we will use group actions to investigate questions regarding symmetry in Western music. Musicians refer to the interval between pitches whose frequencies occur in a 2:1 ratio as an *octave*. For example, in the classic song *Somewhere Over the Rainbow*, the pitches for *Some* and *where* are an octave apart. By humming these pitches, we can hear that they sound good together. In fact, they sound so good together that musicians use the same name for both. For instance, the frequencies 440 Hz and 880 Hz are both musical pitches, or *notes*, labeled as A.

Western musicians divide octaves into 12 distinct equally spaced notes, each separated from its nearest neighbors by a *semitone*. The reasons behind the decisions to partition octaves into 12 equally spaced notes and to use the same name for notes separated by one or more exact octaves is interesting. An accessible and excellent introduction to some of the history, physics, and mathematics related to these decisions can be found in [13].

All notes separated by one or more exact octaves form a *pitch class*. Thus, Western music partitions notes into 12 distinct pitch classes, which are labeled as follows.

$$\left\{ C, C^\sharp = D^\flat, D, D^\sharp = E^\flat, E, F, F^\sharp = G^\flat, G, G^\sharp = A^\flat, A, A^\sharp = B^\flat, B \right\}$$

This is in the same manner in which $\mathbb{Z}_{12}$ partitions $\mathbb{Z}$ into 12 distinct equivalence classes. These equivalence classes will serve as our mathematical model for Western music, as illustrated in the portion of a piano keyboard shown in Figure 13.1.

---

[1]The first draft of this chapter was written by Vicky W. Klima (klimavw@appstate.edu), Appalachian State University, Boone, North Carolina.

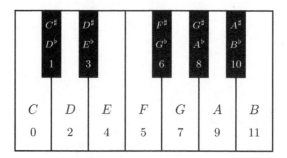

Figure 13.1   Mathematical model for Western music, $\mathbb{Z}_{12}$.

For labeling notes, a sharp ($\sharp$) indicates an increase of one semitone, while a flat ($\flat$) indicates a decrease of one semitone. Since, for example, the note one semitone above $C = 0$ is identical to the note one semitone below $D = 2$, then $C^{\sharp} = D^{\flat} = 1$.

We begin this chapter in Section 13.1 with a discussion of pitch class segments, a mathematical tool used by music theorists to study structural properties of music. Pitch class segments are simply sets of numbers, with each number representing a pitch class. If we think of the numbers in a pitch class segment as unordered and played simultaneously, the segment can be viewed as representing a chord. If we think of the numbers in a pitch class segment as ordered and played successively, the segment can be viewed as representing a scale. In Section 13.1 we also introduce two important operations on pitch class segments—transposition, which is addition of a fixed number of semitones to each element in a segment, and inversion, which is reflection of each element in a segment about a fixed axis.

In Section 13.2 we discuss major and minor diatonic scales. This will include a discussion of the circle of fifths, which is used by music theorists to visualize relationships between scales. The white keys on a piano keyboard beginning at any C and going up one octave form the C major diatonic scale, and every major diatonic scale is a transposition of C major. The white keys beginning at any A and going up one octave form the A minor diatonic scale, and every minor diatonic scale is a transposition of A minor. The following is from *The Harvard Dictionary of Music* [27], a standard reference readers inexperienced with studying music may find useful:

> In tonal music, from the late 17th century on, only two [diatonic scales] are at work: major and minor. Since major and minor scales can be formed on any of the twelve available pitch classes, any given example is named for its starting pitch or tonic, e.g., C major, G minor. A composition based primarily on a given scale is said to be in the key of that scale, e.g., C major, G minor. An

important property of these scales is that the twelve different starting points or tonics produce twelve different scales, no two of which contain precisely the same pitch classes and each of which is uniquely and precisely related to all of the others.

In Section 13.3 we discuss major and minor triads, called *consonant triads* since they are pleasant sounding. Consonant triads are defined in terms of the musical intervals of the perfect fifth (seven semitones), the major third (four semitones), and the minor third (three semitones). A major triad contains a root and notes a major third and perfect fifth above the root, while a minor triad contains a root and notes a minor third and perfect fifth above the root. From *The Harvard Dictionary of Music* [27]:

> The concepts of triad and triad inversion have been central to discussions of tonality since the 18th century, the first explicit theoretical accounts of them dating from around 1600. ... The major and minor triads themselves occur regularly in English music beginning in the 13th century. They are prominent in continental music from the 15th century on, in part because of an increasing preference for four voice texture, combined with the fact that the major and minor triads are the only consonant simultaneities of more than two pitches that can be produced from the intervals regarded as consonant at the time.

Transposition and inversion are key to the study of triadic music, and so the action of D_{12} on pitch class segments representing triads will play a notable role in our investigations. We will also use the Cayley graph of $\mathbb{Z}_{12}$ with generating set $\{3, 4, 7\}$ in Section 13.6 to note relationships between major and minor triads. Many of these ideas were introduced in [8].

In Section 13.7 we address modern twelve-tone composition. From *The Harvard Dictionary of Music* [27]:

> [Twelve-tone composition] is most commonly applied to music by Arnold Schoenberg and his followers. ... In Schoenberg's twelve-tone music, the twelve pitches of the chromatic scale are ordered into a row, or series, that provides the basic pitch structure (or Grundgestalt) for a given composition and is thus an essential element in the work's fundamental conception. ... The entire pitch structure of the composition is then derived from the row, including its melodic, contrapuntal, and harmonic features. ... In addition to the principal or prime form of the row (designated p), new forms may be derived through three basic operations: the row can be reversed (the retrograde, designated R), inverted (the inversion, I), and both reversed and inverted

(the retrograde-inversion, RI). Moreover, each of these four basic versions may be transposed to begin on a different pitch.

Mathematically, we will consider a twelve-tone row as a permutation of the 12 pitch classes, viewing transposition and inversion as left actions of S_{12} on itself, and retrograde as a right action of S_{12} on itself. The ideas in this section were drawn mainly from [13] and [18].

13.1   Introduction

We can think of each element in $\mathbb{Z}_{12}$ as representing the name of the corresponding pitch class, and as giving the number of semitones the pitch class is above C. The musical clock shown in Figure 13.2 gives a visual representation of these dual meanings.

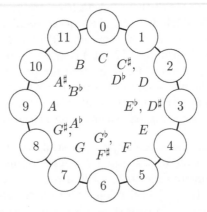

Figure 13.2   The musical clock.

The element $1 \in \mathbb{Z}_{12}$ represents the $C^\sharp$ pitch class, but also a musical interval of one semitone—one arc on the clock. This fits well with our use of $\mathbb{Z}_{12}$ to model pitch classes and their relationships with one another. For example, $(5+10) \bmod 12 = 3$, and if we start at the 5 (or F) pitch class and add 10 semitones (i.e., move 10 positions clockwise on the musical clock), we end on the 3 (or $D^\sharp$) pitch class. The musical clock is essentially a visual representation of $\mathbb{Z}_{12}$ with some additional labels. On the clock each pitch class is represented by one number, but some have more than one name. Musically, a sharp ($\sharp$) indicates a move up by one semitone and a flat ($\flat$) indicates a move down by one semitone. However, moving up from $C = 0$ by one semitone and moving down from $D = 2$ by one semitone both end in the same place. Thus, $C^\sharp = D^\flat = 1$. When a pitch class can be represented by two different names, the names are said to be *enharmonically equivalent*, and can be used interchangeably.

As we consider symmetry in Western music, we will consider the dihedral group D_{12} (with clockwise rotations as positive) acting on the musical clock. We will label the 12 rotations in D_{12} as $T_n = T^n$ for $n = 0, 1, \ldots, 11$, where T is a clockwise rotation of $30°$, and the 12 reflections in D_{12} as $I_n = T^n I_0$, where I_0 is reflection across the vertical. This nonstandard notation comes from music theory, where rotation is known as *transposition*[2] and reflection as *inversion*. Musical transposition and inversion can also operate componentwise on ordered sets of pitch classes known as *pitch class segments*. For example, $T_3(\langle 0, 4, 7 \rangle) = \langle 3, 7, 10 \rangle$ and $I_5(\langle 0, 4, 7 \rangle) = \langle 5, 1, 10 \rangle$, as shown in the Figure 13.3.

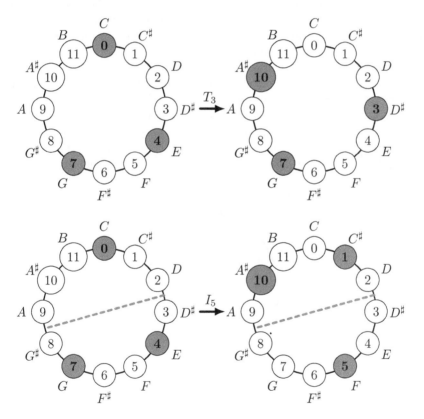

Figure 13.3 Musical actions of T_3 and I_5.

---

[2]In this chapter, we will use the word *transposition* only in its musical context, and refer to permutations that are a single cycle of length two as *2-cycles*.

13.2   Group Actions and Scales

In Western tonal music, composers generally write selections in a *key*, meaning they select at least most of the notes not from all 12 pitch classes, but from a seven-note *diatonic scale*. Each diatonic scale has a *home pitch* around which the melody and harmony are based, and in a tonal work this home pitch is the first note of the major or minor scale used as the pitch collection. This first note of the scale is called the *tonic* and is used to name the scale. The remaining notes are then played in clockwise order around the musical clock. For example, the C *major* scale consists of the white notes on a traditional piano keyboard, $\langle C, D, E, F, G, A, B \rangle = \langle 0, 2, 4, 5, 7, 9, 11 \rangle$. The k *major* scale is the image of C major under the action of T_k. If we rotate the notes in the C major scale on the musical clock, we see that these notes are not fixed under any of the 12 rotations, and thus by Lemma 11.3 there are 12 distinct major diatonic scales. When a diatonic scale is played on an instrument, the pitch class representatives are chosen so that the notes played remain within the octave above the tonic.

The C major scale is shown in Figure 13.4, with its tonic C shaded and the intervals between the notes in the scale emphasized. As the scale goes from C around and back to C, intervals occur in the sequence 2-2-1-2-2-2-1, called the *step intervals* for the scale. Every major diatonic scale is a rotation of C major, and thus has the same step intervals as C major. These step intervals are often used to define a major diatonic scale.

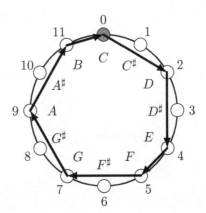

Figure 13.4   Step intervals for the C major scale.

Minor diatonic scales have step intervals 2-1-2-2-1-2-2. If we change our starting point in Figure 13.4 from $C = 0$ to $A = 9$, we see a scale with minor diatonic step intervals. Thus the pitch classes in the C major scale,

$\langle C, D, E, F, G, A, B \rangle = \langle 0, 2, 4, 5, 7, 9, 11 \rangle$, are exactly the pitch classes in the A minor scale, $\langle A, B, C, D, E, F, G \rangle = \langle 9, 11, 0, 2, 4, 5, 7 \rangle$, just arranged in a different order indicating a different tonal center. Musicians refer to such scales as *relative* major and minor scales.

We have viewed major and minor diatonic scales as pitch class segments and step intervals, both based on the musical clock. A third view of these scales requires us to change the way we picture $\mathbb{Z}_{12}$. The group $\mathbb{Z}_{12}$ is cyclic with generators 1, 5, 7, and 11. If we look at the elements around the musical clock in Figure 13.2 starting with 1 and moving clockwise, we see the elements of $\mathbb{Z}_{12}$ in the order in which they are generated by 1. Doing the same with the modified musical clock in Figure 13.5, we see the elements of $\mathbb{Z}_{12}$ in the order in which they are generated by 7. In this figure, note that the pitches in the C major (and A minor) diatonic scale, which are shaded, are connected. Musicians call Figure 13.5 the *circle of fifths*, named for the fact that the musical interval of seven semitones is called a *perfect fifth*. Every connected region of seven notes on the circle of fifths forms both the k major scale and its relative t minor scale, where k is the second clockwise note in the region and t is the fifth. The distance between k and t is $3 \cdot 7 \bmod 12 = 9$ semitones, exactly the distance between C and A that we can observe in the musical clock in Figure 13.2.

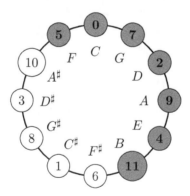

Figure 13.5   The circle of fifths, $\mathbb{Z}_{12} = \langle 7 \rangle$.

13.3   Group Actions and Chords

While a scale is melodic with its pitches played in succession, a chord is harmonic with its pitches played in unison. In this section we focus our attention on the smallest chords, the *triads*, containing only three pitches. A major triad consists of a root around which the harmony is based, and notes four and seven semitones above the root. A minor triad also contains a root and a note seven semitones above the root, with the remaining note three semitones above the root instead of four. We saw the interval of seven semitones at the end of Section 13.2, the perfect fifth which is also the interval between the first and fifth notes in a major (or minor) diatonic scale. A four-semitone interval is called a *major third*, and appears as the interval between the first and third notes in a major diatonic scale. Similarly, a three-semitone interval is called a *minor third*, and appears as the interval between the first and third notes in a minor diatonic scale.

As suggested in [24], we can visualize the notes in chords and the intervals between them using a chord polygon whose vertices are the notes in the chords. We consider the length of a side of a polygon to be the length, in semitones, of the arc between its vertices. Applying each of the elements of D_{12} to $\langle 0, 4, 7 \rangle$ yields all of the major and minor triads, which together form the *consonant* triads. Two examples are shown in Figure 13.6.

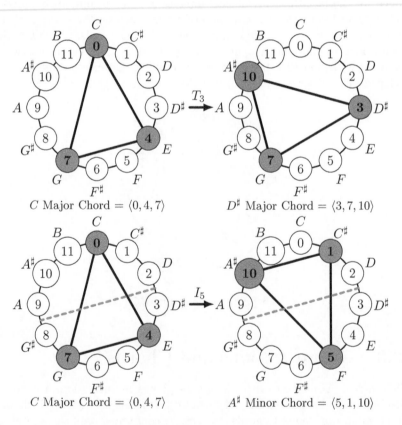

Figure 13.6   Musical actions of T_3 and I_5 on chord polygons.

Our use of pitch class segments to notate chords in Figure 13.6 is a bit artificial—pitches in chords are not ordered, but played in unison. However, the notation allows us to define T_n and I_n via a componentwise action, and helps us keep track of the roots of the chords. Every major triad is the image of C major $= \langle 0, 4, 7 \rangle$ under the action of a rotation, and the root of any major chord formed in this manner will be the first note in its pitch class segment. Minor chords appear as the image of C major under the action of a reflection. The three-semitone interval in the C major chord occurs between the second and third notes. This is exactly the interval between the root and the second note in a minor chord. Thus, when written as a pitch class segment, the root of a minor chord appears as the last pitch class in the segment. In general, we have the following algebraic expressions for the k major and t minor triads.

$$k \text{ major triad} = \langle k, (k+4) \bmod 12, (k+7) \bmod 12 \rangle$$
$$t \text{ minor triad} = \langle (t+7) \bmod 12, (t+3) \bmod 12, t \rangle$$

The C major chord is only fixed in D_{12} by the identity, giving 24 consonant triads, 12 major and 12 minor.

Determining the number of different types of triads depends on what we use as our symmetry group. Two choices make sense. We could either use the subgroup of rotations in D_{12}, in which case major and minor triads would be considered as two different types, or we could use all of D_{12}, in which case all consonant triads would be considered as the same type. In Example 13.1, we will take the first approach, and determine the number of distinct triads up to transposition (i.e., considering two triads as the same if one could be rotated to match the other). We leave it as an exercise to determine the number of distinct triads up to transposition and inversion (i.e., considering two triads as the same if one could be rotated and/or reflected to match the other).

Example 13.1 In this example we will find the number of distinct triads up to transposition. Our first step is to determine the cycle index for the rotations of the musical clock. Keeping the notation of the musical clock, we can represent T, the $30°$ clockwise rotation, as the permutation $T = (0, 1, 2, 3, 4, 5, 6, 7, 8, 9, 10, 11)$. All other rotations are powers of T. For example, the square of T is $T_2 = T^2 = (0, 2, 4, 6, 8, 10)(1, 3, 5, 7, 9, 11)$. Thus, T_2 contributes the term x_6^2 to the cycle index. The first 6-cycle in this representation for T_2 contains the elements of the cyclic subgroup of $\mathbb{Z}_{12}$ generated by 2, since by applying T twice we are consistently adding two to each of the previous entries in our cycle. In general, the first cycle in the disjoint cycle representation of $T_k = T^k$ will contain the elements of the subgroup $\langle k \rangle = \langle \gcd(k, 12) \rangle$ of $\mathbb{Z}_{12}$. Each of the remaining cycles in the disjoint cycle representation of T_k will be cosets of this subgroup. Thus, for each k, T_k contains $\gcd(k, 12)$ cycles, each of length $12/\gcd(k, 12)$. This gives the following full cycle index.

$$f(x_1, x_2, x_3, \ldots, x_{12}) = \frac{1}{12}\left(x_1^{12} + x_2^6 + 2x_3^4 + 2x_4^3 + 2x_6^2 + 4x_{12}\right)$$

Coloring the notes in our chord gray and coloring all other notes white, and labeling gray notes as g and white notes as 1, yields the following pattern inventory.

$$
\begin{aligned}
f(1 + &g, 1 + g^2, 1 + g^3, \ldots, 1 + g^{12}) \\
= \ & 1 + g + 6g^2 + 19g^3 + 43g^4 + 66g^5 + 80g^6 + 66g^7 + 43g^8 + 19g^9 \\
& + 6g^{10} + g^{11} + g^{12}
\end{aligned}
$$

This indicates that, up to transposition, there are 19 different types of three-note chords, two of which are the major and minor triads.   □

13.4   Group Actions and Chords with Maple

In this section, we will show how Maple can be used to find the cycle index for the rotations of the musical clock. To do this, we will use the user-written function **cicn**, which we have written separately from this Maple session and saved as the text file cicn.mpl. To use this function, we must first read it into our Maple session as follows.

```
> read "cicn.mpl";
```

The following command then illustrates how the function **cicn** can be used. Specifically, by entering the following command, we find the cycle index for the rotations of the musical clock in terms of the variable x.

```
> f := cicn(12, x);
```

$$f := \frac{1}{12}x_1^{12} + \frac{1}{12}x_2^6 + \frac{1}{6}x_3^4 + \frac{1}{6}x_4^3 + \frac{1}{6}x_6^2 + \frac{1}{3}x_{12}$$

We can then use the Maple **unapply** function as follows to convert this expression for f into a function.

```
> f := unapply(f, seq(x[i], i=1..12));
```

$$
\begin{aligned}
f := \ & (x\_1, \ x\_2, \ x\_3, \ x\_4, \ x\_5, \ x\_6, \ x\_7, \ x\_8, \ x\_9, \ x\_10, \ x\_11, \ x\_12) \\
& \rightarrow \frac{1}{12}x\_1^{12} + \frac{1}{12}x\_2^6 + \frac{1}{6}x\_3^4 + \frac{1}{6}x\_4^3 + \frac{1}{6}x\_6^2 + \frac{1}{3}x\_12
\end{aligned}
$$

With chord notes colored gray and labeled as g, and all other notes colored white and labeled as 1, the following command gives the resulting pattern inventory.

```
>  pinv := simplify(f(seq(1+g^i, i=1..12)));
```

$$pinv := g^{12} + g^{11} + 6\,g^{10} + 19\,g^9 + 43\,g^8 + 66\,g^7 + 80\,g^6 + 66\,g^5 + 43\,g^4$$
$$+ 19\,g^3 + 6\,g^2 + g + 1$$

From this pattern inventory we can see that up to transposition there are, for example, 19 different types of three-note chords, and 43 different types of four-note chords.

To hear musical simulations of these different chord types, we will use the user-written function **midichords**,[3] which we have written separately from this Maple session and saved as the text file midichords.mpl. To use this function, we must first read it into our Maple session as follows.

```
>  read "midichords.mpl";
```

The following command then illustrates how the function **midichords** can be used. Specifically, the following command creates a file that will play a simulation of an example of each of the 19 different types of three-note chords that exist up to transposition.[4]

```
>  midichords("pinvchords.midi", 0, pinv, 3);
```

$$[[0, 1, 2], [0, 1, 3], [0, 1, 4], [0, 1, 5], [0, 1, 6], [0, 1, 7], [0, 1, 8], [0, 1, 9], [0, 1,$$
$$10], [0, 2, 4], [0, 2, 5], [0, 2, 6], [0, 2, 7], [0, 2, 8], [0, 2, 9], [0, 3, 6], [0, 3, 7],$$
$$[0, 3, 8], [0, 4, 8]]$$

The first parameter in the previous command is the name of a file in *MIDI* (an acronym for *Musical Instrument Digital Interface*) format which will be created to hold the musical simulation, and saved by default in the same directory as the file containing the Maple worksheet from which the command is executed. Some information on how MIDI files are constructed can be found in [39]. Once existing, the MIDI file can be played using a variety of different types of multimedia software. The second parameter in the previous command is a numerical value representing the instrument to be used in the simulation. Examples include $0 =$ acoustic grand piano, $27 =$ electric guitar (clean), $40 =$ violin, and $105 =$ banjo. A complete list of instruments available can be found in [7]. The third parameter in the previous command is the pattern inventory polynomial, and the fourth parameter gives the number of notes in the chords to be created in the simulation. The Maple output from the previous command are the numbers in $\mathbb{Z}_{12}$ for the notes in the chords created in the simulation.

---

[3]The function **midichords** uses the Maple **GroupTheory** package, which was available for the first time in Maple version 18.

[4]Although composers can choose notes from any octave, the notes in all musical simulations produced by **midichords** exist in the middle octave of a standard keyboard.

To find the cycle index for both the rotations and reflections of the musical clock, we can use the user-written function **cidn**, which we have written separately from this Maple session and saved as the text file cidn.mpl. We described and illustrated the use of the function **cidn** in Section 11.5.

13.5   Group Actions and Chords with MAT-LAB

In this section, we will show how MATLAB can be used to find the cycle index for the rotations of the musical clock. To do this, we will use the user-written function **cicn**, which we have written separately from this MATLAB session and saved as the M-file cicn.m. The following commands illustrate how the function **cicn** can be used. Specifically, by entering the following commands, we find the cycle index for the rotations of the musical clock in terms of the variable x.

```
>> syms x
>> cires = cicn(12, x);
```

The **cicn** function returns two entries in a cell array that we have assigned as *cires*. The first entry in *cires* is the resulting cycle index. In the following command we store this cycle index as the variable f.

```
>> f = cires{1};
```

We can then use the MATLAB **pretty** function as follows to display this cycle index in a form that resembles typeset mathematics.

```
>> pretty(f)

   12      6       4       3       2
   x1      x2      x3      x4      x6      x12
  ----  +  ---  +  ---  +  ---  +  ---  +  ---
   12      12      6       6       6       3
```

Next, we will convert this expression for f into a function. To do this, we will use the second entry in *cires*, which is a list of the variables used in the cycle index.

```
>> fvar = cires{2};
>> [fvar{:}]

ans =

[ x1, x2, x3, x4, x5, x6, x7, x8, x9, x10, x11, x12]
```

We can then convert the cycle index stored as f from an expression into a function by entering the following command.

```
>> f = @(x) sym(subs(f, {fvar{:}}, x));
```

With chord notes colored gray and labeled as g, and all other notes colored white and labeled as 1, the following commands give the resulting pattern inventory.

```
>> syms g
>> pinv = expand(f({1+g.^(1:12)}));
>> pretty(pinv)
```

$$g^{12} + g^{11} + 6\,g^{10} + 19\,g^{9} + 43\,g^{8} + 66\,g^{7} + 80\,g^{6}$$

$$+ 66\,g^{5} + 43\,g^{4} + 19\,g^{3} + 6\,g^{2} + g + 1$$

From this pattern inventory we can see that up to transposition there are, for example, 19 different types of three-note chords, and 43 different types of four-note chords.

To hear musical simulations of these different chord types, we will use the user-written function **midichords**, which we have written separately from this MATLAB session and saved as the M-file midichords.m. The following command illustrates how the function **midichords** can be used. Specifically, the following command creates a file that will play a simulation of an example of each of the 19 different types of three-note chords that exist up to transposition.[5]

```
>> chords = midichords('pinvchords.midi', 0, pinv, 3);
```

The first parameter in the previous command is the name of a file in *MIDI* (an acronym for *Musical Instrument Digital Interface*) format which will be created to hold the musical simulation, and saved by default in the same directory as the current MATLAB command window. Some information on how MIDI files are constructed can be found in [39]. Once existing, the MIDI file can be played using a variety of different types of multimedia software. The second parameter in the previous command is a numerical value representing the instrument to be used in the simulation. Examples include 0 = acoustic grand piano, 27 = electric guitar (clean), 40 = violin, and 105 = banjo. A complete list of instruments can be found in [7].

---

[5]Although composers can choose notes from any octave, the notes in all musical simulations produced by **midichords** exist in the middle octave of a standard keyboard.

The third parameter in the previous command is the pattern inventory polynomial, and the fourth parameter gives the number of notes in the chords to be created in the simulation.

The output of the previous command is a cell array stored in the variable *chords*, and contains the numbers in $\mathbb{Z}_{12}$ for the notes in the chords created in the simulation. To display this output efficiently, we will use the user-written functions **pstring** and **stringprint**, which we have written separately from this MATLAB session and saved as the M-files pstring.m and stringprint.m. This **stringprint** function is the same one we introduced and used initially in Section 7.4. The following command displays the numbers in $\mathbb{Z}_{12}$ for the notes in the chords created in the simulation in rows with 51 characters each.

```
>> stringprint(pstring(chords), 51)
```

```
[[0, 1, 2],[0, 1, 3],[0, 1, 4],[0, 1, 5],[0, 1, 6],
[0, 1, 7],[0, 1, 8],[0, 1, 9],[0, 1, 10],[0, 2, 4],
[0, 2, 5],[0, 2, 6],[0, 2, 7],[0, 2, 8],[0, 2, 9],[
0, 3, 6],[0, 3, 7],[0, 3, 8],[0, 4, 8]]
```

To find the cycle index for both the rotations and reflections of the musical clock, we can use the user-written function **cidn**, which we have written separately from this MATLAB session and saved as the M-file cidm.m. We described and illustrated the use of the function **cidn** in Section 11.6.

13.6   Cayley Graphs for $\mathbb{Z}_{12}$

Cayley graphs can be used to help visualize the structure of a group. In this section we will see how several Cayley graphs of $\mathbb{Z}_{12}$ highlight certain musical structures.

Definition 13.1 *Given a finite group G with generating set $S \subseteq G - \{e\}$, the Cayley graph $Cay(G, S)$ has as its vertices the elements of G, with an edge joining vertex g_1 to vertex g_2 if and only if $g_2 = g_1 s$ for some $s \in S$. We can color the graph by assigning a color to each $s \in S$.*

Figure 13.7 shows two Cayley graphs of $\mathbb{Z}_{12}$. The graph $Cay(\mathbb{Z}_{12}, \{1\})$ on the left is a directed version of our original musical clock, and the graph $Cay(\mathbb{Z}_{12}, \{7\})$ on the right is a directed version of the circle of fifths.

If vertices are close to each other on the musical clock $Cay(\mathbb{Z}_{12}, \{1\})$, then the corresponding pitch classes will be separated by a small number of semitones. To address the issue of closeness of vertices on the circle of fifths $Cay(\mathbb{Z}_{12}, \{7\})$, we can think of vertices as representing not the

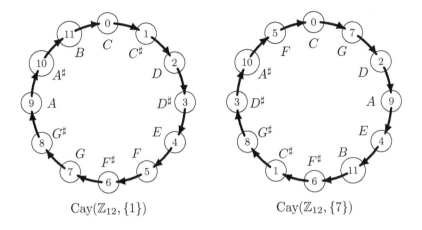

Figure 13.7 Two Cayley graphs of $\mathbb{Z}_{12}$.

names of notes, but rather the names of major keys.[6] We saw in Section 13.2 that the major keys are the connected regions of seven notes in the circle of fifths, each named for the second clockwise note in the region. Thus the collections of notes in keys named for adjacent vertices agree in all but one note. Furthermore, moving from the k major key to the adjacent $(k+7) \bmod 12$ major key, we would drop the $(k-7) \bmod 12$ note and gain the $(k+6 \cdot 7) \bmod 12$ note. These notes differ by only $(k+6 \cdot 7 - (k-7)) \bmod 12 = 7 \cdot 7 \bmod 12 = 1$ semitone. Thus, closeness in the circle of fifths represents closeness in key.

In addition to musical scales and keys, we have also considered chords, focusing on the consonant triads. Recall that major triads contain a root and notes four and seven semitones above the root, while minor triads contain a root and notes three and seven semitones above the root. Thus we should see some of their structure by looking at the Cayley graph of $\mathbb{Z}_{12}$ with generating set $S = \{3, 4, 7\}$, as shown in Figure 13.8. The solid black edges in this figure connect vertices via the generator $s = 4$, the dashed black edges via the generator $s = 3$, and the dashed gray edges via the generator $s = 7$. Although vertices 2, 6, and 10 appear in both the top and bottom rows of the graph, each 2 is the same 2, each 6 is the same 6, and each 10 is the same 10. One can imagine the graph as having the bottom and top rows glued together such that vertices with the same labels lie on top of one another. The left and right sides of the parallelogram should be glued together in the same manner, thus forming a twisted torus.

---

[6]When thinking of a major or minor diatonic scale as a collection of notes used to compose music, it is common to call the notes a *key* rather than a *scale*.

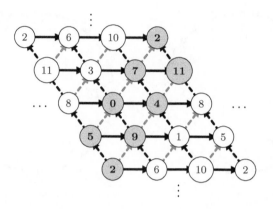

Figure 13.8 $\text{Cay}(\mathbb{Z}_{12}, \{3, 4, 7\})$.

The undirected version of $\text{Cay}(\mathbb{Z}_{12}, \{3, 4, 7\})$ is known as the *Tonnetz*, and was developed by music theorists to study triadic post-tonal music, as described in [8]. On the graph, upward-pointing triangles represent major triads, while downward-pointing triangles represent minor triads. Note that if we reflect the C major chord $\langle 0, 4, 7 \rangle$ across its solid black edge, we obtain the minor chord containing notes 0, 4, and 9, which in our pitch class segment notation is $\langle 4, 0, 9 \rangle$, the A minor triad.[7] Musically, reflection across a solid black edge, which we will call R, yields the *relative* minor chord. Algebraically, this is as follows, with operations done modulo 12.

$$R(\langle x, y, z \rangle) = \begin{cases} \langle y, x, z + 2 \rangle & \text{if } \langle x, y, z \rangle \text{ is a major triad} \\ \langle y, x, z + 10 \rangle & \text{if } \langle x, y, z \rangle \text{ is a minor triad} \end{cases}$$

If we reflect the C major chord $\langle 0, 4, 7 \rangle$ across its dashed black line, we obtain the E minor chord $\langle 11, 7, 4 \rangle$. Musically, reflection across a dashed black edge, which we will call L, is referred to as a *leading tone exchange*. Algebraically, this is as follows, again with operations done modulo 12.

$$L(\langle x, y, z \rangle) = \begin{cases} \langle x + 11, z, y \rangle & \text{if } \langle x, y, z \rangle \text{ is a major triad} \\ \langle x + 1, z, y \rangle & \text{if } \langle x, y, z \rangle \text{ is a minor triad} \end{cases}$$

If we reflect the C major chord $\langle 0, 4, 7 \rangle$ across its dashed gray edge, we obtain the C minor chord $\langle 7, 3, 0 \rangle$. Musically, reflection across a dashed gray edge, which we will call P, sends a chord to its parallel chord of the opposite

---

[7]Recall that for a minor chord the root is listed last with the note a perfect fifth above the root listed first.

type (major to minor and vice versa). The chords are called *parallel* since, while of opposite type, they have the same root and take their name from the same note. Expressing this algebraically is left as an exercise.

We now consider whether these new symmetries in the Tonnetz could be related to our familiar musical clock symmetries T_n and I_n. The following theorem determines this by characterizing the group generated by R and L. An alternative proof of this theorem can be found in [8]. While this theorem is about $\langle R, L \rangle$, this group is also generated by R, L, and P, since $P \in \langle R, L \rangle$, an explanation for which is left as an exercise. In the literature, for example [8] and [9], this group is referred to as the *P-L-R* group.

Theorem 13.2 *The group generated by the functions R and L described above is isomorphic to D_{12}.*

Proof. First note that every dihedral group can be generated by a rotation and a reflection. In general the converse is true as well. More specifically, if a group is generated by elements a of order n (e.g., a rotation) and b of order two (e.g., a reflection) with $(ab)^2 = e$ (i.e., if reflecting and rotating followed by reflecting and rotating again have the net effect of the identity), then the group is the dihedral group of order $2n$. For a precise explanation of this, see [10]. Let $b = R$. Since R is a reflection in the Tonnetz, it must be of order two. Next, observe that $L = LRR$, so $\langle L, R \rangle = \langle LR, R \rangle$. We will show that $a = LR$ is of order 12. Like T_n and I_n, L and R are defined by their actions on the set of consonant triads. First consider the following action of LR on the set of major triads, with operations done modulo 12.

$$
\begin{aligned}
LR(\langle x, x+4, x+7 \rangle) &= L(\langle x+4, x, x+9 \rangle) \\
&= \langle x+5, x+9, x \rangle \\
&= \langle x+5, x+9, x+12 \rangle \\
&= T_5(\langle x, x+4, x+7 \rangle)
\end{aligned}
$$

It can similarly be shown and is left as an exercise that $LR = T_7$ when acting on the minor triads. Thus, over its entire domain LR is a rotation that generates the rotational subgroup of D_{12}, and so is of order 12. All that is left is to show $(ab)^2 = e$. However, $(ab)^2 = (LRR)^2 = L^2 = e$. □

Finally, we can use the Cayley graph in Figure 13.8 to connect the ideas of major and minor diatonic scales, the subject of Section 13.2, with those of major and minor triads, the subject of Section 13.3. The notes in the C major diatonic scale, which are shaded in Figure 13.8, form a connected subgraph that indicates how this scale can be built from consonant triads. Every major diatonic scale is a rotation of the C major scale on the circle of fifths, and thus each major diatonic scale is simply a shift of the C major scale through one dashed gray edge ($s = 7$) on the graph. Each minor diatonic scale is relative to a major scale, and so the graph shows how to build the minor diatonic scales from consonant triads as well. If one shades any collection of eight notes in the shape of the graph, the notes will be those in the k major scale and relative t minor scale, where k is the second pitch class in the upper band of four pitch classes and t is the second pitch class from the left in the lower band.

13.7   Twelve-Tone Rows

Western tonal music centers its harmony and melody around a tone or tonic, leading to music that is played in a certain key, or collection of notes, from a seven-note diatonic scale. Most popular music, including what is typically played on the radio, is tonal in nature. In the late 1800s some composers, such as Richard Wagner, began to write music without a tonal center, although their music still used triadic elements (the consonant triads in Section 13.3), and has thus been called *triadic post-tonal*. In the early 1900s a collection of composers led by Arnold Schoenberg began to write music that was neither tonal nor triadic, and developed a method of musical composition centered around a *twelve-tone* row.

A twelve-tone row is an ordering of all 12 pitch classes. Figure 13.9 shows a selection from Schoenberg's 1936 Violin Concerto (Op. 36) in Schoenberg's own handwriting in which he uses the following twelve-tone row.

$$p = \langle A, A^\sharp, D^\sharp, B, E, F^\sharp, C, C^\sharp, G, G^\sharp, D, F \rangle$$
$$= \langle 9, 10, 3, 11, 4, 6, 0, 1, 7, 8, 2, 5 \rangle$$

Figure 13.9   Twelve-tone row in Schoenberg's Violin Concerto (Op. 36).

Composers using a twelve-tone row developed continuity by repeating the row or related versions of it throughout the composition. We will call the original version of a twelve-tone row its *prime* form. The three most common related versions of a twelve-tone row are its *transposed*, *retrograde*, and *inverted* forms. Transposition is our familiar operation $T_k = T^k$, where T represents a 30° clockwise rotation on the musical clock. Retrograde (R)

reverses the order of the notes in the pitch class segment, and inversion (I) reflects the notes across the axis of symmetry in the musical clock passing through the first note. Figure 13.10 shows some examples of these operations as they act on the twelve-tone row p introduced in Figure 13.9.

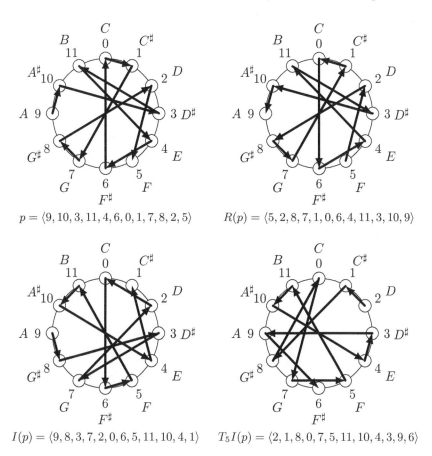

$$p = \langle 9, 10, 3, 11, 4, 6, 0, 1, 7, 8, 2, 5 \rangle$$

$$R(p) = \langle 5, 2, 8, 7, 1, 0, 6, 4, 11, 3, 10, 9 \rangle$$

$$I(p) = \langle 9, 8, 3, 7, 2, 0, 6, 5, 11, 10, 4, 1 \rangle$$

$$T_5 I(p) = \langle 2, 1, 8, 0, 7, 5, 11, 10, 4, 3, 9, 6 \rangle$$

Figure 13.10   Actions of R, I, and T_5 on twelve-tone row p.

Composers using a twelve-tone row could use any combination of transposition, retrograde, and inversion to obtain a related version. For example, Schoenberg's violin concerto contains the original prime form, $p = T_0(p)$, as well as $T_5 I(p)$, $R(p)$, and $T_5 I R(p)$. A careful mathematical analysis of this concerto can be found in [13].

As observed in [18], we can think of each twelve-tone row as representing a permutation on the set $\{0, 1, 2, \ldots, 11\}$. For example, in the prime form

$p = \langle 9, 10, 3, 11, 4, 6, 0, 1, 7, 8, 2, 5 \rangle$ used in Schoenberg's violin concerto, the
0th position is assigned the 9th note, the 9th position the 8th note, the 8th
position the 7th note, and so on, yielding the permutation representation
$p = (0, 9, 8, 7, 1, 10, 2, 3, 11, 5, 6)$.

Transposition, retrograde, and inversion each reorder a twelve-tone row
in a prescribed manner, and we can think of these operations in terms
of permutations. For example, when we transpose the prime form of the
twelve-tone row used in Schoenberg's violin concerto by one semitone, we
want to assign the 0th position the 10th note instead of the 9th, the 9th
position the 9th note instead of the 8th, and so on. Thus, $T(p) = \sigma_T p$,
where $\sigma_T = (0, 1, 2, 3, 4, 5, 6, 7, 8, 9, 10, 11)$. Similarly, when we apply the
retrograde transformation to p, we want to exchange what was in the 0th
and 11th positions, what was in the 1st and 10th positions, and so on.
Thus, $R(p) = p\sigma_R$, where $\sigma_R = (0, 11)(1, 10)(2, 9)(3, 8)(4, 7)(5, 6)$.

Inversion is slightly more complicated. We will first determine how we
could reflect a prime row across the axis of symmetry in the musical clock
passing through 0. We will then transpose the result to be the correct
inversion. To reflect $p = (0, 9, 8, 7, 1, 10, 2, 3, 11, 5, 6)$ across the axis of
symmetry passing through 0, instead of assigning the 0th position the 9th
note, we would assign it the mirror image of the 9th note, which is the
3rd note. Similarly, instead of assigning the 9th position the 8th note, we
would assign it the mirror image of the 8th note, which is the 4th note.
In full, if $\sigma_I = (1, 11)(2, 10)(3, 9)(4, 8)(5, 7)$, then $\sigma_I p$ will give the row
reflected across the axis of symmetry passing through 0. Since $I(p)$ should
reflect across the axis of symmetry passing through p_0 (the initial note in
p), all that is left is to rotate $\sigma_I p$ so that its initial note is p_0, which could
be done by applying T a total of $2p_0$ times. Thus, $I(p) = \sigma_T^{2p_0} \sigma_I p$. We
can simplify $I(p)$ further by observing that $\sigma_I = \sigma_T \sigma_R$, so that in general
$I(p) = \sigma_T^{2p_0} \sigma_T \sigma_R p = \sigma_T^{2p_0+1} \sigma_R p$.

The number of related versions of a prime twelve-tone row p depends
on p itself, but we can find a general upper bound on the number of related
versions. We saw that both T and I transform p via left-multiplication; for
T we left-multiply by σ_T, and for I by $\sigma_T^{2p_0+1} \sigma_R$. Thus any version of p
related to p by a sequence of transpositions and inversions must be in the
right coset $\langle \sigma_T, \sigma_R \rangle p$, where $\langle \sigma_T, \sigma_R \rangle$ is the subgroup generated by σ_T and
σ_R. Note that σ_T is of order 12, σ_R is of order 2, and $(\sigma_T \sigma_R)^2 = \sigma_I^2 = e$.
Thus $\langle \sigma_T, \sigma_R \rangle$ is isomorphic to D_{12}, and there are at most 24 elements in
the right coset $\langle \sigma_T, \sigma_R \rangle p$.[8] Retrograde acts via right-multiplication by σ_R,
and so each application of retrograde, no matter where it is in our sequence
of Ts, Is, and Rs, will give exactly one copy of σ_R on the right of the
related version. Since σ_R is of order 2, this leaves only 2 possibilities—an

---

[8]We would meet this bound if and only if p had no rotational or reflective symmetries.

even number of applications of retrograde forcing the related version into the right coset $\langle \sigma_T, \sigma_R \rangle\, p$, or an odd number of applications of retrograde forcing the related version into the right coset $\langle \sigma_T, \sigma_R \rangle\, p\sigma_R$. This new right coset also has at most 24 elements, giving at most $24 + 24 = 48$ total related versions. The fact that $\langle T, I, R \rangle$ is actually isomorphic to $D_{12} \times \mathbb{Z}_2$, which clearly contains 48 elements, is left as an exercise.

Before seeing a few examples, we introduce a final key fact about permutations in the following lemma.

Lemma 13.3 *If α and β are permutations, then α and $\beta\alpha\beta^{-1}$ (α "conjugated by" β) will have the same disjoint cycle structure (i.e., when expressed as products of disjoint cycles). Moreover, if α and γ are permutations with the same disjoint cycle structure, then there must exist a permutation β with $\beta\alpha\beta^{-1} = \gamma$.*

Proof. Let α and β be permutations expressed as products of disjoint cycles, and suppose $\alpha(a_1) = a_2$. Then the following is true.

$$\beta\alpha\beta^{-1}(\beta(a_1)) = \beta\alpha\left(\beta^{-1}(\beta(a_1))\right) = \beta(\alpha(a_1)) = \beta(a_2)$$

Since β is a bijection, each cycle $(a_1, a_2, \ldots)$ in the disjoint cycle representation of α will correspond to a cycle $(\beta(a_1), \beta(a_2), \ldots)$ in the disjoint cycle representation of $\beta\alpha\beta^{-1}$. Thus α and $\beta\alpha\beta^{-1}$ will have the same disjoint cycle structure. Also, if α and γ have the same disjoint cycle structure, then for each cycle $(a_1, a_2, \ldots)$ in the disjoint cycle representation of α, there must exist a corresponding cycle $(g_1, g_2, \ldots)$ in the disjoint cycle representation of γ. If we define β by $\beta(a_1) = g_1$, $\beta(a_2) = g_2$, and so on for all cycles in the disjoint cycle representation of α, then $\beta\alpha\beta^{-1} = \gamma$. □

Example 13.2 For $\sigma_R = (0,11)(1,10)(2,9)(3,8)(4,7)(5,6)$, Lemma 13.3 gives the following.

$$\begin{aligned} \sigma_R \sigma_T \sigma_R^{-1} &= \sigma_R(0,1,2,3,4,5,6,7,8,9,10,11)\sigma_R^{-1} \\ &= (11,10,9,8,7,6,5,4,3,2,1,0) \\ &= \sigma_T^{-1} \end{aligned}$$

□

A twelve-tone row can exhibit many types of symmetries. In Example 13.3 we consider rows that are their own transposed retrograde, known as *palindromic* rows, an example of which is $\langle 5,8,7,6,10,9,3,4,0,1,2,11 \rangle$, which was used in Anton Webern's Chamber Symphony (Op. 21).

Example 13.3 In this example, we will determine the number of palindromic twelve-tone rows. Suppose p is its own transposed retrograde. Then the following will be true about p for some positive integer k.

$$p = \sigma_T^k p \sigma_R$$
$$\Rightarrow \quad p\sigma_R^{-1}p^{-1} = \sigma_T^k$$
$$\Rightarrow \quad p\sigma_R p^{-1} = \sigma_T^k$$

Thus σ_T^k is a conjugate of σ_R, and so by Lemma 13.3 these permutations must have the same disjoint cycle structure, both containing six 2-cycles. As p ranges over all of S_{12}, the number of distinct $p\sigma_R p^{-1}$ will equal the number elements in S_{12} with the same disjoint cycle structure as σ_R, which is indicated by the partition $P = [0, 6, 0, 0, 0, 0, 0, 0, 0, 0, 0, 0]$. Thus there are 12! possible choices for p, and by Theorem 12.2 only $\dfrac{12!}{2^6 6!}$ possibilities for $p\sigma_R p^{-1}$. Therefore there are $12! \cdot \left(\dfrac{12!}{2^6 6!} \right)^{-1} = 2^6 \cdot 6! = 46{,}080$ possible choices for p that will correspond to each $p\sigma_R p^{-1}$. Since we have observed that if p is palindromic, then $p\sigma_R p^{-1}$ is completely determined, there must be 46,080 palindromic twelve-tone rows. $\qquad \square$

Suppose we apply some combination of transpositions, retrogrades, and inversions, say X, to a palindromic twelve-tone row $p = \sigma_T^s p \sigma_R$. We can use the fact that $\sigma_R \sigma_T = \sigma_T^{-1} \sigma_R$ to express $X(p)$ as $\sigma_T^k \sigma_R^\ell p \sigma_R^m$ for some integers k, ℓ, and m with $\ell, m \in \{0, 1\}$. We then have the following.

$$
X(p) = \begin{cases} \sigma_T^k p \sigma_R^m & \text{if } \ell = 0 \\ \sigma_T^k \sigma_R p \sigma_R^m & \text{if } \ell = 1 \end{cases}
$$

$$
= \begin{cases} \sigma_T^k \left(\sigma_T^s p \sigma_R \right) \sigma_R^m & \text{if } \ell = 0 \\ \sigma_T^k \sigma_R \left(\sigma_T^s p \sigma_R \right) \sigma_R^m & \text{if } \ell = 1 \end{cases}
$$

$$
= \begin{cases} \sigma_T^s \sigma_T^k p \sigma_R^m \sigma_R & \text{if } \ell = 0 \\ \sigma_T^k \sigma_T^{-s} \sigma_R p \sigma_R^m \sigma_R & \text{if } \ell = 1 \end{cases}
$$

$$
= \begin{cases} \sigma_T^s \left(\sigma_T^k p \sigma_R^m \right) \sigma_R & \text{if } \ell = 0 \\ \sigma_T^{-s} \left(\sigma_T^k \sigma_R p \sigma_R^m \right) \sigma_R & \text{if } \ell = 1 \end{cases}
$$

$$
= \begin{cases} \sigma_T^s X(p) \sigma_R & \text{if } \ell = 0 \\ \sigma_T^{-s} X(p) \sigma_R & \text{if } \ell = 1 \end{cases}
$$

Thus $X(p)$ is palindromic as well. This means if we consider one palindromic row equivalent to another if it can be transformed to the other via some combination of transpositions, retrogrades, and inversions, any equivalence class containing a palindromic row must contain only palindromic rows. We consider the problem of determining the total number of equivalence classes containing only palindromic rows in Example 13.4.

Example 13.4 In this example we will determine the number of distinct twelve-tone rows up to any combination of transpositions, retrogrades, and inversions. Recall that transposition and inversion act on a given twelve-tone row p via left multiplication by σ_T^k and $\sigma_T^{2p_0+1}\sigma_R$, respectively, while retrograde acts on p via right multiplication by σ_R. Thus, we see that the number of distinct twelve-tone rows is the number of distinct sets $\langle\sigma_T,\sigma_R\rangle\,p\,\langle\sigma_R\rangle = \langle\sigma_T,\sigma_R\rangle\,p\,\{e,\sigma_R\} = \langle\sigma_T,\sigma_R\rangle\,p\cup\langle\sigma_T,\sigma_R\rangle\,p\sigma_R$ as p ranges over all of S_{12}. First, if $\langle\sigma_T,\sigma_R\rangle\,p = \langle\sigma_T,\sigma_R\rangle\,p\sigma_R$, then $p\sigma_Rp^{-1}\in\langle\sigma_T,\sigma_R\rangle$. To find the number of twelve-tone rows that fall into this case, consider that there are 12! possible twelve-tone rows, and by Theorem 12.2 only $\dfrac{12!}{2^6 6!}$ conjugates of σ_R. So $2^6 6!$ twelve-tone rows correspond to each unique conjugate of σ_R in $\langle\sigma_T,\sigma_R\rangle$. Each conjugate of σ_R in $\langle\sigma_T,\sigma_R\rangle$ must have the same disjoint cycle structure as σ_R—six 2-cycles. We have seen that $\langle\sigma_T,\sigma_R\rangle$ is isomorphic to D_{12}, and thus the only permutations in $\langle\sigma_T,\sigma_R\rangle$ that consist of six 2-cycles are T^6 and $T^k\sigma_R$ for $k\in\{0,2,4,6,8,10\}$ (reflections across the lines connecting the midpoints of the arcs between consecutive notes on the musical clock and the consecutive notes directly across the musical clock), making a total of seven conjugates of σ_R in $\langle\sigma_T,\sigma_R\rangle$, and thus $7\cdot 2^6 6! = 322{,}560$ twelve-tone rows in this case.[9] We are concerned with finding the number of distinct sets $\langle\sigma_T,\sigma_R\rangle\,p\cup\langle\sigma_T,\sigma_R\rangle\,p\sigma_R$ as p ranges over all of S_{12}. However, in this case $\langle\sigma_T,\sigma_R\rangle\,p = \langle\sigma_T,\sigma_R\rangle\,p\sigma_R$, and so we need to count the number of distinct $\langle\sigma_T,\sigma_R\rangle\,p$ where p is one of the $7\cdot 2^6 6!$ twelve-tone rows. Each right coset of $\langle\sigma_T,\sigma_R\rangle$ contains 24 elements, giving $\dfrac{7\cdot 2^6 6!}{24} = 13{,}440$ distinct right cosets, $\dfrac{2^6 6!}{24} = 1920$ of which are palindromic. Now, suppose $\langle\sigma_T,\sigma_R\rangle\,p\neq\langle\sigma_T,\sigma_R\rangle\,p\sigma_R$. We know that all of the remaining $12! - 7\cdot 2^6 6!$ twelve-tone rows fall into this case. Right cosets of a group are either equal or disjoint, and so $\langle\sigma_T,\sigma_R\rangle\,p\cap\langle\sigma_T,\sigma_R\rangle\,p\sigma_R = \emptyset$. Therefore, each $\langle\sigma_T,\sigma_R\rangle\,p\cup\langle\sigma_T,\sigma_R\rangle\,p\sigma_R$ contains $24+24 = 48$ elements, making $\dfrac{12! - 7\cdot 2^6 6!}{48} = 9{,}972{,}480$ distinct twelve-tone rows. This gives a total of $9{,}972{,}480 + 13{,}440 = 9{,}985{,}920$ distinct twelve-tone rows up to any combination of transpositions, retrogrades, and inversions, only 1920 (or 0.000192%) of which are palindromic. □

Many of the ideas in this section can be explored in more detail in [18]. In this article the authors argue that while composers using a twelve-tone system tend to choose symmetric row classes relatively infrequently, for example only 5% of the time for Schoenberg, symmetric row classes actually make up only 0.13% of the 9,985,920 possible row classes, indicating a likely strong preference for these among such composers.

---

[9]Note that the $2^6 6! = 46{,}080$ of these that correspond to T^6 are the palindromic rows that we counted in Example 13.3.

13.8   Twelve-Tone Rows with Maple

In this section, we will show how Maple can be used to apply any combination of transpositions, retrogrades, and inversions to a twelve-tone row. As our example, we will use the following twelve-tone prime row p described in Section 13.7.

$$p = \langle A, A^\sharp, D^\sharp, B, E, F^\sharp, C, C^\sharp, G, G^\sharp, D, F \rangle$$

We begin by entering the numerical representation of the notes in p.

```
>  p := [9, 10, 3, 11, 4, 6, 0, 1, 7, 8, 2, 5];
```
$$p := [9, 10, 3, 11, 4, 6, 0, 1, 7, 8, 2, 5]$$

Next we convert this numerical representation of p into a permutation. To perform this conversion, we will use the user-written function **notetoperm**, which we have written separately from this Maple session and saved as the text file notetoperm.mpl. To use this function, we must first read it into our Maple session as follows.

```
>  read "notetoperm.mpl";
```

The following command then illustrates how the function **notetoperm** can be used. Specifically, by entering the following command, we convert the numerical representation of p into a permutation, with the output represented as a list of lists.

```
>  p := notetoperm(p);
```
$$p := [[0, 9, 8, 7, 1, 10, 2, 3, 11, 5, 6]]$$

To apply a transposition to a twelve-tone row, we will use the user-written function **Tn**, which we have written separately from this Maple session and saved as the text file Tn.mpl. To use this function, we must first read it into our Maple session as follows.

```
>  read "Tn.mpl";
```

The following command then illustrates how the function **Tn** can be used. Specifically, by entering the following command, we apply the first power of T to p. The second parameter in this command is the permutation representation of p, and the output is the permutation representation of the result.

```
>  tp1 := Tn(1, p);
```
$$tp1 := [[0, 10, 3], [1, 11, 6], [2, 4, 5, 7]]$$

To convert the permutation representation of a twelve-tone row back into its sequential numerical representation, we will use the user-written function **permtonote**, which we have written separately from this Maple session and saved as the text file permtonote.mpl. To use this function, we must first read it into our Maple session as follows.

>   read "permtonote.mpl";

The following command then illustrates how the function **permtonote** can be used. Specifically, by entering the following command, we convert the permutation representation of *tp1* back into its sequential numerical representation.

>   permtonote(tp1);
$$[10, 11, 4, 0, 5, 7, 1, 2, 8, 9, 3, 6]$$

By using an integer larger than one as the first parameter in the **Tn** function, we can apply higher powers of T to a permutation. For example, in the following commands we apply the fifth power of T to p, showing the result both as a permutation and in sequential numerical form.

>   tp5 := Tn(5, p);
$$tp5 := [[0, 2, 8], [1, 3, 4, 9], [5, 11, 10, 7, 6]]$$

>   permtonote(tp5);
$$[2, 3, 8, 4, 9, 11, 5, 6, 0, 1, 7, 10]$$

To apply a retrograde to a twelve-tone row, we will use the user-written function **Ret**, which we have written separately from this Maple session and saved as the text file Ret.mpl. To use this function, we must first read it into our Maple session as follows.

>   read "Ret.mpl";

The following command then illustrates how **Ret** can be used. Specifically, by entering the following command, we apply the first power of R to p. The second parameter in this command is the permutation representation of p, and the output is the permutation representation of the result.

>   rp := Ret(1, p);
$$rp := [[0, 5], [1, 2, 8, 11, 9, 3, 7, 4]]$$

In the next command we convert the permutation representation of the previous result into its sequential numerical representation.

>   permtonote(rp);
$$[5, 2, 8, 7, 1, 0, 6, 4, 11, 3, 10, 9]$$

By using an integer larger than one as the first parameter in the **Ret** function, we can apply higher powers of R to a permutation. For example, in the following commands we apply the second power of R to p. Since R^2 is equal to the identity, we obtain the original permutation representation of p.

```
>  Ret(2, p);
```
$$[[0, 9, 8, 7, 1, 10, 2, 3, 11, 5, 6]]$$

To apply an inversion to a twelve-tone row, we will use the user-written function **Inv**, which we have written separately from this Maple session and saved as the text file Inv.mpl. To use this function, we must first read it into our Maple session as follows.

```
>  read "Inv.mpl";
```

The following command illustrates how **Inv** can be used. Specifically, by entering the following command, we apply the first power of an inversion to p. The third parameter in this command is the permutation representation of p, and the output is the permutation representation of the result.

```
>  ip0 := Inv(1, 0, p);
```
$$ip0 := [[0, 3, 1, 2, 9, 4, 8, 5, 6], [7, 11]]$$

The second parameter in the previous command specifies the number of one of the notes through which the axis of symmetry for the reflection will occur. Since this parameter is *0*, the inversion reflects across the axis of symmetry connecting notes 0 and 6. In the next command we convert the previous result into its sequential numerical representation.

```
>  permtonote(ip0);
```
$$[3, 2, 9, 1, 8, 6, 0, 11, 5, 4, 10, 7]$$

By using an integer larger than one as the first parameter in the **Inv** function, we can apply higher powers of inversions to a permutation. For example, in the following commands we apply the fifth power of an inversion to p, reflecting each time across the axis of symmetry connecting notes 2 and 8. Since the second power of an inversion is equal to the identity, the result is the same as if we had applied the first power of the inversion across the same axis of symmetry.

```
>  ip2 := Inv(5, 2, p);
```
$$ip2 := [[0, 7, 3, 5, 10, 2, 1, 6, 4], [8, 9]]$$

```
>  permtonote(ip2);
```
$$[7, 6, 1, 5, 0, 10, 4, 3, 9, 8, 2, 11]$$

To hear musical simulations of the results of these transformations, we will use the user-written function **midinotes**, which we have written separately from this Maple session and saved as the text file midinotes.mpl. To use this function, we must first read it into our Maple session as follows.

```
> read "midinotes.mpl";
```

The following command then illustrates how the function **midinotes** can be used. Specifically, by entering the following command, we create a file containing a musical simulation of the notes in the prime form of p.

```
> midinotes("m1.midi", permtonote(p), 40, 0);
```

The first parameter in the previous command is the name of a file in MIDI format in which the musical simulation will be stored. This file will be created and stored by default in the same directory as the file containing the Maple worksheet from which it is executed. The second parameter is the permutation representation of the sequence of notes to be played. The third parameter is a numerical value designating an instrument to be used in the simulation. In this command we use 40 for a violin. The final parameter gives the option of including a pause every 12 notes in the simulation by using a 1 for this parameter, or playing the notes continuously by using any other number for this parameter. The next two commands create files containing violin simulations of the notes that result from applying T and T^5 to p.

```
> midinotes("m2.midi", permtonote(tp1), 40, 0);
```

```
> midinotes("m3.midi", permtonote(tp5), 40, 0);
```

The next command creates a file containing a violin simulation of the notes that result from applying R to p.

```
> midinotes("m4.midi", permtonote(rp), 40, 0);
```

The next two commands create files containing violin simulations of the notes that result from applying one inversion to p across the axes of symmetry connecting notes 0 and 6, and connecting notes 2 and 8, respectively.

```
> midinotes("m5.midi", permtonote(ip0), 40, 0);
```

```
> midinotes("m6.midi", permtonote(ip2), 40, 0);
```

Simulations can be constructed for the application of any combination of transpositions, retrogrades, and inversions. The following three commands demonstrate how we can apply the transformation T^3R to p, showing the

result both as a permutation and in sequential numerical form, and create
a musical simulation of the result.

```
>   t3r := Tn(3, Ret(1, p));
```
$$t3r := [[0, 8, 2, 11], [1, 5, 3, 10], [6, 9]]$$

```
>   permtonote(t3r);
```
$$[8, 5, 11, 10, 4, 3, 9, 7, 2, 6, 1, 0]$$

```
>   midinotes("m7.midi", permtonote(t3r), 40, 0);
```

The next two commands demonstrate how we can apply the transformation
TIR to *p* with the inversion across the axis of symmetry connecting notes
0 and 6, and create a musical simulation of the result.

```
>   tir := Tn(1, Inv(1, 0, Ret(1, p)));
```
$$tir := [[0, 8, 2, 5, 1, 11, 4], [3, 6, 7, 9, 10]]$$

```
>   midinotes("m8.midi", permtonote(tir), 40, 0);
```

The following sequence of commands demonstrates how a collection of
nested **for** loops can be used to apply all 48 possible combinations of transpositions, retrogrades, and inversions to *p*. The variable *eG* is used to store
identifying representations for the combinations and the variable *G* holds
the resulting notes for these transformations.

```
>   eG := []:

>   G := []:

>   for i from 0 to 1 do

>       for j from 0 to 1 do

>           for k from 0 to 11 do

>               eG := [op(eG), sort(T^k*I^j*R^i, order =

                tdeg(T, I, R))];

>               G := [op(G), op(permtonote(Tn(k, Inv(j, 0,

                Ret(i, p)))))];

>           od:

>       od:

>   od:
```

By entering the following commands we can see the results for the variables eG and G.

```
> eG;
```

$[1, T, T^2, T^3, T^4, T^5, T^6, T^7, T^8, T^9, T^{10}, T^{11}, I, TI, T^2I, T^3I, T^4I, T^5I,$
$\quad T^6I, T^7I, T^8I, T^9I, T^{10}I, T^{11}I, R, TR, T^2R, T^3R, T^4R, T^5R, T^6R,$
$\quad T^7R, T^8R, T^9R, T^{10}R, T^{11}R, IR, TIR, T^2IR, T^3IR, T^4IR, T^5IR,$
$\quad T^6IR, T^7IR, T^8IR, T^9IR, T^{10}IR, T^{11}IR]$

```
> G;
```

$[9, 10, 3, 11, 4, 6, 0, 1, 7, 8, 2, 5, 10, 11, 4, 0, 5, 7, 1, 2, 8, 9, 3, 6, 11, 0, 5, 1, 6,$
$\quad 8, 2, 3, 9, 10, 4, 7, 0, 1, 6, 2, 7, 9, 3, 4, 10, 11, 5, 8, 1, 2, 7, 3, 8, 10, 4, 5, 11,$
$\quad 0, 6, 9, 2, 3, 8, 4, 9, 11, 5, 6, 0, 1, 7, 10, 3, 4, 9, 5, 10, 0, 6, 7, 1, 2, 8, 11, 4,$
$\quad 5, 10, 6, 11, 1, 7, 8, 2, 3, 9, 0, 5, 6, 11, 7, 0, 2, 8, 9, 3, 4, 10, 1, 6, 7, 0, 8, 1,$
$\quad 3, 9, 10, 4, 5, 11, 2, 7, 8, 1, 9, 2, 4, 10, 11, 5, 6, 0, 3, 8, 9, 2, 10, 3, 5, 11, 0,$
$\quad 6, 7, 1, 4, 3, 2, 9, 1, 8, 6, 0, 11, 5, 4, 10, 7, 4, 3, 10, 2, 9, 7, 1, 0, 6, 5, 11, 8,$
$\quad 5, 4, 11, 3, 10, 8, 2, 1, 7, 6, 0, 9, 6, 5, 0, 4, 11, 9, 3, 2, 8, 7, 1, 10, 7, 6, 1, 5,$
$\quad 0, 10, 4, 3, 9, 8, 2, 11, 8, 7, 2, 6, 1, 11, 5, 4, 10, 9, 3, 0, 9, 8, 3, 7, 2, 0, 6, 5,$
$\quad 11, 10, 4, 1, 10, 9, 4, 8, 3, 1, 7, 6, 0, 11, 5, 2, 11, 10, 5, 9, 4, 2, 8, 7, 1, 0, 6,$
$\quad 3, 0, 11, 6, 10, 5, 3, 9, 8, 2, 1, 7, 4, 1, 0, 7, 11, 6, 4, 10, 9, 3, 2, 8, 5, 2, 1, 8,$
$\quad 0, 7, 5, 11, 10, 4, 3, 9, 6, 5, 2, 8, 7, 1, 0, 6, 4, 11, 3, 10, 9, 6, 3, 9, 8, 2, 1, 7,$
$\quad 5, 0, 4, 11, 10, 7, 4, 10, 9, 3, 2, 8, 6, 1, 5, 0, 11, 8, 5, 11, 10, 4, 3, 9, 7, 2, 6,$
$\quad 1, 0, 9, 6, 0, 11, 5, 4, 10, 8, 3, 7, 2, 1, 10, 7, 1, 0, 6, 5, 11, 9, 4, 8, 3, 2, 11, 8,$
$\quad 2, 1, 7, 6, 0, 10, 5, 9, 4, 3, 0, 9, 3, 2, 8, 7, 1, 11, 6, 10, 5, 4, 1, 10, 4, 3, 9, 8,$
$\quad 2, 0, 7, 11, 6, 5, 2, 11, 5, 4, 10, 9, 3, 1, 8, 0, 7, 6, 3, 0, 6, 5, 11, 10, 4, 2, 9, 1,$
$\quad 8, 7, 4, 1, 7, 6, 0, 11, 5, 3, 10, 2, 9, 8, 7, 10, 4, 5, 11, 0, 6, 8, 1, 9, 2, 3, 8, 11,$
$\quad 5, 6, 0, 1, 7, 9, 2, 10, 3, 4, 9, 0, 6, 7, 1, 2, 8, 10, 3, 11, 4, 5, 10, 1, 7, 8, 2, 3,$
$\quad 9, 11, 4, 0, 5, 6, 11, 2, 8, 9, 3, 4, 10, 0, 5, 1, 6, 7, 0, 3, 9, 10, 4, 5, 11, 1, 6, 2,$
$\quad 7, 8, 1, 4, 10, 11, 5, 6, 0, 2, 7, 3, 8, 9, 2, 5, 11, 0, 6, 7, 1, 3, 8, 4, 9, 10, 3, 6,$
$\quad 0, 1, 7, 8, 2, 4, 9, 5, 10, 11, 4, 7, 1, 2, 8, 9, 3, 5, 10, 6, 11, 0, 5, 8, 2, 3, 9, 10,$
$\quad 4, 6, 11, 7, 0, 1, 6, 9, 3, 4, 10, 11, 5, 7, 0, 8, 1, 2]$

The following final command creates a musical simulation for the notes produced by all 48 possible combinations of transpositions, retrogrades, and inversions applied to p.[10] The third parameter in this command is *0* to designate an acoustic grand piano as the instrument, and the final parameter is *1* to cause a pause to be included every 12 notes in the simulation.

```
> midinotes("m9.midi", G, 0, 1);
```

---

[10]Although composers can choose notes from any octave, the notes in all musical simulations produced by **midinotes** exist in the middle octave of a standard keyboard.

13.9   Twelve-Tone Rows with MATLAB

In this section, we will show how MATLAB can be used to apply any combination of transpositions, retrogrades, and inversions to a twelve-tone row. As our example, we will use the following twelve-tone prime row p described in Section 13.7.

$$p = \langle A, A^\sharp, D^\sharp, B, E, F^\sharp, C, C^\sharp, G, G^\sharp, D, F \rangle$$

We begin by entering the numerical representation of the notes in p.

```
>> p = [9, 10, 3, 11, 4, 6, 0, 1, 7, 8, 2, 5]

p =

     9   10    3   11    4    6    0    1    7    8    2    5
```

Next we convert this numerical representation of p into a permutation. To perform this conversion, we will use the user-written function **notetoperm**, which we have written separately from this MATLAB session and saved as the M-file notetoperm.m. The following command illustrates how the function **notetoperm** can be used. Specifically, by entering the following command, we convert the numerical representation of p into a permutation.

```
>> p = notetoperm(p);
```

The previous command causes each cycle of the resulting permutation to be stored as a cell array element. To display the complete permutation efficiently, we will use the user-written function **pstring**, which we have written separately from this MATLAB session and saved as the M-file pstring.m, and introduced and used initially in Section 13.5. The next command causes the permutation representation for p to be displayed.

```
>> pstring(p)

ans =

[[0, 9, 8, 7, 1, 10, 2, 3, 11, 5, 6]]
```

To apply a transposition to a twelve-tone row, we will use the user-written function **Tn**, which we have written separately from this MATLAB session and saved as the M-file Tn.m. The following command illustrates how the function **Tn** can be used. Specifically, by entering the following command, we apply the first power of T to p. The second parameter in this command is the permutation representation of p, and the output is the permutation representation of the result.

```
>> tp1 = Tn(1, p);
>> pstring(tp1)

ans =

[[0, 10, 3],[2, 4, 5, 7],[1, 11, 6]]
```

To convert the permutation representation of a twelve-tone row back into its sequential numerical representation, we will use the user-written function **permtonote**, which we have written separately from this MATLAB session and saved as the M-file permtonote.m. The next command illustrates how the function **permtonote** can be used. Specifically, by entering the following command, we convert the permutation representation of *tp1* back into its sequential numerical representation.

```
>> permtonote(tp1)

ans =

    10   11    4    0    5    7    1    2    8    9    3    6
```

By using an integer larger than one as the first parameter in the **Tn** function, we can apply higher powers of T to a permutation. For example, in the following commands we apply the fifth power of T to p, showing the result both as a permutation and in sequential numerical form.

```
>> tp5 = Tn(5, p);
>> pstring(tp5)

ans =

[[0, 2, 8],[1, 3, 4, 9],[5, 11, 10, 7, 6]]

>> permtonote(tp5)

ans =

     2    3    8    4    9   11    5    6    0    1    7   10
```

To apply a retrograde to a twelve-tone row, we will use the user-written function **Ret**, which we have written separately from this MATLAB session and saved as the M-file Ret.m. The following command illustrates how the function **Ret** can be used. Specifically, by entering the following command, we apply the first power of R to p. The second parameter in this command is the permutation representation of p, and the output is the permutation representation of the result.

```
>> rp = Ret(1, p);
>> pstring(rp)

ans =

[[0, 5],[1, 2, 8, 11, 9, 3, 7, 4]]
```

In the next command we convert the permutation representation of the previous result into its sequential numerical representation.

```
>> permtonote(rp)

ans =
```

| 5 | 2 | 8 | 7 | 1 | 0 | 6 | 4 | 11 | 3 | 10 | 9 |

By using an integer larger than one as the first parameter in the **Ret** function, we can apply higher powers of R to a permutation. For example, in the following commands we apply the second power of R to p. Since R^2 is equal to the identity, we obtain the original permutation representation of p.

```
>> pstring(Ret(2, p))

ans =

[[0, 9, 8, 7, 1, 10, 2, 3, 11, 5, 6]]
```

To apply an inversion to a twelve-tone row, we will use the user-written function **Inv**, which we have written separately from this MATLAB session and saved as the M-file Inv.m. The following command illustrates how the function **Inv** can be used. Specifically, by entering the following command, we apply the first power of an inversion to p. The third parameter in this command is the permutation representation of p, and the output is the permutation representation of the result.

```
>> ip0 = Inv(1, 0, p);
>> pstring(ip0)

ans =

[[0, 3, 1, 2, 9, 4, 8, 5, 6],[7, 11]]
```

The second parameter in the previous **Inv** command specifies the number of one of the notes through which the axis of symmetry for the reflection will occur. Since this parameter is *0*, the inversion reflects across the axis of symmetry connecting notes 0 and 6. In the next command we convert the previous result into its sequential numerical representation.

```
>> permtonote(ip0)

ans =
```

```
    3   2   9   1   8   6   0   11   5   4   10   7
```

By using an integer larger than one as the first parameter in the **Inv** function, we can apply higher powers of inversions to a permutation. For example, in the following commands we apply the fifth power of an inversion to *p*, reflecting each time across the axis of symmetry connecting notes 2 and 8. Since the second power of an inversion is equal to the identity, the result is the same as if we had applied the first power of the inversion across the same axis of symmetry.

```
>> ip2 = Inv(5, 2, p);
>> pstring(ip2)

ans =

[[0, 7, 3, 5, 10, 2, 1, 6, 4],[8, 9]]

>> permtonote(ip2)

ans =
```

```
    7   6   1   5   0   10   4   3   9   8   2   11
```

To hear musical simulations of the results of these transformations, we will use the user-written function **midinotes**, which we have written separately from this MATLAB session and saved as the M-file midinotes.m. The following command illustrates how the function **midinotes** can be used. Specifically, by entering the following command, we create a file containing a musical simulation of the notes in the prime form of *p*.

```
>> midinotes('m1.midi', permtonote(p), 40, 0);
```

The first parameter in the previous command is the name of a file in MIDI format in which the musical simulation will be stored. This file will be created and stored by default in the same directory as specified in the current

MATLAB command window. The second parameter is the permutation representation of the sequence of notes to be played. The third parameter is a numerical value designating an instrument to be used in the simulation. In this command we use *40* for a violin. The final parameter gives the option of including a pause every 12 notes in the simulation by using a *1* for this parameter, or playing the notes continuously by using any other number for this parameter. The next two commands create files containing violin simulations of the notes that result from applying T and T^5 to p.

```
>> midinotes('m2.midi', permtonote(tp1), 40, 0);
>> midinotes('m3.midi', permtonote(tp5), 40, 0);
```

The next command creates a file containing a violin simulation of the notes that result from applying R to p.

```
>> midinotes('m4.midi', permtonote(rp), 40, 0);
```

The next two commands create files containing violin simulations of the notes that result from applying one inversion to p across the axes of symmetry connecting notes 0 and 6, and connecting notes 2 and 8, respectively.

```
>> midinotes('m5.midi', permtonote(ip0), 40, 0);
>> midinotes('m6.midi', permtonote(ip2), 40, 0);
```

Simulations can be constructed for the application of any combination of transpositions, retrogrades, and inversions. The following four commands demonstrate how we can apply the transformation T^3R to p, showing the result both as a permutation and in sequential numerical form, and create a musical simulation of the result.

```
>> t3r = Tn(3, Ret(1, p));
>> pstring(t3r)

ans =

[[0, 8, 2, 11],[1, 5, 3, 10],[6, 9]]

>> permtonote(t3r)

ans =

     8   5   11   10   4   3   9   7   2   6   1   0

>> midinotes('m7.midi', permtonote(t3r), 40, 0);
```

The next three commands demonstrate how we can apply the transformation *TIR* to *p* with the inversion across the axis of symmetry connecting notes 0 and 6, and create a musical simulation of the result.

```
>> tir = Tn(1, Inv(1, 0, Ret(1, p)));
>> pstring(tir)

ans =

[[0, 8, 2, 5, 1, 11, 4],[3, 6, 7, 9, 10]]

>> midinotes('m8.midi', permtonote(tir), 40, 0);
```

The following commands demonstrate how to apply all 48 possible combinations of transpositions, retrogrades, and inversions to *p*. The variable *eG* is used to store identifying representations for the combinations and the variable *G* holds the resulting notes for these transformations.

```
>> syms T I R
>> G = [];
>> ct = 0;
>> for i = 0:1
       for j = 0:1
           for k = 0:11
               ct = ct+1;
               s = strsplit(char(T^k*I^j*R^i), '*');
               eG{ct} = [s{numel(s)}, s{1:numel(s)-1}, ' '];
               G = [G, permtonote(Tn(k, Inv(j, 0, ...
               Ret(i, p))))];
           end
       end
   end
```

The following commands show the results for the variables *eG* and *G*.

```
>> [eG{:}]

ans =

1    T   T^2   T^3   T^4   T^5   T^6   T^7   T^8   T^9   T^10   T^11   I
TI   T^2I   T^3I   T^4I   T^5I   T^6I   T^7I   T^8I   T^9I   T^10I
T^11I   R   TR   T^2R   T^3R   T^4R   T^5R   T^6R   T^7R   T^8R
T^9R   T^10R   T^11R   RI   TIR   T^2IR   T^3IR   T^4IR   T^5IR
T^6IR   T^7IR   T^8IR   T^9IR   T^10IR   T^11IR
```

```
>> sym(G)

ans =

[ 9, 10, 3, 11, 4, 6, 0, 1, 7, 8, 2, 5, 10, 11, 4, 0, 5, 7,
  1, 2, 8, 9, 3, 6, 11, 0, 5, 1, 6, 8, 2, 3, 9, 10, 4, 7,
  0, 1, 6, 2, 4, 9, 3, 7, 10, 11, 5, 8, 1, 2, 7, 3, 8, 10,
  4, 5, 11, 0, 6, 9, 2, 3, 8, 4, 9, 11, 5, 6, 0, 1, 7, 10,
  3, 4, 9, 5, 10, 0, 6, 7, 1, 2, 8, 11, 4, 1, 10, 6, 11, 5,
  7, 8, 2, 3, 9, 0, 5, 6, 11, 7, 0, 2, 8, 9, 3, 4, 10, 1,
  6, 7, 0, 8, 1, 3, 9, 10, 4, 5, 11, 2, 7, 8, 1, 9, 2, 4,
  10, 11, 5, 6, 0, 3, 8, 9, 2, 10, 3, 5, 11, 0, 6, 7, 1, 4,
  3, 2, 9, 1, 8, 6, 0, 11, 5, 4, 10, 7, 4, 3, 10, 2, 9, 7,
  1, 0, 6, 5, 11, 8, 5, 4, 11, 3, 10, 8, 2, 1, 7, 6, 0, 9,
  6, 5, 0, 4, 11, 9, 3, 2, 8, 7, 1, 10, 7, 6, 1, 5, 0, 10,
  4, 3, 9, 8, 2, 11, 8, 7, 2, 6, 1, 11, 5, 4, 10, 9, 3, 0,
  9, 8, 3, 7, 2, 0, 6, 5, 11, 10, 4, 1, 10, 9, 4, 8, 3, 1,
  6, 7, 0, 11, 5, 2, 11, 10, 5, 9, 4, 2, 8, 7, 1, 0, 6, 3,
  0, 11, 6, 10, 5, 3, 9, 8, 2, 1, 7, 4, 1, 0, 7, 11, 6, 4,
  10, 9, 3, 2, 8, 5, 2, 1, 8, 0, 7, 5, 11, 10, 4, 3, 9, 6,
  5, 2, 8, 7, 1, 0, 6, 4, 11, 3, 10, 9, 6, 3, 9, 8, 2, 1,
  7, 5, 0, 4, 10, 11, 7, 4, 10, 9, 3, 2, 8, 6, 1, 5, 0, 11,
  8, 5, 11, 10, 4, 3, 9, 7, 2, 6, 1, 0, 9, 6, 0, 11, 4, 5,
  10, 8, 3, 7, 2, 1, 10, 7, 1, 0, 6, 5, 11, 9, 4, 8, 3, 2,
  11, 8, 2, 1, 7, 6, 0, 10, 5, 9, 4, 3, 0, 9, 2, 3, 8, 7,
  1, 11, 6, 10, 5, 4, 1, 10, 4, 3, 9, 8, 2, 0, 7, 11, 6, 5,
  2, 11, 5, 4, 10, 9, 3, 1, 8, 0, 7, 6, 3, 0, 6, 5, 11, 10,
  4, 2, 9, 1, 8, 7, 4, 1, 7, 6, 0, 11, 5, 3, 10, 2, 9, 8,
  7, 10, 4, 5, 11, 0, 6, 8, 1, 9, 2, 3, 8, 11, 5, 6, 0, 1,
  7, 9, 2, 10, 3, 4, 9, 0, 6, 7, 1, 2, 8, 10, 3, 11, 4, 5,
  10, 1, 7, 8, 2, 3, 9, 11, 4, 0, 5, 6, 11, 2, 8, 9, 3, 4,
  10, 0, 5, 1, 6, 7, 0, 3, 9, 10, 4, 5, 11, 1, 6, 2, 7, 8,
  1, 4, 10, 11, 5, 6, 0, 2, 7, 3, 8, 9, 2, 5, 11, 0, 6, 7,
  1, 3, 8, 4, 9, 10, 3, 6, 0, 1, 4, 8, 2, 7, 9, 5, 10, 11,
  4, 7, 1, 2, 8, 9, 3, 5, 10, 6, 11, 0, 5, 8, 2, 3, 9, 10,
  4, 6, 11, 7, 0, 1, 6, 9, 3, 4, 10, 11, 5, 7, 0, 8, 1, 2]
```

The following final command creates a musical simulation for the notes produced by all 48 possible combinations of transpositions, retrogrades, and inversions applied to p.[11] The third parameter in this command is 0 to designate an acoustic grand piano as the instrument, and the final parameter

---

[11] Although composers can choose notes from any octave, the notes in all musical simulations produced by **midinotes** exist in the middle octave of a standard keyboard.

is *1* to cause a pause to be included every 12 notes in the simulation.

```
>> midinotes('m9.midi', G, 0, 1);
```

Exercises

1. The scale $\langle C, D, D^\sharp, F, F^\sharp, G^\sharp, A, B \rangle = \langle 0, 2, 3, 5, 6, 8, 9, 11 \rangle$ with step intervals 2-1-2-1-2-1-2-1 is an example of an *octatonic* scale. How many distinct octatonic scales exist up to transposition?

2. A *fully diminished* seventh chord contains four notes with step intervals 3-3-3-3. How many distinct fully diminished seventh chords exist up to transposition? How many exist up to transposition and inversion?

3. How many distinct three-note chords exist up to transposition and inversion?

4. How many distinct four-note chords exist up to transposition and inversion?

5. Prove that the major and minor diatonic scales are the unique seven-note scales containing the maximum number of pairs of notes separated by a perfect fifth.

6. In the circle of fifths, two adjacent vertices label keys that differ in only one note, with the note gained exactly one semitone above the note lost. In his explorations of microtonal pitch systems in [1], Balzano refers to this property as the $F \to F^\sharp$ property. Prove algebraically that for connected regions of notes on the circle of fifths, only regions of seven notes have the $F \to F^\sharp$ property.

7. As noted in Exercise 6, for connected regions of notes on the circle of fifths, only regions of seven notes have the $F \to F^\sharp$ property. Connected regions of five notes on the circle of fifths, called *pentatonic* scales, have a similar property. Prove algebraically that for a 30° rotation of a connected region of five notes on the circle of fifths, the note gained will be exactly one semitone below the note lost.

8. Complete the proof of Theorem 13.2 by showing that $LR = T_7$ when acting on the minor triads.

9. Consider a function P that acts on the set of all consonant triads by sending a consonant triad to its parallel triad of the opposite type but with the same name. For example $P(C \text{ major}) = C$ minor, or,

$P(\langle 0, 4, 7\rangle) = \langle 7, 3, 0\rangle$. Like the relative chord function R and leading tone exchange function L, the action of P can be described in terms of the Tonnetz and algebraically. Give such descriptions for P.

10. Music theorists have used the group generated by the function P in Exercise 9, the leading tone exchange function L, and the relative chord function R to study triadic post-tonal music. Show that the group generated by P, L, and R is equal to the group generated by only L and R, which we have seen is isomorphic to D_{12}.

11. Let T, R, and I be the transposition, retrograde, and inversion functions that act on twelve-tone rows as described in Section 13.7.

    (a) Show that $T_n R(p) = R T_n(p)$ for all twelve-tone rows p.

    (b) Show that $IR(p) = RI(p)$ for all twelve-tone rows p.

    (c) Show that $IT_n(p) = T_{-n}I(p)$ for all twelve-tone rows p.

    (d) Prove that $\{T_0, T_5 I, R, T_5 IR\}$, which are the transformations in Schoenberg's 1936 Violin Concerto (Op. 36), forms a subgroup of the group of all transpositions, inversions, and reflections of twelve-tone rows. Is this subgroup cyclic or non-cyclic?

    (e) Prove that the group $\langle T, I, R\rangle$ is isomorphic to $D_{12} \times \mathbb{Z}_2$.

12. Let p be a twelve-tone row. Prove that $p \neq X(p)$ for any transformation X composed of transpositions and inversions but not retrogrades. Similarly, prove that $p \neq X(p)$ for any transformation X composed of retrogrades and inversions but not transpositions.

13. How many non-palindromic twelve-tone row classes exhibit symmetries? (Note that by Exercise 12, these would have to be symmetries involving transpositions, retrogrades, and inversions.) How many row classes contain non-symmetric rows, or rows that cannot be transformed onto themselves via any combination of transpositions, retrogrades, and inversions?

Computer Exercises

14. For Western music with octaves divided into 12 distinct pitch classes, create a file that will play a simulation of an example of each of the different types of four-note chords that exist up to transposition.

15. Consider Western music with octaves divided into 12 distinct pitch classes, and the full group D_{12} of both rotations and reflections acting on the musical clock for this system.

    (a) Determine the cycle index for this system.

(b) Determine the number of different types of three-note and five-note chords that exist for this system.

(c) Create files that will play musical simulations of an example of each of the different types of three-note and five-note chords that exist for this system.

16. The modern Arab system of musical tuning is based upon dividing octaves into 24 distinct pitch classes. Consider the subgroup of rotations in D_{24} acting on the musical clock for this system.

    (a) Determine the cycle index for this system.

    (b) Determine the number of different types of three-note and five-note chords that exist for this system.

17. Repeat Exercise 16 with the full group D_{24} of both rotations and reflections acting on the musical clock for the system.

18. The *31 equal temperament* of musical tuning is based upon dividing octaves into 31 distinct pitch classes. Consider the subgroup of rotations in D_{31} acting on the musical clock for this system.

    (a) Determine the cycle index for this system.

    (b) Determine the number of different types of three-note and five-note chords that exist for this system.

19. Repeat Exercise 18 with the full group D_{31} of both rotations and reflections acting on the musical clock for the system.

20. Create files that will play simulations of the transformations of the twelve-tone prime row $p = \langle B, A^\sharp, D, D^\sharp, G, F^\sharp, G^\sharp, E, F, C, C^\sharp, A \rangle$.[12]

    (a) T

    (b) T^4

    (c) I

    (d) R

    (e) $T^6 R$

    (f) $T^7 IR$

21. Create a file that will play a musical simulation of the notes resulting from applying all 48 possible combinations of transpositions, retrogrades, and inversions to the twelve-tone row p in Exercise 20.

---

[12] From Anton Webern's 1934 Concerto for Nine Instruments (Op. 24).

22. Create files that will play simulations of the transformations of the twelve-tone prime row $p = \langle G, A^\sharp, D, F^\sharp, A, C, E, G^\sharp, B, C^\sharp, D^\sharp, F \rangle$.[13]

   (a) T
   (b) T^5
   (c) I
   (d) R
   (e) $T^9 I$
   (f) $T^{10} I R$

23. Create a file that will play a musical simulation of the notes resulting from applying all 48 possible combinations of transpositions, retrogrades, and inversions to the twelve-tone row p in Exercise 22.

Research Exercises

24. Investigate how musicians convert a piece of music from one key to another. Can you relate this to our mathematical discussions of major and minor diatonic scales?

25. As we noted, Western musicians divide octaves into 12 distinct notes, and in Western tonal music composers generally select at least most of their notes not from all 12 notes, but from a seven-note diatonic scale. Investigate how other cultures make such divisions.

26. In [8], the duality of the group generated by the parallel, retrograde, and leading tone exchange actions and the group generated by the set of all transpositions and inversions is discussed. Write a summary of this discussion with musical examples.

27. In [24], *interval vectors* are used to study chords and their transformations. Investigate how the author uses geometry to show that up to transposition and inversion, only two all-interval chords exist.

28. In [28], Burnside's Theorem is used to count n-tone rows. Write a summary of this discussion as it pertains to twelve-tone rows.

29. Milton Babbitt (1916–2011) used the methods of twelve-tone composition to arrange the pitch classes in a piece of music, and applied similar mathematical precision to his arrangement of rhythm and timbre. Investigate Babbitt's methods, including how his life and music were influenced by (or perhaps themselves influenced) mathematics.

---

[13]From Alban Berg's 1935 Violin Concerto.

Bibliography

[1] G. Balzano. The group-theoretic description of 12-fold and microtonal pitch systems. *Computer Music Journal*, 4(4):66–84, 1980.

[2] T. H. Barr. *Invitation to Cryptology*. Prentice Hall, 2002.

[3] D. Bressoud. *Factorization and Primality Testing*. Springer-Verlag, U.T.M., 1989.

[4] R. Brualdi. *Introductory Combinatorics, 4th ed.* Prentice Hall, 2004.

[5] Certicom Corporation. *Current public-key cryptographic systems*, 1997. Certicom whitepaper.

[6] Certicom Corporation. *Elliptic curve groups over F_{2m}*, 1997. Elliptic curve cryptography tutorial.

[7] C. E. Chow. CS525 Multimedia Computing and Communications. Available at http://cs.uccs.edu/~cs525/midi/stdpatc2.jpg, 2014.

[8] A. S. Crans, T. M. Fiore, and R. Satyendra. Musical actions of dihedral groups. *American Mathematical Monthly*, 116(6):479–495, 2009.

[9] A. Forte. *The Structure of Atonal Music*. Yale University Press, 1973.

[10] J. Gallian. *Contemporary Abstract Algebra, Eighth Edition*. Cengage Learning, 2012.

[11] D. Guichard. Counting non-isomorphic graphs with Maple. *MapleTech*, 26(8):52–56, 1992.

[12] F. Harary and E. Palmer. *Graphical Enumeration*. Academic Press, 1973.

[13] L. Harkleroad. *The Math Behind the Music*. Cambridge University Press, 2006.

[14] I. N. Herstein. *Abstract Algebra*. Macmillan, 1986.

[15] L. S. Hill. Cryptography in an algebraic alphabet. *American Mathematical Monthly*, 36:306–312, 1929.

[16] L. S. Hill. Concerning certain linear transformation apparatus of cryptography. *American Mathematical Monthly*, 38:135–154, 1931.

[17] T. Hungerford. *Algebra*. Springer-Verlag, G.T.M., 1989.

[18] D. J. Hunter and P. T. von Hippel. How rare is symmetry in musical 12-tone rows? *American Mathematical Monthly*, 110(2):124–132, 2003.

[19] R. E. Klima. Applying the Diffie-Hellman key exchange to RSA. *UMAP*, 20(1):21–28, 1999.

[20] N. Koblitz. *A Course in Number Theory and Cryptography*. Springer-Verlag, G.T.M., 1987.

[21] J. Levine. Variable matrix substitution in algebraic cryptography. *American Mathematical Monthly*, 65:170–179, 1958.

[22] R. Lidl and H. Neiderreiter. *Introduction to Finite Fields and their Applications*. Cambridge University Press, 1986.

[23] G. Mackiw. *Applications of Abstract Algebra*. John Wiley & Sons, 1985.

[24] B. McCartin. Prelude to musical geometry. *The College Mathematics Journal*, 29(5):354–370, 1998.

[25] B. C. Murray. *Journey into Space*. Norton, 1989.

[26] O. Pretzel. *Error Correcting Codes and Finite Fields*. Oxford University Press, 1992.

[27] D. M. Randel. *The Harvard Dictionary of Music*. Belknap Press, 2003.

[28] D. L. Reiner. Enumeration in music theory. *American Mathematical Monthly*, 92(1):51–54, 1985.

[29] F. Roberts. *Applied Combinatorics*. Prentice Hall, 1984.

[30] K. Rosen. *Elementary Number Theory and its Applications*. Addison-Wesley, 1988.

[31] H. R. Ryser. Combinatorial mathematics. *The Mathematical Association of America*, 1(14), 1963. The Carus Mathematical Monographs.

[32] J. Savard. The Advanced Encryption Standard (Rijndael). Available at http://www.quadibloc.com/crypto/co040401.htm, 2005.

[33] N. P. Sigmon. Applications of Maple to algebraic cryptography. *Mathematics and Computer Education*, 31(3):220–229, 1997.

[34] N. P. Sigmon and E. L. Stitzinger. Applications of Maple to Reed-Solomon codes. *MapleTech*, 3(3):53–59, 1996.

[35] D. Stinson. *Cryptography, Theory and Practice, 3rd ed.* CRC Press, 2006.

[36] W. Trappe and L. C. Washington. *Introduction to Cryptography with Coding Theory, 2nd ed.* Prentice Hall, 2006.

[37] A. Tucker. Pólya's enumeration formula by example. *Mathematics Magazine*, 47(5):248–256, 1974.

[38] G. A. Walker. *Introduction to Abstract Algebra.* Random House, 1987.

[39] D. White. Skytopia: Crash course on the standard MIDI specification (SMF). Available at http://www.skytopia.com/project/articles/midi.html, 2014.

[40] S. B. Wicker and V. K. Bhargava, Eds. *Reed-Solomon Codes and Their Applications.* IEEE Inc., 1994.

Hints or Answers for Selected Exercises

Chapter 1

1. In order for $(\mathbb{Z}, \cdot)$ with normal integer multiplication to be a group, every element in $\mathbb{Z}$ would have to have a multiplicative inverse in $\mathbb{Z}$.

5. (a) $\alpha \circ \beta = \begin{pmatrix} 1 & 2 & 3 & 4 & 5 & 6 \\ 2 & 4 & 5 & 3 & 6 & 1 \end{pmatrix}$

   (b) $\beta = (142)(365)$

   (c) $\alpha \circ \gamma$ is even

   (d) $\beta^{-1} = (124)(356)$

   (e) One possible answer for β is $\beta = (12)(14)(35)(36)$.

7. (1), (12345), (13524), (14253), (15432), $(25)(34)$, $(13)(45)$, $(15)(24)$, $(12)(35)$, $(14)(23)$

11. Suppose H is a subgroup of a cyclic group G, and a is a cyclic generator for G. Also, suppose j is the smallest positive integer for which $a^j \in H$. Use the fact that $\mathbb{Z}$ is a Euclidean domain to show that a^j is a cyclic generator for H.

12. (a) The order is five.

    (e) The order is six.

13. The distinct left cosets of A_4 in S_4 are $(1)A_4$ and $(12)A_4$.

17. Suppose $\alpha \in S_n$ and $\beta \in A_n$, and show that $\alpha^{-1}\beta\alpha \in A_n$.

21. The kernel is the set of all $n \times n$ matrices over $\mathbb{R}$ whose determinant is 1.

25. Suppose that I is an ideal of F, and I contains at least one nonzero element in F. Use the fact that I is an ideal of F to show that I must contain every element in F.

31. The irreducible elements are the prime integers, their negatives, and the units.

33. (a) The polynomial is primitive, and gives the following table of field elements that correspond to powers of x.

Power	Field Element
x^1	x
x^2	$x+1$
x^3	$2x+1$
x^4	2
x^5	$2x$
x^6	$2x+2$
x^7	$x+2$
x^8	1

(c) The polynomial is not primitive.

34. (a) The polynomial is primitive, and gives the following table of field elements that correspond to powers of x.

Power	Field Element
x^1	x
x^2	x^2
x^3	$x+1$
x^4	x^2+x
x^5	x^2+x+1
x^6	x^2+1
x^7	1

(c) The polynomial is primitive, and gives the following table of field elements that correspond to powers of x.

Power	Field Element	Power	Field Element
x^1	x	x^9	x^3+x
x^2	x^2	x^{10}	x^2+x+1
x^3	x^3	x^{11}	x^3+x^2+x
x^4	$x+1$	x^{12}	x^3+x^2+x+1
x^5	x^2+x	x^{13}	x^3+x^2+1
x^6	x^3+x^2	x^{14}	x^3+1
x^7	x^3+x+1	x^{15}	1
x^8	x^2+1		

35. (b) Try evaluating $f(x)$ at some inputs from $\mathbb{Z}_{11}$.

41. $\gcd(2272, 716) = 4 = 2272(52) + 716(-165)$

47. Since 127 is prime, this will require a primitive polynomial of degree 1 in $\mathbb{Z}_{127}[x]$.

Chapter 2

2. Use Theorem 2.5 with $t = 4$.

8. For one block design use Proposition 2.8 with $t = 2$, $p = 13$, and $n = 1$, and for the other use Proposition 2.9 with $t = 3$, $p = 13$, and $n = 1$. With the cyclic generator 2 for $\mathbb{Z}_{13}^*$, Proposition 2.8 gives the initial blocks $D_0 = \{1, 3, 9\}$ and $D_1 = \{2, 6, 5\}$, which yield a block design with parameters $(13, 26, 6, 3, 1)$.

10. Use Proposition 2.9 with $t = 6$, $p = 5$, and $n = 2$. With the cyclic generator x for the nonzero elements in a finite field of order 25, Proposition 2.9 gives 6 initial blocks, each of which contains 4 elements. The first two initial blocks are $D_0 = \{x^0, x^6, x^{12}, x^{18}\}$ and $D_1 = \{x^1, x^7, x^{13}, x^{19}\}$. The parameters for the resulting block design are $(25, 150, 24, 4, 3)$, which indicate that 150 test-drivers will be needed, each vehicle will be evaluated 24 times, each test-driver will evaluate four vehicles, and every possible pair of vehicles will be evaluated by the same test-driver exactly three times.

12. It is possible. Use Proposition 2.8 to construct a block design in which the objects are the teams and each block contains the teams that should show up at the field on a given evening.

14. For $v = 529 = 23^2$, Proposition 2.8 can be used with $t = 88$, and Proposition 2.9 with $t = 132$.

15. Consider the matrix product $\begin{bmatrix} H & H \\ H & -H \end{bmatrix} \cdot \begin{bmatrix} H & H \\ H & -H \end{bmatrix}^T$.

24. For one block design use Proposition 2.8 with $t = 20$, $p = 11$, and $n = 2$, and for the other use Proposition 2.9 with $t = 30$, $p = 11$, and $n = 2$.

25. For one block design use Theorem 2.5 with $t = 32$, and for the other use Proposition 2.8 with $t = 21$, $p = 127$, and $n = 1$. Theorem 2.5 gives a $(127, 127, 63, 63, 31)$ block design.

Chapter 3

1. (a) One possible vector is (10000000).

   (c) One possible vector is (01111011).

3. (a) There are 49 non-codewords that are guaranteed to be uniquely correctable.

4. The vector (001011010001010) corrects to (001011010010110).

5. The Reed-Muller code with $k = 2$ resulting from H_8 satisfies the stated requirements. The vector (01000110) corrects to (01100110).

6. It is not possible.

8. One possible answer is the following generator matrix G and parity check matrix H.

$$G = \begin{bmatrix} 1 & 1 & 1 & 0 & 0 & 0 & 1 & 1 & 1 \\ 0 & 0 & 0 & 1 & 1 & 1 & 1 & 1 & 1 \end{bmatrix}$$

$$H = \begin{bmatrix} 0 & 0 & 0 & 1 & 1 & 0 & 0 & 0 & 0 \\ 1 & 0 & 1 & 0 & 0 & 0 & 0 & 0 & 0 \\ 1 & 1 & 0 & 0 & 0 & 0 & 0 & 0 & 0 \\ 0 & 0 & 0 & 1 & 0 & 1 & 0 & 0 & 0 \\ 1 & 0 & 0 & 1 & 0 & 0 & 0 & 0 & 1 \\ 1 & 0 & 0 & 1 & 0 & 0 & 0 & 1 & 0 \\ 1 & 0 & 0 & 1 & 0 & 0 & 1 & 0 & 0 \end{bmatrix}$$

11. The vector (011111011111110) corrects to (011111010111110).

15. The vector (11101) corrects to (11100), and (10101) cannot be corrected.

17. Recall that all Hamming codes are one-error correcting.

20. Show that the inequality in Theorem 3.1 is an equality.

23. A code with the Hamming distance function is a metric space. Use the triangle inequality to show that $d(x, z) \le d(x, y) + d(y, z)$.

24. It is an equivalence relation.

25. The Hadamard code with $k = 8$ resulting from H_{32} satisfies the stated requirements.

Chapter 4

1. The generator polynomial is $p(x)$, which gives a $[7, 4]$ code with 16 codewords. One codeword of maximum possible degree is $x^3 g(x)$.

2. The generator polynomial is the same as the generator polynomial in Example 4.2.

4. Refer to Example 4.3, and consider the BCH code that results from only the first four powers of a. The resulting code has 128 codewords and is two-error correcting.

6. The generator polynomial is $x^{10} + x^9 + x^8 + x^6 + x^5 + x^2 + 1$, which gives a $[15, 5]$ BCH code with 32 codewords.

7. (a) The vector corrects to (1001110).

8. (a) The vector corrects to (110001001101011).

   (c) The vector corrects to (100001110110010).

10. (a) The vector corrects to (111101011001000).

    (c) The vector is already a codeword.

11. (a) The vector corrects to (000111011001010).

    (c) The vector corrects to (111011001010000).

12. (a) The following is the parity check matrix.

$$\begin{bmatrix} 1 & a & a^2 & a^2+1 & a^2+a+1 & a+1 & a^2+a \\ 1 & a^2 & a^2+a+1 & a^2+a & a & a^2+1 & a+1 \end{bmatrix}$$

Since multiplying this matrix by the transpose of the received vector in Exercise 7(a) results in a nonzero vector, and multiplying this matrix by the transpose of the codeword to which this vector corrects results in the zero vector, then this parity check matrix "works" for these vectors.

15. Use the fact that $x_1^2 + x_2^2 + \cdots + x_r^2 = (x_1 + x_2 + \cdots + x_r)^2$ over $\mathbb{Z}_2$ for any $x_1, x_2, \ldots, x_r$, since all cross terms in the binomial expansion of the expression on the right will contain a factor of 2.

16. Use the fact that by Exercise 15, we know $p(a^2) = p(a)^2$.

19. The generator polynomial $g(x)$ gives a $[63, 45]$ code with 2^{45} codewords. One codeword of maximum possible degree is $x^{44} g(x)$.

Chapter 5

1. There are 16^{11} codewords.

2. One codeword of maximum possible degree is $a^2 x^{10} g(x)$.

3. The vector converts into a polynomial which can be verified to be a codeword by showing it is a multiple of $g(x)$.

5. (a) The generator polynomial is $g(x) = x^4 + a^3 x^3 + x^2 + ax + a^3$.

   (b) One codeword is $(a^4 x + a^5) g(x) = a^4 x^5 + a^4 x^4 + a^2 x^3 + a^2 x + a$, which converts into the vector (0100010000010110110000).

7. (a) The polynomial corrects to $a^5 x^6 + a^6 x^4 + ax^2 + a^6 x + a^5$.

   (c) The polynomial is already a codeword.

9. (a) The polynomial corrects to $a^7 x^{12} + a^2 x^{11} + a^8 x^{10} + a^6 x^9 + x^8 + ax^7 + a^{10} x^6 + a^6 x^5 + a^4 x^4 + a^7 x^3 + ax^2 + a^6 x + a$.

   (c) The polynomial corrects to $a^8 x^{14} + a^{12} x^{13} + a^3 x^{12} + a^8 x^{11} + a^{12} x^{10} + a^3 x^9 + a^8 x^8 + a^{12} x^7 + a^3 x^6 + a^4 x^5 + a^4 x^4 + a^8 x^3 + a^{12} x^2 + a^{11} x + a^{13}$.

11. Let $m = 2^n - 1$, and follow the proof of Theorem 4.2, noting that $r(x) = b_0 + b_1 x + \cdots + b_{m-1} x^{m-1} \in \mathbb{Z}_2[x]/(p(x))$.

17. There are 256^{223} codewords, each corresponding to a binary vector of length $255 \cdot 8 = 2040$. The maximum error burst length guaranteed to be correctable is $8 \cdot 15 + 1 = 121$.

Chapter 6

1. The ciphertext is HALWVWUDWHJB.

4. The ciphertext is CHFWICOQWJWQOCIWFHC.

7. (c) Superencryption using an affine cipher with encryption function $f(x) = ax + b_1$ followed by a shift cipher with encryption function $f(x) = x + b_2$ would not increase security, since it would be equivalent to using a single affine cipher with encryption function $f(x) = (ax + b_1) + b_2 = ax + (b_1 + b_2)$.

9. On average you would have to try half of the $12 \cdot 26 = 312$ possible pairs of keys. This would take $\frac{312}{2} \cdot 3 = 468$ minutes, or $\frac{468}{60} = 7.8$ hours.

12. (a) There are 72 possible values for a and 95 possible values for b, giving $72 \cdot 95 = 6840$ possible pairs (a, b) of keys.

13. (a) The ciphertext is GCLYNDJLWLWFWJOY.

15. The key matrix for the system is $A = \begin{bmatrix} 5 & 21 \\ 9 & 12 \end{bmatrix}$.

16. (a) The ciphertext is JXHPGXPUNKBJ.

17. (b) The ciphertext is HFXLKQOOFS.

18. Use Theorem 1.16.

20. Consider the matrix product K^2.

30. (a) The plaintext ABORT MISSION PROCEED WITH SECONDARY ORDERS AA encrypts to KNNVZCEIEIEKKXIRXFBXHHNKXUNYUUFQLKSWXQZA.

35. (a) The ciphertext is DCPRBLJOAGFXBDOEGPMM.

Chapter 7

1. (a) The ciphertext is TWSPLULTVLDZACRPO.

3. (a) The ciphertext is TFFRRDDVJUTFCUK.

5. (a) The index of coincidence is $\frac{3670}{242 \cdot 241} = 0.0629$.

7. (b) The estimate for the length of the keyword is 0.7570, so the keyword should be one letter long. This is as expected, since the ciphertext was formed using a monoalphabetic cipher, which uses the same correspondences to encrypt each plaintext letter.

11. (a) The ciphertext is FKMOABZBAVHJEUPUI.

Chapter 8

1. (a) The ciphertext is 0, 222, 222, 0, 128, 175, 99, 192, 201, 222, 201, 99, 0, 216, 175.

2. The decryption exponent is 41.

7. A total of 22 multiplications would be required.

12. Use Lemmas 8.2 and 8.3.

18. (a) Consider what would happen if $P^{b_1} \bmod n_1$ was greater than n_2.

26. The smallest base to which 3215031751 is not a pseudoprime is 151.

Chapter 9

1. $y = 16$, $z = 5$

4. $w = 2x + 1$

5. (a) There are 13 ordered pairs in E.

    (c) $(1, 6) + (1, 6) = (3, 8)$

6. $(0, 1) + (14, 0) = (14, 0)$

7. (a) There are 11 ordered pairs in E.

9. $y = (0, 1)$, $z = (8, 2)$

12. $w = (7, 4)$

16. $y = 3x^6 + 2x^5 + 4x^4 + 2x^3 + x$, $z = x^5 + x^4 + 4x^2 + 4x + 2$

18. (f) To explain why every element in E except $\mathcal{O}$ must be a generator for E, use the fact that E has prime order.

Chapter 10[1]

2. The input into the S-box transformation would be $x^3 + x^2 + x$.

3. $K_2 = \begin{bmatrix} 01001010 & 11110100 & 01001010 & 11110000 \\ 01111011 & 00010000 & 01111111 & 00001000 \\ 11010001 & 10111111 & 11000011 & 10111111 \\ 00001011 & 11111111 & 00000001 & 11111111 \end{bmatrix}$

5. (b) $K_0 = \begin{bmatrix} 00010011 & 00000100 & 00010010 & 00010001 \\ 00000111 & 00000000 & 00000000 & 00011000 \\ 00000100 & 00010110 & 00001101 & 00010011 \\ 00010010 & 00000000 & 00000110 & 00000111 \end{bmatrix}$

  (d) $K_2 = \begin{bmatrix} 11011011 & 11011111 & 11001101 & 11011100 \\ 01100100 & 01100100 & 01100100 & 01111100 \\ 01100111 & 01110001 & 01111100 & 01101111 \\ 01101001 & 01101001 & 01101111 & 01101000 \end{bmatrix}$

6. (a) $A = \begin{bmatrix} 00010001 & 00000100 & 00010001 & 00000100 \\ 00001110 & 00001011 & 00001110 & 00001011 \\ 00000010 & 00001011 & 00000010 & 00001011 \\ 00000111 & 00000100 & 00000111 & 00000100 \end{bmatrix}$

---

[1]Each of the matrices in this section is expressed with the polynomial entries converted into the bytes of which they can be viewed as representations.

$$(c)\ K_1 = \begin{bmatrix} 11010011 & 11011111 & 11010010 & 11011010 \\ 11011000 & 11010000 & 11000011 & 11001110 \\ 01101001 & 01100010 & 01101100 & 01111110 \\ 00111100 & 00111100 & 00110000 & 00111010 \end{bmatrix}$$

$$(e)\ A_0 = \begin{bmatrix} 00010100 & 00001000 & 00011100 & 00001100 \\ 00011111 & 00000011 & 00011101 & 00000110 \\ 00001100 & 00000000 & 00001100 & 00011001 \\ 00001011 & 00000100 & 00001011 & 00001110 \end{bmatrix}$$

$$(g)\ C_1 = \begin{bmatrix} 11111010 & 00110000 & 10011100 & 11111110 \\ 01111011 & 10100100 & 01101111 & 11000000 \\ 11111110 & 11010100 & 11111110 & 01100011 \\ 10101011 & 00101011 & 11110010 & 00101011 \end{bmatrix}$$

$$(i)\ A_1 = \begin{bmatrix} 11100100 & 10110111 & 01001100 & 00101110 \\ 01100110 & 11111111 & 01101010 & 00100101 \\ 11101001 & 00111000 & 01110101 & 11111011 \\ 11100001 & 01001010 & 11100001 & 11010110 \end{bmatrix}$$

$$8.\ A_2 = \begin{bmatrix} 01110000 & 01000110 & 11110010 & 10110010 \\ 00010011 & 11001000 & 01110010 & 00110110 \\ 11011100 & 00110101 & 00010101 & 00111100 \\ 00111101 & 00000011 & 00101011 & 10101101 \end{bmatrix}$$

$$9.\ (b)\ C_1 = \begin{bmatrix} 10011100 & 01110110 & 01110111 & 01100011 \\ 10101011 & 01100111 & 01111101 & 11111110 \\ 11110010 & 00101011 & 11010111 & 01011001 \\ 01100011 & 10101011 & 01110110 & 00101011 \end{bmatrix}$$

$$(d)\ A_0 = \begin{bmatrix} 00011100 & 00001111 & 00000010 & 00000000 \\ 00001100 & 00001110 & 00001010 & 00010011 \\ 00001101 & 00010101 & 00000100 & 00001011 \\ 00001110 & 00001111 & 00001011 & 00000000 \end{bmatrix}$$

11. Show that $SS^{-1} = I$ over $\mathbb{Z}_2$.

13. (b) Starting with A_1, apply **invmixcolumn, invaddrkey, byte-sub, invshiftrow**, and **addrkey** in order to recover the plaintext matrix.

Chapter 11

1. (c) The cycle index is $f(x_1, x_2, x_3) = \frac{1}{6}\left(x_1^3 + 3x_1x_2 + 2x_3\right)$, and you could construct $f(2,2,2) = 4$ distinct necklaces.

(e) The pattern inventory of X_2 is $\mathcal{R}^3 + \mathcal{W}^3 + \mathcal{B}^3 + \mathcal{R}^2\mathcal{W} + \mathcal{R}^2\mathcal{B} + \mathcal{W}^2\mathcal{R} + \mathcal{W}^2\mathcal{B} + \mathcal{B}^2\mathcal{R} + \mathcal{B}^2\mathcal{W} + \mathcal{R}\mathcal{W}\mathcal{B}$.

3. (a) You could construct 8 distinct necklaces.

   (b) You could construct 2 distinct necklaces with two red beads and three white beads.

6. You could construct 10504 nonequivalent buildings.

8. See Example 11.5.

12. Use Burnside's Theorem.

15. There are 3984 distinct equivalence classes.

17. You could construct 108 distinct necklaces with three red beads, two white beads, two blue beads, and one yellow bead.

19. (a) You could construct 20346485 distinct watches.

Chapter 12

1. As sets, the partitions of 4 are $\{1,1,1,1\}$, $\{1,1,2\}$, $\{2,2\}$, $\{1,3\}$, and $\{4\}$. As ordered lists of the number of times that i appears in these sets for $i = 1,2,3,4$, these partitions are $[4,0,0,0]$, $[2,1,0,0]$, $[0,2,0,0]$, $[1,0,1,0]$, and $[0,0,0,1]$, respectively.

3. $f(x_1, x_2, x_3, x_4) = \frac{1}{24}\left(x_1^4 + 6x_1^2 x_2 + 3x_2^2 + 8x_1 x_3 + 6x_4\right)$

5. There are 15 nonequivalent colorings.

7. There should be a total of 64 figures.

9. The cycle index is $f(x_1, x_2, x_3, x_4) = \frac{1}{24}(x_1^6 + 9x_1^2 x_2^2 + 8x_3^2 + 6x_2 x_4)$. There are 11 nonequivalent undirected graphs with four vertices. The pattern inventory is $1 + \mathcal{E} + 2\mathcal{E}^2 + 3\mathcal{E}^3 + 2\mathcal{E}^4 + \mathcal{E}^5 + \mathcal{E}^6$.

16. There are 120 nonequivalent colorings.

18. There are 1044 nonequivalent undirected graphs.

20. There are 131 nonequivalent undirected graphs containing 12 edges.

Chapter 13

1. There are 3 distinct octatonic scales up to transposition.

3. There are 12 distinct three-note chords up to transposition and inversion.

6. Note that a connected region of m notes on the circle of fifths will consist of notes of the form $k, k + 7, k + 2 \cdot 7, \ldots, k + (m - 1) \cdot 7$ in $\mathbb{Z}_{12}$. Try rotating this sequence of notes $30°$ on the circle of fifths, subtracting the formula for the note lost from the formula for the note gained, and setting the result equal to 1.

7. Note that a connected region of five notes on the circle of fifths will consist of notes of the form $k, k + 7, k + 2 \cdot 7, k + 3 \cdot 7, k + 4 \cdot 7$ in $\mathbb{Z}_{12}$.

10. Consider $R(LR)^3$.

11. (a) $T_n R(p) = T_n(p \sigma_R) = \sigma_T^n p \sigma_R = R(\sigma_T^n p) = RT_n(p)$

(c) First use the facts that $(\sigma_T \sigma_R)^2 = \sigma_I^2 = e$ and $(\sigma_R)^2 = e$ to show that $\sigma_T^n = \sigma_R \sigma_T^{-n} \sigma_R$.

(d) Use parts (a), (b), and (c) and also the facts that $I^2 = T_0$ and $R^2 = T_0$ to show that the set is closed.

12. If $p = X(p)$ for a transformation X composed of transpositions and inversions but not retrogrades, then $T_k I(p) = p$ for some integer k. This gives the following.

$$\sigma_T^k \sigma_T^{2p_0+1} \sigma_R p = p$$
$$\Rightarrow \quad \sigma_T^{2p_0+k+1} \sigma_R = p p^{-1}$$
$$\Rightarrow \quad \sigma_T^{2p_0+k+1} \sigma_R = e$$

Thus, $\sigma_T^{2p_0+k+1} = \sigma_R$. However, this is impossible since there is no power of σ_T that is equal to σ_R.

18. (a) $f(x_1, \ldots, x_{31}) = \frac{1}{31}(x_1^{31} + 30x_{31})$

(b) There are 145 distinct three-note chords, and 5481 distinct five-note chords.

19. (a) $f(x_1, \ldots, x_{31}) = \frac{1}{62}(x_1^{31} + 30x_{31} + 31x_1 x_2^{15})$

(b) There are 80 distinct three-note chords, and 2793 distinct five-note chords.

Index